Lecture Notes in Control and Information Sciences

Edited by M. Thoma and A. Wyner

Vol. 81: Stochastic Optimization
Proceedings of the International Conference, Kiew, 1984
Edited by I. Arkin, A. Shiraev, R. Wets
X, 754 pages, 1986.

Vol. 82: Analysis and Algorithms of Optimization Problems
Edited by K. Malanowski, K. Mizukami
VIII, 240 pages, 1986.

Vol. 83: Analysis and Optimization of Systems
Proceedings of the Seventh International Conference of Analysis and Optimization of Systems
Antiba, June 26-27, 1986
Edited by A. Bensoussan, J. L. Lions
XVI, 901 pages, 1986.

Vol. 84: System Modelling and Optimization
Proceedings of the 12th IFIP Conference
Budapest, Hungary, September 2–6, 1985
Edited by A. Prékopa, J. Szelezsán, B. Strazicky
XII, 1046 pages, 1986.

Vol. 85: Stochastic Processes in Underwater Acoustics
Edited by Charles R. Baker
V, 205 pages, 1986.

Vol. 86: Time Series and Linear Systems
Edited by Sergio Bittanti
XVII, 243 pages, 1986.

Vol. 87: Recent Advances in System Modelling and Optimization
Proceedings of the IFIP-WG 7/1 Working Conference
Santiago, Chile, August 27-31, 1984
Edited by L. Contesse, R. Correa, A. Weintraub
IV, 199 pages, 1987.

Vol. 88: Bruce A. Francis
A Course in H_∞ Control Theory
XI, 156 pages, 1987.

Vol. 88: Bruce A. Francis
A Course in H_∞ Control Theory
X, 150 pages, 1987.
Corrected - 1st printing 1987

Vol. 89: G. K. H. Pang/A. G. J. McFarlane
An Expert System Approach to Computer-Aided Design of Multivariable Systems
XII, 223 pages, 1987.

Vol. 90: Singular Perturbations and Asymptotic Analysis in Control Systems
Edited by P. Kokotovic, A. Bensoussan, G. Blankenship
VI, 419 pages, 1987.

Vol. 91 Stochastic Modelling and Filtering
Proceedings of the IFIP-WG 7/1 Working Conference
Rome, Italy, Decembre 10-14, 1984
Edited by A. Germani
IV, 209 pages, 1987.

Vol. 92: L. T. Grujić, A. A. Martynyuk, M. Ribbens-Pavella
Large-Scale Systems Stability Under Structural and Singular Perturbations
XV, 366 pages, 1987.

Vol. 93: K. Malanowski
Stability of Solutions to Convex Problems of Optimization
IX, 137 pages, 1987.

Vol. 94: H. Krishna
Computational Complexity of Bilinear Forms
Algebraic Coding Theory and Applications to Digital Communication Systems
XVIII, 166 pages, 1987.

Vol. 95: Optimal Control
Proceedings of the Conference on Optimal Control and Variational Calculus
Oberwolfach, West-Germany, June 15-21, 1986
Edited by R. Bulirsch, A. Miele, J. Stoer and K. H. Well
XII, 321 pages, 1987.

Vol. 96: H. J. Engelbert/W. Schmidt
Stochastic Differential Systems
Proceedings of the IFIP-WG 7/1 Working Conference
Eisenach, GDR, April 6-13, 1986
XII, 381 pages, 1987.

For information about Vols. 1–80 please contact your bookseller or Springer-Verlag.

Lecture Notes in Control and Information Sciences

Edited by M. Thoma and A. Wyner

151

J. M. Skowronski, H. Flashner, R. S. Guttalu (Eds.)

Mechanics and Control

Proceedings of the 3rd Workshop on Control Mechanics in Honor of the 65th Birthday of George Leitmann
January 22-24, 1990, University of Southern California

Springer-Verlag Berlin Heidelberg GmbH

Editors
Prof. J. M. Skowronski
Prof. H. Flashner
Prof. R. S. Guttalu

Mechanical Engineering
University of Southern California
Los Angeles
CA 90089-1453
USA

ISBN 978-3-540-53517-1 ISBN 978-3-540-46752-6 (eBook)
DOI 10.1007/978-3-540-46752-6

Originally published by Springer-Verlag Berlin Heidelberg New York in 1991.

61/3020-543210 Printed on acid-free paper.

CONTENTS

INTRODUCTION

With the present use of high speed and high precision machines and vehicles working in uncertain environment, adequate modelling with untruncated nonlinearity of characteristics becomes essential. The control programs designed must also be robust against uncertainty, thus signal adaptive. This asks for both, *developing control theory of nonlinear mechanical systems and use of nonlinear mechanics when investigating control systems.*

The three Workshops on Control Mechanics held annually since 1988 at the University of Southern California made substantial contribution to the above set of topics and have established a tradition in the interface between mechanics and control. The Workshops gained considerable international significance, with participants coming from USA, Canada, Australia, UK, Sweden, France, Germany and USSR.

The Third Workshop presented in this Volume is also special by producing papers dedicated to G. Leitmann, one of the founders of the study on dynamics of uncertain systems, and thus emphasising this recently rapidly developing discipline. The three major sets of topics presented deal successively with the general control theory of uncertain systems, with new theoretic methods in control, and with applications to control of robotic manipulators, aircraft control and flexible structures, all well integrated.

The first group includes papers on stabilization (Ryan, Soldatos – Corless – Leitmann, Madani Esfahani – Hui – Zak, Zufiria – Guttalu, Pak – Shieh), design of robust controllers (Kang – Horowitz – Leitmann, Dolphus – Schmitendorf, Blackwell, Chen – Chen, Hamano, Bojadziev – Skowronski), reduced dynamics (Corless, Ardema – Cooper), chaos (Vincent – Yu, Unal) and stochastic analysis (Blaquiere, Meerkov, Jabbari).

The control methodology is represented by direct works on Liapunov formalism (Olas), use of Newton's Law (Stadler) and Hamiltonian modelling (Flashner – Skowronski), global optimal control (Galperin), controllability (Stadler) and stochastic controllability (Blaquiere).

Finally the applications include, as mentioned, the design of manipulator controllers (Ahmad – Mrad, Dolphus – Schmitendorf, Flashner – Skowronski, Kang – Horowitz – Leitmann), aircraft control (Miele, Gazit – Gutmann, Merz) and flexible structures (Ahmard – Mrad, Parks – Pak, Perez – Nwokah).

The basic feature of all papers is the Liapunov formalism applied to control of dynamical systems with untruncated nonlinearity and subject to uncertain perturbations.

We would like to thank the Authors and all participants for their continuing patronage of the Workshop and hope that this Volume may contribute to development of the nonlinear control methodology.

Thanks are in order to the Dean of Engineering, L. Silvermann and the Department of Mechanical Engineering, particularly Head, Professor G. Shiflett and Administrative Officer Mrs J. Givens, of the University of Southern California for their hospitality and intensive help in organizing the Workshops.

J.M. Skowronski
H. Flashner
R.S. Guttalu

Los Angeles, June 1990.

IN HONOR OF GEORGE LEITMANN

A. Miele
Aero-Astronautics Group
Rice University
Houston, Texas 77251-1892

It is a privilege for me to have been invited to give these opening remarks in honor of George Leitmann, who has contributed so much to our area of research. I believe that I can speak with knowledge, because I have known George for at least 30 years and I have been aware of his work for almost 40 years. He has been one of the main engines of the advances that have occurred in optimal control, differential games, and engineering applications. His work has been impressive, original, varied, and above all thorough. His professional achievements and his influence have been considerable. But let me proceed in order.

Born on May 24, 1925 in Vienna (Austria), but conceived in Venice (Italy), the only child of Josef Leitmann and Stella Fisher, George Leitmann attended grammar school and then the Realgymnasium (technical preparatory school). He emigrated to the United States in April 1940, accompanied by his mother and his two grandmothers.

He attended the technical high school in Queens, New York. After graduation, he joined the US Army in February 1944, arriving in the European theater of operations in the fall of 1944, where he served in the reconnaissance unit of a Combat Engineer Battalion attached to the French First Army. For his combat service, France awarded him the "Croix de Guerre avec Palmes" and Belgium awarded him the "Décoration Fourragère". In 1945, he was transferred to the Counterintelligence Corps (CIC), becoming its youngest special agent. After a brief period of service at the

Nuremberg War Trials, George was discharged in May 1946 and began college studies at Columbia University. There, he received the BS and MA Degrees in Physics in 1949 and 1950.

In the fall of 1950, George accepted a position as Physicist at the US Naval Ordnance Test Station, China Lake, California, where he eventually became Head of the Aeroballistics Section. During the seven years at China Lake, there was a two-year parenthesis to obtain the PhD Degree in Engineering Science at the University of California at Berkeley (UCB); more importantly, he met and in 1955 married Nancy Lloyd. The Leitmanns' first child, Josef, was born at China Lake; he now works for the World Bank and is completing the PhD Degree in Regional Planning at UCB. In 1957, George joined the Mechanical Engineering Faculty at UCB as an Assistant Professor; he was promoted to Associate Professor in 1959 and to Professor in 1963. The Leitmanns' second child, Elaine, was born in 1959; both Elaine and her husband Bob graduated from the University of California at Davis and now live in Napa Valley.

In the early 1950's, while still at China Lake, George Leitmann together with Pierre St. Amand wrote the first proposal for a US satellite. George's work on rocket ballistics led him first into flight mechanics and then into trajectory optimization. His early aerospace papers were of considerable interest. In particular, this statement applies to his treatment of the optimal burning program for a sounding rocket, which he presented at the Astronautical Congress in Rome in 1956.It also applies to his treatment of the optimal trajectories for a rocket in a vacuum and a uniform gravitational field; for this problem, he proved that the occurrence of singular arcs is to be excluded.

After joining UCB, George became interested in the foundations of optimal control theory. In the early 1960's, in collaboration with Professor Austin Blaquière of the University of Paris, he developed a geometric approach to optimal control, which was later extended to two-person, zero-sum games. Then, in the late 1960's, George and his students generalized this approach to many-person, nonzero-sum games; they also made significant contributions to sufficiency theorems in optimal control and game theory.

In subsequent years, George steadily extended his interest in applications to include not only flight mechanics and astrodynamics, but also economics, decision making, pursuit-evasion, collision avoidance, bargaining, and so on. As George has often said, the constructive interaction with many outstanding graduate students and some 60 postdoctoral fellows played a major role in this work.

Since the early 1970's, his main interest has been the control of uncertain systems in a deterministic setting, rather than in a stochastic setting or a fuzzy setting. His work in differential games led him naturally into treating control under uncertainty as a game against nature, an approach which is still widely used in the Soviet Union. Today, together with his graduate students and postdoctoral fellows, George continues working on the theory of the deterministic control of uncertain systems, with applications ranging from robotics to river pollution control, and from aircraft control to active control of seismically excited structures.

George Leitmann's contributions are summarized in some 200 publications, including eleven self-authored and coauthored books. However, his favorite publication is his translation from German into English of the "Mantle of Dreams", written by the Hungarian poet Bela Balas, who was the librettist for the composer Bela Bartok.

In addition to this active technical life, George has been busy serving his university and his profession. Among his many university assignments, the following must be mentioned: Chairman of the Applied Mechanics Division; University Ombudsman (during some particularly troubled years, 1968-70); Chairman of the Academic Senate Committee on Privilege and Tenure; Associate Dean for Graduate Affairs; Associate Dean for Academic Affairs; Acting Dean of the College of Engineering; Chairman of the University Research Expeditions Program; Cochairman of the UCB Research Center Executive Committee.

He has been a consulting staff scientist for the Lockheed Company and a consultant to the Martin Corporation, the Aerojet General Corporation, the Princeton Guggenheim Laboratory, and others. He is Editor of the Journal of Mathematical Analysis and Applications; also, he is a member of the editorial boards of Optimal Control Applications and Methods, Rivista di Matematica Pura ed Applicata, and Dynamics and Stability of Systems; and he is an Associate Editor of the Journal of Optimization Theory and Applications (JOTA). Concerning George's contributions to the well being of JOTA, I can speak with first-hand knowledge: I have sought his advice on innumerable technical matters, and his keen judgment has helped me in resolving many delicate problems.

In addition to the above consulting and editorial activities, George has been a long-time member of the Scientific Council of the International Center for

Mechanical Sciences in Udine, Italy; a member of the Advisory Board of the Alexander Von Humboldt Foundation in Germany; and a member of the Mathematics of Control Committee of IFAC. He was Chairman of the NSF Engineering Fellowship Panel, and currently serves on the NRC Advisory Committee on Scholarships and Fellowships.

He has been honored by his peers numerous times. A Fellow of the AIAA, he has received the Pendray Aerospace Literature Award and the Mechanics and Control of Flight Award. He is the recipient of the Levy Medal of the Franklin Institute and the Medal of the University of Liège. He is a member of the National Academy of Engineering, a foreign member of the Academy of Sciences of Bologna, a foreign member of the Academy of Engineering of Argentina, as well as a member of the International Academy of Astronautics. A recipient of the Von Humboldt Senior Scientist Award, he also holds honorary doctorates from the Technical University of Vienna and the University of Paris.

Of course, George Leitmann is more than an academician. His extramural interests are varied. An avid swimmer, he feels guilty if he does not cover at least a kilometer every day. He has an interest in art and art history, and he collects art; while Nancy is disposed toward modern art, George has reverted, as he puts it, to earlier art, in particular 19th century paintings and bronzes, but also middle-eastern bronzes, precolumbian ceramics, and Bohemian crystal. A knowledgeable imbiber of wine, he owns a retreat in Napa Valley, and he is a member of the Order of the Knights of the Vine.

The year 1990 is truly a banner year in the life of George Leitmann. It marks his 35th wedding anniversary, his

40th year in California, his 50th year in the USA, and his 65th birthday. On the occasion of the forthcoming birthday, I conclude these opening remarks by saying: "George, you are a most esteemed colleague and a dear personal friend. I admire you not only for your professional achievements, but also for the depth of your human qualities. I salute you with respect and affection."

GEORGE LEITMANN

A FEW WORDS FOR THE 65 th ANNIVERSARY OF George LEITMANN

at the dinner of the third workshop on Control Mechanics : Sunday night, January 21, 1990.

Speaker : Austin Blaquiere

It is a great honour and a great pleasure for me to speak at this dinner. I am endebted to the organizers of this meeting, in particular to my old friend Janislaw Skowronski. When he spoke to me of dedicating the Third Workshop on Control Mechanics to George, for his 65th-anniversary, I had a great enthusiasm and I knew that this enthusiasm was to be shared by many people as George has so many friends in the world.

It is good to see to-night the "family" celebrating this event which is at the same time a private one and one of the highlights of our scientific community.

The academic career of George is rich by its containt and paved with many prestigious honors :

it began with physics at Columbia University, where he graduated with honors in 1949 and obtained a master's degree a year later. Then, he worked as a research scientist (with emphasis on rocket motion and guidance) for the Naval Ordnance Test Station at China Lake. In 1954, he resumed graduate studies, with a Navy fellowship, at the University of California at Berkeley. He was awarded the PhD Degree in Engineering Science in 1956 and joined the Berkeley faculty the following year.

He has served on numerous academic committees and boards . From 1981 to 85 he has been Dean for Graduate Affairs, and is currently Dean for Academic Affairs.

Many prestigious honors have been bestowed on George, including the Pendray Award and the Mechanics and Control of Flight Award of the AIAA, the Levy Medal of the Franklin Institute, the Medal of the University of Liège, and a von Humboldt Senior Scientist Award.

George is a member of the National Academy of Engineering, Foreign Member of the Academy of Sciences of Bologna.

In the past months he has been constituted Docteur Honoris Causa of one of the most prestigious Universities of Paris, regarding Mathematics, namely the University Paris 9 - Dauphine. Prof. Jean-Pierre Aubin from that University sends his greetings and regrets not to be with us to-night. The celebration is expected in May. The precise date will be given by the Rector.

The influence which George's contributions have had and continue to have is evident from many factors : the results of his work are in wide use, his theories are taught in many universities both in the States and abroad. Books of his have been translated into Russian, Polish, Japanese and German. His pupils occupy prominent positions in academia, industry and government.

If I had to briefly characterize the whole research of George, I would say that it is concerned with every possibility of improving the relations between human beings, and their relations with their surroundings. Thus, his research parallels his way of life.

It is not the place and the time, I believe, to speak further of George's scientific achievement, since we will enjoy three most-welcome days of meeting. To-night, let me praise a quarter of century of friendship. If many people know his work, but few know from personal experience that he is generous, perceptive, diplomatic, witty with a touch of poetry. He is a man whose character has been seasoned by a broad range of experiences

He was born in 1925 to Josef and Stella Leitmann in Vienna, which was his home until 1940. He and his mother were obliged by the fury of nazism to leave their native land for the United States.

In February 1944, following high school graduation, George enlisted in the Army. He was assigned as an NCO to the 286th Combat Engineer Battalion Reconnaissance Squadron and subsequently transferred to Europe. He was in combat from

December 1944 to May 1945 and received the *French Croix de Guerre* and the *Belgian Fourragère decoration* . In August 1945, he was assigned to the CIC as a special agent, and he served at the Nuremberg Trials from February to May 1946, when he was discharged.

He is one of these most courageous men who, in the Army of the United States of America, delivered Europe and the world from nazism and restored our freedom.

To have him as a collegue and a friend has been one of the highlights of my life.

Anniversaries are a time for remembrance. As I try to put some order in my memories, in the center of so many priviledged instants of time, one question comes to my mind:

"*well, George, you are a dear friend, what makes you so rare?*"
Let me greet one of your outstanding characteristics which sum up, I believe, several other ones, namely

you are trustable hundred-per-cent.

This is true for your friendship, also for the advices you give regarding practical problems, psychological ones, diplomatical ones, scientific ones, and others. Surely, you have not the solution for all of them, but, in all the cases, you give the best of yourself for helping to approach a reasonable, possibly approximate, solution, and the one you help can be sure that this solution or this approximation is the best one.

Democrite has said : *All of what exists in the Universe is due to hazard and necessity.* I guess that you always strive to avoid hazard. More precisely, you strive to enclose the hazard in a deterministic circle, not by a statistical description, but by dealing with bounds on uncertain factors. You exhibit the best you can do under the conditions laid down by uncertainty. You thus reach necessity in the sense of Democrite which corresponds also, I believe, in this sentence, to the wide-spread belief that nature operates at the minimum cost : it does just what is necessary.This is a statement of efficiency which is at the basis of optimal control.

Said in other words, one has the feeling that the premonitory sentence of Democrite could be put as an exergue to your achievement.

You put everything at its right place, in the proper light.

You know that you are not an oracle, and that the outward world and the life itself, though plausible, are only relative convenient operational hypotheses. This gives you the proper distance for facing the problems. As you are mindful of others, you place yourself at their level and on their ground, you listen, strive to understand and to help in the most direct way : this gives more price and efficiency to your words.

Here is one reason why so many people, your peers, your pupils and your students, so often ask for your advice. Let me recall that you have been appointed UCB's first Ombudsman in 1968, a post that you held during some of the university's most trying days of on-campus political activism and student unrest.

Everyone who has worked with you knows how stimulating it is. I often recall memories of my first stay in Berkeley, in 1964. Your office was close to mine in the provisional building of Mechanical Engineering on the campus. Your rigour, the soundness of your views, added to your friendly welcome, exerted a great attraction on me as they exert a great attraction on every people you meet. This has been the starting point of a long collaboration and of an ever lasting friendship between us and our families.

We have travelled in many countries, somewhat, like the strolling brothers riding the sacred Himalayan cow, in the Mantle of Dreams, a book of the Poet Béla Balàzs which you have so beautifuly translated. Our steps led us in many places, some of which seem to belong also to the realm of dreams : the observatory of Ulugh Beg, the tomb of Tamerlane, the blue domes of the mistress Bibi-Khanym, and several others. (These places were not yet reached by Perestroïka).

Here, let me greet the patience of our spouses Nancy and Paulette.

Now, those who have the great priviledge of having been welcome in the private life of your family know how, you and Nancy who herself received an MA from Boston University, are hospitable, generous, perceptive, mindful of the life of your city, of the States, and of what is going on abroad. You both are not only interested but deeply concerned with arts, in particular with music, painting, theater and poetry. You are recognized in your community as wisdom patrons of the arts.

Your children, Josef, a UC-Berkeley and a Harvard graduate, and married daughter Elaine, a UC-Davis graduate, now in the Napa Valley, are well prepared for entering the 21th century and to tread in the royal road you have shown to them.

Let me extend all my friendly wishes to all of you, and the wishes of all of us, in this great event : Happy Birthday and long life George, for you and your family!

ADAPTIVE CONTROL OF FLEXIBLE JOINT ROBOTS DERIVED FROM ARM ENERGY CONSIDERATIONS

Shaheen AHMAD, Fouad MRAD

Real-time Robot Control Laboratory
School of Electrical Engineering
Purdue University
West Lafayette, IN 47907 - 0501
U S A

1. Abstract *Almost all industrial robots exhibit joint flexibility due to mechanical compliance of their gear boxes. In this paper we outline the design of three controllers for flexible joint robots. Two of the three controllers are suitable for parameter adaptation, the candidate Lyapunov functions for these two controllers are derived from arm energy considerations.*

The desired actuator trajectory in a flexible joint robot is dependent not only on the desired kinematic trajectory of the link but also on the link dynamics. Unfortunately, link dynamic parameters are unknown in most cases, as a result the desired actuator trajectory is also unknown. To overcome this difficulty, a number of control schemes require the use of link acceleration and link jerk feedback. In this paper we describe three control schemes for flexible joint robots which do not use link jerk or acceleration. One of the controllers is suitable for trajectory tracking when the robot parameters are known in advance. The other two control laws are derived from candidate Lyapunov functions which resemble the energy of the arm deviating from the desired trajectory. Trajectory tracking and adaptation of robot arm parameters are possible with two of the controllers described in this paper. Our control schemes do not require the numerical differentiation of the velocity signal, or the inversion of the inertial matrices. Simulations are presented to verify the validity of the control scheme. The superiority of the proposed scheme over existing rigid robot adaptive schemes is also illustrated through simulation.

2. Introduction

Many of today's rigid robots are driven by actuators with high gear ratios, the load due to the arm at the actuator is reduced by a factor of n_g, where, $n_g > 1$, is the gear ratio. In fact, inertia of the arm experienced by the actuator is reduced by $(1/n_g^2)$, and as the actuator acceleration is n_g times the joint acceleration, the overall load is reduced by $(1/n_g)$. Thus the load experienced by robots with high gear ratios are dominated by actuator dynamics, link dynamics are secondary. Recent trend is towards high-technology direct-drive robots. Here, the actuators are directly connected to the links and the lack of high gear ratios and increasing demand for high-speed operation, requires the control system to compensate for the dominant nonlinear link dynamics. Thus the presence of high gear ratios reduces the effective load experienced at the motors but at slower robot operations, and the absence of gearing adds to the complexity of the control problem. Robots which move fast (apparently with reasonable manufacturing cycle times) and or carry large loads have additional problems. It is experimentally found that most gearing systems are compliant, as a result, actuators are connected to the robot links through effectively flexible shafts. Experimental evidence also indicates that joint flexibility should be accounted for in both modeling and control of manipulators [1] [14] [4]. The presence of joint flexibility in the direct-drive high-speed actuators can be modeled by a "linear" torsional spring. This flexibility may be attractive in some practical applications when the robot must make contact with an unknown surface.

Numerous techniques to control Flexible Joint Robots have been suggested [14], [2], [3], [6], [12], [4]. One approach is based on the idea of feedback linearization, which requires the measurement of joint acceleration and jerk to be used in the feedback loop [2], [12]. Another method is based on the concept of reduced order system and requires the restriction of the system to a suitable integral manifold in the state space [6].

We derive three controllers which drives the FJR to track a desired trajectory . Similar to the work on rigidly jointed robots [10],[11], [7], our controller design starts by selecting a candidate Lyapunov function which is similar to the energy of the FJR. Our control scheme does not require link jerk, or acceleration feedback or the inversion of the inertia matrix, in addition parameter adaptation is easily accommodated for two of the three controllers.

At this time, the only adaptive control scheme for flexible joint robots that we are aware of that uses position and velocity feedback is the one derived from singular perturbation arguments by Ghorbel, Spong and Hung [4]. In order to derive an adaptive scheme from a singular perturbation argument, several assumptions are necessary, these include sufficient joint stiffness and that it is possible to ignore the higher order terms in the singular perturbation expansion. Assumptions such as these are not necessary in our derivations.

An important problem in adaptive control is that of parameter convergence, providing a sufficiently rich tracking signal has sometimes been assumed to be adequate conditions for parameter convergence. However tracking a persistently exciting trajectory does not mean that all of the unknown parameters of a certain manipulator can be estimated. In general, the maximum number of parameters that may be estimated depends on the trajectory used for estimation and on the kinematic structure of the manipulator. These unknown parameters could be categorized as uniquely identifiable, identifiable in linear combinations only, or unidentifiable. Typically, only those dynamic parameters that affect the force/torque equations of at least one joint can be identified.

The organization of the remainder of this paper is now described. Section #3 and #4 summarize the dynamics and trajectory model of the manipulator. A trajectory tracking controller which does not use link jerk and acceleration is described in section #5, it is assumed that the manipulator parameters are known in advance. Section #6 explores the use of the arm's energy as possible Lyapunov function candidates. The energy based Lyapunov functions derived in section #6 are used to derive adaptive control schemes in section #7. Simulation results are given in section #8, conclusions are presented in section #9.

3. Manipulator Models

Experimental investigations of industrial robots with harmonic drive transmission and other forms of gearing indicate that joint flexibility contributes significantly to the overall dynamics of the system [1], [13]. The dynamic equations of the flexible joint robots are given as:

$$\tau = D_m \ddot{q}_m + B_m \dot{q}_m + K_s (q_m - q) \qquad (1)$$

$$0 = D(q)\ddot{q} + C(q,\dot{q})\dot{q} + g(q) + K_s (q - q_m) \qquad (2)$$

where, an n-link manipulator becomes a 2n-degrees of freedom system:

D_m : Diagonal motor inertia matrix $\in \mathbb{R}^{n\times n}$
B_m : Diagonal motor damping matrix $\in \mathbb{R}^{n\times n}$
K_s : Diagonal drive shaft stiffness matrix $\in \mathbb{R}^{n\times n}$
q_m : Vector of sensed motor angles $\in \mathbb{R}^{n\times 1}$
$D(q)$: Link inertia matrix $\in \mathbb{R}^{n\times n}$
$C(q,\dot{q})$: Centrifugal and coriolis terms matrix $\in \mathbb{R}^{n\times n}$
$g(q)$: Gravitational vector term $\in \mathbb{R}^{n\times 1}$
q : Vector of link joint angles $\in \mathbb{R}^{n\times 1}$

Matrices D_m, B_m, K_s, are all positive definite matrices. Further, $D(q)$ is symmetric, positive definite and both $D(q)$ and $D^{-1}(q)$ are both bounded as function of q [13], [4]. When K_s tends toward infinity, the robot is considered to have rigid joints (i.e. $q = q_m$). The dynamic equations which represent the rigidly jointed robot, with the same inertial and coriolis matrices as the FJR defined above, are:

$$\tau = [D(q) + D_m]\ddot{q} + [C(q,\dot{q}) + B_m]\dot{q} + g(q) \qquad (3)$$

Some properties of the rigid model concerning the inertia matrix, coriolis and centrifugal force matrix were discussed by Koditschek. Those properties remain valid for the flexible model [4]. The first most important property shows that $D(q)$ and $C(q,\dot{q})$ are not independent, but the matrix $(\dot{D} - 2C)$ is skew symmetric, this can be easily derived from the Lagrangian formulation of the manipulator dynamics (see Appendix A). The second property confirms that the individual terms of the right hand side of equation (2), excluding the $K_s(q - q_m)$ term, could be represented by a linear relationship between a suitably selected set of unknown manipulator and load parameters [11], [4], [13], in other words equation (2) could be rewritten as:

$$0 = Y(\ddot{q},\dot{q},q)\, P + K_s\, (q - q_m) \qquad (4)$$

where $Y(\ddot{q},\dot{q},q) \in \mathbb{R}^{n\times r}$, is called the regressor matrix of known functions, and $P \in \mathbb{R}^{r\times 1}$ is a vector of unknown parameters.

4. Trajectory Model

Let $q_d(t) \in C^4$ denote a desired link trajectory in which case $q_d(t), \dot{q}_d(t), \ddot{q}_d(t), \dddot{q}_d(t)$ are all bounded and continuously differentiable. The set of desired motor trajectory can be derived using equation (4). The diagonal stiffness matrix, $K_s \in \mathbb{R}^{n\times n}$, can be written as $K_s = \text{Diag}[k_{si}]$, where $k_{si} > 0$, for $i = 1,2,\ldots,n$, represents the spring constant of the ith drive shaft. Since all of these constants are positive and K_s is a diagonal matrix, as a result matrix K_s is invertible and positive definite.

We assume the link parameters and the load handled by the end effector are time invariant, i.e.

$$P = \text{Constant vector, thus, } \dot{P} = \ddot{P} = 0 \tag{5}$$

The above assumption is valid in a large class of applications. The desired motor trajectory may now be computed as follows:

$$q_{md}(t) = K_s^{-1} Y(\ddot{q}_d, \dot{q}_d, q_d)P + q_d(t) \tag{6}$$

$$\dot{q}_{md}(t) = K_s^{-1} \dot{Y}(\ddot{q}_d, \dot{q}_d, q_d)P + \dot{q}_d(t) \tag{7}$$

$$\ddot{q}_{md}(t) = K_s^{-1} \ddot{Y}(\ddot{q}_d, \dot{q}_d, q_d)P + \ddot{q}_d(t) \tag{8}$$

The subscript "d" is used to denote the desired trajectory.

Notice that the desired motor trajectory $q_{md}(t)$, $\dot{q}_{md}(t)$ and $\ddot{q}_{md}(t)$ are dependent on the desired link trajectory $q_d(t)$, $\dot{q}_d(t)$ and $\ddot{q}_d(t)$ and also on the unknown parameters P and the link dynamics represented by Y_d, $\dot{Y}_d$ and $\ddot{Y}_d$. This makes it difficult to design a control law which utilizes the desired motor position and velocity.

Using equations (6), (7) and (8), removing subscripts d, and using equation (1) and (2), we can rewrite equation (1) in-link coordinates q as:

$$\tau = D_m K_S^{-1} D(q) q^{(4)} + N(q, q^{(1)}, q^{(2)}, q^{(3)}) \tag{8a}$$

$$= Y^*(q, q^{(1)}, q^{(2)}, q^{(3)}, q^{(4)}) P^* \tag{8b}$$

where, $N(.,.,.,.) \in \mathbb{R}^n$,is a nonlinear function and $q^{(i)}$ is the ith time derivative of q. From the structure of equation (8a) we can see that the FJR can be stabilized by feeding back a nonlinear function of the link position, velocity, acceleration and jerk. Notice that the fourth order dynamics in the link coordinates can also be written in the regressor matrix form in terms of some suitably selected vector of unknown parameters P^*.

5. Control of the Flexible Joint Robot when the Arm Parameters are Known

An adaptive controller for the FJR can be derived if measurements of q, $q^{(1)}$, $q^{(2)}$ and $q^{(3)}$ are available. Generally it is difficult to measure acceleration and jerk and it is desirable to design control schemes which only require the use of link position and velocity and motor position and velocity. In this section we will show that we can derive such a controller from

Lyapunov's second method if the arm's dynamic parameters are known in advance. First notice that if the arm parameters are known, then the acceleration and jerk of the joints can be obtained directly from the link and motor position and velocity measurements.

$$q^{(2)} = D^{-1}\{K_s(q_m - q) - C(q, q^{(1)})q^{(1)} - g(q)\} \tag{9a}$$

$$\text{and } q^{(3)} = \frac{d}{dt}q^{(2)} = \mathcal{F}(q, q^{(1)}, q_m, q_m^{(1)}) \tag{9b}$$

Having shown that $q^{(2)}$ and $q^{(3)}$ can be obtained from velocity and position feedback, we will now assume that they are available.

Given, $q_d(t) \in C^4$, is the desired link trajectory, we can define tracking error $e(t) = q_d(t) - q(t)$, and a composite error vector $\eta(t) \in \mathbb{R}^n$, such that $\eta(t) = e^{(3)} + \Lambda_2 e^{(2)} + \Lambda_1 e^{(1)} + \Lambda_0 e$, where $\Lambda_0, \Lambda_1, \Lambda_2 \in \mathbb{R}^{n\times n}$ are gain matrices. We will also define, $\mu(t) = \Lambda_2 e^{(3)} + \Lambda_1 e^{(2)} + \Lambda_0 e^{(1)}$.

Theorem #1

The following control torque applied to the dynamical system given in (8a):

$$\tau = N(q, q^{(1)}, q^{(2)}, q^{(3)}) + \beta C(q, q^{(1)})\eta + K_D\eta + \beta D(\mu + q_d^{(4)}) \tag{10}$$

ensures $e(t) \rightarrow 0$ as $t \rightarrow \infty$ for an appropriate choice of matrices $\Lambda_0, \Lambda_1, \Lambda_2$ and $K_D \in \mathbb{R}^{n\times n}$, given $\beta = D_m K_s^{-1}$. Furthermore, $\tau = \tau(q, q^{(1)}, q_m, q_m^{(1)})$.

Proof of Theorem #1:

Consider the Lyapunov function

$$V = \frac{1}{2}\eta^t D\eta$$

Then using the fact $(\dot{D} - 2C)$ is skew symmetric (see Appendix A), we obtain,

$$\begin{aligned} \dot{V} &= \eta^t(D\dot{\eta} + \frac{1}{2}\dot{D}\eta) - \frac{1}{2}\eta^t(\dot{D} - 2C(q, q^{(1)}))\eta \\ &= \eta^t\beta^{-1}(\beta D(q_d^{(4)} + \mu) + N + \beta C(q, q^{(1)})\eta - \tau) \end{aligned} \tag{12}$$

If we set τ as given in the above (10), we have:

$$\dot{V} = -\eta^t\beta^{-1}K_D\eta < 0 \tag{13}$$

As $\beta = D_m K_s^{-1}$ is a positive definite diagonal matrix, in which case we can set K_D as a positive definite diagonal matrix such that $\beta^{-1}K_D$ is positive definite. This ensures that $\eta(t) \rightarrow 0$ as $t \rightarrow \infty$, therefore for an appropriate choice of the matrices $\Lambda_0, \Lambda_1, \Lambda_2$ such that the eigenvalues of $(s^3 I + \Lambda_2 s^2 + \Lambda_1 s + \Lambda_0)$ are in the LHP, ensures that, $e(t) \rightarrow 0$, as, $t \rightarrow \infty$.

Notice that as $q^{(3)}$ and $q^{(2)}$ can be expressed in terms of link and motor positions and velocities, as shown in equations (9a) and (9b), $\tau(q, q^{(1)}, q^{(2)}, q^{(3)}) = \tau(q, q_m^{(1)}, q, q^{(1)})$. ###

Notice when the arm parameters are uncertain, we cannot calculate the link acceleration and jerk (as in (9a) and (9b)). As a result, this scheme is not suitable for adaptation. In the next two sections of this paper we derive adaptive controllers based on the arm energy consideration.

6. Selection of an Energy-based Lyapunov function

The total energy of the robot arm is E, it is the sum of the kinetic and potential energies of the actuator and linkages:

$$E=\tfrac{1}{2}\dot{q}_m^t D_m \dot{q}_m+\tfrac{1}{2}\dot{q}^t D\dot{q}+\tfrac{1}{2}(q-q_m)^t K_s(q-q_m)+\Phi(q) \tag{14}$$

where, $\Phi(q)$ is the gravitational potential energy of the linkage. Then the power input to the FJR is through the actuator and is given as:

$$\frac{dE}{dt}=(\tau_m-B_m\dot{q}_m)^t\dot{q}_m \tag{15}$$

Notice that when $\Phi(q) = 0$, $E(\dot{q}_m,\dot{q},q_m,q)$ becomes a quadratic in q, $\dot{q}$, q_m, and $\dot{q}_m$. Notice also, if we set $\tau_m = B_m\dot{q}_m-\Omega\dot{q}_m$, then

$$\frac{dE}{dt}=-\dot{q}_m^t\Omega\dot{q}_m \le 0 \tag{16}$$

where, $q_m \in \mathbb{R}^n$, and $\Omega \in \mathbb{R}^{n\times n} >0$ is a positive definite matrix .

We can conclude that, with an appropriate rate feedback, we may track a static joint trajectory. This exposition shows why most FJR with appropriate velocity feedback can be stabilized. This exposition indicates that we may select a Lyapunov function similar to E given in (14), and we may stabilize the FJR along a nonstatic link trajectory by suitable position and velocity feedback.

Excluding the potential energy of the FJR, the energy of the robot arm along a prespecified trajectory is:

$$E(t)=\tfrac{1}{2}\dot{q}_d^t D\dot{q}_d+\tfrac{1}{2}(q_d-q_{md})^t K_s(q_d-q_{md})+\tfrac{1}{2}\dot{q}_{md}^t D_m\dot{q}_{md} \tag{17}$$

Likewise, the energy of the FJR which causes the deviation from the desired trajectory is given as:

$$V(t)=\tfrac{1}{2}\dot{e}^t D\dot{e}+\tfrac{1}{2}(e-e_m)^t K_s(e-e_m)+\tfrac{1}{2}\dot{e}_m^t D_m\dot{e}_m \tag{18}$$

where, we defined the error terms as: $e = (q_d - q)$ and $e_m = (q_{md} - q_m)$.

Throughout the trajectory it is desired to have $\frac{dV}{dt} < 0$, furthermore $\dot{V}(t)$ and $V(t)$ should be dependent on e and e_m as well as $\dot{e}_m$ and $\dot{e}$. We can make $V(t)$ dependent on $\dot{e}$, e, $\dot{e}_m$ and e_m by selecting:

$$V(t)=\tfrac{1}{2}\dot{e}_m^t D_m\dot{e}_m+\tfrac{1}{2}\dot{e}^t D\dot{e}+\tfrac{1}{2}(e-e_m)^t K_s(e-e_m)$$

$$+\tfrac{1}{2}e^t K_p e+\tfrac{1}{2}e_m^t K_{pm} e_m \tag{19}$$

where, $K_p \in \mathbb{R}^{n\times n}$, $K_{pm} \in \mathbb{R}^{n\times n}$ are some positive definite gain matrices. The derivation of τ_m to make $\dot{V}(t) < 0$ and quadratically dependent on the variables e, e_m, $\dot{e}$ and $\dot{e}_m$ will be addressed in the next section.

7. Control and Adaptation Law Design

As the dynamic parameters of the arm are unknown and assumed to be time invariant, we can define the parameter error vector as $e_p = \tilde{P} - P$, where $\tilde{P}$ is the estimated parameter vector. Notice $\dot{e}_p = \dot{\tilde{P}}$, as $\dot{P} = 0$. Based on the estimated value of the parameter vector $\tilde{P}$, we obtain an estimate of the desired motor position as $\tilde{q}_{md}$ using equation (6). Similarly, we can compute the estimated motor velocity and acceleration. We can define the following motor error as $\tilde{e}_m = (\tilde{q}_{md} - q_m)$. Similar terms for $\dot{\tilde{e}}_m$ and $\ddot{\tilde{e}}_m$ can be defined. Based on the above Lyapunov function (19), we can find the energy of the trajectory deviating from the desired trajectory as:

$$V(t) = \tfrac{1}{2}\dot{\tilde{e}}_m^t D_m \dot{\tilde{e}}_m + \tfrac{1}{2}\dot{e}^t D \dot{e} + \tfrac{1}{2}(e - \tilde{e}_m)^t K_s (e - \tilde{e}_m)$$

$$+ \tfrac{1}{2} e^t K_p e + \tfrac{1}{2}\tilde{e}_m^t K_{pm} \tilde{e}_m + \tfrac{1}{2} e_p^t M e_p \tag{20}$$

The last term in (20) is added to account for parameter adaptation, where $K_p, K_{pm} \in \mathbb{R}^{n \times n}$, and $M \in \mathbb{R}^{r \times r}$ are some positive gain matrices.
For convenience let us define:

$$D(q)\ddot{q}_d + C(q,\dot{q})\dot{q}_d + g(q) = \Psi(\ddot{q}_d, \dot{q}_d, \dot{q}, q)P \tag{21}$$

where, $\Psi \in \mathbb{R}^{n \times r}$, and

$$Y_d = Y(\ddot{q}_d, \dot{q}_d, q_d) \tag{22}$$

where, $Y_d \in \mathbb{R}^{n \times r}$.

Theorem #2:

If the control torque is bounded such that $\|\tau\| < \tau_{max}$. The system given by the dynamical model (1) and (2), subjected to the below control and adaptation laws, results in bounded trajectory tracking error.

$$\tau = D_m \ddot{\tilde{q}}_{md} + B_m \dot{q}_m + K_s(\tilde{q}_{md} - q_d) + K_{pm}\tilde{e}_m + K_{dm}\dot{\tilde{e}}_m$$

$$+ \frac{\dot{\tilde{e}}_m}{\|\dot{\tilde{e}}_m\|^2}\left[\dot{e}^t((\Psi - Y_d)\tilde{P} + K_p e) + \dot{e}^t K_d \dot{e} + e^t K_p e + \tilde{e}_m^t K_{pm} \tilde{e}_m\right] \tag{23}$$

where, $K_{dm}, K_d \in \mathbb{R}^{n \times n}$ are some positive gain matrices. The parameter adaptation law is given by:

$$\dot{\tilde{P}}(t) = M^{-1}\Psi^t(\ddot{q}_d, \dot{q}_d, \dot{q}, q)\, \dot{e} \tag{24}$$

The trajectory tracking error is bounded in a set which is given by:

$$\alpha_1 \|\tilde{e}_m\| + \|\dot{e}\|(\alpha_2 \|e\| + \alpha_3(t))/\|\dot{\tilde{e}}_m\| + \alpha_4(t) < \tau_{max} \tag{25}$$

where α_1, α_2, α_3, α_4 are some positive scalars, and $\|\cdot\|$ is the Euclidean vector norm.

Proof of Theorem 2:

Differentiation of the positive Lyapunov function candidate V(t) in equation (20) yields the following:

$$\dot{V}(t)=\dot{\tilde{e}}_m^t[D_m\ddot{\tilde{e}}_m+K_s(\tilde{e}_m-e)+K_{pm}\tilde{e}_m]$$

$$+\dot{e}^t[D\ddot{e}+\tfrac{1}{2}\dot{D}\dot{e}+K_pe+K_s(e-\tilde{e}_m)]+e_p^tM\dot{e}_p-\dot{e}^t(\tfrac{1}{2}\dot{D}-C)\dot{e} \tag{26}$$

In order to simplify equation (26), we have subtracted the term $\dot{e}^t(\frac{1}{2}\dot{D}-C)\dot{e}=0$, see (Appendix A). Simplifying equation (26) and substituting the dynamic equation of the FJR given by (1) and (2), we have

$$\dot{V}(t)=\dot{\tilde{e}}_m^t\{D_m\ddot{\tilde{q}}_{md}+K_s(\tilde{q}_{md}-q_d)+K_{pm}\tilde{e}_m-[D_m\ddot{q}_m+K_s(q_m-q)]\}$$

$$+\dot{e}^t\{D\ddot{q}_d+C\dot{q}_d+K_s(q_d-\tilde{q}_{md})-[D\ddot{q}+C\dot{q}+K_s(q-q_m)+g(q)]$$

$$+K_pe+g(q)\}+e_p^tM\dot{e}_p$$

$$=\dot{\tilde{e}}_m^t[D_m\ddot{\tilde{q}}_{md}+K_s(\tilde{q}_{md}-q_d)+K_{pm}\tilde{e}_m+B_m\dot{q}_m-\tau]$$

$$+\dot{e}^t[D\ddot{q}_d+C\dot{q}_d+K_s(q_d-\tilde{q}_{md})+K_pe+g(q)]+e_p^tM\dot{e}_p \tag{27}$$

Then by substituting the controller (23) into (27), using the definition of Ψ given by equation (21), and by using the fact that, $K_s(q_d-\tilde{q}_{md})=-Y_d\tilde{P}$, derived from (6), we get:

$$\dot{V}(t)=\dot{\tilde{e}}_m^t\{-K_{dm}\dot{\tilde{e}}_m-\frac{\dot{\tilde{e}}_m}{\|\dot{\tilde{e}}_m\|^2}[\dot{e}^t((\Psi-Y_d)\tilde{P}+K_pe)+\dot{e}^tK_d\dot{e}+e^tK_pe+\tilde{e}_m^tK_{pm}\tilde{e}_m]\}$$

$$+\dot{e}^t[\Psi P-Y_d\tilde{P}+K_pe]+e_p^tM\dot{e}_p$$

$$=-\dot{\tilde{e}}_m^tK_{dm}\dot{\tilde{e}}_m-\dot{e}^tK_d\dot{e}-e^tK_pe-\tilde{e}_m^tK_{pm}\tilde{e}_m-\dot{e}^t\Psi e_p+e_p^tM\dot{e}_p$$

$$=-\dot{\tilde{e}}_m^tK_{dm}\dot{\tilde{e}}_m-\dot{e}^tK_d\dot{e}-e^tK_pe-\tilde{e}_m^tK_{pm}\tilde{e}_m+e_p^t[M\dot{e}_p-\Psi^t\dot{e}] \tag{28}$$

Since, $\dot{e}_p=\dot{\tilde{P}}-\dot{P}$, and as, $\dot{P}=0$ (robot arm parameters are time invariant), we can substitute the adaptation law (24) into (28) and the final expression for the derivative of the Lyapunov function is given as:

$$\dot{V}(t)=-\dot{\tilde{e}}_m^tK_{dm}\dot{\tilde{e}}_m-\dot{e}^tK_d\dot{e}-e^tK_pe-\tilde{e}_m^tK_{pm}\tilde{e}_m<0 \tag{29}$$

Which guarantees the convergence of $\dot{\tilde{e}}_m$, $\dot{e}$, $\tilde{e}_m$, and e as time goes to infinity.

The problem that can arise with this controller is that as $\|\dot{\tilde{e}}_m\|\rightarrow 0$, large torques are required to maintain the manipulator along the desired trajectory . As this is impractical , let us assume that the available joint torques are bounded , i.e. $\|\tau\|<\tau_{max}$, in which case we have $\dot{V}(t)<0$, if:

$$\|\gamma(t)\|+\|K_{pm}\|\|\tilde{e}_m\|+\frac{\|\dot{e}\|}{\|\dot{\tilde{e}}_m\|}[\|\rho(t)\|+\|K_p\|\|e\|] < \tau_{max} \tag{30a}$$

$$\text{where, } \gamma(t)=D_m\ddot{\tilde{q}}_{md}+B_m\dot{q}_m+K_s(\tilde{q}_{md}-q_d) \tag{30b}$$

$$\rho(t)=\Psi(\ddot{q}_d,\dot{q}_d,\dot{q},q)\tilde{P}-K_s(\tilde{q}_{md}-q_d) \tag{30c}$$

Let us define the following positive constants: $\alpha_1=\|K_{pm}\|$, $\alpha_2=\|K_p\|$, and positive scalars $\alpha_3(t)=\|\rho(t)\|$, and $\alpha_4(t)=\|\gamma(t)\|$.
The ultimate bound on the trajectory errors is then given by the set S defined as:

$$S=\{\, e(t),\dot{e}(t),\tilde{e}_m(t),\dot{\tilde{e}}_m(t) \mid \alpha_1\|\tilde{e}_m\|+\|\dot{e}\|(\alpha_2\|e\|+\alpha_3(t))/\|\dot{\tilde{e}}_m\|+\alpha_4(t) = \tau_{max} \,\} \tag{31}$$

###

Notice that for a particular trajectory an upperbound on $\alpha_3(t)$ and $\alpha_4(t)$ can be found, this allows one to find an upperbound on trajectory error for all time.

In order to reduce excessive torque demands as $\|\dot{\tilde{e}}_m\|\to 0$, we make use of the structural reduction in the system to propose a secondary controller. Let $\Lambda\in\mathbb{R}^{n\times n}$ be some positive diagonal matrix, then we let

$$\Gamma(\ddot{q}_d,\dot{q}_d,\dot{q},q)P=D(q)[\ddot{q}_d+\Lambda\dot{e}]+C(q,\dot{q})[\dot{q}_d+\Lambda e]+g(q) \tag{32}$$

where, $\Gamma\in\mathbb{R}^{n\times r}$. Furthermore let us define the following variables:
$s\in\mathbb{R}^n$, $s=(\dot{e}+\Lambda e)$. Let us also define a region where, $\ddot{e}_{mi}=\ddot{\tilde{q}}_{mdi}-\ddot{q}_{mi}$, as:

$$\mu_{min}(i) \le \ddot{e}_{mi} \le \mu_{max}(i) \quad \text{for } i=1,2,\dots,n \tag{33}$$

where, $\mu_{min}(i)$, and $\mu_{max}(i)$ are real scalars. Let us also set vector $\lambda=(\lambda_1,\dots\lambda_n)^t\in\mathbb{R}^n$ be defined such that:

$$\lambda_i = \tfrac{1}{2}D_{mi}\{Sgn(s_i)[\mu_{min}(i)-\mu_{max}(i)]+\mu_{min}(i)+\mu_{max}(i)\} \tag{34}$$

for $i=1,2,\dots,n$.

$$\text{where, } Sgn(s_i)=\begin{cases}+1 & \text{if } s_i>0\\ -1 & \text{if } s_i<0\\ 0 & \text{if } s_i=0\end{cases} \tag{35}$$

Theorem #3:

If, $\|\dot{\tilde{e}}_m\|^2<\epsilon>0$, we employ the below control law (36) and adaptation law (37) in addition (when $\|\dot{\tilde{e}}_m\|^2>\epsilon$), we employ the control and adaptation law described in theorem 2, then the norm of the system tracking error will decrease to the order of $O(\sqrt{\epsilon})$. The second stage control law is given by:

$$\tau=\Gamma(\ddot{q}_d,\dot{q}_d,\dot{q},q)\tilde{P}+D_m\ddot{\tilde{q}}_{md}+B_m\dot{q}_m+K_d s-\lambda \tag{36}$$

where, K_{dm}, $K_d\in\mathbb{R}^{n\times n}$ are some positive definite gain matrices. The second stage adaptation law is given by

$$\dot{\tilde{P}}(t)=M^{-1}\Gamma^t(\ddot{q}_d,\dot{q}_d,\dot{q},q)[\dot{e}+\Lambda e] \tag{37}$$

Proof of theorem 3:

Let us consider the system when $||\dot{\tilde{e}}_m||^2 \leq \epsilon$. If ϵ is suitably small then at this stage the motor is tracking the estimated actuator trajectory in velocity, but a steady state error may exist between the actual and desired motor position.

Notice now as $||\dot{\tilde{e}}_m||^2 \rightarrow 0$, **a structural reduction in the system is** apparent as the "Lyapunov" function V(t) in (20) resembles that of a rigid robot, as the first term is approximately zero. We exploit this property in the secondary control. The dynamic equations (1) and (2) can be added to obtain a single system equation:

$$\tau = D(q)\ddot{q}+C(q,\dot{q})\dot{q}+g(q)+D_m\ddot{q}_m+B_m\dot{q}_m \tag{38}$$

Consider now the Lyapunov function candidate W(t):

$$W(t)=\tfrac{1}{2}(\dot{e}+\Lambda e)^tD(\dot{e}+\Lambda e) + \tfrac{1}{2}e_p^tMe_p \tag{39}$$

Differentiating W(t) with respect to time, substituting Γ given by (32), and using the dynamic equation (38) leads us to:

$$\dot{W}(t)=s^t[D(\ddot{e}+\Lambda\dot{e})+\tfrac{1}{2}\dot{D}(\dot{e}+\Lambda e)] + e_p^tM\dot{e}_p$$

$$=s^t[D\ddot{q}_d+D\Lambda\dot{e}+C(q,\dot{q})(\dot{q}_d+\Lambda e)-(D\ddot{q}+C(q,\dot{q})\dot{q})] + e_p^tM\dot{e}_p$$

$$=s^t[D(\ddot{q}_d+\Lambda\dot{e})+C(q,\dot{q})(\dot{q}_d+\Lambda e)+g(q)$$

$$+D_m\ddot{q}_m+B_m\dot{q}_m-\tau] + e_p^tM\dot{e}_p$$

$$=s^t[\Gamma P +D_m\ddot{\tilde{q}}_{md}-D_m\ddot{\tilde{e}}_m+B_m\dot{q}_m-\tau] + e_p^tM\dot{e}_p \tag{40}$$

Substituting for τ from equation (36) into (40) yields,

$$\dot{W}(t)=s^t[-\Gamma\, e_p -K_ds-D_m\ddot{\tilde{e}}_m+\lambda] + e_p^tM\dot{e}_p$$

$$=-s^t\Gamma e_p -s^tK_ds -s^t[D_m\ddot{\tilde{e}}_m-\lambda]+ e_p^tM\dot{e}_p$$

$$= -s^tK_ds -s^t[D_m\ddot{\tilde{e}}_m-\lambda]+e_p^t[M\dot{e}_p-\Gamma^ts] \tag{41}$$

As, $\dot{e}_p(t) = \dot{\tilde{P}}(t)$, since $\dot{P}=0$, now let us substitute $\dot{\tilde{P}}(t)$ given by equation (37) into (41), it yields:

$$\dot{W}(t)= -s^tK_ds -s^t[D_m\ddot{\tilde{e}}_m-\lambda]+ e_p^t[MM^{-1}\Gamma^ts-\Gamma^ts]$$

$$= -s^tK_ds-\sum_{i=1}^{n}s_i(D_{mi}\ddot{\tilde{e}}_{mi}-\lambda_i) < 0 \tag{42}$$

A substitution for the values of λ_i's from equation (34) guarantees that $\dot{W}(t)$ is upper bounded by zero and decreases for any nonzero $(s=\dot{e}+\Lambda e)$, and s converges to zero with time going to infinity for positive definite gain matrices Λ, and K_d. Consequently, this implies that both $\dot{e}(t)$ and $e(t)$ decreases to zero as time goes to infinity providing the controller stays in the secondary controller. If however the controller switches back to the primary controller the link and motor errors decrease towards zero until $\|\dot{\tilde{e}}_m\|^2=\epsilon$. Therefore the dual control scheme will result in the norm of the tracking error $(\sqrt{\|e\|^2+\|\tilde{e}_m\|^2+\|\dot{\tilde{e}}_m\|^2+\|\dot{e}\|^2})$ to decrease to the order of $O(\sqrt{\epsilon})$. ###

8. Simulation Results For A Two-Link Planar FJR.

We now describe the computer simulation for a two-link planar manipulator with revolute joints (see Figure 1). The linkage are composed of two identically uniform beams which are infinitely rigid, with actuators mounted at the joints (see Figure 2). We assume that the load carried by the end-effector is a part of the second link. From equations (1) and (2), the dynamic equations of the two link manipulator are given as:

$$\begin{bmatrix}\tau_1\\ \tau_2\end{bmatrix}=\begin{bmatrix}d_{m1} & 0\\ 0 & d_{m2}\end{bmatrix}\begin{bmatrix}\ddot{q}_{m1}\\ \ddot{q}_{m2}\end{bmatrix}+\begin{bmatrix}b_{m1} & 0\\ 0 & b_{m2}\end{bmatrix}\begin{bmatrix}\dot{q}_{m1}\\ \dot{q}_{m2}\end{bmatrix}+\begin{bmatrix}k_{s1} & 0\\ 0 & k_{s2}\end{bmatrix}\begin{bmatrix}q_{m1}-q_1\\ q_{m2}-q_2\end{bmatrix} \tag{43}$$

and,

$$\begin{bmatrix}0\\ 0\end{bmatrix}=\begin{bmatrix}d_{11} & d_{12}\\ d_{21} & d_{22}\end{bmatrix}\begin{bmatrix}\ddot{q}_1\\ \ddot{q}_2\end{bmatrix}+\begin{bmatrix}c_{11} & c_{12}\\ c_{21} & c_{22}\end{bmatrix}\begin{bmatrix}\dot{q}_1\\ \dot{q}_2\end{bmatrix}+\begin{bmatrix}g_1(q)\\ g_2(q)\end{bmatrix}+\begin{bmatrix}k_{s1} & 0\\ 0 & k_{s2}\end{bmatrix}\begin{bmatrix}q_1-q_{m1}\\ q_2-q_{m2}\end{bmatrix} \tag{44}$$

where the coefficients can be derived from the Lagragian formulation (similar to that in Paul's book, [8]). Notice that $g\in\mathbb{R}^1$ is the gravitational acceleration and it is assumed to be $9.81\,ms^{-2}$.

$d_{11}=I_1+I_2+(m_1+4m_2)l_1^2+m_2l_2^2+4m_2l_1l_2\cos(q_2)$
$d_{12}=d_{21}=I_2+m_2l_2^2+2m_2l_1l_2\cos(q_2)$
$d_{22}=I_2+m_2l_2^2$
$c_{11}=-4m_2l_1l_2\dot{q}_2\sin(q_2)$
$c_{12}=-2m_2l_1l_2\dot{q}_2\sin(q_2)$
$c_{21}=2m_2l_1l_2\dot{q}_1\sin(q_2)$
$c_{22}=0$, and $I_j=.33m_jl_j^2+.01m_jl_j^2$ for $j=1,2$
$g_1(q)=g[(m_1+2m_2)l_1\cos(q_1)+m_2l_2\cos(q_1+q_2)]$

$$g_2(q)=gm_2l_2\cos(q_1+q_2) \tag{45}$$

For notational convenience let us define

$$S_1=\sin q_1 \quad C_1=\cos q_1 \quad C_{12}=\cos(q_1+q_2)$$

$$S_2=\sin q_2 \quad C_2=\cos q_2 \quad S_{12}=\sin(q_1+q_2)$$

We can rewrite the manipulator dynamics in the regressor form with the unknown parameters appearing linearly as:

$$\begin{bmatrix}0\\0\end{bmatrix}=\begin{bmatrix}\ddot{q}_1 & (\ddot{q}_1+\ddot{q}_2) & (\alpha C_2-\beta S_2) & gC_1 & gC_{12}\\ 0 & (\ddot{q}_1+\ddot{q}_2) & (\ddot{q}_1C_2+\dot{q}_1^2S_2) & 0 & gC_{12}\end{bmatrix}\begin{bmatrix}p_1\\p_2\\p_3\\p_4\\p_5\end{bmatrix}$$

$$+\begin{bmatrix}k_{s1} & 0\\ 0 & k_{s2}\end{bmatrix}\begin{bmatrix}q_1-q_{m1}\\ q_2-q_{m2}\end{bmatrix} \tag{46}$$

$$\text{where, } \alpha = 2\ddot{q}_1 + \ddot{q}_2 \text{ and } \beta = 2\dot{q}_1\dot{q}_2 + \dot{q}_2^2 \tag{47}$$

Furthermore, the unknown parameter vector P is given as:

$$P=\begin{bmatrix}p_1\\p_2\\p_3\\p_4\\p_5\end{bmatrix}=\begin{bmatrix}I_1+m_1l_1^2+4m_2l_1^2\\ I_2+m_2l_2^2\\ 2m_2l_1l_2\\ l_1(m_1+2m_2)\\ m_2l_2\end{bmatrix} \tag{48}$$

Therefore the vector functions of unknown parameters $P \in \mathbb{R}^5$ and the regressor matrix $Y(\ddot{q},\dot{q},q) \in \mathbb{R}^{2\times5}$ are well defined. After choosing the desired links trajectory, we use equations (6), (7) and (8) to derive the desired motor trajectory.

The control law given in equation (23) and (36) and the adaptation laws given (24) and (37) were used with the following definitions of Ψ and Γ:

$$\Psi=\begin{bmatrix}\ddot{q}_{1d} & (\ddot{q}_{1d}+\ddot{q}_{2d}) & (\alpha_1C_2-\beta_1S_2) & gC_1 & gC_{12}\\ 0 & (\ddot{q}_{1d}+\ddot{q}_{2d}) & (\ddot{q}_{1d}C_2+\dot{q}_{1d}\dot{q}_1S_2) & 0 & gC_{12}\end{bmatrix} \tag{49}$$

$$\text{where, } \alpha_1 = 2\ddot{q}_{1d} + \ddot{q}_{2d} \text{ and } \beta_1 = 2\dot{q}_{1d}\dot{q}_2 + \dot{q}_{2d}\dot{q}_2 \tag{50}$$

and

$$\Gamma=\begin{bmatrix}(\ddot{q}_{1d}+a) & \gamma_1 & (\alpha_2C_2-\beta_2S_2) & gC_1 & gC_{12}\\ 0 & \gamma_2 & (\ddot{q}_{1d}+a)C_2+(\dot{q}_{1d}+c)\dot{q}_1S_2 & 0 & gC_{12}\end{bmatrix} \tag{51}$$

$$\text{where, } \alpha_2 = 2(\ddot{q}_{1d}+a)+\ddot{q}_{2d}+b\ ,\ \beta_2 = 2(\dot{q}_{1d}+c)\dot{q}_2 + (\dot{q}_{2d}+d)\dot{q}_2$$

$$\text{and } \gamma_1=(\ddot{q}_{1d}+\ddot{q}_{2d}+a+b)\ ,\ \gamma_2=(\ddot{q}_{1d}+\ddot{q}_{2d}+a+b) \tag{52}$$

assuming a, b, c, and d are derived from:

$$\Lambda\dot{e}=\begin{bmatrix}a\\b\end{bmatrix} \text{ and } \Lambda e=\begin{bmatrix}c\\d\end{bmatrix} \tag{53}$$

We selected a sampling period of 10×10^{-3} seconds corresponding to a servo rate of 100 Hz. We selected, $\epsilon = 0.5$, and the second controller (36) was activated when, $\|\dot{\tilde{e}}_m\|^2 < \epsilon$. The value of ϵ is quite large and it was selected to ensure small torque demand in the first level controller

(23). The following bounds were used in the definition of λ_i's $-2 \leq \|\ddot{e}_{mi}\| \leq 2$, for i =1,2. Table S-1 shows the numerical values of the parameter vector P, and the known motors parameters.

parameter	value
$K_{s1}=k_{s2}$	50 N m rad^{-1}
$D_{m1}=d_{m2}$	.05 kg m^2
$D_{m1}=b_{m2}$	.05 N m sec rad^{-1}
P_1	1.66
P_2	0.42
P_3	0.63
P_4	3.75
P_5	1.25
$l_1=l_2$	.25 m
$m_1=m_2$	5 kg

Table S-1 : Actual parameters values

Four different cases were simulated to show the improvement obtained over current adaptive control schemes for robotic manipulators. The need for adaptive control is also illustrated through simulations. Simulations for the controller described in equation(10), when parameters of the arm are known are not presented in this paper.

As seen from table (S-1), the robot considered here has extremely flexible joints. A load of 5kg, when the arm is fully extended and parallel to the horizontal plane, results in the inner joint q_1 to deflect by 1rad (or 57.3°). Current industrial robots have joint stiffness in excess of several hundred $Nmrad^{-1}$. Notice also, this manipulator is not light and each link has a weight of 5kg.

In the below simulations, we assume the manipulator is initially at rest with $q_1 = -90°$, and $q_2 = 0°$. The desired trajectory is given by:

$$q_{1d}(t) = [-\frac{\pi}{2} + 0.3\sin(\pi t)]\text{rad}. \quad (54)$$

$$q_{2d}(t) = [-0.3 + 0.3\cos(\pi t)]\text{rad}. \quad (55)$$

Case #1:

In order to show that current rigid robot adaptive schemes are ineffective when applied to FJR, we applied the elegant adaptive control schemes suggested by Slotine and Li [11] to the FJR described in table (S-1). As this controller was derived on the assumption that the joints are rigid, equation (3) was used for the rigid robot model and the rigid robot control law was:

$$\tau=\Gamma(\ddot{q}_d,\dot{q}_d,\dot{q},q)\tilde{P}+D_m(\ddot{q}_d+\Lambda e)+B_m\dot{q}+K_d(\dot{e}+\Lambda e) \quad (56)$$

and the adaptation law was:

$$\dot{\tilde{P}}(t)=M^{-1}\Gamma^t(\ddot{q}_d,\dot{q}_d,\dot{q},q)[\dot{e}+\Lambda e] \quad (57)$$

The controller gains were found to be

$$K_p = K_{pm} = \begin{bmatrix} 0.25 & 0 \\ 0 & 0.25 \end{bmatrix} \quad K_d = K_{dm} = \begin{bmatrix} 1 & 0 \\ 0 & 1 \end{bmatrix} \tag{58}$$

$$\Lambda = \begin{bmatrix} 0.25 & 0 \\ 0 & 0.25 \end{bmatrix} \quad M = \begin{bmatrix} 1 & 0 \\ 0 & 1 \end{bmatrix} \tag{59}$$

The response of the manipulator to Slotine and Li's adaptive control law is shown in Figures (3a, 3b, and 3c). Figure 3a shows the link angle responses, the motor responses are shown in Figure 3b, and parameter $\tilde{P}_1$ is shown in Figure 3c. Notice all the parameters behave similarly to $\tilde{P}_1$, shown in Figure 3c. From Figures 3a and 3b, it can be seen that unacceptable link and motor responses are obtained before the system goes unstable. Figure 3c shows that the parameters vary wildly before diverging.

We expect that all other rigid robot adaptive control schemes would also produce unstable responses when applied to control FJR's with such low joint stiffness. These simulations indicate clearly the need to develop new adaptive control schemes for the FJR. Note that the rigid control law (56) gives acceptable responses for very large joint stiffness, eg with $K_{si}=6000\text{Nmrad}^{-1}$.

Case #2:

In order to show the effectiveness of the adaptive controller given by equations(23) and (24) we applied the control scheme to the FJR described in table S-1. Here we set $|\tau_1|<25\text{Nm}$ and $|\tau_2|<25\text{Nm}$. The controller gains were selected as shown below:

$$K_p = K_{pm} = \begin{bmatrix} 10 & 0 \\ 0 & 10 \end{bmatrix} \quad K_d = K_{dm} = \begin{bmatrix} 3 & 0 \\ 0 & 3 \end{bmatrix} \tag{60}$$

$$\Lambda = \begin{bmatrix} 2 & 0 \\ 0 & 2 \end{bmatrix} \quad M = \begin{bmatrix} 2 & 0 \\ 0 & 2 \end{bmatrix} \tag{61}$$

The response of the motors, joint angles and parameters are shown in Figures 4a, 4b and 4c respectively. Notice that stable responses are obtained , parameters do not diverge and small steady state errors in joint#1 and in joint#2 are observed. This one level adaptive scheme is clearly more effective than the rigid control schemes.

Case #3:

In order to show the need for adaptation and the effectiveness of the dual level control scheme, we applied the control scheme given by equations (23) and (36). We assumed the parameter vector $\tilde{P}=[2,1,1,3.5,1]$. The parameters are different from their actual values given in table S-1. The response of the FJR to the control scheme without the adaptation is given in Figures (5a, and 5b). We can see the tracking errors of the links in Figure 5a, and the tracking errors of the motors in Figure 5b. Notice that the scheme given in equations (23) and (36) is more effective in tracking the FJR trajectory than Slotine and Li's rigid adaptive scheme, which gave unstable responses. The controller gains used were the same as those given in Case #2. We can see a significant steady state errors develop, clearly this is undesirable in many applications. In order to compensate for the steady state tracking error, it is desirable to employ an adaptive control scheme. Notice, even if $\tilde{P}$ was determined such that, $\tilde{P}-P=0$, the need for adaptation is not eliminated as the robot may pick up unknown loads and therefore alter the P vector. This would once again result in steady state tracking error.

Case #4:

In order to show the effectiveness of the adaptive control scheme given by equations (23) through (36), we applied our two level control scheme to the FJR. The response is given in Figures (6a, 6b, and 6c). Figure 6a shows the responses for the links, while Figure 6b shows the responses of the motors, and Figure 6c shows the estimates of the parameters. We can see that the motor and the link tracking errors go to zero. The parameters also do not diverge, although they do not converge to their exact actual values, they oscillate about their true values. The controller gain matrices given in equations (60) and (61) were used for this case. Clearly, the response of the manipulator to the adaptive FJR scheme described in this paper is significantly better than applying rigid robot adaptive schemes as seen in case #1. Notice also the the two level adaptive scheme has superior performance over the non-adaptive control law simulated in case #3 and the single level adaptive controller simulated in case #2. The non-adaptive controller developed significant steady state errors while the single level adaptive controller with torque constraints developed some steady state tracking errors. Extensive simulations show that the steady state tracking error developed by this two level controller is quite small and the error is mainly in the motor coordinates. Notice also the behavior of the parameters, they vary slowly about their nominal values with the two-level scheme, whereas they do not appear to track their nominal values as accurately in the one-level scheme for the given simulation time.

9. Conclusion

In this paper we have presented several schemes to control flexible joint robots. Two of them are adaptive control schemes, they were derived without employing linearization techniques, link acceleration or jerk measurements, and inertia matrix inversion. Adaptive controllers for the FJR were derived using Lyapunov's second method. From the simulation results, it is clear that the improvement in the tracking and parameter estimation is significant over rigid robot adaptive schemes. Therefore it is necessary to account for joint flexibility effects when deriving control schemes for industrial robots with such form of compliance.

It is obvious that some correction scheme could be added to the derived adaptation law to improve the robustness of our controller in the presence of bounded disturbances or unmodeled dynamics [5]. Experimental work will be necessary to demonstrate the practicality of our scheme.

It is important to point out that most industrial robots use feedback sensors mounted on the actuator and in order to compensate for joint flexibility additional sensors must be mounted to measure the joint angles and velocities.

10. Acknowledgement

We would like to thank Professor's Martin Corless and Stanislaw Żak and Dr.Mehdi Madani-Esfahani for initial expository discussions on the area of flexibly jointed systems and Lyapunov stability theory.

Appendix A : To show [$\dot{D}$ − 2C] is Skew Symmetric.

Here we will show that D(q) and C(q,$\dot{q}$) are not independent, but the matrix ($\dot{D}$−2C) is skew symmetric [10], [4], [13]. This can be easily derived from the Lagrangian formulation of the manipulator dynamics. In order for a square matrix W to be skew-symmetric, we need $W^t = -W$. From equation (2), we can represent the $(kj)^{th}$ element of C(q,$\dot{q}$) by

$$c_{kj} = \sum_{i=1}^{n} [\frac{\partial d_{kj}}{\partial q_i} - ½\frac{\partial d_{ij}}{\partial q_k}]\, \dot{q}_i \tag{A1}$$

where, d_{kj} is the $(kj)^{th}$ element of the inertia matrix D(q). Now, by interchanging the (i,j) indices and using the symmetry property of D (q), we note:

$$\sum_{i,j}^{n} \frac{\partial d_{kj}}{\partial q_i} = ½\sum_{i,j}^{n} [\frac{\partial d_{kj}}{\partial q_i} + \frac{\partial d_{ki}}{\partial q_j}] \tag{A2}$$

Therefore, we can substitute (A2) into (A1), and:

$$c_{kj} = \sum_{i=1}^{n} ½[\frac{\partial d_{kj}}{\partial q_i} + \frac{\partial d_{ki}}{\partial q_j} - \frac{\partial d_{ij}}{\partial q_k}]\, \dot{q}_i \tag{A3}$$

Let, W(q,$\dot{q}$)=[$\dot{D}$(q)−2C(q,$\dot{q}$)], then the $(jk)^{th}$ element of W is:

$$w_{kj} = \dot{d}_{kj} - 2c_{kj}$$

$$= \sum_{i=1}^{n} [\frac{\partial d_{kj}}{\partial q_i} - (\frac{\partial d_{kj}}{\partial q_i} + \frac{\partial d_{ki}}{\partial q_j} - \frac{\partial d_{ij}}{\partial q_k})]\, \dot{q}_i$$

$$= \sum_{i=1}^{n} [\frac{\partial d_{ij}}{\partial q_k} - \frac{\partial d_{ki}}{\partial q_j}]\, \dot{q}_i \tag{A4}$$

Since D(q) is symmetric, it is clear from (A4) that, $w_{kj} = -w_{jk}$. Therefore, W(q,$\dot{q}$) is skew symmetric, furthermore the diagonal entries of W are zero as:

$$w_{jj} = \sum_{i=1}^{n} [\frac{\partial d_{ij}}{\partial q_j} - \frac{\partial d_{ji}}{\partial q_j}]\, \dot{q}_i = 0 \tag{A5}$$

Again by the symmetry property of D(q), (A5) is straight forward.

Now we can conclude that W(q,$\dot{q}$)=[$\dot{D}$(q)−2C(q,$\dot{q}$)] is skew symmetric with zero diagonal entries, which yields

$$\dot{q}^t\, [\dot{D}(q) - 2C(q,\dot{q})]\, \dot{q} = 0 \tag{A6}$$

12. References

[1] S. Ahmad, "Analysis of Robot Drive Train Errors, their Static Effects and their Compensations," *IEEE Journal of Robotics and Automation*, Vol. 4, No. 2, April 1988.

[2] A. De-Luca, "Dynamic Control of Robots with Joint Elasticity," *Proceedings of 1988 IEEE International Conference on Robotics and Automation*, Philadelphia, PA, pp. 152-158, April 1988.

[3] L. Fu and K. Chen, "Nonlinear Adaptive Motion Control for a Manipulator with Flexible Joints," *Proceedings of 1989 IEEE International Conference on Robotics and Automation*, Phoenix, AZ, pp. 1201-1206, May 1989.

[4] F. Ghorbel, W. Spong and J. Hung, "Adaptive Control of Flexible Joint Manipulators," *Proceedings of 1989 IEEE International Conference on Robotics and Automation*, Phoenix, AZ, pp. 1188-1193, May 1989.

[5] P. Ioannou, "Robust Adaptive Controller with Zero Residual Tracking Errors," *IEEE Trans. Automatic Control*, Vol. 31, No. 8, pp. 773-776, Aug. 1986.

[6] K. Khorasani, "Robust Adaptive Stabilization of Flexible Joint Manipulators," *Proceedings of 1989 IEEE International Conference on Robotics and Automation*, Phoenix, AZ, pp. 1194-1199, May 1989.

[7] D. Koditschek, "Lyapunov Analysis of Robot Motion," *Tutorial Workshop of the 1987 IEEE International Conference on Robotics and Automation*, Raleigh, NC, pp. 1B-3-1 to 1B-3-56, March 1987.

[8] R. Paul, *Robot Manipulators: Mathematics, Programming and Control*, MIT Press, 1981.

[9] S. Sastry, P. Hsu, M. Bodson and B. Paden, "Adaptive Identification and Control for Manipulators without Using Joint Accelerations," *Proceedings of 1987 IEEE International Conference on Robotics and Automation*, Raleigh, NC, pp. 1210-1215, March 1987.

[10] J. Slotine and W. Li, "Adaptive Manipulator Control: A Case Study," *IEEE Trans. Automatic Control*, Vol. 33, No. 11, pp. 995-1003, Nov. 1988.

[11] J. Slotine and W. Li, "On the Adaptive Control of Robot Manipulators," *Int. J. Robotics Research*, Vol. 6, No. 3, pp. 49-59, Fall 1987.

[12] M. Spong, K. Khorasani and P. Kokotovic, "An Integral Manifold Approach to the Feedback Control of Flexible Joint Robots," *IEEE Journal of Robotics and Automation*, Vol. RA-3, No. 4, pp. 291-300, Aug. 1987.

[13] M. Spong, "Modeling and Control of Elastic Joint Robots," *Transactions of the ASME*, Vol. 109, pp. 310-319, Dec. 1987.

[14] G. Widmann and S. Ahmad, "Control of Robots with Flexible Joints," *Proceedings of 1987 IEEE International Conference on Robotics and Automation*, Raleigh, NC, pp. 1561-1566, March 1987.

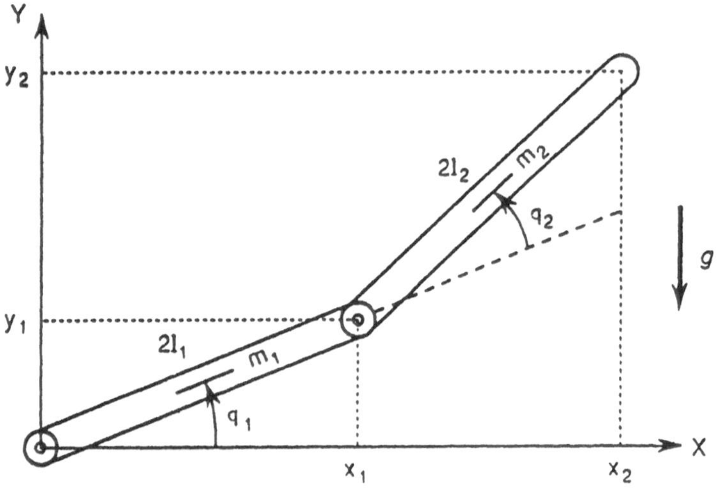

Figure 1. The Two Link Planar Manipulator.

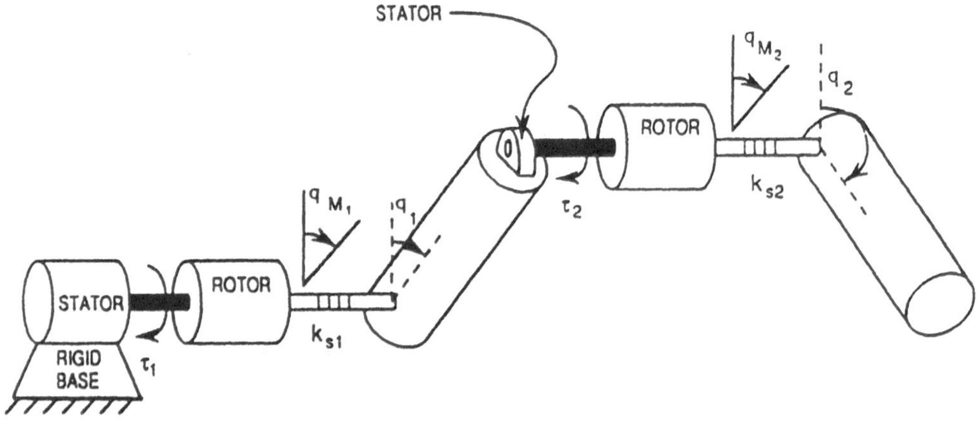

Figure 2. Conceptual Diagram of Two Link Compliant Manipulator.

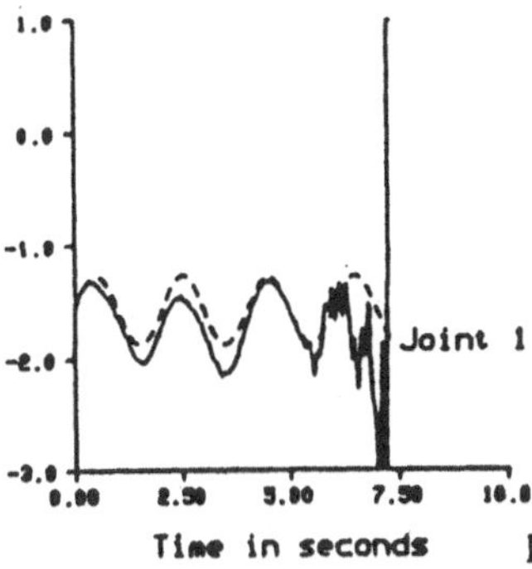

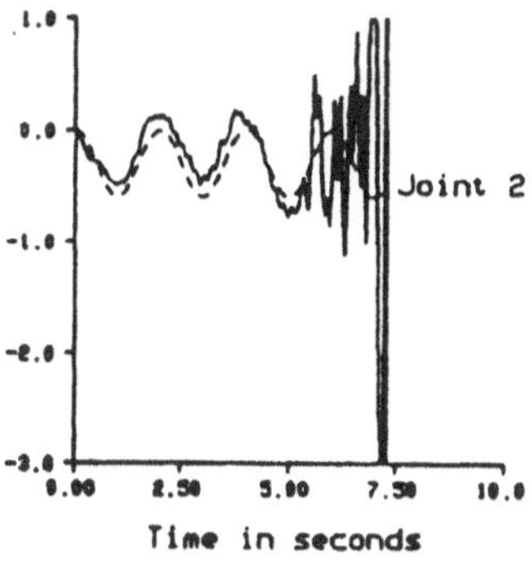

Figure 3a

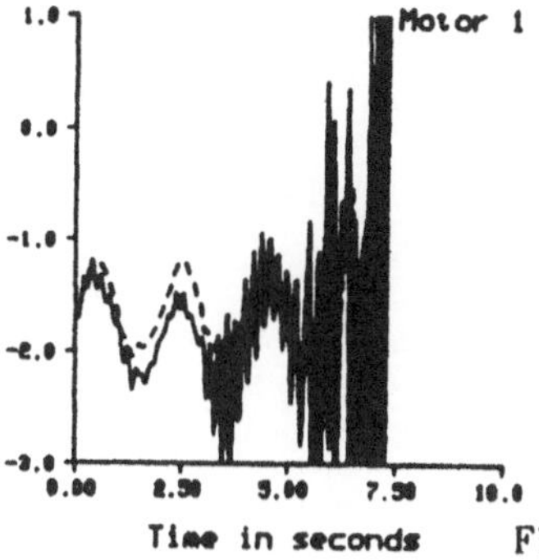

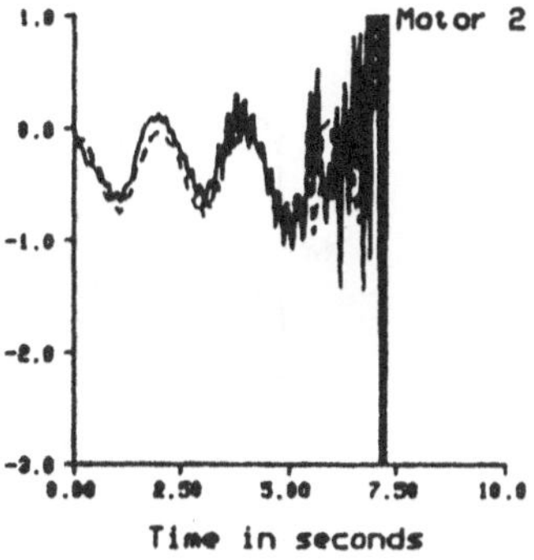

Figure 3b

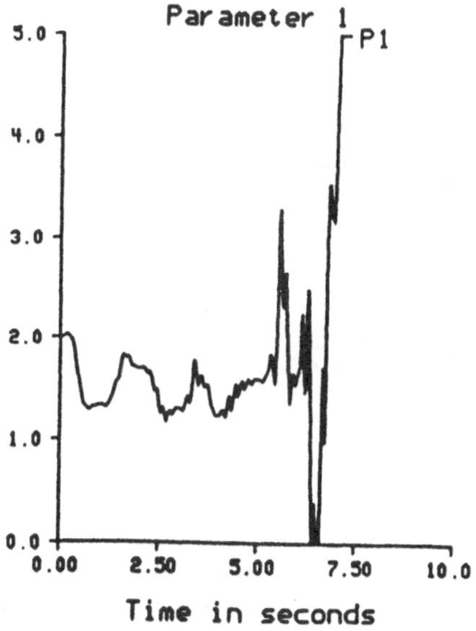

Figure 3c

Case#1: Current rigid robot adaptive scheme applied to FJR.

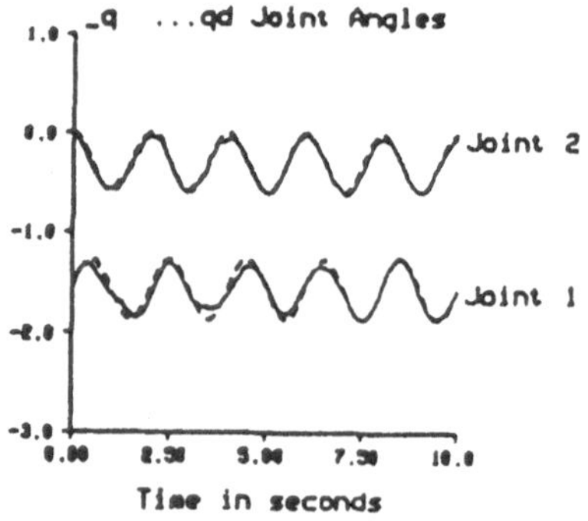

Figure 4a

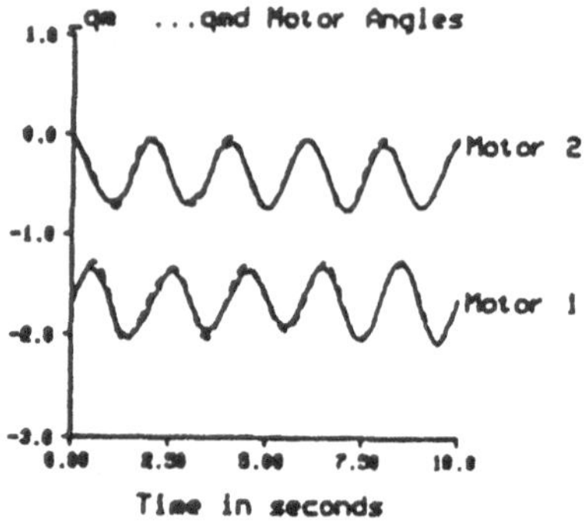

Figure 4b

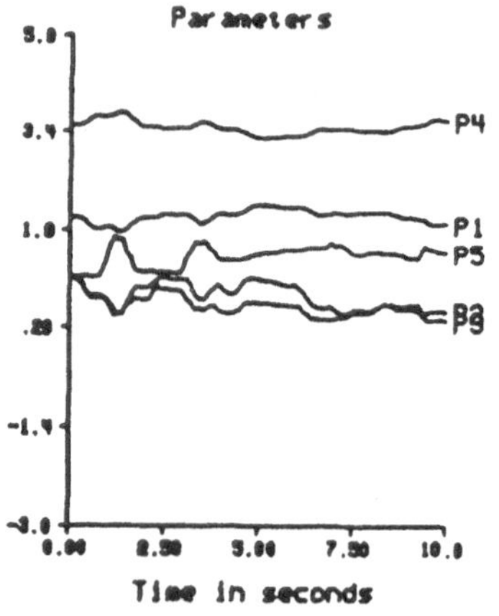

Figure 4c

Case#2: Derived one-level FJR adaptive control scheme.

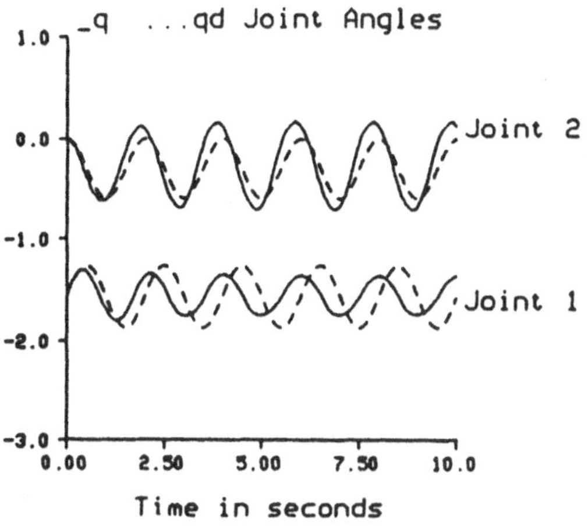

Figure 5a

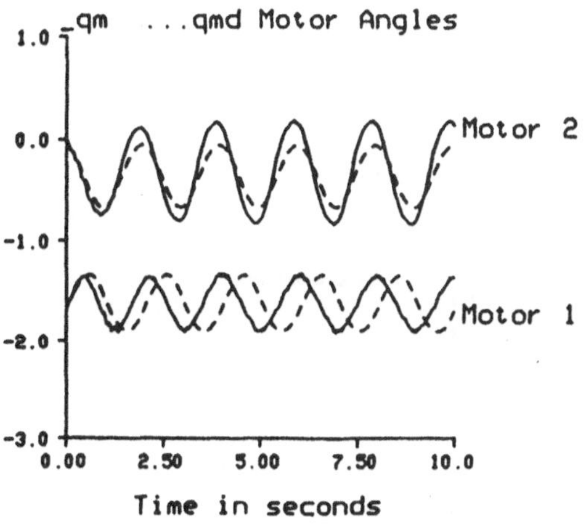

Figure 5b

Case#3: Derived two-level FJR control scheme without adaptation.

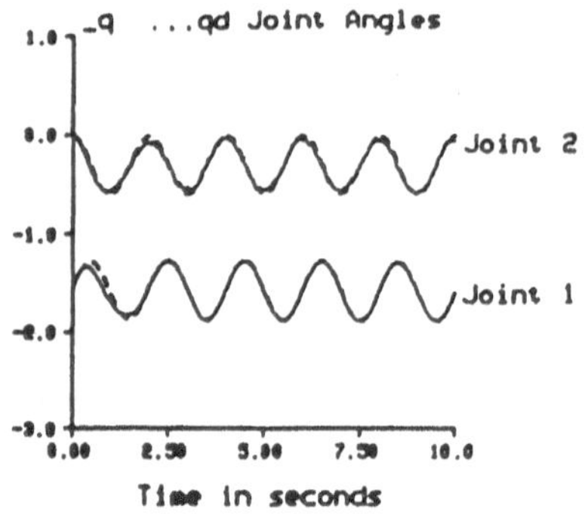

Figure 6a

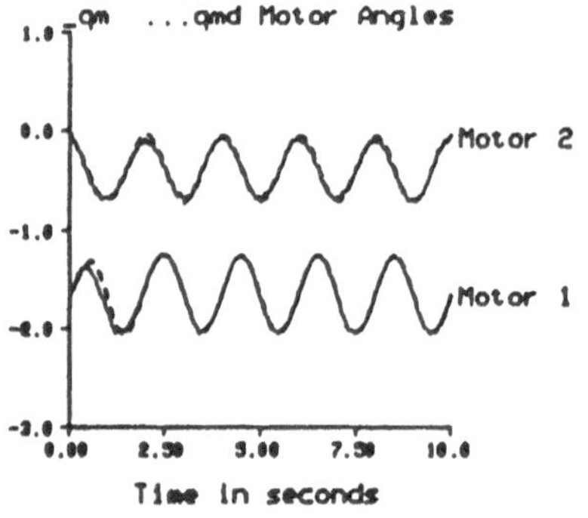

Figure 6b

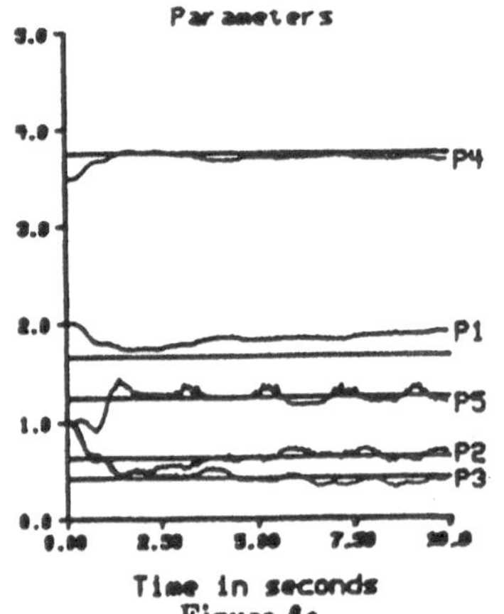

Figure 6c

Case#4: Derived two-level FJR adaptive control scheme.

PERTURBATION METHOD FOR IMPROVED TIME–OPTIMAL CONTROL OF DISK DRIVES

Mark D. Ardema and Evert Cooper
Santa Clara University, Santa Clara, CA. 95053, U.S.A.

1. Introduction

One of the most important performance goals of a disk file actuator is rapid access time. The disk file actuator as incorporated within its magnetic head/disk assembly is a high–order, flexible dynamic system with unpredictable behavior due to manufacturing tolerances and temperature variations. The controller must operate within demanding limits of response time, power consumption, and storage capacity. Although the goal of rapid access time argues for a time optimal control law, the high system order necessitates approximation techniques. Further, the desire to maintain near time–optimality in the presence of temperature fluctuations and time–varying unmodeled dynamic effects motivates the need for an adaptive approach. The disk drive control problem is summarized in Reference 1.

Figure 1 shows the over–all architecture of a possible disk drive actuator controller. The time–optimal controller computes a control signal in real–time based on a reduced–order reference model of the true system. This control signal is augmented by a signal produced by an adaptive control loop. Only the optimal controller and its associated reference model is of interest in this paper.

It is highly desirable to construct a reference model closely representative of the true plant to minimize the efforts of the adaptive loop and to make the performance near–optimal. The main problem is that the system order must be severely limited. Although there are general algorithms for computing time–optimal controls for high–order systems (Ref. 2) and approximate methods for third–order systems (Ref. 3), these are impractical for our purpose, and analysis must be restricted to second order systems.

An attractive method for dealing with prohibitively high–order systems is to use singular perturbation theory. This method separates a dynamic system into reduced–order subsystems of different time scales, and provides a way to synthesize a near–optimal controller by combining the two reduced system controllers (see, for example, Ref. 4 and 5). This approach has been investigated for linear time–optimal systems in References 5 and 6. It is found that the optimal control of the full system is much like

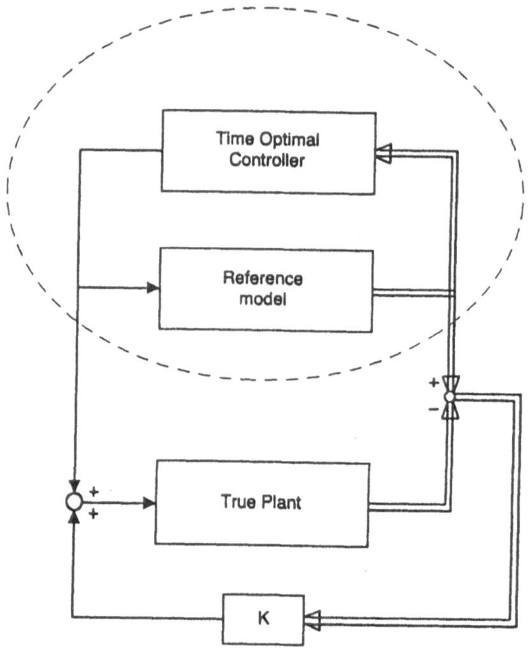

Figure 1. Schematic of disk drive actuator control system.

the optimal control of the reduced system (perturbation parameter set to zero) except that there are rapid control switches near the end of the process to bring the dynamics neglected by the reduced solution to rest. Reference 5 contains an example similar to the specific problem we consider in this paper.

In this paper, we analyze a representative third-order time optimal linear system as a singular perturbation problem. The perturbation parameter appears in such a way as to make the reduced problem a simple double integrator, for which the time optimal control is well-known and easily implementable. We then consider the boundary layer problem to obtain corrections to the reduced solution. Of particular interest is determination of the additional control switch time that appears in the third-order system but not in the reduced system. The analysis is restricted to zero order terms in the perturbation parameter.

Finally, we simulate the performance of the singular perturbation open-loop controller for a variety of system parameters. As expected, the controller performs very well if the boundary layers are sufficiently small.

Our ultimate goal in this research is to synthesize a closed-loop controller based on a first-order asymptotic analysis. The present paper is a necessary first step in the construction of the first-order controller.

2. Problem Formulation and Reduced Problem

As a representative problem, consider the system

$$\begin{aligned} \dot{x} &= y \\ \dot{y} &= z \\ \epsilon \dot{z} &= -z + u \end{aligned} \tag{2.1}$$

We wish to transfer the system from one rest position to another, the latter taken to be the state space origin,

$$\begin{aligned} x(0) &= x_o, & x(t_f) &= 0 \\ y(0) &= 0, & y(t_f) &= 0 \\ z(0) &= 0, & z(t_f) &= 0 \end{aligned} \tag{2.2}$$

while minimizing the transfer time:

$$J = \int_0^{t_f} dt \tag{2.3}$$

The control u is bounded, $|\, u \,| \leq 1$.

The necessary conditions for optimal control are easily stated. For this purpose, form the Hamiltonian function

$$H = -1 + y\lambda_x + z\lambda_y + (-z + u)\lambda_z \tag{2.4}$$

and the adjoint equations

$$\begin{aligned} \dot{\lambda}_x &= 0 \\ \dot{\lambda}_y &= -\lambda_x \\ \epsilon \dot{\lambda}_z &= -\lambda_y + \lambda_z \end{aligned} \tag{2.5}$$

The Maximum Principle (Ref. 7) then gives the optimal control as

$$u = sgn \lambda_z \tag{2.6}$$

It may be shown that for this system there is no unbounded control and that there are at most two control switches between bounded values of control (Ref. 7). From (2.6) these switches occur at the zeros of λ_z. The adjoint equations (2.5) may be integrated to give

$$\begin{aligned}\lambda_x &= C_1 \\ \lambda_y &= -C_1 t + C_2 \\ \lambda_z &= -C_1 t + C_2 - \epsilon C_1 + C_3 e^{t/\epsilon}\end{aligned} \tag{2.7}$$

Thus the control switch times depend on the constants C_1, C_2, and C_3, but determining these constants is nontrivial.

To obtain the reduced problem associated with system (2.1), set $\epsilon = 0$.

$$\begin{aligned}\dot{x}_r &= y_r \\ \dot{y}_r &= u_r \\ z_r &= u_r\end{aligned} \tag{2.8}$$

with

$$\begin{aligned}x_r(0) &= x_0, \qquad & x_r(t_{f_r}) &= 0 \\ y_r(0) &= 0, \qquad & y_r(t_{f_r}) &= 0\end{aligned} \tag{2.9}$$

and

$$J = \int_0^{t_{f_r}} dt \tag{2.10}$$

The solution to this problem is well-known. Assuming that $x_0 < 0$, there is one switch, at $t = t_{s_r}$ from $u = +1$ to $u = -1$. The solution for $0 \le t \le t_{s_r}$ is

$$\begin{aligned}x_r &= t^2/2 + x_0 \\ y_r &= t \\ u &= 1\end{aligned} \tag{2.11}$$

and for $t_{s_r} \le t \le t_{f_r}$,

$$\begin{aligned}x_r &= -(2\sqrt{-x_0} - t)^2/2 \\ y_r &= 2\sqrt{-x_0} - t \\ u &= -1\end{aligned} \tag{2.13}$$

where the switch time and final time are given by

$$\begin{aligned}t_{s_r} &= \sqrt{-x_0} \\ t_{f_r} &= 2\sqrt{-x_0}\end{aligned} \tag{2.14}$$

The adjoint variables, $0 \le t \le t_{f_r}$ are found to be

$$\begin{aligned} \lambda_{x_r} &= \frac{1}{\sqrt{-x_0}} \\ \lambda_{y_r} &= 1 - \frac{t}{\sqrt{-x_0}} \end{aligned} \tag{2.15}$$

The switching function of the reduced problem is λ_{y_r}, as given by (2.15b). Comparing this switching function with that of the full problem, (2.7c), shows that the difference between the two consists of two boundary–layer type terms. It will be seen later, that to zero order the first of these terms is negligible and the second is important only near $t = t_f$.

3. Asymptotic Analysis and Control Law

The asymptotic analysis proceeds by dividing the motion into five segments as follows: (1) an initial boundary layer in which the z state variable rapidly and asymptotically approaches its equilibrium value, 1; (2) an outer region ending at the first switch time; (3) an interior boundary layer beginning at the first switch time in which z approaches its new equilibrium value, -1; (4) an outer region ending at the second switch time; and (5) a terminal boundary layer.

To zero order, the boundary layer motions, which are asymptotically stable, take place in zero time while the slow variables remain frozen at their boundary values. Consequently, the zero–order solution for the slow variables is exactly the same as for the reduced problem. Further, the first control switch time, t_s, and the final time, t_f, are the same as well. The only effect of the boundary layer motions, to zero–order, is to introduce a second control switch at a time, t_{s_2}, an order ϵ before t_f.

To derive an expression for t_{s_2}, we consider the terminal boundary layer. The boundary layer is obtained by stretching time–to–go by introducing a new independent variable $\alpha = (t_f - t)/\epsilon$ in (2.1) and (2.5) and then setting $\epsilon = 0$ and $u = 1$:

$$\begin{aligned} \frac{dx_b}{d\alpha} &= 0 & \frac{d\lambda_{x_b}}{d\alpha} &= 0 \\ \frac{dy_b}{d\alpha} &= 0 & \frac{d\lambda_{y_b}}{d\alpha} &= 0 \\ \frac{dz_b}{d\alpha} &= z_b - 1 & \frac{d\lambda_{z_b}}{d\alpha} &= \lambda_{z_b} - \lambda_{z_b} \end{aligned} \tag{3.1}$$

The boundary conditions are

$$x_b(0) = 0, \quad y_b(0) = 0, \quad z_b(0) = 0 \tag{3.2}$$

and the solution is

$$\begin{aligned} x_b &= 0, \qquad y_b = 0, \qquad z_b = 1 - e^{\alpha} \\ \lambda_{x_b} &= C_4, \quad \lambda_{y_b} = C_5, \quad \lambda_{z_b} = C_6 e^{-\alpha} + C_5 \end{aligned} \tag{3.3}$$

The constants of integration C_4, C_5, and C_6 are found by matching with the second outer solution. The result is, for λ_z,

$$\lambda_{z_b} = 2e^{-\alpha} - 1 \tag{3.4}$$

Setting this to zero gives the second switch time, t_{s_2},

$$\lambda_{z_b}(t_{s_2}) = 2e^{-\alpha_{s_2}} - 1 = 0 \tag{3.5}$$

Thus $\alpha_{s_2} = ln2$ and, solving for t_{s_2},

$$t_{s_2} = 2\sqrt{-x_0} - \epsilon ln2 \tag{3.6}$$

The zero–order, open–loop control algorithm is now easily stated. First, precompute $t_{s_1}(= t_{s_r})$, $t_f(= t_{f_r})$, and t_{s_2} from (2.14) and (3.6). Then begin the process with control $u = 1$, and when $t = t_{s_1}$, switch the control to $u = -1$. When $t = t_{s_2}$, switch the control back to $u = +1$ and end the process at $t = t_f$.

An alternative to the open–loop controller just stated, would be to construct a zero–order composite representation of the switching function λ_z. To this end, recall that the switching function in the second outer region is given by (2.15b);

$$\lambda_{z_0} = \lambda_{y_r} = 1 - \frac{t}{\sqrt{-x_0}} \tag{3.7}$$

and that in the terminal boundary layer it is given by (3.4). Taking limits

$$\lim_{t \to t_f} \lambda_{z_0} = -1 = \lim_{\alpha \to \infty} \lambda_{z_b} \tag{3.8}$$

shows that these functions match and that the common part is -1. Consequently, the composite function is

$$\lambda_{z_c} = \lambda_{z_0} + \lambda_{z_b} - \text{common part}$$
$$\lambda_{z_c} = 1 - \frac{t}{\sqrt{-x_0}} + \frac{2e^{t/\epsilon}}{e^{2\sqrt{-x_0}/\epsilon}} \tag{3.9}$$

The two switch time t_{s_1} and t_{s_2} are then the roots of the equation:

$$0 = 1 - \frac{t_s}{\sqrt{-x_o}} + \frac{2e^{t_s/\epsilon}}{e^{2\sqrt{-x_o}/\epsilon}} \tag{3.10}$$

The zero–order solution will be asymptotically valid if the difference between the value of the fast variable z and its equilibrium value is asymptotically negligible just prior to the switch times. Consider, for example, $t \in [0, t_{s_1}]$. In this interval, z is formed by stretching the independent variable by $\tau = t/\epsilon$, setting $\epsilon = 0$ and $u = 1$, and applying initial condition (2.2c). The result is

$$z_b = 1 - e^{-\tau} \tag{3.11}$$

This must be near equilibrium at $t = t_{s_1}$, that is

$$1 - z_b(t_{s_1}) = e^{-\sqrt{-x_0}/\epsilon} < < 1$$

or

$$e^{\sqrt{-x_0}/\epsilon} > > 1 \tag{3.12}$$

This shows that either $\mid x_0 \mid$ must be large, or ϵ must be small, or both, compared with 1.

4. Numerical Examples

In this section, we apply the open–loop, zero–order control algorithm to a simulation program that numerically integrates (2.1) subject to the initial conditions of (2.2).

Figures 2 and 3 show the time–histories of the state variables x, y, and z for the case of $x_0 = -1$ and two different values of ϵ. For $\epsilon = 1.0$, Figure 2, (3.12) is clearly violated and, further, $t_f - t_{s_2}$ is not small. Consequently, good performance can not be expected and in fact all state variables show considerable error at t_f. For $\epsilon = 0.1$, Figure 3, (3.12) is satisfied and $t_f - t_{s_2}$ is small. This results in good performance, although there is some error in $y(t_f)$. First order corrections may be expected to greatly reduce this error.

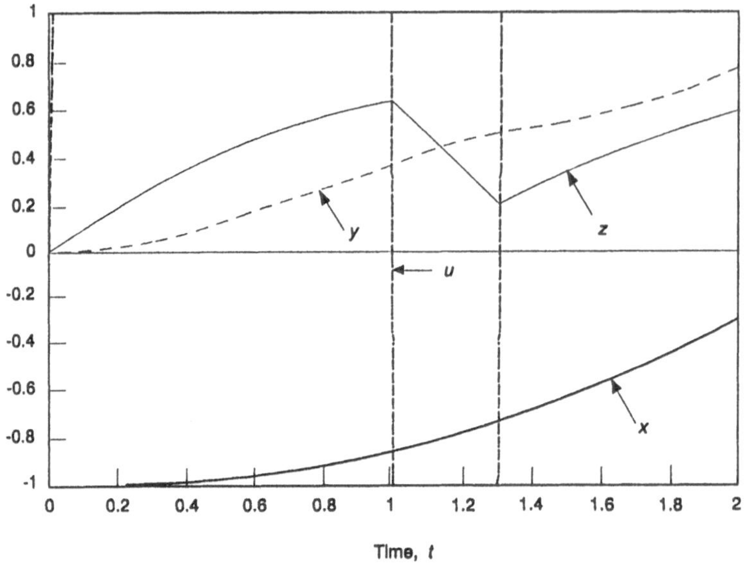

Figure 2. Time histories of state variables for $x_0 = -1$, $\epsilon = 1.0$.

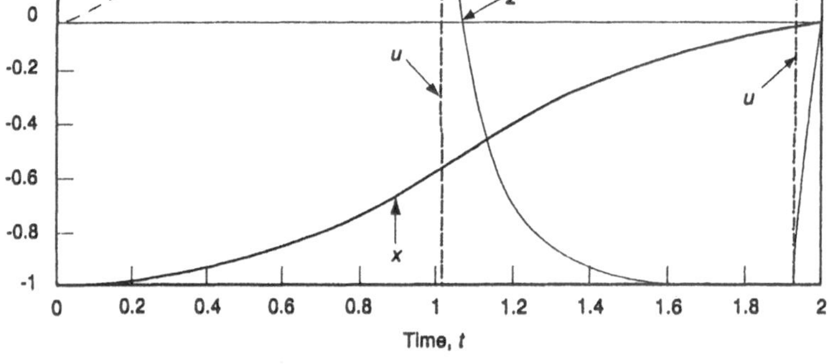

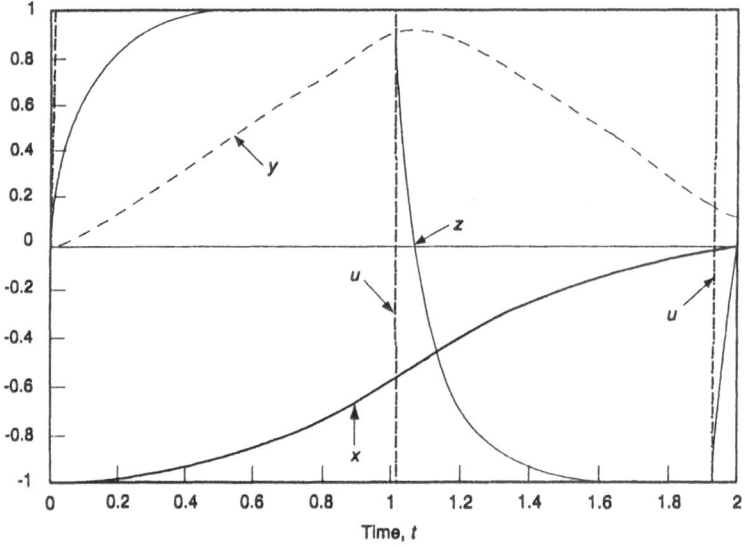

Figure 3. Time histories of state variables for $x_0 = -1$, $\epsilon = 0.1$.

The final values of the state variables as a function of ϵ are shown on Figure 4 for the case $x_0 = -1$. It is seen that all variables tend to zero as ϵ tends to zero, confirming the asymptotic validity of the algorithm.

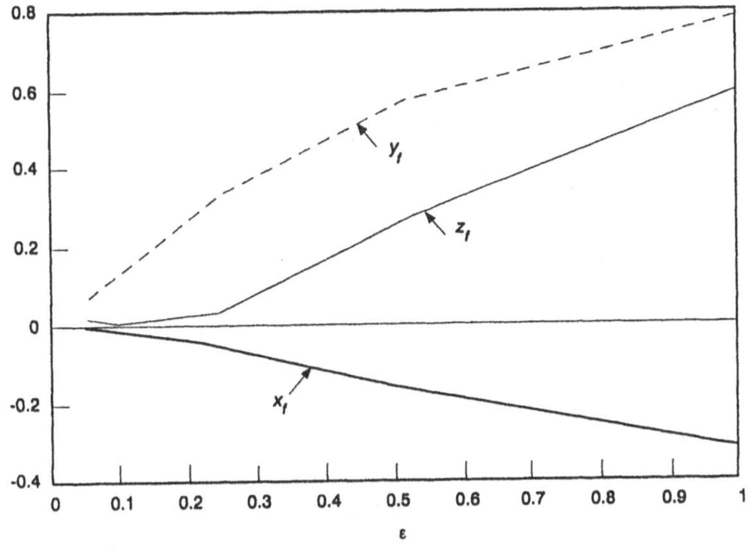

Figure 4. Effect of ϵ for x_0 = -1.

The case of a change in displacement of 5 units (Figs. 5, 6, and 7) is considered next. For $\epsilon = 1.0$ (Fig. 5), performance has improved relative to the case of a change in displacement of one unit (Fig. 2), but final values are still unacceptably large due to not satisfying (3.12) to a sufficient degree. For $\epsilon = 0.1$ (Fig. 6), the performance of the algorithm is quite good. Figure 7 shows the asymptotic convergence as $\epsilon \to 0$ for $x_0 = -5$.

The error in final position, $x(t_f)$, is shown as a function of ϵ for the two displacement changes in Figure 8.

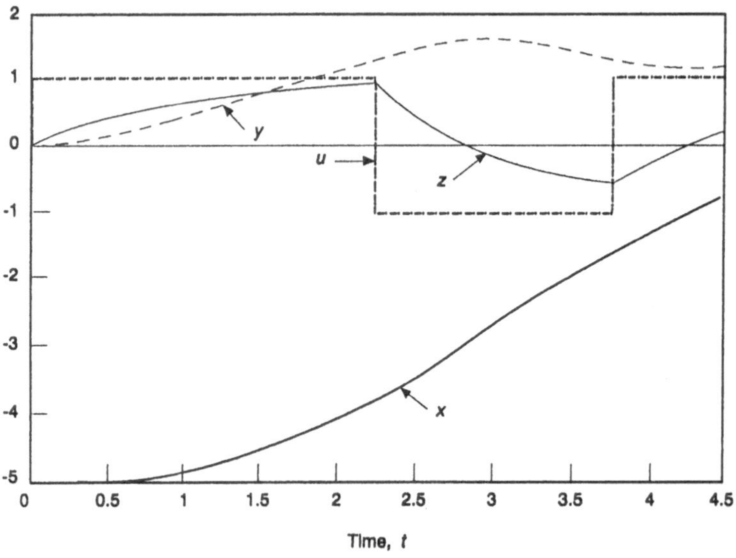

Figure 5. Time histories of state variables for x_0 = -5, ϵ = 1.0.

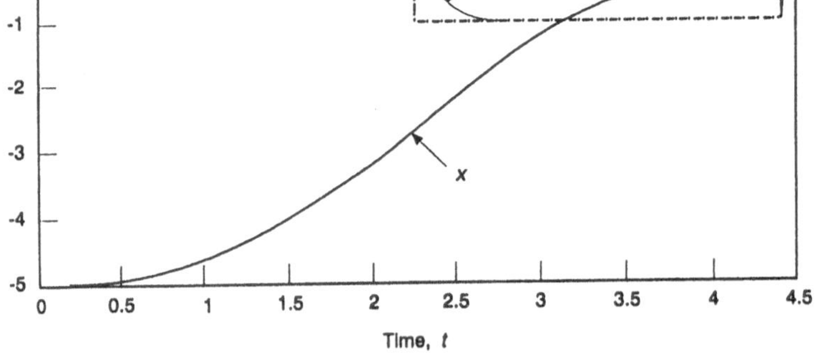

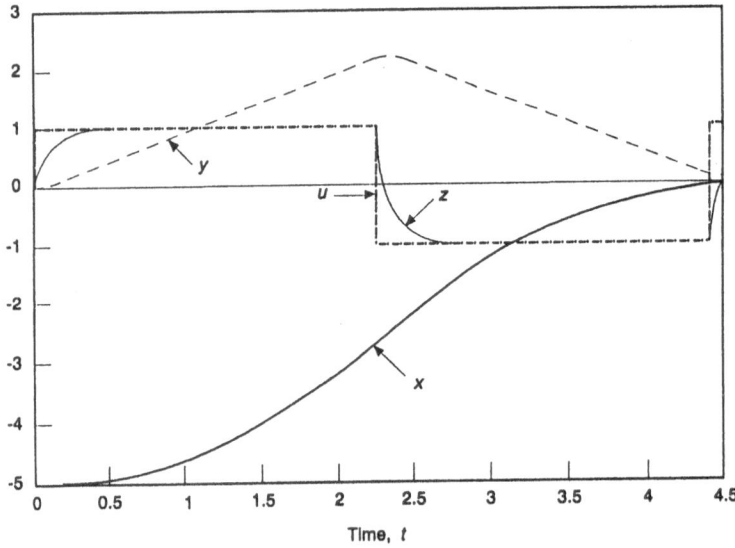

Figure 6. Time histories of state variables for x_0 = -5, ϵ = 0.1.

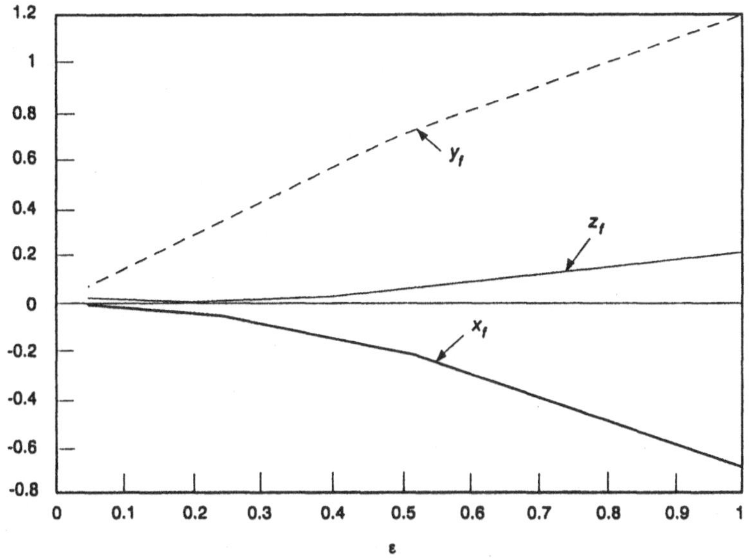

Figure 7. Effect of ϵ for $x_0 = -5$.

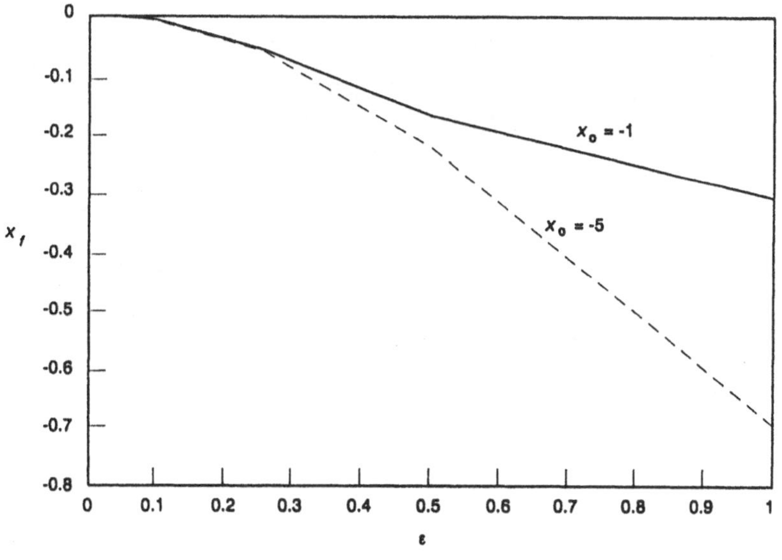

Figure 8. Error in final position.

In Figures 9 and 10 we investigate using the zero–order composite switching function (3.9) to determine control switch times t_{s_1} and t_{s_2}. Figure 9 ($x_0 = -1$) shows that for ϵ greater than about 0.35, λ_{z_c} has no zeros and hence no switches are predicted, an obvious error. For ϵ less than about 0.20, however, the first switch time is predicted quite accurately and the second one is close to t_f as required. Figure 10 ($x_0 = -5$) shows that good prediction of switch times is obtained for ϵ less than about 0.50. It is apparent that for ϵ sufficiently small the zeros of the zero–order composite switching function λ_{z_c} correctly predict the control switch times. The first switch time is near the time predicted by the reduced solution and the second occurs before the final time by an increment proportional to ϵ.

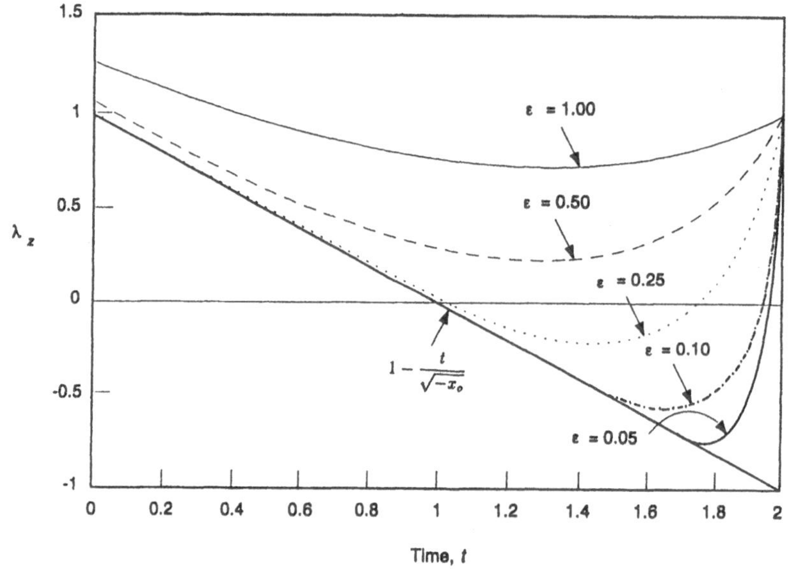

Figure 9. Composite switching function for $x_0 = -1$.

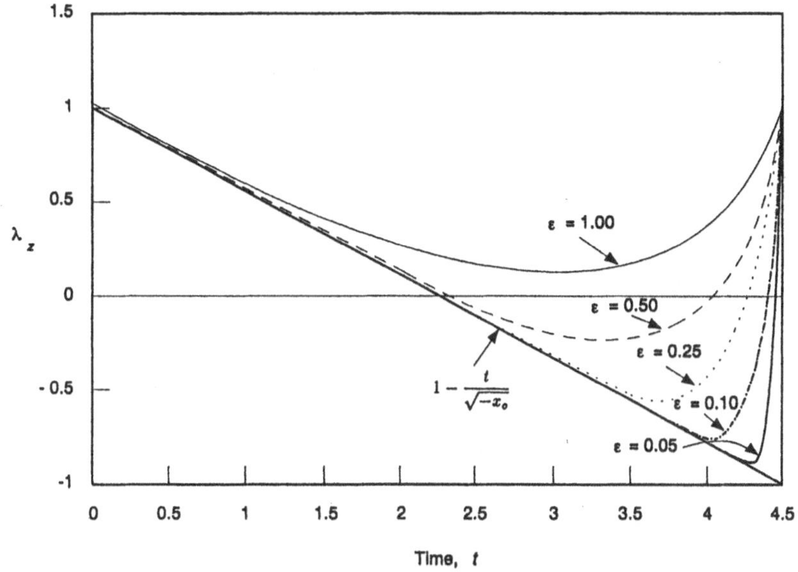

Figure 10. Composite switching function. for x_0 = -5.

5. References

1. Cooper, E., "Minimizing Power Dissipation in a Disk File Actuator," IEEE Trans. on Magnetics, Vol. 24, No. 3, May 1988.
2. Yastreboff, M., "Synthesis of Time–Optimal Control by Time Interval Adjustment,"IEEE Trans. Auto. Control, Dec. 1969, pp. 707.
3. Kassam, S.A., Thomas, J.B., and McCrumm, J.D., "Implementation of Sub–Optimal Control for a Third–Order System,"Comput. Elect. Engng., Vol. 2, 1975, pp. 307.
4. Ardema, M.D., "An Introduction to Singular Perturbations in Nonlinear Optimal Control,"Singular Perturbations in Systems and Control, M.D. Ardema, ed., International Centre for Mechanical Sciences, Courses and Lectures No. 280, 1983.
5. Kokotovic, P.V., Khalil, H.K., and O'Reilly, J., Singular Perturbation Methods in Control: Analysis and Design, Academic Press, 1986.
6. Kokotovic. P.V. and Haddad, A.H., "Controllability and Time–Optimal Control of Systems with Slow and Fast Modes," IEEE Trans. Auto Control, Feb. 1975, pp. 111.
7. Leitmann, G., The Calculus of Variations and Optimal Control, Plenum, 1981.

ROBUSTNESS ANALYSIS
IN THE TIME DOMAIN AND OUTPUT SPACE

C. C. Blackwell

Center for Dynamic Systems and Control Studies

Department of Mechanical Engineering

University of Texas - Arlington

P. O. Box 19023

Arlington, Texas 76019

Abstract. *The results of efforts to extend the Direct Principle of Lyapunov to stability analysis in controlled output space are described. The motivation is to apply the DPL in the space of the actual variables of concern in the system performance, thus avoiding conservatism and in some cases gaining other advantages. The details of the application of the procedure are developed, and an example is presented.*

INTRODUCTION

Much of the recent robust control research and development effort has been devoted to the reduction of conservatism in the robustness analysis procedure (e.g., Yedavalli, 1989 and previous work, Barmish, 1985, Black, 1990, and Blackwell, 1990). This problem can occur in both H_∞ analysis results and in results of the use of the Direct Principal of Lyapunov (DPL). Robustness analysis based on the

DPL has emphasized finding an ultimate boundedness attractor in the state space of the control system. The boundary of this region is parameterized by a critical value of the Lyapunov function being used as the basis of the robustness analysis (and possibly the controller synthesis). In mechanical systems control problems as well as those of all other types of physical systems, it is often true that the range of values of all the states is not of concern. Instead, the range of values of a vector of *controlled outputs*, typically of smaller dimensionality than the state, is of concern. This scenario is quite natural to deal with in H_∞ (Francis, 1988), but has been awkward for treatment in the time domain with the DPL. To find an ultimate boundedness attractor in the controlled output space presents the possibility of reduced conservatism with respect to this potential source, thus rendering the DPL perspective as natural for such problems as is H_∞. A discussion of the background, some indirect treatments, and current results of direct treatment of this problem are presented.

The bases of this discussion are (1) the work of Leitmann (e.g., 1981) and others in the use of the Direct Principle of Lyapunov (DPL, more frequently called the Direct or Second "Method") to synthesize controllers which yield practical stability (PS), and (2) the practical motivation to define robustness in terms which characterize the performance requirements of a control system but which are mathematically compatible with the DPL and PS. The content of the sequel presumes that the reader is familiar with the contents of Leitmann (1981). The same symbology is used, and the same qualifications, mathematical definitions and restrictions apply. This obviates the need for restatement of many necessary but already well established important details of the theory. Developments similar to Leitmann (1981) can be found in Corless and Leitmann (1983), and Barmish and Leitmann (1982).

A Simple Uncertain System Representation.

In the literature, one can find state space oriented treatments of a large spectrum of types of uncertainty and a large spectrum of combinations of types of uncertainty. The purpose of this paper is to introduce the idea of analysis in the controlled output space and to present the basis upon which it can be done. A very simple representation of an uncertain system is selected as the vehicle for this development in an effort to minimize the potential for distraction and confusion due to the complex detail of the rigorous treatment of combinations of uncertainty. This representation form provides an adequate vehicle for this discussion, and for the demonstration of some results. Consider the following uncertain system representation

$$\dot{x}(t) = A\,x(t) + B\,u(t) + C\,d(t);\ x(t_o) = x_o \tag{1}$$

with controlled output:

$$z = E\,x \tag{2}$$

where

$x \in \Re^{nx}$ (the state vector of the system),

$d(t)\colon \Re \to \Re^{nd} \mid \|d\| \le \rho_d,\ \rho_d \ge 0$ (a possibly time-varying plant disturbance vector),

$u(t)\colon \Re \to \Re^{nu}$ (the control vector),

$z \in \Re^{nz}$ (the controlled output),

$A \in \Re^{nx \times nx}$,

$B \in \Re^{nx \times nu}$,

$C \in \Re^{nx \times nd}$,

$E \in \Re^{nz \times nx}$,

and: (1) A is a stability matrix, (2) E is full rank, (3) u and d are Lebesgue measurable in t. We will refer to this representation as $\mathcal{S}1$.

The Nominal System Representation. Define the *nominal system representation* $\mathcal{S}2$ as the representation $\mathcal{S}1$ with u and d set to zero

$$\dot{x}(t) = A\,x(t);\ x(t_o) = x_o \tag{3}$$

Essential Results From Practical Stability Theory

It is well established (e.g., Kalman and Bertram, 1960) that for such a representation as $\mathcal{S}2$, $\ni$ positive definite quadratic scalar functions

$$\Theta(x,t) = x'P\,x, \text{ where } P \in \Re^{nx \times nx} \tag{4}$$

which are *Lyapunov functions* $V(x,t) = \Theta(x,t)$ of $\mathcal{S}2$ if:

$$P = P' > 0, \tag{5}$$

is a solution of

$$PA + A'P = -\,Q,$$

where

$$Q \in \Re^{nx \times nx} \mid Q = Q' > 0.$$

In this case, $dV(x,t)/dt = \nabla V(x,t)'Ax = -x'Qx < 0 \;\forall\; x \neq 0$, thus establishing the stability of $\mathcal{S}2$ w.r.t. $x = 0$.

It is also well established (Corless and Leitmann, 1983) that although system $\mathcal{S}1$ cannot have a Lyapunov function, (4) can be effectively used as a convergence measure to address the question of guaranteed nearness of approach of *the state of* $\mathcal{S}1$ to the origin. Evaluating $\Theta(x,t)$ and its derivative on the trajectories of $\mathcal{S}1$, one obtains

$$\dot{\Theta} = -\,x'Qx + 2x'PCd \tag{6}$$

One can establish that

$$\dot{\Theta} \leq h[\eta]$$

where

$$h[\eta] = -\,\lambda_{min}[Q]\eta^2 + 2\|PC\|\rho_d\eta \tag{7}$$

$$\eta = \|x\| \tag{8}$$

and

$$\tilde{\eta}_{max} = \eta\colon \max_{\eta}[h[\eta]] \mid h(\eta) = 0 \tag{9}$$

Equations (6) through (9) establish the radius r_b of the smallest ball $\mathfrak{B}[\eta]$ in state space which it can be guaranteed that the state of $\mathcal{S}1$ can approach.

$$r_b = \tilde{\eta}_{max} \tag{10}$$

This ball characterizes the surface $\Theta[x,t] = k_b$, which is the lowest value that we can guarantee $\Theta[x,t]$ will assume. It is characterized by

$$\lambda_{max}[P]\, r_b^2 = k_b \tag{11}$$

and establishes the lower limiting value of the ultimate bound of norm(x), ρ_l, which is

$$\rho_l = r_b \sqrt{\lambda_{max}(P)/\lambda_{min}(P)} \tag{12}$$

By selecting $\Theta^*[x,t] = k^* = k_b + \Delta k$, where $\Delta k \in \Re_+$, one establishes an attractor $\mathcal{A} = \{x \in \Re^{nx} | \; k_b < \Theta[x,t] \le k^*\}$. In the event that $\|x_0\| \in \mathcal{A}$, it is guaranteed that x will remain there, and in the event that $\|x_0\| \notin \mathcal{A}$, it is guaranteed that x will enter $\mathcal{A}$ in finite time T_o^* and remain there (see Leitmann, 1981). The greater Δk, the smaller T_o^*.

We can now present a rigorous definition of robustness in the framework of the DPL.

A Definition of Robustness in State Space

Let X be the closed, compact set of acceptable values of the state. We define the control to be robust to d with respect to X if, in the case that $\|x(t_o)\| \notin$ X, then (a) the control will cause $x \in$ X in a finite period of time T_c $\forall$ $\|d\| \le \rho_d$, and (b) the control will cause $x \in$ X $\forall$ $t \ge t_o + T_c$, and in the case that $\|x(t_o)\| \in$ X, then the control will cause $x \in$ X $\forall$ $t \ge t_o$. We will use the expression: "u is robust to {X; d}." This definition of robustness is intended to be compatible with the definition of practical stability, but is also an attempt to address the highly varied circumstances for which the notion of robustness is useful. The sizes, shapes, and orientations of the the PS attractor (PSA) and X are not likely to be the same.

The desirability of incurring minimal conservatism in the process of establishing the robustness of a control system is self-evident. In the use of PS fundamentals, either of two undesirable outcomes can occur: (1) the PS attractor is smaller than **X**, or (2) robustness cannot be affirmed because the PS attractor is not wholly contained in **X**. It is desirable, therefore, to fashion the PS attractor to be as nearly the same as **X** as possible.

A DEFINITION OF ROBUSTNESS IN OUTPUT SPACE

Let **Z** be the closed compact set of acceptable values of the controlled output z. We difine the control to be robust to d with respect to **Z** if in the case that $z(t_o) \notin \mathbf{Z}$, then (a) the control will cause $z \in \mathbf{Z}$ in a finite period of time T_c $\forall$ $\|d\| \leq \rho_d$, and (b) will cause $z \in \mathbf{Z}$ $\forall$ $t \geq t_o + T_c$, and if $z(t_o) \in \mathbf{Z}$, then the control will cause $z \in \mathbf{Z}$ $\forall$ $t \geq t_o$. We will use the expression: "u is robust to $\{\mathbf{Z};\ d\}$."

<u>A Comment on Observability via the Controlled Output.</u> In some cases, $z = Ex$ may not observe x fully. in such cases, it will be advantageous to decompose x into its real Jordan canonical basis and reconstruct z from the canonical basis vector elements which it observes, and then continue the analysis. For the remainder of the discussion, it will be assumed that this has been done. The results will make it desireable to transform from the state via a similarity transformation to a new state which contains the output $x = Tq$, where $q' = [z'|w]$ where w is the null space conjugate of z. This means that E in the transformed system is always $[I|\ 0]$. One can obtain w via $w = \Phi x$, where

$$\Phi \in \Re^{(nx\ -\ ny)\ \times\ nx} \mid \Phi\Phi' = \mathrm{I},\ E\Phi' = 0. \tag{13}$$

Then a one-to-one transformation of the state

$$q \triangleq \begin{bmatrix} z \\ w \end{bmatrix} = \begin{bmatrix} E \\ \Phi \end{bmatrix} x = Tx \tag{14}$$

will allow mapping from the state to q, which explicitly contains the controlled output vector.

<u>Scaling to Reduce Conservatism.</u> Suppose it is desired to establish that the magnitude of each element of the state vector, x_i, is bounded by $\sigma_i > 0$, or $|x| \leq \sigma$. Then it is useful to transform to a

scaled state s, $z = Ss$, where $S = diag[\sigma]$. This transformation approximates $\mathbf{Z}$ with a hypersphere and avoids the conservatism caused by a distended hyperellipsoid due to the smallest individual bound.

Robustness Analysis Using a Real Canonical Transformation. Šiljak (1986) has advanced the idea of use of real canonical transformations for the purpose of establishing robustness to ignored dynamics. A sketch of this procedure follows. For system $\mathcal{S}1$, $\exists$ a real Jordan transformation $x = M_R r$, where $M_R \in \Re^{nx \times nx}$, and $r \in \Re^{nx}$, where M_R is block diagonal with no more than two by two blocks (see Takahashi, Rabins, and Auslander, 196). As a result of this transformation, (1) becomes

$$\dot{r}(t) = J_R\, r(t) + M_R^{-1} B\, u(t) + M_R^{-1} C\, d(t);\ r(t_o) = r_o \tag{15}$$

This allows the establishment of bounds on the canonical vector elements individually or at most in blocks of two. Since

$$z = EM_R\, x \tag{16}$$

the individual elements in z can be bounded. This can sometimes (but not always) provide a reduction in conservatism.

PRACTICAL STABILITY IN OUTPUT SPACE

The mathematical structure of PS in state space is the paradigm for the development of the results of PS in output space to be presented.

Define a positive definite quadratic function of z

$$\Theta_z(z,t) = z' P_z z \tag{17}$$

where

$$P_z \in \Re^{nz} \mid P_z = P_z' \geq 0 \tag{18}$$

On any trajectory of $\mathcal{S}2$,

$$\dot{\Theta}(z,t) = \nabla_z{}'\dot{z} = \dot{z}'\, P_z\, z + z'\, P_z\, \dot{z} \tag{19}$$

which, in terms of the state, is

$$\dot{\Theta}(z,t) = x'A'E'P_z\, Ex + x'E'\, P_z\, EAx \tag{20}$$

or

$$\dot{\Theta}(z,t) = x'(A'E'P_z\, E + E'P_zEA)x \tag{21}$$

THEOREM: *For system $\mathcal{S}2$, if $\exists$ a $Q_z = Q_z' \geq 0$ and a P_z satisfying (18) $\ni$ Q_z and P_z satisfy*

$$A'E'P_z\, E + E'P_zEA = -\, E'Q_zE \tag{22}$$

Then (17) is a Lyapunov function of system $\mathcal{S}2$ in controlled output space.

Proof: *Using (22) in (21):*

$$\dot{\Theta}(z,t) = \nabla_x'[\Theta]EAx = -\, x'E'Q_zEx = -\, z'Q_zz \tag{23}$$

Thus $\dot{\Theta}(z,t)$ is negative definite. □

Comment: There is no guarantee that a stable A_e will be found for all valid E even though A is a stability matrix.

Equation (22) is not in a convenient form for the search for the pair $[P_z,\ Q_z]$. Multiplying (22) from the left by E and from the right by E' yields

$$EA'E'P_z\, EE' + EE'P_zEAE' = -\, EE'Q_zEE' \tag{24}$$

and multiplying (24) from the left and from the right by $(EE')^{-1}$ results in a standard Lyapunov form

$$(EE')^{-1}EA'E'P_z + P_zEAE'(EE')^{-1} = -\, Q_z \tag{25}$$

One might think of $EAE'(EE')^{-1}$ as an "equivalent state coefficient matrix," A_e, which allows (25) to be written as

$$A_e'\, P_z + P_z\, A_e = -\, Q_z \tag{26}$$

It has been rigorously established that for any $Q_z > 0$, (26) provides a positive definite solution P_z iff A_e is a stability matrix (e.g., Kalman and Bertram, 1960).

Completion of the Robustness Analysis in Output Space. Having established the details of determining P_z and Q_z, we can complete the robustness analysis in output space. The process follows the pattern already presented for the case of the state space. The rate of change of $\Theta[z,t]$ on any trajectory of $\mathcal{V}1$ is

$$\dot{\Theta} = -z'Q_z z + 2z'P_z ECd \tag{27}$$

$$\dot{\Theta} \leq -\lambda_{min}[Q_z]\eta_z^2 + 2\|P_z EC\|\eta_z\rho_d \tag{28}$$

from which

$$r_{b_z} = 2\|P_z EC\|\rho_d / \lambda_{min}[Q_z] \tag{29}$$

and finally

$$\rho_{l_z} = r_{b_z}\sqrt{\lambda_{max}[P_z]/\lambda_{min}[P_z]} \tag{30}$$

EXAMPLES

The vehicle for the following examples is a representation of pitch plane dynamics of the CH-47 (cargo helicopter). It was taken from Doyle and Stein (1981) and extended sufficiently to support the example. The state vector is [horizontal speed, vertical speed, pitch angle rate, pitch angle]', the control vector is [collective, cyclic]', and the disturbance vector is [horizontal gust, vertical gust]'. The plant state coefficient matrix is

$$A = \begin{bmatrix} -.0200 & .0050 & 2.400 & -32.0 \\ -.140 & .4400 & -1.30 & -30.0 \\ .0000 & .0180 & -1.60 & 1.200 \\ .0000 & .000 & 1.000 & .0000 \end{bmatrix}$$

while the control coefficient matrix and the disturbance coefficient matrix are

$$B = \begin{bmatrix} .1400 & -.120 \\ .3600 & -8.60 \\ .3500 & .0090 \\ .0000 & .0000 \end{bmatrix} \text{ and } C = \begin{bmatrix} .0200 & -.005 \\ .1400 & -.440 \\ .0000 & -.018 \\ .0000 & .0000 \end{bmatrix}$$

It is known that $\|d\| \leq \rho_d = 1$.

The plant is not stable. After stabilization via full state feedback $u = Kx$ with criterion

$$K\colon \min_K \int x'x \, dt$$

a controlled system state coefficient matrix

$$A_c = \begin{bmatrix} .0774 & -.0884 & 1.5579 & -35.0811 \\ .1198 & -5.8734 & -1.9290 & -23.2725 \\ .2433 & .0109 & -3.7624 & -7.0488 \\ .0000 & .0000 & 1.0000 & .0000 \end{bmatrix}$$

is obtained. The selection of $Q = I_4$ results in

$$P = \begin{bmatrix} .9573 & -.0159 & -2.3524 & -14.7063 \\ -.0159 & .0854 & .0307 & .0762 \\ -2.3524 & .0307 & 18.5484 & 73.0113 \\ -14.7063 & .0762 & 73.0113 & 346.6934 \end{bmatrix}$$

with eigenvalues $\lambda(P) = [\,362.8164,\ 3.2340,\ .1525,\ .0817]$

Example Part 1: State Bound.

An ultimate bound on η_x is desired. Since $\rho_d = 1$ and $\|PC\| = 1.3494$, $r_b = 2\ \|PC\|\ \rho_d = 2.6988$. This in turn yields

$$\rho_{lx} = r_b \sqrt{\lambda_{max}[P]/\lambda_{min}[P]} = 179.8957$$

Example Part 2: Vertical Speed is the Controlled Output.

Selecting $z = [0, 1, 0, 0]x$ results in $A_e = -5.8734$. Next, taking $Q_z = 1$ results in $P_z = -1/(2A_e) = .0851$ and $r_{b_z} = 2\|P_z EC\| = .0786$. Since P_z is a scalar, $\rho_{l_z} = .0786$

Example Part 3: Horizontal Speed is the Controlled Output.

It is no coincidence that horizontal speed is selected as the controlled output for Part 3 and not Part 2. Since $z = [1, 0, 0, 0]x$, $A_e = .0774$ which is not a stability matrix. A similar result is found for the case in which the pitch angle is selected to be the controlled output.

Example Part 4: Pitch Angle Rate is the Controlled Output.

In this case, $z = [0, 0, 1, 0]x$ and $A_e = -3.7624$. Taking $Q_z = 1$ results in $P_z = .1329$ which gives $\|P_z EC\| = .0024$. Thus $\rho_{l_z} = .0048$.

Example Part 5: Attempt to Bound Horizontal Speed with Less Conservatism.

It may be that horizontal speed can be grouped with another output to obtain a less conservative bound than that obtained in Part 1. It can be confirmed that no pairing is successful. However, let

$$E = \begin{bmatrix} 1 & 0 & 0 & 0 \\ 0 & 0 & 1 & 0 \\ 0 & 0 & 0 & 1 \end{bmatrix}$$

then

$$A_e = \begin{bmatrix} .0774 & 1.5579 & -35.0811 \\ .2433 & -3.7624 & -7.0488 \\ .0000 & 1.0000 & .0000 \end{bmatrix} \quad P_z = \begin{bmatrix} .9488 & -2.3575 & -14.6928 \\ -2.3575 & 18.6110 & 73.1957 \\ -14.6928 & 73.1957 & 346.7659 \end{bmatrix}$$

From the above, $r_{b_z} = 2.634$ and $\rho_{l_z} = 133.5815$. Comparison to the results of Part 1 shows that some conservatism has been eliminated.

CONCLUDING REMARKS

Both sufficient conditions for valid performance of practical stability analysis in output space *and* a procedure for doing so have been presented. Robustness analysis in output space can produce substantially reduced conservatism. At the least, the use of output space robustness analysis can produce a reduction of the bound estimate of a single controlled output by a factor of $\sqrt{n_z / n_x}$.

For any specific case, the equivalent system may not be stable. In such cases, it may be that the true controlled output can be grouped with additional arbitrarily selected outputs to form a ficticious controlled output which is of smaller dimensionality than the state and which will produce a stable equivalent system (the smaller the dimensionality, the better).

Two procedures for reducing conservatism in analyzing robustness in state space are worthwhile. Scaling so that the bound of each uncertain element is unity can be useful, as is the scaling of the desired bound of each controlled output element. Analysis of robustness by making use of a real Jordan canonical transformation can be useful.

There are at least two questions which are relevant to these results. First, can anything be done via feedback control to produce a stable equivalent system state coefficient matrix while maintaining system stability? Second, the requirement that the equivalent system state coefficient matrix be stable is not necessary for the controlled output to be bounded. Are there conditions which are necessary and sufficient?

ACKNOWLEDGEMENTS

The author is endebted to Professor George Leitmann for very helpful suggestions for improving these results.

REFERENCES

Barmish, B. R. (1985). "Necessary and sufficient conditions for quadratic stabilizability of an uncertain system," *J. Opt. Theory and Applic.*, vol. 46, no. 4, pp. 399-408.

Barmish, B. R., and G. Leitmann (1982). "On ultimate boundedness control of uncertain systems in the absence of matching conditions," *IEEE Trans. Auto. Cont.*, vol. AC-27, no. 1, pp. 153-159.

Black, K. W. (1990). "A relaxed mismatch condition for reduced conservatism in Lyapunov stability analysis," *Proc.* 1990 *American Automatic Control Conf.*, San Diego, California, May 23-25.

Blackwell, C. C. (1990). "Synthesis of disturbance attenuating, noise rejecting regulator control via the matrix Riccati Equation," *Proc.* 1990 *American Control Conf.*, San Diego, California. May 23-25.

Corless, M., and G. Leitmann (1981). "Continuous state feedback guaranteeing uniform ultimate boundedness for uncertain dynamic systems," *IEEE Trans. Auto. Cont.*, vol. AC-26, no. 5, pp. 1139-1144.

Doyle, J. C., and G. Stein (1981). "Multivariable feedback design: concepts for a classical/modern synthesis," *IEEE Trans. Auto. Con.*, vol. AC-26, no. 1, pp. 4-16.

Francis, B. A. (1988). *A Course in H_∞ Control Theory*. Springer-Verlag, New York.

Kalman, R. E., and J. E. Bertram (1960a). "Control system analysis and design via the "Second Method of Lyapunov": I. continuous-time systems," *ASME J. Basic Engr.*, vol. 82, no. 2, pp. 371-393.

Leitmann, G. (1981). "On the efficacy of nonlinear control in uncertain linear systems," *ASME J. Dynam. Syst., Meas., and Contr.*, vol. 102, pp. 95-102.

Šiljak, D. D. (1987). "Stability of reduced-order models via vector Liapunov functions," *Proceedings of the* 1987 *American Automatic Control Conference*, Minneapolis, Minnesota, June 10-12, vol. 1, pp. 482-489.

Takahashi, Y., M. J. Rabins, and D. M. Auslander (1970). *Control and Dynamic Systems*, Addison-Wesley Publishing Company. Reading, Massachussetts.

Yedavalli, R. K. (1989). "Robust control design for aerospace applications," *IEEE Trans. Aerosp. and Electronic Sys.*, vol. 25, no. 3, pp. 314-324.

SUFFICIENCY CONDITIONS FOR EXISTENCE OF AN OPTIMAL FEEDBACK CONTROL IN STOCHASTIC MECHANICS[1]

A. Blaquière[2]

Dedicated to George Leitmann

Abstract. In the framework of stochastic mechanics, the following problem is considered : in a set of admissible feedback controls v, with range in E^n, find one minimizing the expectation $E_{sx}\{\int_s^T L(t,\xi(t),v(t,\xi(t)))dt + W_T(\xi(T))\}$ for all $(s,x) \in [0,T)\otimes E^n$, where $L(t,x,\nu) = (1/2)m\nu^2 - U(t,x)$ is the classical action integrand and ξ is a n-dimensional diffusion process (in the weak sense [17]) with drift v and diffusion coefficient D = constant > 0. W_T and U are given real functions . Sufficiency conditions for existence of such an optimal feedback control are given.

1. Introduction

Consider the following familiar problem of calculus of variations in n-dimensional Euclidean space E^n : find a E^n-valued function z on a fixed interval [O, T] minimizing

$$\int_0^T L(t, z(t), z'(t))\, dt + F(z(T)) \tag{1.1}$$

with z(0) = x given.

---

[1] Presented at the THIRD WORKSHOP ON CONTROL MECHANICS in honour of GEORGE LEITMANN, January 22-24, 1990, University of Southern California, Los Angeles, California (USA).

[2] Université Paris 7, Laboratoire d'Automatique Théorique, Tour 14-24, 2 Place Jussieu, 75251 PARIS CEDEX 05 (FRANCE).

For $L(t, x, v) = (1/2)mv^2 - U(t, x)$, the classical action integrand, such a problem appears in classical mechanics as a consequence of Hamilton's least action principle and leads to well known equations of mechanics.

In relation with quantum mechanics, stochastic analogues of (1.1) have been considered by different authors, in particular by Blaquière (see [3]) in the framework of stochastic optimal control theory, and more recently by Guerra-Morato [8], Yasue [14], Blaquière-Marzollo [4] , Kime-Blaquière [5], Blaquière [6], Wakolbinger [19]. Nelson's stochastic mechanics [11] is not directly concerned with variational principles however, as shown in [5] , its mathematical structure is closely related to stochastic optimal control theory which, itself, provides a stochastic analogue of problem (1.1) as a special case when the end times are prescribed.

Nelson's stochastic mechanics postulates the motion of a particle according to a stochastic differential equation

$$d\xi(t) = v(t, \xi(t))dt + \sqrt{2D}\ dw(t) \tag{1.2}$$

with $D = h/4\pi m$, h the Planck's constant, m the mass of the particle.

To account for *time-reversibility* in quantum mechanics, both "forward" and "backward" drifts are considered in stochastic mechanics. This can be done in different ways ;

-by introducing a stochastic analogue of second Newton's law (Nelson [11];

-by introducing a stochastic analogue of the classical action integrand (Guerra-Morato [8], Yasue [14]);

-by considering a pair of stochastic optimal control problems, one "forward" in time, one "backward" in time (Blaquière-Marzollo [4], Kime-Blaquière[5], Blaquière [6]).

In this paper we are concerned with existence of an optimal feedback control for the problems stated in [5]. We have not dealt with time reversal since one may pass from the "forward" to the "backward" problem (and back) by changing t into T-t.

2. Notation

E^n denotes n-dimensional Euclidean space, (0, T) an interval in E^1. S denotes the Cartesian product $(0, T)\otimes E^n$. Definitions of *stochastic process, Brownian motion, standard Brownian motion*, are found in basic books (we shall use in particular [7], [9] [17] .

By Theorem 1.2, p. 38 in Friedman [9], there is a continuous version of a Brownian motion. *From now on when we speak of Brownian motion, it is always tacitly assumed that we speak of a continuous version.*

Let (Ω, F, P) be a probability space, and $\{F_t\}$, for t in some interval I, be an increasing family of sub-σ-algebras of F. We say that a process f on I is an *adapted process* if for each t in I, f(t) is F_t - measurable. A *nonanticipative process* and classes of functions $L^p_w[\alpha, \beta]$ and $M^p_w[\alpha, \beta]$ $(1\le p\le\infty)$ are defined as in Friedman [9], pp. 55, 56.

For given (Ω, F, P), let $b(t,x) = (b_1(t,x), \dots b_n(t,x))$, $\sigma(t,x) = (\sigma_{ij}(t,x))^n_{ij=1}$ and suppose the functions $b_i(t,x)$, $\sigma_{ij}(t,x)$ are measurable in $(t,x)\in[0, T]\otimes E^n$. A solution of the stochastic differential equation

$$d\xi(t) = b(t,\xi(t))dt + \sigma(t,\xi(t))dw(t), \qquad 0\le t\le T, \tag{2.1}$$

with initial data $\xi(s) = x$ ($(s,x) \in [0, T)\otimes E^n$) is to be interpreted as in [9] as a solution of the integral equation

$$\xi(t) = \xi(s) + \int_s^t b(r,\xi(r))dr + \int_s^t \sigma(r,\xi(r))dw(r) . \tag{2.2}$$

Here, w is standard Brownian motion of dimension n. With the vector notation $\xi = (\xi_1, \dots \xi_n)$, we have

$$d\xi_i(t) = b_i(t,\xi(t))dt + \sum_{j=1}^n s_{ij}(t,\xi(t))dw_j(t), \qquad i = 1, \dots n .$$

Note that it is implicitely assumed that $b(t,\xi(t))$ belongs to $L^1_w[0, T]$ and $\sigma(t,\xi(t))$ belongs to $L^2_w[0, T]$.

Let us say that a function $\psi(t,x)$ is of class $C^{1,2}$ if the partial derivatives ψ_t, ψ_x, ψ_{xx} are continuous on $[0, T]\otimes E^n$. For Q an open subset of $E^1\otimes E^n$, let us denote by $C^{1,2}(Q)$ the space of functions $\psi(t,x)$ continuous on Q together with the partial derivatives $\psi_t, \psi_{x_i}, \psi_{x_i x_j}$, $i,j = 1,...n$. We say that ψ satisfies a *polynomial growth condition* on Q if, for some constants D, k, $|\psi(t,x)|\le D(1+|x|^k)$ where $(t,x)\in Q$. Let $C_p^{1,2}(Q)$ denote the class of functions ψ in $C^{1,2}(Q)$ which satisfy a polynomial growth condition on Q. We shall denote the class of functions of x on E^n which are twice continuously differentiable and satisfy a polynomial growth condition by $C_p^2(E^n)$.

The closure and the interior of a set A will be denoted by clos A and int A, respectively.

3. Problem Statement

For some probability space (Ω,F,P), consider the stochastic differential equation

$$d\xi(t) = v(t,\xi(t))dt + \sigma dw(t) \tag{3.1}$$

with initial data $\xi(s) = x \in E^n$, at time $s\in[0, T)$, where w is a standard n-dimentional Brownian motion, and

$$\sigma_{ij} = \sqrt{2D}\,\delta_{ij}$$

where δ is the Kronecker delta, and D is a positive constant. We assume that v belongs to a class of *admissible control functions* defined as follows:

Definition 3.A. A *feedback control law* v (the term feedback refers to the fact that the control is a function of the state $\xi(t)$) is *admissible* if v is a Borel measurable function from $[0, T]\otimes E^n$ into E^n, such that

(a) For each (s,x) in $[0, T)\otimes E^n$ there exists a probability P_{sx}^v on the measurable space (Ω,F), and a Brownian motion with respect to P_{sx}^v - say w_{sx} - such that

$$d\xi(t) = v(t,\xi(t))dt + \sqrt{2D}\ dw_{sx}(t), \qquad s\le t\le T, \qquad (3.2)$$

with $\xi(s) = x$, have a solution, unique in probability law; and

(b) For each $k>0$, $E_{sx}^v |\xi(t)|^k$ is bounded for $s\le t\le T$ and

(c) $$E_{sx}^v \int_s^T |v(t,\xi(t))|^2\, dt < \infty \quad ;$$

the bound in (b) may depend on (s,x).

We denote by V the class of all admissible feedback control laws.

Now, for $(t,x) \in$ clos S and $\nu \in E^n$, let

$$L(t,x,\nu) = (1/2)m\nu^2 + V(t,x) \qquad (3.3)$$

where V is continuous on clos S, and let $W_T : E^n \to R_+$ (R_+ denoting nonnegative real numbers) be continuous and assume

$$|V(t,x)| \le C(1 + |x|)^k$$
$$W_T(x) \le C(1 + |x|)^k \qquad (3.4)$$

for some constants C, k.

We define a *cost function*

$$J(s,x,v) = E_{sx}^v \{ \int_s^T L(t,\xi(t),v(t,\xi(t)))dt + W_T(\xi(T)) \} .$$

Conditions (b), (c) and (3.4) insure that $J(s,x,v)$ is finite for each $v \in V$. Further assumptions about V will be made later as needed. The case where

$\xi(T)$ is required to belong to a prescribed subset of E^n will be considered elsewhere.

Now let the optimal control problem be as follows : *find a feedback control* v^* *which minimizes* $J(s,x,v)$ *among all feedback controls* $v \in V$.

The following Verification Theorem gives sufficient conditions for existence of a minimizing v^*.

Theorem 3.B Let $W(s,x)$ be a solution of the dynamic programming equation

$$0 = \frac{\partial W}{\partial s} + \min_{\nu \in E^n} [D\Delta W + \nu \cdot \text{grad } W + (1/2)m\nu^2 + V(s,x)] \qquad (3.5)$$

$$(s,x) \in S ,$$

with boundary data

$$W(T,x) = W_T(x), \quad x \in E^n, \qquad (3.6)$$

such that W is in $C_p^{1,2}(S)$ and continuous on clos S . Then,

(a) $W(s,x) \leq J(s,x,v)$ for any admissible feedback control v and any initial data $(s,x) \in S$.

(b) If v^* is an admissible feedback control such that

$$D\Delta W + v^*(s,x).\text{grad } W + (1/2)m(v^*(s,x))^2 + V(s,x)$$
$$= \min_{\nu \in E^n} [D\Delta W + \nu.\text{grad } W + (1/2)m\nu^2 + V(s,x)] \qquad (3.7)$$

for all $(s,x) \in S$, then $W(s,x) = J(s,x,v^*)$ for all $(s,x) \in S$. Thus, v^* is optimal.

Though our condition (c) in Definition 3.A is weaker than the corresponding one of Fleming-Rishel [7], p. 156, due to the special form (3.3) of $L(t,x,\nu)$, the proof of Theorem 3.B is similar to the one of Theorem 4.1 of that Reference.

The proof of existence of an optimal feedback control for the problem stated above, under assumptions which will be made later, will comprise two steps :

first, the proof of existence of a W in $C_p^{1,2}(S)$ and continuous on clos S, satisfying (3.5) and (3.6); and

secondly, the proof of existence of an admissible v* satisfying (3.7).

We shall need a preliminary observation, namely

4. A Logarithmic (or Exponential) Transformation

For the time being, let us assume that there exist functions W and v* satisfying the hypotheses of the Verification Theorem. Thus

$$\frac{\partial W}{\partial s} = - D\, \Delta W + (1/2m)(\text{grad } W)^2 - V \qquad (4.1)$$

$$mv^* = - \text{grad } W, \qquad (4.2)$$

for all $(s,x) \in S$.

We now perform the logarithmic (or exponential) change of variable on (4.1), (4.2); that is, we let

$$\rho(s,x) = \exp\,(- W(s,x)/2mD), \qquad (s,x) \in \text{clos } S. \qquad (4.3)$$

(ρ can be defined on clos S since W is continuous on clos S).

Thus

$$\rho(T,x) = \exp\,(- W_T(x)/2mD\,) = \rho_T(x). \qquad (4.4)$$

A noteworthy property of this transformation has been observed in [1] and further used by different authors (see [20]), namely, equation (4.1) becomes

$$\frac{\partial \rho}{\partial s} = - D\, \Delta\rho + (V/2mD)\, \rho\,, \qquad (s,x) \in S, \qquad (4.5)$$

and, from (4.2) we have

$$v^*(s,x) = (2D/\rho(s,x))\text{ grad } \rho(s,x)\;, \qquad (s,x)\in S\,. \qquad (4.6)$$

Further

$$v^*(T,x) = (2D/\rho_T(x)) \text{ grad } \rho_T(x), \quad x \in E^n, \qquad (4.7)$$

if the partial derivatives exist.

Finding a solution to the equation

$$\frac{\partial\phi}{\partial s} = -D\,\Delta\phi + (V/2mD)\,\phi, \qquad (s,x) \in S, \qquad (4.8)$$

with the terminal condition

$$\phi(T,x) = \rho_T(x) \qquad \text{on } E^n, \qquad (4.9)$$

constitutes a *Cauchy problem.*

Let $Q = (0, \Theta)\otimes E^n$, with $T<\Theta$, and concerning V let us introduce

(A1) $V : \text{clos } Q \to R_+$ *is bounded and continuous on the strip* clos Q *and satisfies a Hölder condition with respect to* x *on* E^n.

A weaker assumption, with Q replaced in (A1) by S, would be enough for asserting that there exists a unique continuous and bounded solution ϕ to the Cauchy problem (4.8),(4.9), with

$$\phi(s,x) = \int_{E^n} p(s,x;t,y)\,\phi(t,y)\,dy, \qquad s<t\leq T, \qquad (4.10)$$

where p denotes the fundamental solution of (4.8) (see Il'in, Kalashnikov, Oleinik [18] . The introduction of Q will be needed later.

5. Existence of a Well-behaved W

The proof of Theorem 5.A has been given in [5].

Assume :

(A2) V *is uniformly Hölder continuous in* (s,x) *in compact subsets of* clos Q .

Theorem 5.A If $W_T : E^n \to R_+$, continuous on E^n, satisfies

$$W_T(x) \leq C(1 + x^2) \qquad (5.1)$$

for some C, and if V satisfies (A1) and (A2), then there exists W in $C_2^{1,2}(S)$ and continuous on clos S, satisfying (3.5) and (3.6), with $W(s,x) \geq 0$ for all $(s,x) \in \text{clos } S$.

Assume :

(A3) *Under* (A1) *and* (A2), $W_T(x)$ $(x \in E^n)$ *is given by*

$$\exp(-W_T(x)/2mD) = \int_{E^n} p(T,x;\Theta,y)\, g(y)\, dy \qquad (5.2)$$

where $g : E^n \to [0, 1]$ *is a continuous function with compact support.*

By a slight change in the proof of Theorem 5.A, one obtains

Theorem 5.B Let (A1)-(A3) hold. Then there exists a nonnegative function W on clos S satisfying (3.5), (3.6) and

$$W \in C^{1,2} \cap C_2^{1,2}(S), \qquad (5.3)$$

$$W_T \in C_2^2(E^n) \quad . \qquad (5.4)$$

6. Existence of an Optimal Feedback Control

Define v^* : clos S $\to E^n$ by

$$v^*(s,x) = (-1/m) \text{ grad } W(s,x) , \quad (s,x) \in \text{clos } S. \qquad (6.1)$$

We wish to prove that v^* is an optimal feedback control for the problem stated in §3.

Since v^* satisfies

$$D\,\Delta W + v^*(s,x).\text{grad } W + (1/2)m(v^*(s,x))^2 + V(s,x) =$$
$$+\min_{\nu \in E^n} [D\Delta W + \nu \,.\, \text{grad } W + (1/2)m\nu^2 + V(s,x)] \qquad (6.2)$$
$$\text{for all } (s,x) \in \text{clos } S,$$

Theorem 3.B tells us that *we have only to prove that* v^* *is an admissible feedback control.* For that purpose, we will have to prove a few lemmas.

Lemma 6.A Let (A1)-(A3) hold. Then v^* is a Borel measurable function from clos S into E^n , such that, for each (s,x) in $[0, T)\otimes E^n$ there exists a probability P_{sx} on the measurable space (Ω,F), and a n-dimensional

Brownian motion with respect to P_{sx} - say w_{sx} - and a nonanticipative continuous process ξ, such that

$$\xi(t) = x + \int_s^t v^*(r,\xi(r))dr + \sqrt{2D}\,(w_{sx}(t) - w_{sx}(s)), \qquad (6.3)$$

$$s \le t \le T.$$

Proof. Since, by (5.3) and Definition (6.1), v^* is continuous on clos S, v^* is Borel measurable.

Let $w°(t)$ $(s \le t \le T)$ be a standard n-dimensional Brownian motion with respect to the probability space $(\Omega,F,P°)$. Let $\{F_t\}$ be an increasing family of sub-σ-algebras of F, such that (i) $F(w°(r),\ s \le r \le t)$ is a subset of F_t and (ii) $F(w°(\lambda+t)-w°(t),\ 0 \le \lambda \le T-t)$ is independent of F_t for all $t \in [s, T]$.

$w°$ has the following properties :

1. $w°$ is a continuous process, and therefore it is a separable process.
2. $w°$ is a measurable process, since any continuous process is measurable.
3. By condition (i) above, $w°(t)$ is F_t-measurable for each $t \in [s, T]$; that is, it is adapted to $\{F_t\}$.

Therefore $w°$ is a nonanticipative process with respect to $\{F_t\}$ in the sense of Friedman [9], pp. 55, 56.

4. By the continuity of $w°$, $w°$ is in $L^p_{w°}[s, T]$, $1 \le p \le \infty$.

Consider the system of stochastic differential equations

$$d\eta(t) = \sqrt{2D}\, dw°(t), \qquad s \le t \le T, \qquad (6.4)$$
$$\eta(s) = x, \qquad (6.5)$$

on $(\Omega,F,P°)$.

Clearly, η has the same properties 1-4 above as $w°$. Therefore, η *is a continuous nonanticipative process.*

Since, by (5.3), W is of class $C^{1,2}$, we deduce from the Ito stochastic differential rule

$$W(t,\eta(t)) - W(s,x) = \int_s^t \frac{\partial W}{\partial t}(r,\eta(r))dr + \int_s^t \text{grad } W(r,\eta(r)) \, d\eta(r) + D\int_s^t \Delta W(r,\eta(r) \, dr , \tag{6.6}$$

$0 \leq s \leq t \leq T$, with probability 1 .

Otherwise, we have from (3.5)

$$\frac{\partial W}{\partial s} = - D \,\Delta W + (1/2m)\, (\text{grad } W)^2 - V \tag{6.7}$$

from which we deduce

$$\int_s^t V(r,\eta(r)) \, dr = - \int_s^t \frac{\partial W}{\partial t}(r,\eta(r)) \, dr + (1/2m)\int_s^t [\text{grad } W(r,\eta(r))]^2 \, dr - - D\int_s^t \Delta W(r,\eta(r)) \, dr , \tag{6.8}$$

$0 \leq s \leq t \leq T$, with probability 1 .

By the continuity of V and of the other functions involved, and by the continuity of the process η , all the integrals in (6.8) are finite with probability 1.

By adding (6.6) and (6.8) we obtain

$$\int_s^t V(r,\eta(r)) \, dr + W(t,\eta(t)) - W(s,x) = = \int_s^t (\text{grad } W(r,\eta(r)) \, d\eta(r) + (1/2m) \int_s^t [\text{grad } W(r,\eta(r))]^2 \, dr , \tag{6.9}$$

$0 \leq s \leq t \leq T$, with probability 1 .

Let

$$\theta(t) = (-1/m\sqrt{2D}) \text{ grad } W(t,\eta(t)) = (1/\sqrt{2D}) \, v^*(t,\eta(t)) , \tag{6.10}$$

$$\zeta_s^t(\theta) = \int_s^t \theta(r) \, dw^\circ(r) - (1/2) \int_s^t |\theta(r)|^2 \, dr . \tag{6.11}$$

Since v^* is continuous on clos S , and η has the same properties 1-4 as $w°$, the stochastic process θ enjoys the same properties 1-4 as $w°$. Therefore, θ *is a continuous nonanticipative process,* and $\theta \in L^p_{w°}[s, T]$, $1 \le p \le \infty$.

From (6.9)-(6.11) we deduce, by taking the exponentials of both sides of (6.9) after division by -2mD,

$$\exp \zeta_s^t(\theta) = \frac{\exp(-W(t,\eta(t))/2mD)}{\exp(-W(s,x)/2mD)} \exp\left(-(1/2mD) \int_s^t V(r,\eta(r))dr \right). \quad (6.12)$$

$0 \le s \le t \le T$, with probability 1 .

Since $\rho = \exp(-W/2mD)$ is the (unique) solution of the Cauchy problem (4.8),(4.9), by Theorem 5.3 of Friedman [9], p. 148, we have

$$\rho(s,x) = E° \rho_T(\eta(T)) \exp\left(-(1/2mD) \int_s^T V(r,\eta(r))dr \right),$$

where $E°$ denotes expectation with respect to $P°$; that is

$$\exp(-W(s,x)/2mD) = \int_\Omega \exp(-W(T,\eta(T))/2mD)$$

$$\exp\left[-(1/2mD)\int_s^T V(r,\eta(r))dr\right] dP°(\omega) \quad (6.13)$$

From (6.12) and (6.13) we deduce

$$\int_\Omega \exp \zeta_s^T(\theta)\, dP°(\omega) = 1 . \quad (6.14)$$

(One can also prove directly that $\exp \zeta_s^t(\theta)$ is a martingale, whence (6.14) follows).

Let P_{sx} be absolutely continuous with respect to $P°$, with

$$d P_{sx}(\omega) = \exp \zeta_s^T(\theta)\, dP°(\omega), \quad (6.15)$$

$$w_{sx}(t) = w°(t) - \int_s^t \theta(r)\, dr . \quad (6.16)$$

Then, since θ is a function in $L^2_{w°}[s, T]$, and since, by (6.14) and (6.15),

$$P_{sx}(\Omega) = 1 ,$$

by the theorem of Girsanov (Friedman [9], p. 156), $w_{sx}(t)$ $(s\le t\le T)$ is an n-dimensional Brownian motion with respect to the probability space (Ω,F,P_{sx}).

From (6.4) and (6.5)

$$\eta(t) - x = \sqrt{2D}\ (w^\circ(t) - w^\circ(s)) \ . \tag{6.17}$$

From (6.16) and the definition of θ

$$w^\circ(t) = w_{sx}(t) - (1/m\sqrt{2D}) \int_s^t \text{grad } W(r,\eta(r))\, dr =$$

$$= w_{sx}(t) + (1/\sqrt{2D}) \int_s^t v^*(r,\eta(r))\, dr \ . \tag{6.18}$$

From (6.17),(6.18) and $w_{sx}(s) = w^\circ(s)$, we obtain

$$\eta(t) = x + \int_s^t v^*(r,\eta(r))\, dr + \sqrt{2D}\ (w_{sx}(t) - w_{sx}(s)), \tag{6.19}$$

$$s \le t \le T,$$

and since we have seen that η is a continuous nonanticipative process, Lemma 6.A is proved, with $\xi = \eta$.

Note that (6.19) means that η is a solution of the stochastic differential equation

$$d\eta(t) = v^*(t,\eta(t))\, dt + \sqrt{2D}\ dw_{sx}(t)\ , \qquad s \le t \le T\ , \tag{6.20}$$

with the initial data

$$\eta(s) = x\ . \tag{6.21}$$

η and w_{sx} are continuous nonanticipative processes with respect to the probability space (Ω,F,P_{sx}).

Of course, by (6.10), $v^*(t,\eta(t))$ like $\theta(t)$ $(s\le t\le T)$ is in $L^p_{w^\circ}[s, T]$, $1 \le p \le \infty$, in particular in $L^1_{w^\circ}[s, T]$. By weakening the conditions on $\{F_t\}$, as in Friedman [9], p.165, since P_{sx} is absolutely continuous with respect to P° one can see easily that this implies that $v^*(t,\eta(t))$ is in $L^1_{w_{sx}}[s, T]$ as required.

Lemma 6.B Let A(1)-A(3) hold. Then (6.20), (6.21) have a unique solution.

Proof. Suppose $\xi'(t)$ and $\xi''(t)$ $(s \le t \le T)$ are two solutions. Then

$$\xi'(t) = x + \int_s^t v^*(r,\xi'(r))\, dr + \sqrt{2D}\, (w_{sx}(t) - w_{sx}(s)) ,$$

$$\xi''(t) = x + \int_s^t v^*(r,\xi''(r))\, dr + \sqrt{2D}\, (w_{sx}(t) - w_{sx}(s)) ,$$

and, accordingly,

$$\xi'(t) - \xi''(t) = \int_s^t [f'(r) - f''(r)]\, dr ,$$

where $f'(r) = v^*(r,\xi'(r))$, $f''(r) = v^*(r,\xi''(r))$.

By Cauchy-Schwarz inequality

$$|\xi'(t) - \xi''(t)|^2 \le (t-s) \int_s^t | f'(r) - f''(r)|^2\, dr \le (T-s) \int_s^t |f'(r) - f''(r)|^2\, dr . \quad (6.22)$$

Now, let G_k be an open ball in E^n with center O and radius k with $x \in G_k$.

For $\omega \in \Omega$, let $\tau'^k(\omega)$ (resp. $\tau''^k(\omega)$) denote the smallest time t $(0 \le s \le t \le T)$ such that $\xi'(t,\omega) \notin G_k$ (resp. $\xi''(t,\omega) \notin G_k$) , or $\tau'^k(\omega) = T$ (resp. $\tau''^k = T$) if $\xi'(t,\omega) \in G_k$ (resp. $\xi''(t,\omega) \in G_k$) for all $s \le t \le T$.

$\tau'^k(\omega)$ (resp. $\tau''^k(\omega)$) is called the *exit time for* $\xi'(t,\omega)$ (resp. $\xi''(t,\omega)$) *from* $(0, T)\otimes G_k$. Note that if $\tau'^k(\omega) = T$ (resp. $\tau''^k(\omega) = T$) , then $|\xi'(t,\omega)|<k$ (resp. $|\xi''(t,\omega)|<k$) for all $s \le t \le T$.

Let

$$\Omega_k = \{ \omega : \quad \tau'^k(\omega) = \tau''_k(\omega) = T \} .$$

From (6.22) we deduce

$$\int_{\Omega_k} |\xi'(t) - \xi''(t)|^2 \, dP_{sx}(\omega) \leq (T-s)\int_{\Omega_k} [\int_s^t |f'(r) - f''(r)|^2 \, dr] \, dP_{sx}(\omega) . \tag{6.23}$$

All expectations and integrals in (6.23) exist since ξ' and ξ'' are bounded by k for all $\omega \in \Omega_k$, and v^* being continuous is bounded on $[0, T]\otimes(\text{clos } G_k)$.

By Fubini's theorem

$$\int_{\Omega_k} |\xi'(t) - \xi''(t)|^2 \, dP_{sx} \leq (T-s)\int_s^t dr \, [\int_{\Omega_k} |f'(r) - f''(r)|^2 dP_{sx}(\omega)]. \tag{6.24}$$

$v^*(t,y) = (-1/m) \text{ grad } W(t,y)$, being continuously differentiable with respect to y_i, $i = 1, \ldots n$, on the open convex set $(0, T)\otimes G_k$, with these derivatives bounded on $[0, T]\otimes(\text{clos } G_k)$, is Lipschitzian on $(0, T)\otimes G_k$.

Therefore

$$|f'(r) - f''(r)| \leq K_k |\xi'(r) - \xi''(r)| \qquad \text{for } \omega \in \Omega_k ,$$

where the constant K_k may depend on G_k.

Substitution into (6.24) result in

$$\int_{\Omega_k} |\xi'(t) - \xi''(t)|^2 \, dP_{sx}(\omega) \leq C_k \int_s^t dr \, [\int_{\Omega_k} |\xi'(r) - \xi''(r)|^2 \, dP_{sx}(\omega) \quad , \tag{6.25}$$

where the constant C_k may depend on G_k.

Let

$$h(t) = \int_{\Omega_k} |\xi'(t) - \xi''(t)|^2 \, dP_{sx}(\omega) .$$

By (6.25), the function h satisfies

$$h(t) \leq C_k \int_s^t h(r) \, dr , \qquad h(s) = 0. \tag{6.26}$$

$\xi'(.,\omega)$ and $\xi''(.,\omega)$ being P_{sx}-almost surely continuous, and bounded for $\omega \in \Omega_k$, by the Lebesgue's theorem of dominated convergence, h is continuous on [s, T]. Further h is nonnegative. Therefore, by Gronwall's inequality , (6.26) implies

$$h(t) \equiv 0 .$$

From the definition of h, this in turn implies that

$$P_{sx}\{ \omega : \quad \omega \in \Omega_k , \quad \xi'(t,\omega) = \xi''(t,\omega) \quad \text{for all } s \le t \le T \} = P_{sx}(\Omega_k) . \quad (6.27)$$

In other words, (almost surely Ω_k , P_{sx}) ξ' and ξ" have the same sample functions.

E^n is the countable union of sets G_k , k = 1, 2, ... Sample functions $\xi'(.,\omega)$ and $\xi''(.,\omega)$ are continuous for all $\omega \in \Omega - N_o$, where N_o is a subset of Ω of P_{sx}-measure 0 . Because of this continuity and of (6.27), for any $\omega \in \Omega - N_o$, there is a $k \in \{ 1, 2, ... \}$ and a subset Ω_k of Ω containing ω, and a subset N_k of Ω_k of P_{sx}-measure 0, such that for all $\omega' \in \Omega_k - N_k$

$$\xi'(.,\omega') = \xi''(.,\omega') .$$

Let $N = \cup_{k=1,2,...} N_k$. Since N is the countable union of disjoint sets N_1, $N_{k+1}-N_k$ (k=1, 2, ...) each of which has P_{sx}-measure 0, $P_{sx}(N) = 0$. Further, for $\omega \in \Omega - N_o$:

$$\text{either } \xi'(.,\omega) = \xi''(.,\omega) \quad \text{or} \quad \omega \in N .$$

Therefore

$$P_{sx} \{ \omega : \quad \xi'(t,\omega) = \xi''(t,\omega) \quad \text{for all } s \le t \le T \} = 1 ,$$

which concludes the proof of Lemma 6.B.

<u>Lemma 6.C</u> Let (A1)-(A3) hold. Then (6.20), (6.21) have a unique solution in probability law.

Lemma 6.C is a straightforward consequence of Lemma 6.B.

Lemma 6.D Let (A1)-(A3) hold. Then, for each (s,x) in $[0, T)\otimes E^n$ and for each k>0, $E_{sx}|\eta(t)|^k$ is bounded for $s \le t \le T$.

Lemma 6.D follows from the estimate

$$E^\circ|\eta(t)|^k \le C_k(1 + E^\circ|\eta(s)|^k), \qquad s \le t \le T,$$

where the constant C_k depends on k and T-s (Flemming-Rishel [7], p.119), together with $\eta(s) = x$, and the fact that P_{sx} is absolutely continuous with respect to P°.

Lemma 6.E Let (A1)-(A3) hold. Then, for each (s,x) in $[0, T)\otimes E^n$

$$E_{sx}\int_s^T |v^*(t,\eta(t))|^2 \, dt < \infty .$$

Proof. By substituting (6.20) into (6.9) we obtain

$$\int_s^t V(r,\eta(r))\,dr + W(t,\eta(t)) - W(s,x) =$$

$$= \int_s^t \text{grad } W(r,\eta(r))\, v^*(r,\eta(r))\,dr + \sqrt{2D}\int_s^t \text{grad } W(r,\eta(r))\, dw_{sx}(r) +$$

$$+ (1/2m)\int_s^t [\text{grad } W(r,\eta(r))]^2\,dr. \qquad (6.28)$$

$$0 \le s \le t \le T, \qquad \text{a.s. in } P_{sx} \text{ (or in } P^\circ).$$

Then, in view of (6.1), (6.28) rewrites

$$\int_s^t V(r,\eta(r))\,dr + W(t,\eta(t)) - W(s,x) =$$

$$= (-m\sqrt{2D})\int_s^t v^*(r,\eta(r))\,dw_{sx}(r) - \int_s^t (1/2)m\,|v^*(r,\eta(r))|^2\,dr \qquad (6.29)$$

$$0 \le s \le t \le T, \qquad \text{a.s. in } P_{sx} \text{ (or in } P^\circ).$$

Let G_k be an open ball in E^n with center O and radius k, with $x \in G_k$, and let $\tau(k,\omega)$ denote the exit time for $\eta(t,\omega)$ from G_k. Write relation (6.29) for $t = \tau(k,\omega)$:

$$\int_s^{\tau(k)} V(r,\eta(r))\ dr + W[\tau(k),\eta(\tau(k))] - W(s,x) =$$

$$= (-m\sqrt{2D} \int_s^{\tau(k)} v^*(r,\eta(r))\ \ dw_{sx}(r) - \int_s^{\tau(k)} (1/2)m\ |v^*(r,\eta(r))|^2\ dr\ , \quad (6.30)$$

for P_{sx}-almost all ω .

By (5.3), $v^* = (-1/m)$ grad W is continuous on clos S and, accordingly, it is bounded on the compact set $[0, T]\otimes(\text{clos } G_k)$; that is ,

$$|v^*(t,y)| \le K, \qquad \text{for } (t,y) \in [0, T]\otimes(\text{clos } G_k)\ ,$$

for some K.

It follows that the function u defined by

$$u(t) = v^*(t,\eta(t)), \qquad s \le t \le T, \qquad (6.31)$$

satisfies :

$$\int_\Omega [\int_s^{\tau(k)} |u(t,\omega)|^2\ dt\ ]\ dP_{sx}(\omega) \le K^2(T-s).$$

In other words

$$u \in M^2_{w_{sx}}[s,\ \tau(k)]\ ,$$

in Friedman's notation [9], pp. 56 and 72.

Therefore, from Theorem 3.10 in Friedman [9], p. 72 (see also Fleming-Rishel [7], p. 124)

$$E_{sx} \int_s^{\tau(k)} u(r)\ \ dw_{sx}(r) = 0\ .$$

By taking expectations in (6.30), we obtain

$$W(s,x) = E_{sx}\{ \int_s^{\tau(k)} [(1/2)m\ |v^*(r,\eta(r))|^2 + V(r,\eta(r))]\ dr + W(\tau(k),\eta(\tau(k))\ \}. \quad (6.32)$$

All expectations and integrals in (6.32) exist since $(r,\eta(r))$ and $(\tau(k),\eta(\tau(k))$ lie in the compact set $[0, T]\otimes(\text{clos } G_k)$.

We first use (6.32) to prove that, under (A1)-(A3),

$$E_{sx}\ [\int_s^{\tau(k)} (1/2)m\ |v^*(r,\eta(r))|^2\ dr\ ] \qquad (6.33)$$

is bounded above by a constant *not depending on* k.

From (A1) according to which V is bounded on clos S , and from Theorem 5.B according to which W is nonnegative on clos S ,

$$E_{sx} \{ \int_s^{\tau(k)} V(r,\eta(r))\, dr + W[\tau(k),\eta(\tau(k))] \}$$

in the right-hand side of (6.32), is *bounded below on* clos S *by a constant not depending on* k. Therefore, by (6.32), for given (s,x) in $[0, T)\otimes E^n$, (6.33) *is bounded above by a constant* M° *not depending on* k; that is,

$$0 \le E_{sx} [\int_s^{\tau(k)} (1/2)m\, |v^*(r,\eta(r))|^2\, dr] \le M^\circ. \qquad (6.34)$$

Now let $k\uparrow\infty$.

Because of the continuity of the η-process, for P_{sx}-almost every $\omega \in \Omega$, we have

$$0 \le \int_s^{\tau(k)} (1/2)m\, |v^*(r,\eta(r))|^2\, dr \;\uparrow\; \int_s^T (1/2)m\, |v^*(r,\eta(r))|^2\, dr\ .$$

Therefore, by the monotone convergence theorem

$$E_{sx} [\int_s^{\tau(k)} (1/2)m\, |v^*(r,\eta(r))|^2\, dr \;\uparrow\; E_{sx} [\int_s^T (1/2)m\, |v^*(r)\eta(r))|^2\, dr.$$

By (6.34), it follows that

$$E_{sx} [\int_s^T (1/2)m\, |v^*(r,\eta(r))|^2\, dr \;\le\; M^\circ$$

which ends the proof of Lemma 6.E.

(6.2) and Lemmas 6.A, 6.C-6.E result in

<u>Theorem 6.F</u> Let (A1)-(A3) hold. Then there exists an optimal feedback control for the problem stated in §3.

References

1. A. Blaquiere, Liens entre la théorie géométrique des processus optimaux et la mécanique ondulatoire, *C.R. Acad. Sc. Paris, Série A.*, Vol. 262 (1966), pp. 539-595.
2. A. Blaquiere, Interprétation d'un coefficient de diffusion complexe en mécanique ondulatoire, *C.R. Acad. Sc. Paris, Série A*, Vol. 268 (1969), pp. 1304-1306.
3. A. Blaquiere, System Theory : A new approach to wave mechanics, *J. Optim. Theor. Appl.*, 32, 4 (1980), pp. 463-478.

4. A. Blaquiere and A. Marzollo, Introduction à la théorie moderne de l'optimisation et à certains de ses aspects fondamentaux en physique, in *La pensée physique contemporaine*, edited by S. Diner, D. Fargue, G. Lochak, Editions Augustin Fresnel, Moulidars, 1982 (Proceedings of a workshop organized by La Fondation Louis de Broglie in 1980).

5. K. Kime and A. Blaquiere, From two Stochastic Optimal Control Problems to the Schrödinger Equation, in *Modeling and Control of Systems in Engineering, Quantum Mechanics, Economics and Biosciences.* Lecture Notes in Control and Information Sciences 121. Springer-Verlag Berlin Heidelberg 1989.

6. A. Blaquiere, Girsanov Transformation and two Stochastic Optimal Control Problems. The Schrödinger System and Related Controllability Results, in *Modeling and Control of Systems in Engineering, Quantum Mechanics, Economics and Biosciences.* Lecture Notes in Control and Information Sciences 121. Springer-Verlag Berlin Heidelberg 1989.

7. W. Fleming and R. Rishel, Deterministic and Stochastic Optimal Control, Springer-Verlag, Berlin, 1975.

8. F. Guerra and L. Morato, Quantization of dynamical systems and stochastic control theory, *Physical Review D*, 27, 8 (1983), pp. 1774-1786.

9. A. Friedman, Stochastic Differential Equations and Applications, Vol. I, Academic Press, New York, 1975.

10. S. Mitter, Non-linear Filtering and Stochastic Mechanics, Stochastic Systems : The Mathematics of Filtering and Identification with Applications, Proc. NATO Advanced Study Institute, Les Arcs, Savoie, France 1980, Reidel, Dordrecht, 1981.

11. E. Nelson, Derivation of the Schrödinger Equation from Newtonian Mechanics, *Physical Review*, 150, 4 (1966), pp. 1079-1085.
12. E. Nelson, Dynamical Theories of Brownian Motion, Princeton U.P., Princeton, NJ, 1967.

13. L. Papiez, Stochastic optimal control and quantum mechanics, *J. Math. Phys.* 23, 6 (1982), pp. 1017-1019.

14. K. Yasue, Quantum mechanics and stochastic control theory, *J. Math. Phys.* 22, 5 (1981), pp. 1010-1020.
15. K. Yasue, Stochastic Calculus of Variations, *J. Func. Analysis*, 41 (1981) pp. 327-340.

16. J.C. Zambrini, Variational processes and stochastic versions of mechanics, *J. Math. Phys.*, 27, 9 (1986), pp. 2307-2330.

17. A. Bensoussan, Stochastic Control by Functional Analysis Methods, North-Holland Publishing Company, 1982.

18. A.M. Il'in, A.S. Kalashnikov, and O.A. Oleinik, Linear Equations of the Second Order of Parabolic Type, in *Russian Mathematical Surveys*, edited by K.A. Hirsch, Vol. XVII, Macmillan and Co. Ltd, London, 1962.

19. A. Wakolbinger, A Simplified Variational Characterisation of Schrödinger Processes, *J. Math. Phys.*, 1989.

20. W. Fleming, Logarithmic transformations with applications in probability and stochastic control (abstract), in *Modeling and Control of Systems in Engineering, Quantum Mechanics, Economics and Biosciences.* Lecture Notes in Control and Information Sciences 121. Springer-Verlag Berlin Heidelberg 1989.

AVOIDANCE CONTROL MECHANICS FOR FOOD-CHAIN MODELS SUBJECT TO UNCERTAINTIES

G. Bojadziev
Department of Mathematics and Statistics
Simon Fraser University
Burnaby, B.C. V5A 1S6
Canada

J. Skowronski
Department of Mechanical Engineering
University of Southern California
Los Angeles, CA U.S. 90089-1453

ABSTRACT

We consider a general food-chain system subject to uncertain disturbances. A Liapunov design methodology is used to establish control growth policies of qualitative nature aimed to restrict the fluctuations of the size of the chain elements (populations, consumers, resources), hence to facilitate transition from growth (decay) to manageable population levels.

1. INTRODUCTION

The classical models in population dynamics based on generalization of the Lotka-Volterra predator-prey model do not take into consideration that in reality ecological and also economical systems are continually disturbed by unpredictable forces due to changes of the size of the interacting populations (consumers, resources), harvesting, climatic conditions, pollution, diseases, migration. It leads to undesirable large fluctuations of the size of the populations. This may trigger in general two types of responses (see Bojadziev [1]).

(i) The populations (some or all) forming the system may change abruptly their behaviour or behavioural policy in order to control in some desired fashion the growth or decay of the system by damping the fluctuations, thus ensuring the existence of an acceptable size of the populations (internal control). The controlling populations can apply the internal control to manage the growth or decay of their own members (*self control*) in order to control the behaviour of the entire system. A typical example is given by Bojadziev and Skowronski [2] where a predator (consumer)-prey (food or resource) system is considered. In this work a controlling factor adjusts the number of predators so that a reasonable level of both populations is maintained. A case of self-control in R^4 is considered by Bojadziev [3]. Also, the controlling populations can apply the internal control to some other populations participating in the system (*interacting control*). The internal control can be a combination of both the above, self and interacting control.

(ii) Human influence from outside the system may be imposed to restrict the growth or decay of the populations. For example the resource manager, outside of the system, may wish to achieve more stable productivity level or economic security for the enterprise (*external control*). Goh et al [4] and Vincent [5] have presented various optimal pest management policies (*external control*) for Lotka-Volterra predator-prey system in R^2.

We discuss growth (decay) restricting policies concerning internal and external control from *qualitative* point of view. The rationale is that ecological and economical systems having behaviour based on some kind of policy of qualitative nature could overcome for certain time the effect of unpredictable disturbances and respond quickly to new changes of circumstances after a decision has been implemented. Qualitative control ensures flexible behaviour which is compatible with the nature of an evolutionary process.

The control objective in our paper is to find a behavioural policy that results in damping large fluctuations in populations participating as elements in a food-chain system subjected to uncertain disturbances. In nature food supply and energy transfer often take the form of chains. We consider *open-end chains*. The number of elements, L, measures the length of the chain.

With proper adjustments we use a Liapunov-design methodology developed for avoidance control of general systems by Leitmann and Skowronski [6], Corless et al [7], and Corless and Leitmann [8]. Avoidance control can be considered as one aspect of differential games of kind (qualitative games) (see Blaquire and Gerard [9]). The food-chains under consideration allow us to use a specific

Liapunov function. We deal with avoidance control policies aimed to achieve coexistence by restricting the population growth (decay) in food-chains to a manageable level (neither too large nor too small), from point of view of participating populations (internal control) or managers outside the chain (external control). The use of avoidance instead of its counterparts of ultimate boundedness practical stability, target controllability, etc. allows for larger sets of operation.

2. PROBLEM STATEMENT

The dynamics of the general class of food-chain systems with length $L = 2n$ under study is modelled by the vector equation

$$\dot{x}(t) = f(x(t)) + \Phi(x(t), u(t), w(t)) , \tag{2.1}$$

where $t \in R_+$ is the time variable, $x(t) \in \Delta \subset R_+^{2n}$ is the state (population) vector bounded in the population space Δ with Int $\Delta \neq \phi$. The components of the vector function $f: \Delta \to R^{2n}$ are given by

$$\begin{aligned} f_1(x) &= x_1\left(\alpha_1 - \frac{\beta_1}{\gamma_1} x_2\right) , \\ f_{2k}(x) &= x_{2k}\left(- \alpha_{2k} + \frac{\beta_{2k-1}}{\gamma_{2k}} x_{2k-1} - \frac{\beta_{2k}}{\gamma_{2k}} x_{2k+1}\right) , \\ f_{2k+1}(x) &= x_{2k+1}\left(- \alpha_{2k+1} + \frac{\beta_{2k}}{\gamma_{2k+1}} x_{2k} - \frac{\beta_{2k+1}}{\gamma_{2k+1}} x_{2k+2}\right) , \\ f_{2n}(x) &= x_{2n}\left(- \alpha_{2n} + \frac{\beta_{2n-1}}{\gamma_{2n}} x_{2n-1}\right) \\ &\qquad k=1,\ldots,n-1 , \end{aligned} \tag{2.2}$$

where α_s, γ_s, $s=1,\ldots,2n$, and β_s, $s=1,\ldots,2n-1$, are positive constants; α_s is the growth rate coefficient, β_s is the interaction coefficient, and γ_s is the trophic weight factor; γ_j/γ_i expresses the gain-loss ratio when population i interacts with population j.

The vector $u(t) \in U \subset R^m$ is the control input; the vector

$w(t) \in W \subset R^p$ is a uncertain input bounded in the band W, hence the vector function $\Phi: \Delta \times U \times W \to R^{2n}$ is uncertain. The uncertain elements w and Φ reflect the presence of disturbances whose nature was described in Section 1. We suppose that $f(\cdot) + \Phi(\cdot)$ belongs to a known class of functions $\mathcal{K}$. We consider $u(t)$ to be given by a set-valued, memoryless state feedback controller

$$u(t) \in p(t,x(t)) , \tag{2.3}$$

in an admissible class $\mathcal{P}$.

The system (2.1) with (2.2) subject to an uncertain input $w(t)$ and control $u(t)$ can be presented in the form

$$\dot{x}(t) \in K(t,x(t)) , \tag{2.4}$$

with

$$K(t,x) \triangleq \{f(x) + \Phi(x,u,w): u \in p(t,x(t)) , w \in W\} . \tag{2.5}$$

Equations (2.4) - (2.5) describe a generalized dynamical system (see Roxin [10] and Fillipov [11]). By solutions of (2.4)-(2.5) we mean an absolutely continuous function $x(\cdot): R_+ \to \Delta$, which satisfies (2.4)-(2.5) almost everywhere on R_+.

For $\Phi \equiv 0$, the model (2.1) reduces to the uncontrolled and undisturbed (UCUD) model

$$\dot{x}(t) = f(x(t)) , \tag{2.6}$$

which represents a general food-chain of Lotka-Volterra type.

The meaning the s-th component Φ_s of Φ, is that for $\Phi_s > 0$ the s-th population x_s (consumer) is enhanced (increasing returns), and for $\Phi_s < 0$ it is dampered (diminishing returns). The specific way in which Φ is selected determines the nature of the control (internal, external).

3. THE EQUILIBRIUM AND LIAPUNOV FUNCTION

From $f(x(t)) = 0$ with (2.2) we get for the coordinates of the nontrivial equilibrium $E^0(x^0) \in R^{2n}$ of the UCUD model

$$x_2^0 = \frac{\alpha_1\gamma_1}{\beta_1}, \qquad x_{2n-1}^0 = \frac{\alpha_{2n}\gamma_{2n}}{\beta_{2n-1}},$$

$$x_{2k-1}^0 = \frac{\alpha_{2k}\gamma_{2k} + \beta_{2k}x_{2k+1}^0}{\beta_{2k-1}}, \qquad k=1,\dots,n-2, \tag{3.1}$$

$$x_{2k+2}^0 = \frac{-\alpha_{2k+1}\gamma_{2k+1} + \beta_{2k}x_{2k}^0}{\beta_{2k+1}}, \qquad k=1,\dots,n-1 .$$

Reality requires that $E^0 \in \text{Int } R_+^{2n}$, the interior of the positive (population) cone. Since $x_{2n-1}^0 > 0$, it follows from (3.1) that $x_{2k-1}^0 > 0$, k-1,...,n-1. Also from (3.1) we see that $x_2^0 > 0$. However, in order to secure that $x_{2k+2}^0 > 0$, k=2,...,n-1, we assume that $x_{2k}^0 > \alpha_{2k+1}\gamma_{2k+1}/\beta_{2k}$. For example from $x_2^0 = \alpha_1\gamma_1/\beta_1 > \alpha_3\gamma_3/\beta_2$ it follows that $x_4^0 > 0$, etc.

The model (2.6)-(2.2) has the Volterra function

$$V(x) = \sum_{s=1}^{2n} \gamma_s x_s^0 \left(\frac{x_s}{x_s^0} - \ell n \frac{x_s}{x_s^0} - 1 \right), \tag{3.2}$$

continuous on $\text{Int } R_+^{2n}$, which is actually a Liapunov function with the following properties.

(i) The minimum of $V(\mathbf{x})$ is attained at the equilibrium $E^0(\mathbf{x}^0)$ given by (3.1); $\min V(\mathbf{x}) = V(\mathbf{x}^0) = 0$;

(ii) The function $V(\mathbf{x})$ is monotone increasing about E^0 (has the "nesting property"); $V(\mathbf{x}) \to \infty$ as $\|\mathbf{x}\| \to 0, \infty$;

$$\text{(iii)} \quad \frac{dV(\mathbf{x})}{dt} = \nabla V(\mathbf{x})\mathbf{f}(\mathbf{x}) = 0, \tag{3.3}$$

where the components of $\mathbf{f}$ are given by (2.2). From here it follows that the equilibrium $E^0(\mathbf{x}^0)$ is stable.

The UCUD model (2.6)-(2.2) has a first integral

$$V(\mathbf{x}) = h, \qquad h = \text{const} > 0, \tag{3.4}$$

which represents a family of level surfaces $\mathcal{V}_h$ in $V \times \Delta \subset R_+^{2n+1}$. The orthogonal projection of $\mathcal{V}_h$ onto Δ generates 2n dimensional hypersurfaces $\mathcal{H}_h$ in Δ which are closed, do not intersect, contain inside the equilibrium E^0, and accommodate orbits of (2.6)-(2.2). Note from above, that if

$h_1 < h_2$, the hypersurface $\mathcal{H}_{h_1}$ is inside the hypersurface $\mathcal{H}_{h_2}$, i.e. $\mathcal{H}_{h_1} \cap \mathcal{H}_{h_2} = \phi$ (levels do not cross). This implies that

$$V(\mathbf{x}^1) < V(\mathbf{x}^2) , \qquad \mathbf{x}^s \in \mathcal{H}_{h_s} , \; s=1,2 . \tag{3.5}$$

4. AVOIDANCE CONTROL

Following Ref. [1,2,5,6,7] and using level surfaces we introduce some definitions and a theorem which gives sufficient conditions for avoidance of the food-chain system modelled by (2.1)-(2.2).

Consider the constant vectors $\varepsilon \in \Delta$, $\delta \in \Delta$. To be more specific, but without loosing generality, we assume for the components of ε and δ:

$$0 < \varepsilon_s < \delta_s < x_s^0, \qquad s=1,\dots,2n ; \tag{4.1}$$

we call ε_s *avoidance parameters* and δ_s *safety parameters*. Then from the nesting property (ii) of $V(\cdot)$ and (4.1) it follows

$$V(\delta) = h_\delta < V(\varepsilon) = h_\varepsilon \tag{4.2}$$

and we introduce the *avoidance* set

$$A \triangleq \{\mathbf{x} \in \Delta: V(\mathbf{x}) \geq V(\varepsilon) = h_\varepsilon\} \tag{4.3}$$

with the boundary

$$\partial A = \mathcal{H}_{h_\varepsilon} \triangleq \{\mathbf{x} \in \Delta: V(\mathbf{x}) = h_\varepsilon\} , \tag{4.4}$$

and the *safety* zone

$$S \triangleq \{\mathbf{x} \in \Delta: h_\delta \leq V(\mathbf{x}) < h_\varepsilon\} \tag{4.5}$$

with the boundary

$$\partial S = \mathcal{H}_{h_\delta} \triangleq \{\mathbf{x} \in \Delta: V(\mathbf{x}) = h_\delta\}. \tag{4.6}$$

Moreover we need the *admissible* set

$$\Omega \triangleq \{\mathbf{x} \in \Delta: V(\mathbf{x}) < h_\varepsilon\} = \Delta \backslash A \tag{4.7}$$

and the *desirable* region

$$D \triangleq \{\mathbf{x} \in \Delta: V(\mathbf{x}) < h_\delta\} = \Omega \backslash S. \tag{4.8}$$

It is clear that every continuous solution $x(\cdot)$ which enters A from Ω must pass through S (see [6]).

Definition 1. System (2.4)-(2.5) avoids A iff, for each solution $x(\cdot): R_+ \to \Delta$ of (2.4)-(2.5), $x(0) \notin A \Rightarrow x(t) \notin A \; \forall \; t > 0$.

Definition 2. The set A defined by (4.3) is avoidable by (2.1)-(2.2) iff there exists a function $p(\cdot) \in \mathcal{P}$ such that the system (2.4)-(2.5) avoids A.

Adjusting the results in [6] we now give sufficient conditions for the avoidance of A.

Avoidance Control Theorem. The solution $x(\cdot)$ of the food-chain model (2.1)-(2.2) or the equivalent system (2.4)-(2.5) is controllable in Ω for avoidance of A if there exists a controller $p(\cdot)$ in $\mathcal{P}$ for which

$$\frac{dV}{dt} = \nabla V(x)[f(x) + \Phi(x(t), u(t), w(t)) \le 0 \qquad (4.9)$$

for all $w(t) \in W$ where $V(\cdot)$ is the Liapunov function (3.2).

Proof. Assume that the region A is not avoidable, hence some $x(\cdot)$ enters A. Then there exists $t_1 > 0$ such that $x(t_1) \in A$. Since S is the safety zone for A and $x(\cdot)$ is continuous, there exist $t_2, t_3 \in [0,t_1]$ with $t_2 < t_3$ such that $x(t) \in S$ for $t \in [t_2,t_3)$ and $x(t_3) \in \partial A$. From the nesting property of $V(x)$ it follows that $V(x(t): t \in [t_2,t_3)) < V(x(t_3))$ meaning that the function $V(x)$ is increasing. This contradicts (4.9) which states that $V(x)$ is non-increasing along solutions of (2.1)-(2.2).

5. CONTROL LAW

The inequality (4.9) which gives sufficient conditions for avoidance, with (3.2) and (3.3) after some elaboration reduces to

$$\sum_{s=1}^{2n} \gamma_s x_s^0 \left(\frac{1}{x_s^0} - \frac{1}{x_s} \right) \Phi_s(x(t), u(t), w(t)) \le 0 \; . \qquad (5.1)$$

The relationship (5.1) between the control vector $u(t) \in U$ and the population vector $x(t) \in \Delta$ for all $w \in W$ may be considered a general control condition in an implicit form. It is implied by the following *Control Law*:

Select $u(t)$ such that

$$\Phi_s(x(t),\ u(t),\ w(t)) > 0 \quad \text{for} \quad x_s < x_s^0\ , \quad \forall\ s,\quad \forall\ w,$$

$$\Phi_s(x(t),\ u(t),\ w(t)) < 0 \quad \text{for} \quad x_s > x_s^0, \quad \forall\ s,\ \forall\ w\ . \tag{5.2}$$

Such a selection for $u(t)$ may not be difficult since often in applications Φ_s is a linear function of $u_s(t)$ and $w_s(t)$, i.e.

$$\Phi_s(x(t),\ u(t),\ w(t)) = \Phi_s^{(1)}(x(t))u_s(t) + \Phi_s^{(2)}(x(t))w_s(t)\ . \tag{5.3}$$

The control laws (5.1) and (5.2) provide an avoidance control policy which is essentially a growth (decay) restriction policy. This policy is demonstrated in the next section on a particular selection of the function Φ by the means of numerical simulation.

6. NUMERICAL SIMULATION

Here, we consider a case study whose dynamics is a particular case of (2.1)-(2.2). A fourth order Runge-Kutta numerical method was used with step .01 to integrate the corresponding modelling equation, i.e. to find the solution $x(\cdot)$.

Setting in (2.1)-(2.2) n=1, $\alpha_1 = \alpha_2 = \beta_1 = \gamma_1 = \gamma_2 = 1$; $\Phi_S = u_S + w_S$, $s = 1,2$, we obtain the model

$$\dot{x}_1 = x_1(1 - x_2) + u_1x_1 + w_1\ ,$$

$$\dot{x}_2 = x_2(-\ 1 + x_1) + u_2x_2 + w_2\ , \tag{6.1}$$

This is a Lotka-Volterra model subjected to external control and uncertainties $\Phi_s = u_sx_s + w_s$. The selection of Φ_s is a particular case of (5.3). The terms u_sx_s which contain the control variables u_s can be considered as the rate of application of an ideal pesticide if $u_s < 0$ or the rate of application of a stimulator if $u_s > 0$. The disturbances w_s have an effect equivalent to harvesting if $w_s < 0$ or stocking if $w_s > 0$. We assume that they are bounded,

$$w_s \in [-0.5, 0.5], \quad s=1,2 . \tag{6.2}$$

The equilibrium of the UCUD model (6.1) is $E^0(1,1) \in \text{Int } R_+^2$ and the Liapunov function (3.2) becomes

$$V(x) = \sum_{s=1}^{2} (x_s - \ln x_s - 1) . \tag{6.3}$$

The control law (5.2) now reads

$$\begin{aligned} u_s x_s + w_s &> 0 \quad \text{for} \quad x_s < 1 , \\ u_s x_s + w_s &< 0 \quad \text{for} \quad x_s > 1 . \end{aligned} \tag{6.4}$$

Note that since w_s is unknown within [-.5,.5] (see (6.2)), we consider the worst case when selecting the control against the uncertainties. Thus, from the inequalities (6.4) we obtain

$$\begin{aligned} \min_{w_s} (u_s x_s + w_s) &= u_s x_s - 0.5 > 0 \quad \text{for} \quad x_s < 1, \\ \max_{w_s} (u_s x_s + w_s) &= u_s x_s + 0.5 < 0 \quad \text{for} \quad x_s > 1 . \end{aligned} \tag{6.5}$$

To be more specific we select here from (6.5)

$$\begin{aligned} u_s &= \frac{0.6}{x_s} \quad \text{for} \quad x_s < 1 , \\ u_s &= -\frac{0.6}{x_s} \quad \text{for} \quad x_s > 1 , \end{aligned} \tag{6.6}$$

Further we choose the avoidance and security parameters as follows $\varepsilon_1 = \varepsilon_2 = .4$ and $\delta_1 = \delta_2 = .5$. Then from (4.2) with (6.3) we find $h_\varepsilon = 0.6326$ and $h_\delta = 0.3863$ which specify the boundaries ∂A and ∂S (see (4.4) and (4.6)) presented on Fig. 1. In order to perform the numerical simulation we choose $w_1 = \frac{1}{2} \sin t$ and $w_2 = \frac{1}{2} \cos t$ which is in accordance to (6.2).

Consider the solution of (6.1) with initial populations $x_1(0) = 2$, $x_2(0) = 1$ (measured in nondimensional units) and fixed control $u(0) = (.5, .5)^T$. Considering (6.3) as a test function, at each time step we calculate the components x_1 and x_2 of the solution $\mathbf{x}(\cdot)$ of (6.1) and the value $V[\mathbf{x}(\cdot)]$. As long as

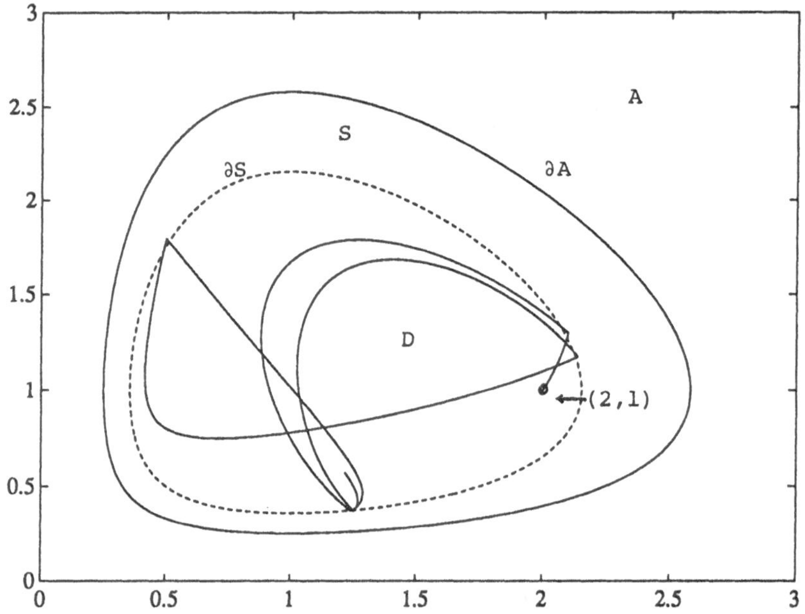

Fig. 1

$V(x(\cdot)) \le h_\delta$ which means that $x(\cdot) \in D$, we continue the calculations. However, at time $t_1 = 0.13$, $V(x[t_1]) = 0.3946 > h_\delta$, hence the solution $x(\cdot)$ has crossed ∂S and is in the state $M(x_1(t_1), x_2(t_1)) \in S$, where $x_1 = 2.0996$, $x_2 = 1.2960$. According to our control policy we change at M (switching point) the control from $u(0)$ to $u = (-0.286, -0.463)$ which is in accordance with the control law (6.6). The solution changes abruptly and enters again the desirable region D.

The avoidance control policy for $t \in [0, 10]$ is displayed on Fig. 1. Actually the solution $x(\cdot)$ enters five times the safety zone S and due to the change of control is pushed back into the desirable region D. The oscillations of x_1 and x_2 versus time are presented on Fig. 2, and the levels of the Liapunov function (6.2) on Fig. 3. Thus, the control policy restricts the growth of the populations to the desirable region D with some exceptions when the response enters into S (hence it is always in the admissible set Ω).

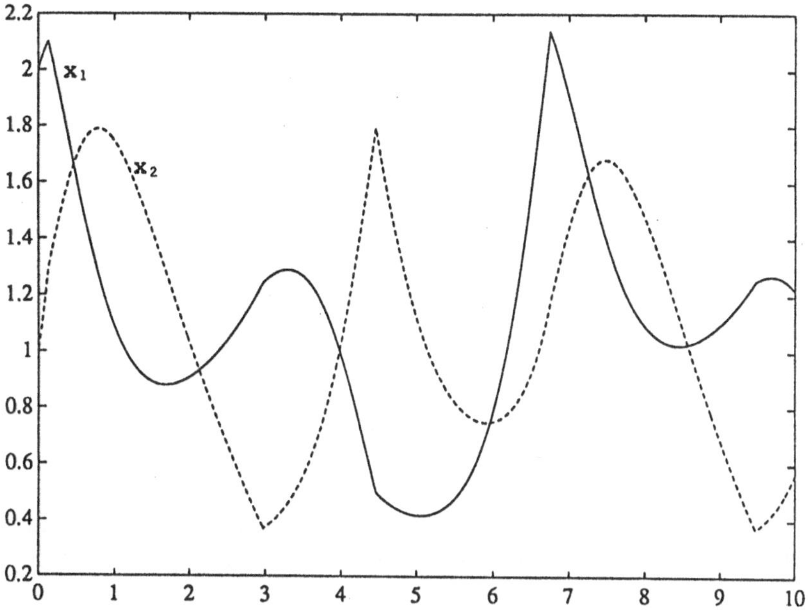
2.2
2
1.8
1.6
1.4
1.2
1
0.8
0.6
0.4
0.2
0 1 2 3 4 5 6 7 8 9 10
x1
x2

Fig. 2

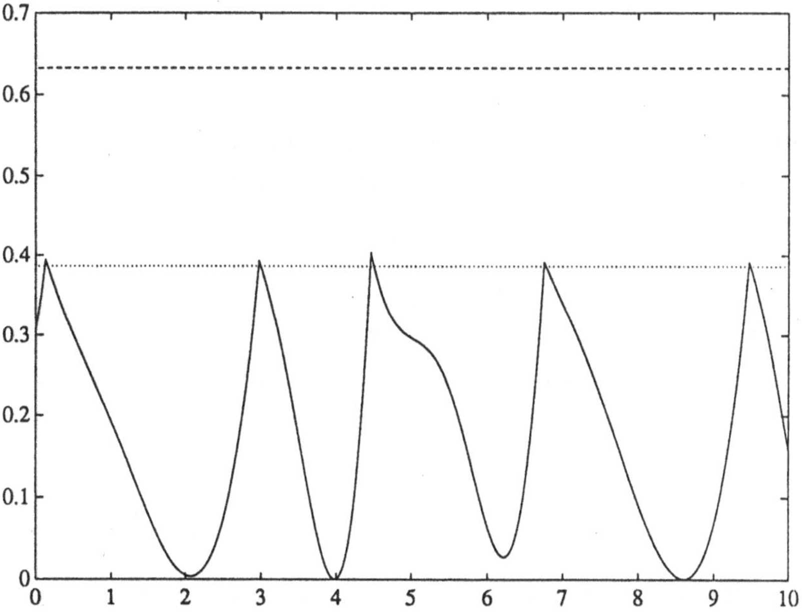
0.7
0.6
0.5
0.4
0.3
0.2
0.1
0
0 1 2 3 4 5 6 7 8 9 10

Fig. 3

Acknowledgement. This work was partly supported by Grant No. 3967 from the Natural Sciences and Engineering Research Council of Canada and a grant from the Centre for Systems Science, Simon Fraser University.

REFERENCES

1. G. Bojadziev, Management of food-chain systems via Liapunov design control. Z. für Operations Research (to appear).

2. G. Bojadziev, J. Skowronski, Controlled food consumption. Methods of Operations Research, vol 49, Hain Verlag bei Athenaum, 499-506 (1985).

3. G. Bojadziev, Self controlled food-chain model of length four. Proceedings of IASTED International Symposium "Adaptive and Knowledge - Based Control and Signal Processing", Honolulu, Aug. 16-18, 1989, 11-13.

4. B.S. Goh, G. Leitmann, T.L. Vincent, Optimal control of a prey-predator system, Math. Biosci. 19, 263-286 (1974).

5. T.L. Vincent, Pest management programs via optimal control theory, Biometrics 31, 1-10 (1975).

6. G. Leitmann, J. Skowronski, Avoidance control. J. of Optimiz. Theory Appl. 23, 581-591 (1977).

7. M. Corless, G. Leitmann, J. Skowronski, Adaptive control for avoidance or evasion in uncertain environment. Comput. Math. Applic. 13, 1-11 (1987).

8. M. Corlesss, G. Leitmann, Adaptive controllers for avoidance or evasion in an uncertain environment: some examples. Computers Math. Applic. 18, 161-170 (1989).

9. A. Blaquière, F. Gérard, On the geometry of optimal strategies in two-person games of kind. J. of Computer and System Sciences, 2 (No. 3), 228-304 (1968).

10. E. Roxin, On generalized dynamical systems defined by a contingent equation. J. Diff. Equations 1, 185-205 (1965).

11. A.F. Filippov, Classical solutions of differential equations with multi-valued right-hand side. SIAM J. Control 5, 200-231 (1967).

ROBUST CONTROL OF UNCERTAIN SYSTEMS WITH TIME-VARYING UNCERTAINTY: A COMPUTER-AIDED SETTING

Y.H. Chen

J.S. Chen

The George W. Woodruff School of Mechanical Engineering

Georgia Institute of Technology

Atlanta, GA 30332, USA

ABSTRACT

The problem of stabilizing a class of linear uncertain systems by using linear state feedback is addressed. The system possesses uncertainty which is time-varying, unknown, but lies within a prescribed bounded set. No statistical information of the uncertainty is imposed. Nor any *matching condition* is required. Necessary and sufficient conditions for quadratic stabilizability are formulated. The controller synthesis and stability analysis can be investigated by a two level optimization process. The result is believed to be practical (in the sense that it can be implemented for non-trivial dynamic systems) and non-conservative.

I. INTRODUCTION

The main problem addressed in this article can be stated as follows: For a class of dynamical systems which possesses time-varying but bounded uncertainty, how to construct a linear feedback control which guarantees stability in the least conservative sense?

The uncertain system proposed here is originated from the setup provided by a series of work by Gutman (1976) and Gutman and Leitmann (1975) where the *worst case* optimal strategy of a player against an uncertain but bounded opponent is considered. The system is later (Gutman 1979, Leitmann 1979) used for a more generic control research issue: stabilization. The major justification of considering such system is that there may often be unknown and *unknowable* characteristics appear in dynamic models. Here the term "unknowable" is referred in both the *deterministic* and *statistical* sense. In the deterministic context, it is possible that one does not know the true value of the uncertainty, which is time-varying, due to the imperfect knowledge of its source. Nor can one learn the uncertainty from the system performance such as the proposal made by adaptive control (e.g., Narendra and Annaswamy 1989). In the statistical context, it is also likely that one is not able to ascertain the probability laws governing the happening of certain key events (note that this is the issue on the difficulty of *acquiring* the information rather than *imposing* certain assumptions).

The stabilization of uncertain systems with unknown, time-varying, but bounded uncertainty has then drawn considerable attention. Numerous criteria have been devised to characterize the uncertainty such that the stability is guaranteed *if* the criteria are satisfied. Very often the criteria are conservative due to their intrinsic feature: to be sufficient but not necessary. For example, Chen (1986) and Leitmann (1981) both require the so-called "matching condition" which restricts the structure of the uncertainty. The condition can be relaxed to a certain extent (Barmish and Leitmann 1982, Chen 1987, Chen and Leitmann 1987) but never released. Other alternatives such as the Riccati setting (Gu and Chen 1989, Petersen and Hollot 1986, Schmitendorf 1988) also fall within this general category. The work by Barmish (1983, 1985) is perhaps the first among those which intend to establish necessary and sufficient conditions of uncertainty for stability. The results by Barmish are rather originative but are, pragmatically speaking, difficult to apply for most of non-trivial dynamic systems. Moreover, the stabilizing controller design is tedious. The work by Khargonekar *et al.* (1990) also provides a necessary and sufficient condition for a class of uncertain systems. While many uncertainties can be expressed as a subset of the uncertainty expression by the "over bounding" technique, the resulting criterion is often extremely conservative. Finally, a recent work by Wei (1990) also tackles the issue on quadratic stabilizability. However, it only deals with single-input and structurally independent case and the result requires some entries of the uncertain system matrix to be sign-invariant.

The major contributions of the present paper can then be made clear in view of the above review. We divide them into two parts. First, necessary and sufficient conditions for quadratic stabilizability by linear feedback control are formulated. Second, using these conditions, a linear feedback control can be designed for the uncertain system via a two level optimization process. The inner level optimum can be reached by only searching the corners of the uncertainty bounding set. In the outer level, the *local* maximum is also the *global* maximum. Hence most of the standard optimization algorithms can be directly applied to find such maximum.

The results we propose here can be applied to various practical systems. The uncertainty bound and the control gain are less conservative in comparing with previous work.

II. BASIC PROBLEM

Consider a class of uncertain dynamic systems described by

$$\dot{x}(t) = [A + \Delta A(t)]x(t) + [B + \Delta B(t)]u(t) \tag{2.1}$$

where $t \in \mathbf{R}$, $x(t) \in \mathbf{R}^n$ is the state, $u(t) \in \mathbf{R}^m$ is the control, and $y(t) \in \mathbf{R}^p$ is the output. The (constant) matrix A denotes the *nominal* part and the matrix $\Delta A(t)$ denotes the *uncertain* part of the system. The matrix B is the nominal part and $\Delta B(t)$ is the uncertain part of the input matrix. The output matrix C is also constant. The functions $\Delta A(\cdot)$ and $\Delta B(\cdot)$ are arbitrary and unknown. Note that they are time-varying and no constraints on the rate of change are imposed. The only assumption imposed on $\Delta A(\cdot)$ and $\Delta B(\cdot)$ is that they are Lebesgue measurable and satisfying

$$(\Delta A(t), \Delta B(t)) \in \Omega \qquad \text{for all } t \geq 0 \tag{2.2}$$

where Ω is a compact convex set in $\mathbf{R}^{n \times (n+m)}$.

Remark 1. In many practical cases, the set Ω is contained in a subspace of $\mathbf{R}^{n \times (n+m)}$. It is then beneficial to express (2.1) as

$$\dot{x}(t) = [A + \sum_{i=1}^{s} q_i(t)A_i]x(t) + [B + \sum_{i=1}^{s} q_i(t)B_i]u(t) \tag{2.3}$$

where $((A_i, B_i), i = 1, 2, \cdots, s)$ are a set of basis vectors (they are actually matrices) of the subspace containing Ω; and

$$q(t) = (q_1(t), q_2(t), \cdots, q_s(t))^T \tag{2.4}$$

is a Lebesgue measurable vector function, satisfying

$$q(t) \in \hat{\Omega} \subset \mathbf{R}^s \qquad \text{for all } t \geq 0 \tag{2.5}$$

Corresponding to the compactness of Ω, $\hat{\Omega}$ is also a compact set in $\mathbf{R}^s$. In many physical systems, (2.3) is often directly available, with $q_i(t)$ being the uncertain factors. Since $\mathbf{R}^{n \times (n+m)}$ is its own subspace, (2.1) and (2.3) are equivalent expressions. Although the

$q_i(t)$'s should be linearly independent, they can be functionally dependent. An example of this is given by

$$\Delta A(t) = \begin{bmatrix} p(t) & p^2(t) \\ p^2(t) & p(t) \end{bmatrix} \qquad |p(t)| \leq 1 \tag{2.6}$$

which can be rewritten as

$$\Delta A(t) = q_1(t) \begin{bmatrix} 1 & 0 \\ 0 & 1 \end{bmatrix} + q_2(t) \begin{bmatrix} 0 & 1 \\ 1 & 0 \end{bmatrix} \tag{2.7}$$

with

$$\hat{\Omega} = \{(p, p^2) : |p| \leq 1\} \tag{2.8}$$

III. LINEAR STATE FEEDBACK CONTROL DESIGN

The following notations are first defined. The set of symmetric $n \times n$ matrices is denoted by S^n. Accordingly, S^n_+ denotes the set of $n \times n$ positive definite matrices. The closure $\bar{S}^n_+$ is the set of positive semi-definite matrices. For a scalar ϵ, $S^n_{\epsilon+}$ is defined as

$$S^n_{\epsilon+} = \{Q : Q - \epsilon I \in S^n_+\} \tag{3.1}$$

It is also true that $S^n_{\epsilon+}$ is a convex cone with the apex at ϵI, i.e., for any $Q_1, Q_2 \in S^n_{\epsilon+}$ and $\alpha > 0$, $0 \leq \beta \leq 1$, then $\beta Q_1 + (1-\beta)Q_2 \in S^n_{\epsilon+}$ and $\epsilon I + \alpha(Q_1 - \epsilon I) \in S^n_{\epsilon+}$. Also notice that for any compact set $C \subset S^n_+$ implies the existence of an $\epsilon > 0$ such that $C \subset S^n_{\epsilon+}$. A linear operator $\mathbf{T}^D_G : S^n \to S^l$ is defined by

$$\mathbf{T}^D_G P = W \qquad \text{if} \qquad -D^T(GP + PG^T)D = W \tag{3.2}$$

We shall abbreviate $\mathbf{T}^I_G$ by $\mathbf{T}_G$. It is important to note that $\mathbf{T}^D$ is itself a linear map:

$$\mathbf{T}^D_{\alpha_1 G_1 + \alpha_2 G_2} = \alpha_1 \mathbf{T}^D_{G_1} + \alpha_2 \mathbf{T}^D_{G_2} \qquad \text{for } \alpha_1, \alpha_2 \in \mathbf{R} \tag{3.3}$$

Definition 1. The dynamic system (2.1) is *quadratically stabilizable* if there exist matrices K and $P (\in S^n_+)$ and a scalar $\alpha > 0$ such that for $u = -Kx$,

$$x^T[P(A + E - (B + F)K) + (A + E - (B + F)K)^T P]x \leq -\alpha \|x\|^2 \tag{3.4}$$

for all $x \in \mathbf{R}^n$, $(E, F) \in \Omega$.

To simplify the derivations, we first consider the case that $\Delta B(t) \equiv 0$ for all $t \geq 0$.

Theorem 1. Let $D \in \mathbf{R}^{n\times(n-m)}$ be such that

$$V = \begin{pmatrix} B^T \\ D^T \end{pmatrix} \tag{3.5}$$

is nonsingular. Let its inverse be partitioned as

$$V^{-1} = (\hat{B} \ \ \hat{D}) \tag{3.6}$$

The system

$$\dot{x}(t) = [A + \Delta A(t)]x(t) + Bu(t) \tag{3.7}$$

is quadratically stabilizable if and only if that there exists a matrix $S \in S_+^{n-m}$ such that

$$\mathbf{T}_{A+E}^{\hat{D}} S = -\hat{D}^T[(A+E)S + S(A+E)^T]\hat{D} \in S_+^{n-m} \text{for all } E \in \Omega. \tag{3.8}$$

Proof: Necessity: If the system (3.7) is quadratically stabilizable, then, according to the definition, there exist K and P such that

$$\Gamma = P(A+E-BK) + (A+E-BK)^T P \in S_-^n \text{for all } E \in \Omega. \tag{3.9}$$

This condition is in fact equivalent to

$$\hat{\Gamma} = V^{-T}P^{-1}\Gamma P^{-1}V^{-1} \in S_-^n \qquad \text{for all } E \in \Omega \tag{3.10}$$

Since $B^T\hat{B} = I$, $B^T\hat{D} = 0$, $D^T\hat{B} = 0$, $D^T\hat{D} = I$, it can be verified that

$$\hat{\Gamma} = \begin{pmatrix} \hat{\Gamma}_{11} & \hat{\Gamma}_{12} \\ \hat{\Gamma}_{12}^T & \hat{\Gamma}_{22} \end{pmatrix} \tag{3.11}$$

where

$$\hat{\Gamma}_{11} = -\mathbf{T}_{A+E}^{\hat{B}} P^{-1} - KP^{-1}\hat{B} - \hat{B}^T P^{-1} K^T \tag{3.12}$$

$$\hat{\Gamma}_{12} = -\hat{B}^T(\mathbf{T}_{A+E}P^{-1})\hat{D} - KP^{-1}\hat{D} \tag{3.13}$$

$$\hat{\Gamma}_{22} = -\mathbf{T}_{A+E}^{\hat{D}} P^{-1} \tag{3.14}$$

The negativeness of the matrix $\hat{\Gamma}$ requires that all the principal submatrices are negative definite (Gantmacher 1959). Therefore we need $\hat{\Gamma}_{22}$ be negative definite, i.e., for $S = P^{-1}$

$$\mathbf{T}_{A+E}^{\hat{D}} \in S_{+}^{n-m} \qquad \text{for all } E \in \Omega \tag{3.15}$$

This ends the necessity part.

Sufficiency: If an $S \in S_{+}^{n}$ exists such that (3.8) holds, let $P = S^{-1}$. Then the system (3.7) is quadratically stabilizable if (3.10) is satisfied (note that Ω is compact, and therefore (3.10) is sufficient for quadratic stabilizability). Partition $\hat{\Gamma}$ as in (3.11). Choose K such that

$$K = \frac{1}{\epsilon} R^{-1} B^T P \tag{3.16}$$

where $R \in S_{+}^{m}$ and $\epsilon > 0$, then

$$\hat{\Gamma}_{11} = -\mathbf{T}_{A+E}^{\hat{B}} P^{-1} - \frac{2}{\epsilon} R^{-1} \tag{3.17}$$

$$\hat{\Gamma}_{12} = -\hat{B}^T (\mathbf{T}_{A+E} P^{-1}) \hat{D} \tag{3.18}$$

$$\hat{\Gamma}_{22} = -\mathbf{T}_{A+E}^{\hat{D}} P^{-1} \tag{3.19}$$

If ϵ is chosen to be sufficiently small, the following can be satisfied:

$$-\mathbf{T}_{A+E}^{\hat{B}} P^{-1} - \frac{2}{\epsilon} R^{-1} + [\hat{B}^T (\mathbf{T}_{A+E} P^{-1}) \hat{D}][\mathbf{T}_{A+E}^{\hat{D}} P^{-1}]^{-1} [\hat{B}^T (\mathbf{T}_{A+E} P^{-1}) \hat{D}]^T \in S_{-}^{m} \tag{3.20}$$

for all $E \in \Omega$. The above condition together with (3.8) imply that

$$\hat{\Gamma}_{22} \in S_{-}^{n-m} \tag{3.21}$$

and

$$\hat{\Gamma}_{11} - \hat{\Gamma}_{12} \hat{\Gamma}_{22}^{-1} \hat{\Gamma}_{21} \in S_{-}^{m} \tag{3.22}$$

which, in turn, mean that $\hat{\Gamma} \in S_{-}^{n}$. The sufficiency part is also proved. ◇◇

Remark 2. This necessary and sufficient condition depends strongly on the orthogonal subspace of B (denoted by $\hat{D}$). This is anticipated since $\hat{D}$ governs the subspace where the control u can not reach but the uncertainty E may. If one decomposes E into the matched and mismatched part (e.g., Chen 1987), i.e., choose E_{match} and E_{mis} such that $E = BE_{\text{match}} + E_{\text{mis}}$, then the condition (3.8) is reduced to (note that $B^T\hat{D} = 0$)

$$\mathbf{T}^{\hat{D}}_{A+E_{\text{mis}}} S \in S_+^{n-m} \qquad \text{for all } E \in \Omega \tag{3.23}$$

Remark 3. The stabilizing gain matrix K shown in (3.16) now only depends on ϵ and R once P is found. In view of (3.20), by letting

$$\sigma_B = \sigma_{\max}(\hat{B}) \tag{3.24}$$

$$\sigma_D = \sigma_{\max}(\hat{D}) \tag{3.25}$$

$$\sigma_T = \max_{E\in\Omega} \sigma_{\max}(\mathbf{T}_{A+E} S) \tag{3.26}$$

$$\sigma_R = \sigma_{\max}(R) \tag{3.27}$$

(where $\sigma_{\max}(\cdot)$ denotes the maximum singular value of the designated matrix), one can choose ϵ such that

$$\frac{1}{\epsilon} > \frac{1}{2}\sigma_R \left[\sigma_B^2 \sigma_T + \frac{(\sigma_B \sigma_T \sigma_D)^2}{\eta}\right] \tag{3.28}$$

The above theorem needs to be put in a more tractable form for any practical application. This is so since the condition (3.8) refers to every point E in Ω, which is difficult to check in practice.

Lemma 1. If $U, V, W \in S^n$, $0 < \alpha < 1$, and $W = \alpha U + (1-\alpha)V$. Then

$$\lambda_{\min}(W) \geq \alpha\lambda_{\min}(U) + (1-\alpha)\lambda_{\min}(V) \tag{3.29}$$

where $\lambda_{\min}(\cdot)$ denotes the minimum eigenvalue of the designated matrix. Therefore

$$\lambda_{\min}(W) \geq \min\{\lambda_{\min}(U), \lambda_{\min}(V)\} \tag{3.30}$$

Proof:

$$\begin{aligned}\lambda_{\min}(W) &= \min_{\|x\|=1} x^T W x \\ &= \min_{\|x\|=1} x^T[\alpha U + (1-\alpha)V]x \\ &= \min_{\|x\|=1} [\alpha x^T U x + (1-\alpha)x^T V x] \\ &\geq \alpha \min_{\|x\|=1} x^T U x + (1-\alpha) \min_{\|x\|=1} x^T V x \\ &= \alpha\lambda_{\min}(U) + (1-\alpha)\lambda_{\min}(V)\end{aligned} \tag{3.31}$$

Eq. (3.30) then follows. ◇◇

Definition 2. Given a set C and a point $p \in C$. Suppose for any line segment which contains p and is contained in C, p is one of its end points. Then p is called a *protruded point.*

The following theorem narrows down the search domain of E.

Theorem 2. Let $S \in S^n$, $\hat{D} \in \mathbf{R}^{n\times(n-m)}$, $A \in \mathbf{R}^{n\times n}$ and Ω be compact and convex. Let

$$\gamma = \min_{E\in\Omega} \lambda_{\min}[\mathbf{T}^{\hat{D}}_{A+E}(S)] \tag{3.32}$$

Then γ can be reached by a protruded point of Ω.

Proof: Since Ω is compact, the minimum γ exists and can be reached by some point $E_0 \in \Omega$. Let H_0 be an affine subspace of $R^{n\times n}$ of minimum dimension such the $\Omega \subset H_0$. Also let $\Omega_0 = \Omega$. If E_0 is a protruded point, then the proof is completed. Otherwise, there exists a line segment L, such that $E_0 \in L$, $L \subset \Omega$, and both ends of the line segment are on the boundary of Ω. Since $\mathbf{T}^{\hat{D}}_{A+E}S$ is linear with respect to E (recall (3.3)), according to (3.30), the minimum can be reached by one of the end points of the line segment. Let this minimum end point be E_1. If E_1 is a protruded point, then the proof is again completed. Otherwise, let H_1 be a hyperplane (Luenberger 1969) of H_0 passing through all the line segments which contain E_1 but do not end at E_1. Let $\Omega_1 = H_1 \cap \Omega_0$ then all the protruded points of Ω_1 are also the protruded points of Ω_0. Repeat the same process on the triple $\{\Omega_1, H_1, E_1\}$, either a protruded point is found or a new triple $\{\Omega_2, H_2, E_2\}$ can be produced. The search domain is again one dimension less. Since Ω is within a finite dimensional space, the process will lead to the protruded point which determines the minimum. ◇◇

Remark 4. Theorem 2 restricts the search domain for the minimum to be only the protruded points. If the bounding set Ω is a hyperpolyhedron, then there are only a finite number of protruded points (i.e., the vertices) and a complete check by computer programming is feasible.

Remark 5. Following similar analysis, it can be shown that Theorem 1 is still valid if "$S \in S^n_+$" is replaced by "$S \in S^n_+$ and $\|S\| = 1$", where $\|\cdot\|$ is any norm of matrices.

Moreover, the matrix norm does not need to be induced (Desoer and Vidyasagar 1975). For later use, we shall propose to choose the "sum of singular value" norm, which reduces to the "trace" for positive semi-definite matrices:

$$\|S\| = \mathrm{tr}(S) \qquad \text{if } S \in \bar{S}_+^n \tag{3.33}$$

This is advantageous since it is then a *linear* function of the matrix.

The following result is practical from computer programming aspect.

Theorem 3. Let

$$\phi(S) = \min_{E\in\Omega} \lambda_{\min}(\mathbf{T}_{A+E}^{\hat{D}} S) \tag{3.34}$$

and

$$\psi_\xi(S) = \min\{\phi(S), \xi\lambda_{\min}(S)\} \tag{3.35}$$

where $\xi > 0$. Define the set W:

$$W = \{S \in \bar{S}_+^n : \mathrm{tr}(S) = 1\} \tag{3.36}$$

Then any *local* maximum of $\psi_\xi(S)$ in W is also the *global* maximum in W.

Proof: Since the operator $\mathrm{tr}(\cdot)$ is linear and the set $\bar{S}_+^n$ is convex, the set W is also convex. We first assume that the point S_1 is a local maximum point of $\psi_\xi(S)$ and the point S_2 is the global maximum point and $S_1 \neq S_2$. Then it must be that $\psi_\xi(S_1) < \psi_\xi(S_2)$. Define

$$\omega(E, S) = \lambda_{\min}(\mathbf{T}_{A+E}^{\hat{D}} S) \tag{3.37}$$

That W be convex implies

$$\alpha S_1 + (1-\alpha)S_2 \in W \qquad \text{for all } 0 < \alpha < 1 \tag{3.38}$$

From Lemma 1,

$$\lambda_{\min}(\alpha S_1 + (1-\alpha)S_2) \geq \alpha\lambda_{\min}(S_1) + (1-\alpha)\lambda_{\min}(S_2) \tag{3.39}$$

Since $\mathbf{T}_{A+E}^{\hat{D}}(\cdot)$ is linear, we hence have

$$\omega(E, \alpha S_1 + (1-\alpha)S_2) \geq \alpha\omega(E, S_1) + (1-\alpha)\omega(E, S_2) \tag{3.40}$$

Taking the minimum on both sides yields

$$\begin{aligned} \min_{E\in\Omega} \omega(E, \alpha S_1 + (1-\alpha)S_2) &\geq \min_{E\in\Omega}[\alpha\omega(E, S_1) + (1-\alpha)\omega(E, S_2)] \\ &\geq \alpha \min_{E\in\Omega} \omega(E, S_1) + (1-\alpha) \min_{E\in\Omega} \omega(E, S_2) \end{aligned} \tag{3.41}$$

In view of (3.34), this implies that

$$\phi(\alpha S_1 + (1-\alpha)S_2) \geq \alpha\phi(S_1) + (1-\alpha)\phi(S_2) \tag{3.42}$$

Combining (3.39) and (3.42) results in

$$\begin{aligned} \psi_\xi(\alpha S_1 + (1-\alpha)S_2) =& \min\{\phi(\alpha S_1 + (1-\alpha)S_2), \xi\lambda_{\min}(\alpha S_1 + (1-\alpha)S_2)\} \\ \geq& \min\{\alpha\phi(S_1) + (1-\alpha)\phi(S_2), \\ & \quad \alpha\xi\lambda_{\min}(S_1) + (1-\alpha)\xi\lambda_{\min}(S_2)\} \\ \geq& \alpha \min\{\phi(S_1), \xi\lambda_{\min}(S_1)\} + (1-\alpha)\min\{\phi(S_2), \xi\lambda_{\min}(S_2)\} \\ =& \alpha\psi_\xi(S_1) + (1-\alpha)\psi_\xi(S_2) \\ >& \psi_\xi(S_1) \end{aligned} \tag{3.43}$$

This contradicts the early assumption that S_1 is the local maximum since α can be made arbitrarily close 1. The proof is completed. ◇◇

Theorem 4. The uncertain system (3.7) is quadratically stabilizable if and only if

$$\eta_\xi = \max_{S\in W} \psi_\xi(S) > 0 \tag{3.44}$$

where ξ is an arbitrary positive scalar.

Proof: Since W is compact and the function $\psi_\xi(\cdot)$ is continuous, the maximum is always defined. If (3.44) holds, let S^* be the maximum point, then (3.8) clearly holds for $S = S^*$. Notice that (3.44) implies $\lambda_{\min}(S^*) > 0$ and therefore $S^* \in S_+^n$. The sufficiency part is proved. For the necessity part, if the system (3.7) is quadratically stabilizable, then (3.8) holds for some $S = \hat{S}$. The matrix $\hat{S}$ can be rescaled to satisfy

$$\mathrm{tr}(\hat{S}) = 1 \tag{3.45}$$

For such an $\hat{S}$,

$$\phi(\hat{S}) > 0 \tag{3.46}$$

and

$$\lambda_{\min}(\hat{S}) > 0 \tag{3.47}$$

Therefore (3.44) holds. The proof is completed. ◇◇

Remark 6. The parameter ξ is chosen to reflect that S and $\mathbf{T}_{A+E}^{\hat{D}} S$ are generally of different scale.

Remark 7. The maximum of $\psi_\xi(S)$ can be searched via standard optimization algorithms (which often provide the *candidates* of maximum through necessary conditions). This is due to the result in Theorem 3. Moreover, the search for $\phi(S)$ (which in turn determines $\psi_\xi(S)$) is only via the protruded points (from Theorem 2). If the bounding set Ω is a hyperpolyhedron, then only a finite number of points need to be checked.

IV. GENERALIZATION TO $\Delta B(t) \not\equiv 0$

All the analysis in Section 3 refers to the special case that $\Delta B(t) \equiv 0$ for all $t \geq 0$. As both $\Delta A(t)$ and $\Delta B(t)$ arise in the uncertain system (2.1), the following result can be applied.

Theorem 5. (Barmish 1983). The uncertain system (2.1) is quadratically stabilizable if and only if the following system is quadratically stabilizable:

$$\frac{d}{dt}\begin{bmatrix} x(t) \\ u(t) \end{bmatrix} = \begin{bmatrix} A + \Delta A(t) & B + \Delta B(t) \\ 0 & 0 \end{bmatrix} \begin{bmatrix} x(t) \\ u(t) \end{bmatrix} + \begin{bmatrix} 0 \\ I \end{bmatrix} v(t) \tag{4.1}$$

Based on this theorem, one only needs to study the system (4.1) as $\Delta B(t) \not\equiv 0$ for quadratic stabilizability. However, if we view (4.1) as an *augmented* system with state variable (x, u), then (4.1) corresponds to (3.7) and the analysis outlined in the last section can be directly applied since no input matrix uncertainty appears in (4.1). For this augmented system, the function $\phi(S)$ is then

$$\phi(S) = \min_{(E,F)\in\Omega} \lambda_{\min}[-(A+E)S_{11} - (B+F)S_{21} - S_{11}(A+E)^T - S_{21}^T(B+F)^T] \tag{4.2}$$

where $S \in R^{(n+m)\times(n+m)}$, and S_{ij}'s are the submatrices of appropriate dimensions. After the maximum point S is found, one can then obtain $\psi_\xi(S)$. The stabilizing control is given by

$$u(t) = S_{21}S_{11}^{-1}x(t) \tag{4.3}$$

$S \in S^n$, $\hat{D} \in \mathbf{R}^{n\times(n-m)}$, $A \in \mathbf{R}^{n\times n}$ and Ω be compact and convex. Note that the matrix P in Definition 1 is equal to S_{11}^{-1}.

If one formulates the augmented system (4.1), the analysis and design procedure for quadratic stabilizability is as follows. Note that one can do this even as $\Delta B(t) \equiv 0$.

Step 1. Define the *new* A, ΔA, and $B = ([0\ I]^T)$ from (4.1). Choose the corresponding D, $\hat{B}$, and $\hat{D}$. Choose the parameter ξ to reflect the proper scaling.

Step 2. Maximize the function $\psi_\xi(S)$. Record the maximizing point S and the maximum value η_ξ. If $\eta_\xi > 0$, the system is declared to be quadratically stabilizable.

Step 3. Choose the control (4.3).

Remark 8. Step 2 involves a two level optimization process. The inner level is the minimization of the function $\phi(S)$. The outer level is the maximization of the function $\psi_\xi(S)$. Note that the *global* maximum can be achieved by Theorem 3.

Remark 9. The bounding set Ω is crucial in evaluating the function ψ_ξ, which involves a minimization process. As mentioned in Remark 1, as Ω is a hyperpolyhedron (this is often the case in practice as, for example, the uncertain system can often be expressed as (2.3) with each uncertain factor bounded by $\underline{q_i} \leq q_i(t) \leq \bar{q}_i$), the minimum can always be reached by one of its vertices according to Theorem 2. It is therefore sufficient to compute the values at the corners and take the minimum one. For other cases, we suggest to bound Ω by a least conservative hyperpolyhedron.

V. ILLUSTRATIONS

Basically, most numerical optimization algorithms utilize the gradient information of an objective function to find new values for the design variable. The design variables may also be subjected to certain constraints. In our case, $\psi_\xi(S)$ is the objective function and that S must belong to the set W is the constraint. Only the elements in the upper triangle of matrix S are our design variables since S needs to be symmetric. Certainly, S must also be a positive semi-definite matrix. This can be assured by constraining the determinant of every principal submatrix of S to be non-negative. Note that we can assure the positive definiteness of S by the argument provided in the proof of Theorem 4 if the maximum point is found and $\eta_\xi > 0$.

In the following examples, the internal penalty method is used to convert the constrained optimization problem to the unconstrained optimization problem (Fox 1971). The search direction for design variables is first determined by the variable metric method(or is called the Davidon-Fletcher-Powell method). Once the search direction is determined, design variables are changed step by step along this direction until the optimal point on this direction is found. Following the aforementioned uni-dimension search, the variable metric method is applied again to find the next search direction.

Other advanced constrained optimization algorithms (e.g., see Luenberger 1984) can also be adopted.

Example 1. Let us consider the following problem (Barmish 1985) where quadratic stabilizability by nonlinear control was considered. The system and input matrices and the magnitude of uncertainties are

$$A(r) = \begin{bmatrix} r & 1 \\ 0 & r \end{bmatrix}, \qquad |r| \leq \bar{r}$$

$$B(s) = \begin{bmatrix} s \\ 1 \end{bmatrix}, \qquad |s| \leq \bar{s}.$$

By using the augmented matrix and the numerical optimization algorithm mentioned above, as $\bar{r} = \bar{s} = .5$, we can find the matrix

$$S = \begin{bmatrix} .2963 & -.1869 & .0207 \\ -.1869 & .1771 & -.1296 \\ .0207 & -.1296 & .5266 \end{bmatrix}$$

satisfying the requirements and the corresponding feedback gain is

$$K = [\,-1.17 \quad -1.97\,].$$

Example 2. The dynamics of a helicopter in a vertical plane for an air speed of 60 knots to 170 knots (Narendra and Tripathi 1973) is considered next. The nominal and uncertain matrices are as following:

$$A = \begin{bmatrix} -0.0366 & 0.0271 & 0.0188 & -0.4555 \\ 0.0482 & -1.01 & 0.0024 & -4.0208 \\ 0.1002 & 0.2855 & -0.707 & 1.3229 \\ 0 & 0 & 1 & 0 \end{bmatrix}$$

$$\Delta A = \begin{bmatrix} 0 & 0 & 0 & 0 \\ 0 & 0 & 0 & 0 \\ 0 & a_{32} & 0 & a_{34} \\ 0 & 0 & 0 & 0 \end{bmatrix}$$

$$B = \begin{bmatrix} 0.4422 & 0.1761 \\ 3.0447 & -7.5922 \\ -5.52 & 4.99 \\ 0 & 0 \end{bmatrix}, \qquad \Delta B = \begin{bmatrix} 0 & 0 \\ b_{21} & 0 \\ 0 & 0 \\ 0 & 0 \end{bmatrix}$$

and $|a_{32}| \leq 0.2192$, $|a_{34}| \leq 1.2031$, $|b_{21}| \leq 2.0673$. In this formulation, x_1, x_2 are horizontal and vertical velocity, x_3, x_4 are pitch rate and pitch angle. Moreover, u_1, u_2 are collective pitch control and longitudinal cyclic pitch control.

Using the augmented matrix and optimization algorithm described above, we can find a matrix

$$S = \begin{bmatrix} .1973 & -.0034 & .0378 & .0422 & .0020 & -.0165 \\ -.0034 & .1912 & .0080 & .0110 & .0606 & .0264 \\ .0378 & .0080 & .1698 & -.0516 & .0504 & .0079 \\ .0422 & .0110 & -.0516 & .1003 & .0160 & -.0364 \\ .0020 & .0606 & .0504 & .0160 & .1829 & -.0058 \\ -.0165 & .0264 & .0079 & -.0364 & -.0058 & .1584 \end{bmatrix}$$

satisfying the requirements and the corresponding feedback gain is

$$K = \begin{bmatrix} -.1640 & .2699 & .4511 & .4308 \\ .0364 & .1692 & -.1066 & -.4519 \end{bmatrix}.$$

It is *smaller* than the K matrix obtained by Schmitendorf (1988), which is

$$K = \begin{bmatrix} -1.0181 & 0.2674 & 1.1123 & 1.7966 \\ 0.9531 & 0.8428 & -0.1412 & -0.7419 \end{bmatrix}.$$

Even if we increase all the bounds of uncertainty by 70%, we can still find a matrix

$$S = \begin{bmatrix} .1829 & -.0212 & -.0399 & .0605 & .0042 & -.0173 \\ -.0212 & .2152 & -.0114 & .0355 & .0668 & .0523 \\ -.0399 & -.0114 & .1229 & -.0444 & .0511 & .0233 \\ .0605 & .0355 & -.0444 & .0505 & .0234 & -.0037 \\ .0042 & .0668 & .0511 & .0234 & .2295 & .0071 \\ -.0173 & .0523 & .0233 & -.0037 & .0071 & .1990 \end{bmatrix}$$

satisfying the requirements. The feedback gain for this case becomes

$$K = \begin{bmatrix} -.272 & .083 & .878 & 1.502 \\ .024 & .279 & .169 & -.150 \end{bmatrix}.$$

Example 3. Next we consider the problem of stabilizing the longitudinal short period mode of a F5E fighter at two operating points. This problem was also solved by Schmitendorf (1988). The system and input matrices and corresponding uncertainties are

$$A = \begin{bmatrix} -0.8251 & 17.76 & 90.245 \\ 0.1734 & -0.7549 & -11.1 \\ 0 & 0 & -250 \end{bmatrix},$$

$$\Delta A = \begin{bmatrix} -0.1645 & -0.35 & 5.905 \\ 0.0914 & -0.0963 & -0.29 \\ 0 & 0 & 0 \end{bmatrix},$$

$$B = \begin{bmatrix} -91.44 \\ 0 \\ 250 \end{bmatrix}, \qquad \Delta B = \begin{bmatrix} -6.34 \\ 0 \\ 0 \end{bmatrix},$$

where x_1 = normal acceleration, x_2 =pitch rate, x_3 =elevator angle, and u =elevator control. Point 1 corresponds to $(A + \Delta A, B + \Delta B)$ and point 2 to $(A - \Delta A, B - \Delta B)$. By means of the same numerical optimization technique, we can find that

$$S = \begin{bmatrix} .3441 & -.0361 & .0343 & .0797 \\ -.0361 & .1154 & .0302 & .0364 \\ .0343 & .0302 & .2974 & .0604 \\ .0797 & .0364 & .0604 & .2431 \end{bmatrix}$$

satisfying the requirements and the corresponding feedback gain is

$$K = \begin{bmatrix} 0.2556 & 0.3595 & 0.1370 \end{bmatrix}$$

which is favorable than the result obtained in Schmitendorf (1988). The feedback gain obtained there is

$$K = \begin{bmatrix} 0.593 & 1.8965 & -0.642 \end{bmatrix}.$$

VI. CONCLUSIONS

We investigate the quadratic stabilizability for uncertain systems. The uncertain system possesses arbitrary, time-varying, but bounded uncertainty. Necessary and sufficient conditions are formulated and are applicable via an optimization setting. The design procedure includes a two level optimization process . The inner level searches for the minimum and can be achieved by checking a finite number of vertices if the bounding set is a hyperpolyhedron. The outer level can reach the global maximum since the set is convex. Further issue along this line such as finding the largest bound of the uncertainty for the stability to remain will be investigated.

ACKNOWLEDGEMENT

The contents of this paper are based on research supported by the National Science Foundation.

REFERENCES

Barmish, B.R., 1983, "Stabilization of uncertain systems via linear control", *IEEE Trans. Automatic Control*, Vol. 28, pp. 848-850.

Barmish, B.R., 1985, "Necessary and sufficient conditions for quadratic stabilizability of an uncertain system", *J. Optimization Theory and Applications*, Vol. 46, No. 4, pp. 399-408.

Barmish, B.R., and Leitmann, G., 1982, "On ultimate boundedness control of uncertain systems in the absence of matching conditions", *IEEE Trans. Automatic Control*, 27, pp. 153-157.

Chen, Y.H., 1986, "On the deterministic performance of uncertain dynamical systems," *International J. of Control*, Vol. 43, No. 5, pp. 1557-1579.

Chen, Y.H., 1987, "On the robustness of mismatched uncertain dynamical systems," *Journal of Dynamic Systems, Measurement, and Control*, Vol. 109, pp. 29-35.

Chen, Y.H., and Leitmann, G., 1987, "Robustness of uncertain systems in the absence of matching assumptions", *International J. Control*, Vol. 45, pp. 1527-1542.

Desoer, C.A., and Vidyasagar, M., 1975, *Feedback Systems: Input-Output Properties*, Academic Press, New York.

Fox, R.L., 1971, *Optimization Methods for Engineering Design*, Addison Wesley, Reading.

Gantmacher, F.R., 1959, *Theory of Matrices*, Vol. 1, Chelsea, New York.

Gu, K., and Chen, Y.H., 1989, "Uncertain systems: new uncertainty characterization for linear control design", *28th IEEE Conference on Decision and Control*, pp. 1508-1512.

Gutman, S., 1976, "Uncertain dynamical systems - a differential game approach", *NASA TMX*, Vol. 73, pp. 135.

Gutman, S., 1979, "Uncertain dynamical systems - a Lyapunov min-max approach", *IEEE Trans. Automatic Control*, Vol. 24, pp. 437-443.

Gutman, S., and Leitmann, G., 1975, "On a class of linear differential games", *J. Optimization Theory and Application*, Vol. 17, pp. 511-522.

Khargonekar, P.P., Petersen, I.R., and Zhou, K., 1990, "Robust stabilization of uncertain linear systems and H^∞ control theory", *IEEE Trans. Automatic Control*, Vol. 35, No. 3, pp. 356-361.

Leitmann, G., 1979, "Guaranteed asymptotic stability for some linear systems with bound -ed uncertainties", *J. Dynamic Systems, Measurement, and Control*, Vol. 101, pp. 212-216.

Leitmann, G., 1981, "On the efficacy of nonlinear control in uncertain linear systems," *J. Dynamical Systems, Measurement, and Control*, Vol. 103, pp. 95-102.

Luenberger, D.G., 1969, *Optimization by Vector Space Methods*, John Wiley, New York.

Luenberger, D.G., 1984, *Linear and Nonlinear Programming*, Second Edition, Addison Wesley, Reading.

Narendra, K.S., and Annaswamy, A.M., 1989, *Stable Adaptive Systems*, Prentice-Hall, New York.

Narendra, K.S., and Tripathi, S.S., 1973, "Identification and optimization of aircraft dynamics", *J. Aircraft*, Vol. 10, pp. 193-199.

Petersen, I.R. and Hollot, C.V., 1986, "A Riccati equation approach to the stabilization of uncertain linear systems," *Automatica*, Vol. 22, No. 4, pp. 397-411.

Schmitendorf, W.E., 1988, "Designing stabilizing controllers for uncertain systems using the Riccati equation approach," *IEEE Trans. Automatic Control*, Vol. 33, No. 4, pp. 376-379.

Wei, K., 1990, "Quadratic stabilizability of linear systems with structural independent time-varying uncertainties", *IEEE Trans. Automatic Control*, Vol. 35, No. 3, pp. 268-277.

ASYMPTOTIC STABILITY OF SINGULARLY PERTURBED SYSTEMS WHICH HAVE MARGINALLY STABLE BOUNDARY LAYER SYSTEMS†

Martin Corless
School of Aeronautics and Astronautics
Purdue University
West Lafayette, Indiana 47907

ABSTRACT

A class of linear singularly perturbed systems with singular perturbation parameter $\mu > 0$ is considered. To assure asymptotic stability of the full order system for $\mu > 0$ sufficiently small, it is customary to require that both the reduced-order system ($\mu = 0$) and the boundary layer system are asymptotically stable. Here we relax the requirement on the boundary layer system to stability (i.e., not necessarily asymptotic stability) and show that, subject to one additional condition, the full order system is asymptotically stable for sufficiently small μ. The result is illustrated by an application in which we consider the stability robustness of a simple feedback-controlled mechanical system w.r.t. to an unmodeled flexibility.

† This paper is based on research supported by the US National Science Foundation under grant MSM-87-06927.

1. INTRODUCTION

In modeling a "real" system, one quite often "neglects" some dynamics in the quest for a manageable model; neglecting dynamics results in a model of lower order than a model in which the "neglected" dynamics are taken into account. This occurs, e.g., if one models a flexible mechanical body as a rigid body. A natural question which arises is whether one can predict the behavior of the higher order model using the lower order one.

One approach to the above question is to consider the higher order system as a singularly perturbed system; see [3,4]. Roughly speaking, a singularly perturbed system is a system whose behavior depends on a scalar parameter μ in such a fashion that for a specific value of the parameter (usually taken to be zero) the order of the system is lower than that for other parameter values. Associated with a singularly perturbed system are two systems of lower order, the reduced-order system ($\mu = 0$) and the associated boundary layer system. One usually tries to determine the behavior of the full order system from the behavior of these two lower order systems.

In this paper, we consider the asymptotic stability of a class of linear singularly perturbed systems with singular perturbation parameter $\mu \geq 0$. To assure asymptotic stability of the full order system for $\mu > 0$ sufficiently small, it is customary to require that both the reduced-order system ($\mu = 0$) and the associated boundary layer system are asymptotically stable; an exception is [1]. Here, as in [1], we relax the requirement on the boundary layer system to stability (i.e., not necessarily asymptotic stability) and show that, subject to one additional condition, the full order system is asymptotically stable for sufficiently small μ. The condition is different from the corresponding condition in [1]. Neither condition implies the other. The proof is based on constructing appropriate Lyapunov functions.

Before proceeding with the general problem statement, we present a simple illustrative application in which we consider the stability robustness of a feedback controlled mechanical system w.r.t. to an unmodeled flexibility.

1.1. Example: Stability Robustness with Respect to an Unmodeled Flexibility

Consider a simple mechanical system consisting of two particles of unit mass connected by a massless bar B; see Figure 1. The system is constrained to move along an inertially fixed line and particle P_2 is subject to a control force u. Using only q_1 and $\dot{q}_2$ as measurements, we wish to construct a feedback controller generating u which assures that the resulting closed-loop system is asymptotically stable about 0.

In designing the controller, we shall model B as a rigid bar of fixed length l_0. We then model B as a flexible bar and determine whether or not the closed-loop flexible model is stable, provided the bar is sufficiently stiff. If stable, we consider the control design to be robust w.r.t. the unmodeled flexibility.

When B is modeled as a rigid bar of fixed length l_0,

$$q_2 = q_1$$

and letting

$$x_1 \triangleq q_1 , \quad x_2 \triangleq \dot{q}_1 ,$$

the motion of the system can be described by

$$\dot{x}_1 = x_2$$
$$\dot{x}_2 = \tfrac{1}{2}u \ .$$

Considering any linear controller of the form

$$u = -l_1 q_1 - l_2 \dot{q}_2$$

where

$$l_1 > 0 , \quad l_2 > 0$$

we obtain the asymptotically stable closed-loop rigid model

$$\dot{x}_1 = x_2 \tag{1.1a}$$
$$\dot{x}_2 = -\tfrac{1}{2} l_1 x_1 - \tfrac{1}{2} l_2 x_2 \ . \tag{1.1b}$$

To investigate robustness w.r.t. the unmodeled flexibility we now model B as a linear spring of undeformed length l_0 and spring constant k and consider behavior for large k. This yields the following equations of motion.

$$\ddot{q}_1 = \ \ k(q_2 - q_1)$$
$$\ddot{q}_2 = -k(q_2 - q_1) - l_1 q_1 - l_2 \dot{q}_2 \ .$$

Choosing x_1 and x_2 as before, letting

$$y \triangleq k(q_2 - q_1)$$
$$\mu \triangleq 1/k$$

the system is described by

$$\dot{x}_1 = x_2 \tag{1.2a}$$

$$\dot{x}_2 = y \tag{1.2b}$$

$$\mu\ddot{y} = -l_1x_1 - l_2x_2 - 2y - \mu l_2\dot{y} \ . \tag{1.2c}$$

Equation (1.2) describes a singularly perturbed system with singular perturbation μ; as μ gets smaller, the flexible bar becomes stiffer.

The reduced-order system associated with (1.2) is obtained by setting $\mu = 0$, i. e.,

$$y = -\tfrac{1}{2}l_1x_1 - \tfrac{1}{2}l_2x_2 \tag{1.3}$$

$$\dot{x}_1 = x_2 \tag{1.4a}$$

$$\dot{x}_2 = -\tfrac{1}{2}l_1x_1 - \tfrac{1}{2}l_2x_2 \ . \tag{1.4b}$$

System (1.4) is exactly the same as the rigid model (1.1); hence it is asymptotically stable.

The boundary layer system associated with (1.2) is obtained by first introducing a new time variable $\tau \triangleq t/\mu^{1/2}$, letting $\eta(\tau) \triangleq x(\mu^{1/2}\tau)$, $\xi(\tau) \triangleq y(\mu^{1/2}\tau)$, and then setting $\mu = 0$ to obtain

$$\dot{\eta} = 0 \tag{1.5}$$

$$\ddot{\xi} = -l_1\eta_1 - l_2\eta_2 - 2\xi \ . \tag{1.6}$$

System (1.6) is only marginally stable. Hence, we cannot use standard results to guarantee asymptotic stability of the full order system (1.2).

2. SYSTEMS UNDER CONSIDERATION

Consider a singularly perturbed system described by

$$\begin{aligned} \dot{x} &= A_{11}(\mu)x + A_{12}(\mu)y + \mu^{1/2}A_{13}(\mu)\dot{y} \ , \\ \mu\ddot{y} &= A_{31}(\mu)x + A_{32}(\mu)y + \mu A_{33}(\mu)\dot{y} \ , \end{aligned} \tag{2.1}$$

where $t \in \mathbb{R}$ is the "time" variable, $x \in \mathbb{R}^n$ and $y \in \mathbb{R}^m$ describe the state of the system, and $\mu \in [0,\infty)$ is the singular perturbation parameter.

2.1. The Reduced-Order System

The reduced-order system associated with (2.1) is obtained by letting $\mu = 0$ in (2.1). Thus, the reduced-order system can be described by

$$\dot{x} = A_{11}(0)x + A_{12}(0)y, \qquad (2.2)$$
$$0 = A_{31}(0)x + A_{32}(0)y.$$

In Section 2.2, we introduce an assumption which implies that $A_{32}(0)$ is invertible. Invoking that assumption here, the reduced-order system can be described by

$$\dot{x} = \overline{A}x, \qquad (2.3)$$
$$y = Hx, \qquad (2.4)$$

where

$$\overline{A} \triangleq A_{11}(0) - A_{12}(0)A_{32}(0)^{-1}A_{31}(0), \qquad (2.5a)$$
$$H \triangleq -A_{32}(0)^{-1}A_{31}(0). \qquad (2.5b)$$

We make the following assumption on the reduced-order system.

Assumption A1. *The reduced-order system (2.3) is asymptotically stable.*

2.2. The Boundary Layer System

The boundary layer system associated with (2.1) is obtained by introducing a new time variable

$$\tau \triangleq \mu^{-1/2}t \qquad (2.6)$$

and considering the behavior of the resulting system for $\mu = 0$. Defining

$$\eta(\tau) \triangleq x(\mu^{1/2}\tau) = x(t) \ , \qquad (2.7)$$
$$\xi(\tau) \triangleq y(\mu^{1/2}\tau) = y(t) \ ,$$

and utilizing (2.1), one obtains

$$\frac{d\eta}{d\tau} = \mu^{1/2}[A_{11}(\mu)\eta + A_{12}(\mu)\xi + A_{13}(\mu)\frac{d\xi}{d\tau}], \qquad (2.8)$$
$$\frac{d^2\xi}{d\tau^2} = A_{31}(\mu)\eta + A_{32}(\mu)\xi + \mu^{1/2}A_{33}(\mu)\frac{d\xi}{d\tau}.$$

Letting $\mu = 0$ in (2.8) yields

$$\frac{d\eta}{d\tau} = 0, \tag{2.9}$$

$$\frac{d^2\xi}{d\tau^2} = A_{31}(0)\eta + A_{32}(0)\xi. \tag{2.10}$$

Equation (2.9) implies that η is constant, i.e.,

$$\eta(\tau) = \eta(0) \quad \forall\ \tau \in \mathbb{R}.$$

With $A_{32}(0)$ invertible, (2.10) has a unique equilibrium solution given by

$$\xi(t) \equiv H\eta(0)\ ,$$

where H is given by (2.5b).

It can readily be shown that system (2.10) is not asymptotically stable about its equilibrium solution. The following assumption assures, among other things, that system (2.10) is marginally stable.

Assumption A2. *There exists a positive-definite, symmetric matrix* $S \in \mathbb{R}^{m\times m}$ *such that*

(i) $SA_{32}(0)$ *is negative-definite and symmetric,*

(ii) $S\overline{C}$ *is negative-definite,*

where

$$\overline{C} \triangleq A_{33}(0) + A_{32}(0)^{-1}A_{31}(0)A_{12}(0). \tag{2.11}$$

Condition (i) of Assumption A2 implies that $A_{32}(0)$ is invertible. It also implies that system (2.10) is marginally stable about its equilibrium solution. Ref. [1] requires that all the eigenvalues of $A_{32}(0)$ be negative and distinct; this is a stronger requirement than (i).

We introduce a final assumption on the dependency of the system matrices on μ.

Assumption A3. *Each matrix-valued function* $A_{ij}(\cdot)$, $i = 1,3$, $j = 1,2,3,$ *is continuous at* 0 *and*

$$\lim_{\mu\to 0} \mu^{-1/2}[A_{31}(\mu) - A_{31}(0)] = 0\ , \tag{2.12}$$
$$\lim_{\mu\to 0} \mu^{-1/2}[A_{32}(\mu) - A_{32}(0)] = 0\ .$$

Note that (2.12) is satisfied if $A_{31}(\cdot)$ and $A_{32}(\cdot)$ are differentiable at 0.

3. A SPECIAL CASE : $A_{31}(0) = 0$

In this section, we consider the class of singularly perturbed systems described by (2.1) which satisfy

$$A_{31}(0) = 0 \ ; \tag{3.1}$$

hence

$$\bar{A} = A_{11}(0) \ , \quad \bar{C} = A_{33}(0) \tag{3.2}$$

and A1 and condition (ii) of A2 reduce to the requirements that $A_{11}(0)$ be asymptotically stable and $SA_{33}(0)$ be negative definite, respectively.

We have the following result.

Theorem 3.1. *Consider a singularly perturbed system described by (2.1) which satisfies Assumptions A1-A3 and (3.1). Then there exists $\mu^* > 0$ such that (2.1) is asymptotically stable for all $\mu \in (0, \mu^*)$.*

Proof. Let $S \in \mathbb{R}^{m\times m}$ be any matrix which assures A2 and choose any scalar $\alpha > 0$ which satisfies(1)

$$SA_{33}(0) < -\alpha S \ . \tag{3.3}$$

Introduce new state vectors $z_1, z_2 \in \mathbb{R}^m$ defined by

$$z_1 \triangleq y \ , \ z_2 \triangleq \mu^{1/2}[\alpha y + \dot{y}] \ . \tag{3.4}$$

Then, utilizing (2.1), we obtain

$$\begin{aligned} \dot{x} &= [B_{11} + \Delta B_{11}(\mu)]x + B_{12}(\mu)z_1 + B_{13}(\mu)z_2 \\ \dot{z}_1 &= -\alpha z_1 + \mu^{-1/2} z_2 \\ \dot{z}_2 &= B_{31}(\mu)x + [\mu^{-1/2}A_{32}(0) + \Delta B_{32}(\mu)]z_1 + [B_{33} + \Delta B_{33}(\mu)]z_2 \end{aligned} \tag{3.5}$$

where

(1) If $M, N \in \mathbb{R}^{m\times m}$, the notation $M < N$ means $x^T M x < x^T N x$ for all $x \in \mathbb{R}^m$, $x \neq 0$.

$$\begin{aligned} B_{11} &= A_{11}(0) , & \Delta B_{11}(\mu) &= A_{11}(\mu) - A_{11}(0), \\ B_{12}(\mu) &= A_{12}(\mu) - \mu^{1/2}\alpha A_{13}(\mu) , & B_{13}(\mu) &= A_{13}(\mu), \end{aligned} \tag{3.6}$$

$$\begin{aligned} B_{31}(\mu) &= \mu^{-1/2} A_{31}(\mu) , \\ \Delta B_{32}(\mu) &= \mu^{-1/2}[A_{32}(\mu) - A_{32}(0)] - \mu^{1/2}\alpha[A_{33}(\mu) + \alpha I] , \\ B_{33} &= A_{33}(0) + \alpha I , \quad \Delta B_{33}(\mu) = A_{33}(\mu) - A_{33}(0) . \end{aligned}$$

Asymptotic stability of (2.1) is equivalent to asymptotic stability of (3.5). Note that, as a consequence of A3 and (3.1),

$$\lim_{\mu\to 0} \Delta B_{11}(\mu) = \lim_{\mu\to 0} B_{31}(\mu) = \lim_{\mu\to 0} \Delta B_{32}(\mu) = \lim_{\mu\to 0} \Delta B_{33}(\mu) = 0 . \tag{3.7}$$

We proceed now to construct a Lyapunov function for (3.5). As a consequence of A1, B_{11} is asymptotically stable; hence there exist positive-definite matrices $P, Q \in \mathbb{R}^{n\times n}$ such that

$$PB_{11} + B_{11}^T P + 2Q = 0 . \tag{3.8}$$

Introduce the positive-definite matrix $R \in \mathbb{R}^{m\times m}$ defined by

$$R \triangleq -SA_{32}(0) . \tag{3.9}$$

As a candidate Lyapunov function for (3.5), we consider the positive-definite function $U_\mu : \mathbb{R}^n \times \mathbb{R}^{2m} \to \mathbb{R}_+$ given by

$$U_\mu(x, z) \triangleq V(x) + k_\mu W(z) \tag{3.10}$$

where

$$z \triangleq \begin{bmatrix} z_1 \\ z_2 \end{bmatrix}$$

$$V(x) \triangleq \tfrac{1}{2} x^T P x , \tag{3.11a}$$

$$W(z) \triangleq \tfrac{1}{2} z_1^T R z_1 + \tfrac{1}{2} z_2^T S z_2 , \tag{3.11b}$$

and the scalar $k_\mu > 0$ will be specified later.

Utilizing (3.6), (3.8) and (3.9) it follows that along any solution of (3.5),

$$\begin{aligned} \frac{dV}{dt} &= -x^T Q x + x^T P \Delta B_{11}(\mu) x + x^T P [B_{12}(\mu) \;\; B_{13}(\mu)] z , \\ \frac{dW}{dt} &= z_2^T S B_{31}(\mu) x - \alpha z_1^T R z_1 + z_2^T S B_{33} z_2 + z_2^T S [\Delta B_{32}(\mu) \;\; \Delta B_{33}(\mu)] z. \end{aligned} \tag{3.12}$$

As a consequence of the choice of α (see (3.3)), we have

$$SB_{33} = SA_{33}(0) + \alpha S < 0 \; ; \tag{3.13}$$

hence we can define positive-definite functions $\phi : \mathbb{R}^n \to \mathbb{R}_+$ and $\psi : \mathbb{R}^{2m} \to \mathbb{R}_+$ by

$$\phi(x) \triangleq (x^T Q x)^{1/2} \; , \tag{3.14a}$$

$$\psi(x) \triangleq (\alpha z_1^T R z_1 - z_2^T S B_{33} z_2)^{1/2} \; . \tag{3.14b}$$

Recalling (3.7), it follows from (3.12)-(3.14) and A3 that there exist non-negative scalars $\Delta a_{11}(\mu)$, $a_{12}(\mu)$, $a_{21}(\mu)$, and $\Delta a_{22}(\mu)$ such that

$$\frac{dV}{dt} \le -[1 - \Delta a_{11}(\mu)]\phi(x)^2 + a_{12}(\mu)\phi(x)\psi(x) \; , \tag{3.15a}$$

$$\frac{dW}{dt} \le a_{21}(\mu)\phi(x)\psi(x) - [1 - \Delta a_{22}(\mu)]\psi(x)^2 \; , \tag{3.15b}$$

and

$$\lim_{\mu\to 0} \Delta a_{11}(\mu) = \lim_{\mu\to 0} a_{21}(\mu) = \lim_{\mu\to 0} \Delta a_{22}(\mu) = 0 \; ; \tag{3.16a}$$

also there exists $\bar{\mu} > 0$, $\bar{\beta} \ge 0$ such that

$$a_{12}(\mu) \le \bar{\beta} \quad \forall \quad \mu \in (0, \bar{\mu}) \, . \tag{3.16b}$$

Recalling (3.10) and utilizing (3.15) we have that along any solution of (3.5),

$$\frac{dU_\mu}{dt} \le - \begin{bmatrix} \phi(x) \\ \psi(x) \end{bmatrix}^T M(\mu) \begin{bmatrix} \phi(x) \\ \psi(x) \end{bmatrix} , \tag{3.17}$$

where

$$M(\mu) \triangleq \begin{bmatrix} 1 - \Delta a_{11}(\mu) & -\tfrac{1}{2}[a_{12}(\mu) + k_\mu a_{21}(\mu)] \\ -\tfrac{1}{2}[a_{12}(\mu) + k_\mu a_{21}(\mu)] & k_\mu[1 - \Delta a_{22}(\mu)] \end{bmatrix} . \tag{3.18}$$

As a consequence of (3.16), there exists $\mu^* > 0$ such that

$$\begin{aligned} 1 - \Delta a_{11}(\mu) &> 0 \; , \\ [1 - \Delta a_{11}(\mu)][1 - \Delta a_{22}(\mu)] - a_{12}(\mu)a_{21}(\mu) &> 0 \; , \end{aligned} \tag{3.19}$$

for all $\mu \in (0, \mu^*)$.

Consider now any $\mu \in (0, \mu^*)$ and choose $k_\mu > 0$ to satisfy

$$\begin{aligned} k_\mu &> a_{12}(\mu)^2/4[1 - \Delta a_{11}(\mu)][1 - \Delta a_{22}(\mu)] && \text{if } a_{21}(\mu) = 0 \; , \\ k_\mu &< 4[1 - \Delta a_{11}(\mu)][1 - \Delta a_{22}(\mu)]/a_{21}(\mu)^2 && \text{if } a_{12}(\mu) = 0 \; , \; a_{21}(\mu) \ne 0 \; , \\ k_\mu &= a_{12}(\mu)/a_{21}(\mu) && \text{if } a_{12}(\mu) \ne 0 \; , \; a_{21}(\mu) \ne 0 \; . \end{aligned}$$

Elementary calculations demonstrate that

$$M(\mu) > 0 \ .$$

It now follows from (3.17) that U_μ is a Lyapunov function for system (3.5) . Thus (3.5) and hence (2.1) is asymptotically stable. □

4. THE MAIN RESULT

4.1. The Main Result

We return now to the class of systems considered in Section 2. The main result of this paper is contained in the following theorem.

Theorem 4.1. *Consider a singularly perturbed system described by (2.1) which satisfies Assumptions A1-A3. Then there exists* $\mu^* > 0$ *such that (2.1) is asymptotically stable for all* $\mu \in (0, \mu^*)$.

Proof. The proof proceeds by introducing a state transformation so that the transformed system satisfies condition (3.1) in addition to A1-A3. Then one applies Theorem 3.1.

Introduce new state variable $e \in \mathbb{R}^m$ defined by

$$e \triangleq y - Hx \tag{4.1}$$

where H is given by (2.5b). Then, utilizing (2.1), a few calculations yield

$$\begin{aligned} \dot{x} &= \tilde{A}_{11}(\mu)x + \tilde{A}_{12}(\mu)e + \mu^{1/2}\tilde{A}_{13}(\mu)\dot{e} \ , \\ \mu\ddot{e} &= \tilde{A}_{31}(\mu)x + \tilde{A}_{32}(\mu)e + \mu\tilde{A}_{33}(\mu)\dot{e} \ , \end{aligned} \tag{4.2}$$

where

$$\begin{aligned} \tilde{A}_{11} &\triangleq [I + \mu^{1/2}\tilde{A}_{13}H][A_{11} + A_{12}H] \ , \\ \tilde{A}_{12} &\triangleq [I + \mu^{1/2}\tilde{A}_{13}H]A_{12} \ , \\ \tilde{A}_{13} &\triangleq A_{13}E^{-1} \ , \\ \tilde{A}_{31} &\triangleq E[A_{31} + A_{32}H] + \mu[-HA_{11}^2 - HA_{11}A_{12} + \tilde{A}_{33}H(A_{11} + A_{12}H)] \ , \\ \tilde{A}_{32} &\triangleq EA_{32} + \mu[-HA_{11} + \tilde{A}_{33}H]A_{12} \ , \\ \tilde{A}_{33} &\triangleq [A_{33} - HA_{12} - \mu^{1/2}H(A_{11}A_{13} + A_{13}A_{33})]E^{-1} , \\ E &\triangleq I - \mu^{1/2}HA_{13} \ . \end{aligned} \tag{4.3}$$

From (4.3) and (2.5b) one can readily show that

$$\tilde{A}_{31}(0) = 0 \ . \tag{4.4}$$

Also, satisfaction of A1-A3 by (2.1) assures that A1-A3 are satisfied with A replaced by $\tilde{A}$. It now follows from Theorem 3.1 that system (4.2) is asymptotically stable for μ sufficiently small. Hence (2.1) is asymptotically stable for sufficiently small μ. □

4.2. The Example Revisited

Recalling the example of Section 1.1, it is readily seen that Assumptions A1 and A3 are satisfied. A simple calculation yields

$$A_{33}(0) = -2 \ ; \ \ \bar{C} = -\tfrac{1}{2} l_2 < 0 \ .$$

Hence, A2 is satisfied (let S = 1) and the feedback-controlled flexible model is asymptotically stable provided the bar is sufficiently stiff.

REFERENCES

[1] CHOW, J.H., ALLEMONG, J.J, AND KOKOTOVIC, P.V, Singular Pertubation Analysis of Systems with Sustained High Frequency Oscillations, *Automatica,* Vol. 14, pp. 271-279, 1978.

[2] CORLESS, M., Stability Robustness of Linear Feedback-Controlled Mechanical Systems in the Presence of a Class of Unmodelled Flexibilities, 27th Conference on Decision and Control, Austin, Texas, 1988.

[3] KOKOTOVIC, P.V., and KHALIL, H.K. (editors), Singular Pertubations in Systems and Control, IEEE Press, 1986.

[4] KOKOTOVIC, P.V., KHALIL, H.K., and O' REILLY, J., Singular Perturbation Methods in Control: Analysis and Design, Academic Press, London, 1986.

[5] SABERI, A., and KHALIL, H.K., Quadratic-Type Lyapunov Functions for Singularly Perturbed Systems, *IEEE Transactions on Automatic Control,* Vol. AC-29, pp. 542-550.

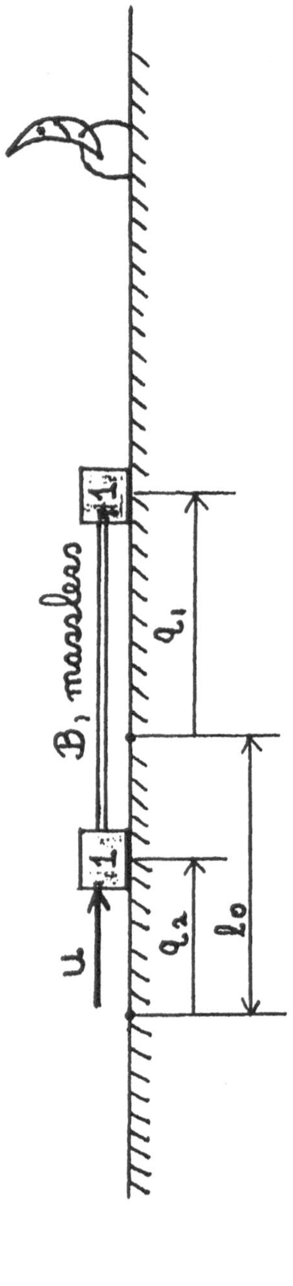

FIGURE 1

ROBOT TRAJECTORY CONTROL:
ROBUST OUTER LOOP DESIGN USING LINEAR CONTROLLER

R.M. Dolphus and W.E. Schmitendorf

Mechanical Engineering
University of California
Irvine, CA 92715

Abstract Trajectory control of robotic manipulators is investigated using a robust linear control law. For a system with unknown but bounded parameters, we implement an inverse dynamics control scheme that consists of a feedback linearizing control (inner loop) based on a nominal system, followed by a robust linear feedback control (outer loop) based on the uncertainty bounds. The control is robust in the sense that the system practically tracks (i.e. follows arbitrarily close to) the desired trajectory of position vs. time for any uncertainty within the allowed range.

I. INTRODUCTION

Inverse dynamics (or computed torque) is one method for controlling the trajectory of a robotic manipulator [1, 2]. The controller is comprised of a nonlinear inner loop designed to cancel nonlinear system dynamics, and a linear outer loop designed to drive tracking errors to zero. Asymptotic tracking of a desired trajectory (position vs. time) is obtained provided the system model is accurate.

If the system model is not precisely known (e.g. if the robot is carrying unknown loads), then a controller whose design is based only on the nominal system may perform

† This research was supported by NSF Grant No. ECS-8805759.

poorly due to inexact cancellation of modeled vs. actual system dynamics. To assure satisfactory performance, we design the outer loop control to be robust over a given range of uncertainties. Using the method described in [3], a linear outer loop control can be chosen such that the robot tracks arbitrarily close to a desired trajectory. Thus we guarantee practical tracking (defined rigorously below). If the uncertainties are zero (the nominal system is equivalent to the actual system) then we achieve asymptotic tracking.

Consider the dynamics of an n-link robot written as

$$M(q,r)\ddot{q} + h(q,\dot{q},r) = u \tag{1}$$

where $q \varepsilon \Re^n$ denotes the generalized coordinates, $u \varepsilon \Re^n$ is the control (applied torques), $M \varepsilon \Re^{n \times n}$ is the inertia matrix, and $h \varepsilon \Re^n$ is a vector of nonlinear terms (e.g. gravity, Coriolis effects). The uncertainties satisfy $r(t) \varepsilon \; R \subset \Re^p$ where R is a compact set and $r(\cdot)$ is assumed to be piecewise continuous. $M(q,\cdot)$ and $h(q,\dot{q},\cdot)$ are assumed to be continuous. We are given a desired trajectory $q^d(\cdot)$ satisfying $q^d(t) \varepsilon Q_1^d \subset \Re^n$ where Q_1^d is a compact set and $q^d(\cdot)$ is assumed to be continuous with piecewise continuous first derivative and piecewise continuous second derivative. Consequently we can define compact sets Q_2^d and Q_3^d such that $\dot{q}^d \varepsilon Q_2^d \subset \Re^n$ and $\ddot{q}^d(t) \varepsilon Q_3^d \subset \Re^n$ for all $t \geq 0$. Functions $r(\cdot)$ and $q^d(\cdot)$ satisfying the above conditions will be called <u>admissible</u>. The inertia matrix M is positive definite and we assume that there exist positive constants $\underline{M}$ and $\bar{M}$ such that

$$\underline{M} \leq \left\| M(q,r)^{-1} \right\| \leq \bar{M} < \infty \tag{2}$$

for all $q \varepsilon \Re^n$ and $r \varepsilon R$.

Let $S_\delta(t) \triangleq \{q : \| q - q^d(t) \| \leq \delta\}$. Given $\delta > 0$, (1) can be rendered <u>ultimately bounded with respect to $S_\delta(t)$</u> if there exists a control $u_\delta(\cdot)$ such that the following condition holds: for all $q_0 \varepsilon \Re^n$, admissible $r(\cdot)$, and admissible $q^d(\cdot)$, there exists a finite $T > 0$ such that $q(t) \varepsilon S_\delta(t)$ for all $t > T$. The control $u_\delta(\cdot)$ is called a <u>δ-tracking control</u> and δ is called the <u>tracking radius</u>. If (1) can be rendered ultimately bounded with respect to $S_\delta(t)$ for all $\delta > 0$ then (1) is said to be <u>practically trackable</u> and the control $u_\delta(\cdot)$ is a <u>practical tracking control</u>.

In the next section, we follow the development of the inverse dynamics equations in the presence of uncertainties as given by [1]. The development results in a nominally linear uncertain system that must be stabilized. The new system is of the same form as the class of systems covered in [3]. Thus we proceed to apply the method from [3] for the outer loop. In [1], a nonlinear outer loop control is used based on the design technique of [4]. In contrast, here we obtain a linear outer loop control.

II. MAIN RESULTS

Consider a fixed nominal set of uncertainties $\hat{r} \varepsilon R$ and define $\hat{M}(q) \triangleq M(q,\hat{r})$ and $\hat{h}(q,\dot{q}) \triangleq h(q,\dot{q},\hat{r})$. Let

$$u = \hat{M}(q)v + \hat{h}(q,\dot{q}) \ . \tag{3}$$

Substituting (3) into (1) and using the invertibility of the inertia matrix M leads to

$$\ddot{q} = v + \left(M^{-1}\hat{M} - I\right)v + M^{-1}\left(\hat{h} - h\right) . \tag{4}$$

Define the following

$$E(q,r) \triangleq M^{-1}(q,r)\,\hat{M}(q) - I \tag{5a}$$

$$\Delta h(q,\dot{q},r) \triangleq \hat{h}(q,\dot{q}) - h(q,\dot{q},r) \tag{5b}$$

$$\eta(q,\dot{q},r) \triangleq E(q,r)v + M^{-1}(q,r)\Delta h(q,\dot{q},r) \quad . \tag{5c}$$

Then (4) becomes

$$\ddot{q} = v + \eta \quad . \tag{6}$$

Introducing the error vector

$$e = \begin{bmatrix} e_1 \\ e_2 \end{bmatrix} \triangleq \begin{bmatrix} q - q^d \\ \dot{q} - \dot{q}^d \end{bmatrix} \tag{7}$$

(6) becomes

$$\dot{e} = Ae + B\left(v + \eta - \ddot{q}^d\right) \tag{8}$$

where

$$A = \begin{bmatrix} O_n & I_n \\ O_n & O_n \end{bmatrix}, \quad B = \begin{bmatrix} O_n \\ I_n \end{bmatrix}.$$

We now choose

$$v = \ddot{q}^d - K_1\left(q - q^d\right) - K_2\left(\dot{q} - \dot{q}^d\right) + \Delta v \tag{9a}$$

$$v = \ddot{q}^d - Ke + \Delta v \tag{9b}$$

where K_1 and K_2 are diagonal matrices with elements consisting of position and velocity gains, e.g., $K_1 = \mathrm{diag}\left[\omega_1^2, \dots \omega_n^2\right]$, $K_2 = \mathrm{diag}\left[2\zeta_1\omega_1, \dots 2\zeta_n\omega_n\right]$, and $K = [K_1 \;\; K_2]$. We combine (8) and (9) to obtain

$$\begin{aligned} \dot{e} = {} & [\bar{A} - BE(q,r)K]e + [B + BE(q,r)]\Delta v \\ & + B\left[E(q,r)\ddot{q}^d + M^{-1}(q,r)\Delta h(q,\dot{q},r)\right] \end{aligned} \tag{10}$$

where $\bar{A} \triangleq A - BK$ is Hurwitz.

The form of (10) is exactly the same as the form of systems considered in [3], which is to say that (10) is nominally linear, has an additive disturbance, and the uncertainties satisfy certain matching conditions. The method in [3] gives a linear control using a non-iterative Riccati equation design technique. Refer to Appendix A for details.

Applying the method to our problem, we define

$$\bar{E} \triangleq \max \left\{ \| E(q,r) \| : q \varepsilon \Re^n, r \varepsilon R \right\} < 1 \tag{11}$$

$$\rho \triangleq \max \left\{ \left\| E(q,r)\ddot{q}^d + M^{-1}(q,r)\Delta h(q,\dot{q},r) \right\| : q \varepsilon Q_1, \dot{q} \varepsilon Q_2, \ddot{q}^d \varepsilon Q_3^d, r \varepsilon R \right\} \tag{12}$$

where Q_1 and Q_2 represent bounds on q and $\dot{q}$. Q_1 and Q_2 can be chosen as sets somewhat larger than Q_1^d and Q_2^d (to account for the larger of the eventual tracking error δ and the norm of the initial error $\|e_0\|$). The bounding sets Q_1 and Q_2 are necessary in (12) to keep ρ finite. Note that this difficulty does not arise in (11); $\hat{M}$ is chosen such that the inequality in (11) is satisfied for all $q \in \Re^n$. There is always a suitable choice for $\hat{M}$ as discussed in [5]. For example, one can always choose

$$\hat{M} = \frac{1}{2}(\bar{M} + \underline{M})I \tag{13}$$

in which case it can be shown that

$$\|E\| = \left\|M^{-1}\hat{M} - I\right\| \le \frac{\bar{M} - \underline{M}}{\bar{M} + \underline{M}} < 1 \tag{14}$$

for all $q \in \Re^n$, $r \in R$.

We proceed with the steps from [3] as follows.

<u>Step 1.</u> Calculate Q_d from

$$Q_d = K^T K \bar{E}^2 \quad . \tag{15}$$

<u>Step 2.</u> For $Q > 0$, solve the Riccati equation for P

$$P\bar{A} + \bar{A}^T P - PBB^T P\left(1 - \bar{E}^2\right) + \frac{Qd}{(1 - \bar{E})^2} + Q = 0 \quad . \tag{16}$$

<u>Step 3.</u> For a given tracking radius δ, choose the scalar γ to satisfy

$$\gamma > \left(\frac{\bar{\lambda}_P}{\underline{\lambda}_P}\right) \frac{\rho^2}{2\underline{\lambda}_Q \delta^2 (1 - \bar{E})} + 1 \tag{17}$$

where $\bar{\lambda}_P(\underline{\lambda}_P)$ is the largest (smallest) eigenvalue of P.

<u>Step 4.</u> The δ-tracking control is then

$$\Delta v = -\gamma B^T P e . \tag{18}$$

One might try different choices for Q in (16) to minimize the gains given in (18). In the absence of uncertainties, $\hat{M} = M$, $\hat{h} = h, E = 0$, and $\Delta h = 0$; hence we can let $\Delta v = 0$ (by choosing a vanishingly small Q), which gives the standard inverse dynamics controller. Figure 1 shows the proposed control scheme (from [1]).

III. EXAMPLES

Example 1. Single Link Robot

A single link robot is shown in Figure 2. The equation of motion is

$$J\ddot{\theta} + mg\ell \sin\theta = u$$

and we use constraints as in [1],

$$5 \le J \le 10$$
$$5 \le mg\ell \le 10.$$

Rewriting the equations using our notation we have

$$q = \theta$$
$$r_1 = J$$
$$r_2 = mg\ell$$
$$R = \{r:\ r_i \varepsilon\, [5, 10]\}.$$

With nominal values of $r_1 = r_2 = 5$, we obtain

$$\hat{M}(q) = 5$$
$$\hat{h}(q,\dot{q}) = 5 \sin(q)$$
$$E(q,r) = M^{-1}\hat{M} - I = 5/r_1 - 1$$
$$\Delta h(q,\dot{q},r) = \hat{h} - h = (5 - r_2)\sin(q)$$
$$A = \begin{bmatrix} 0 & 1 \\ 0 & 0 \end{bmatrix}, \quad B = \begin{bmatrix} 0 \\ 1 \end{bmatrix}$$

Choosing $K_1 = 100$, $K_2 = 20$ (eigenvalues at -10, -10) gives $K = [100 \quad 20]$ and

$$\bar{A} = A - BK = \begin{bmatrix} 0 & 1 \\ -100 & -20 \end{bmatrix} .$$

Assuming $\left|\ddot{q}^d(t)\right| \le 4\,\text{rad/sec}^2$, we obtain

$$\bar{E} = \max\left\{ \left|\frac{5}{r_1} - 1\right| : r_1 \varepsilon [5,10] \right\} = 0.5 < 1$$

$$\rho = \max\left\{ \left|\left(\frac{5}{r_1} - 1\right)\ddot{q}^d + \frac{1}{r_1}(5 - r_2)\sin(q_1)\right| : r_i \varepsilon [5,10],\ \ddot{q}^d \varepsilon [-4,4],\ q \varepsilon [0,2\pi] \right\} = 2.5$$

Algorithm: 1) $Q_d = K^T K\, \bar{E}^2 = \begin{bmatrix} 2500 & 500 \\ 500 & 100 \end{bmatrix}$.

2) With $Q = 10^6 I$, $P = \begin{bmatrix} 1{,}008{,}450 & 1{,}034.8 \\ 1{,}034.8 & 1{,}129.8 \end{bmatrix}$.

3) From (17), $\gamma > \frac{(0.005584)}{\delta^2} + 1$.

4) From (18), $-\gamma B^T P = -\gamma\,[1{,}034.8 \quad 1{,}129.8]$.

Table 1 shows the outer loop feedback matrix for different values of δ. Note that the gains increase as $\frac{1}{\delta^2}$ for smaller δ.

Table 1 Outer Loop Feedback Gains (One Link)

δ	$-(K + \gamma B^T P)$
∞	[-1,134.8 -1,149.8]
0.1	[-1,712.6 -1,780.6]
0.01	[-58,917 -64,237]
0.001	[-5,779,371 -6,309,907]

The system was simulated using the desired trajectory indicated in Figure 3 (maximum acceleration, then switch to maximum deceleration). Figures 4, 5, and 6 show position vs. time, position error vs. time, and control torque vs. time, respectively, for each of three cases. In Case 1, $\Delta \underline{v} = 0$ and there is no uncertainty; tracking error is essentially zero. In Case 2, $\Delta \underline{v} = 0$ but the actual system differs from the nominal system by the maximum amount allowed; tracking error is noticeable. In Case 3, the actual system differs from nominal as in Case 2, but the control from Table 1 associated with $\delta = 0.1$ is used; tracking error is greatly reduced. In fact, for this maneuver, δ actual is $\frac{1}{100}$ of δ guaranteed (0.1), indicating the conservatism of our design technique. Results for the three cases are given in Table 2.

Table 2 Simulation Results (One Link)

Case	$J(\ell\text{bm-ft}^2)$	$mg\ell\left(\frac{\ell\text{bm-ft}^2}{\text{sec}^2}\right)$	outerloop	$\Vert q - q^d \Vert_{max}$ (rad)	$\vert u \vert_{max}$ (ft − ℓbf)
1	5	5	$\Delta v = 0$	0.000,005,3	0.70
2	10	10	$\Delta v = 0$	0.043	1.44
3	10	10	$\delta = 0.1$	0.000,93	1.39

Example 2. Two Link Robot

A two link robot is shown in Figure 2. The equations of motion (from [1]) are

$$M(q)\ddot{q} + h(q,\dot{q}) = u$$

$$M(q) = \begin{bmatrix} (P_1 + P_2 + P_3 + 2P_4 \cos(q_2) + P_5 + P_6) & (P_3 + P_4 \cos(q_2) + P_6) \\ (P_3 + P_4 \cos(q_2) + P_6) & (P_3 + P_6) \end{bmatrix}$$

$$h(q,\dot{q}) = \begin{bmatrix} \left(-P_4 \sin(q_2)\left(2\dot{q}_1\dot{q}_2 + \dot{q}_2^2\right) + (P_7 + P_8)\cos(q_1) + P_9 \cos(q_1 + q_2)\right) \\ \left(P_4 \sin(q_2)\dot{q}_1^2 + P_9 \cos(q_1 + q_2)\right) \end{bmatrix}$$

$$P_1 = m_1 \ell_{c_1}^2 \quad P_2 = m_2 \ell_1^2 \qquad P_3 = m_2 \ell_{c_2}^2 \qquad P_4 = m_2 \ell_1 \ell_{c_2}$$

$$P_5 = J_1 \quad P_6 = J_2 \quad P_7 = m_1 \ell_{c_1} g \quad P_8 = m_2 \ell_1 g \quad P_9 = m_2 \ell_{c_2} g\,.$$

For this example we assume there is a mass m at the end of the arm which can vary as $m \varepsilon [0, 1]$. Choosing other values, we get the following terms.

$$m_1 = 20\ \ell bm \qquad m_2 = (12 + m)\ \ell bm$$

$$\ell_1 = 1.2\ ft \qquad \ell_2 = 1.0\ ft$$

$$\ell c_1 = 0.6\ ft \qquad \ell_{c_2} = \left(\frac{6+m}{12+m}\right) ft$$

$$J_1 = 2.4\ \ell bm \cdot ft^2 \qquad J_2 = \frac{4(m+3)}{(m+12)}\ \ell bm \cdot ft^2$$

We choose $\hat{M}$ and $\hat{h}$ based on the unloaded state $m = 0$. Then A and B are given by

$$A = \begin{bmatrix} 0 & 0 & 1 & 0 \\ 0 & 0 & 0 & 1 \\ 0 & 0 & 0 & 0 \\ 0 & 0 & 0 & 0 \end{bmatrix} \qquad B = \begin{bmatrix} 0 & 0 \\ 0 & 0 \\ 1 & 0 \\ 0 & 1 \end{bmatrix}.$$

For K we use

$$K = \begin{bmatrix} 100 & 0 & 20 & 0 \\ 0 & 144 & 0 & 24 \end{bmatrix}$$

(this gives eigenvalues of -10, -10, -12, -12). We will use the desired trajectories of Figure 3 for both q_1 and q_2 here.

$$\bar{E} = \max\ \left\{ \|E\| : m \varepsilon [0, 1],\ \ q_2 \varepsilon [0, 2\pi] \right\} = .707$$

$$\rho = \max \left\{ \left\| E \ddot{q}^d + M^{-1} \Delta h \right\| :\ m \varepsilon [0,1],\ q_{1,2} \varepsilon [0,2\pi]\ \dot{q}_{1,2} \varepsilon [-4,4],\ \ddot{q}^{\,d}_{1,2} \varepsilon [-4,4] \right\} = 13.2$$

Algorithm:

1) $Q_d = K^T K \bar{E}^2$

2) $Q = 10^6 I$, solve (16) for P

3) From (17), $\gamma > \dfrac{(0.23)}{\delta^2} + 1$

4) Calculate (18)

Table 3 shows the outer loop feedback matrix for different values of δ.

The system was simulated using the desired trajectory indicated in Figure 3 for both angles. Figures 7, 8, and 9 show norm of position error vs. time, control torque 1 vs. time, and control torque 2 vs. time, respectively, for each of three cases.

Table 3 Outer Loop Feedback Gains (Two Link)

δ	$-(K + \gamma B^T P)$
∞	$\begin{bmatrix} -1,368.3 & 0 & -1,398.0 & 0 \\ 0 & -1,380.5 & 0 & -1,394.9 \end{bmatrix}$
0.1	$\begin{bmatrix} -30,216 & 0 & -32,741 & 0 \\ 0 & -29,503 & 0 & -32,577 \end{bmatrix}$
0.01	$\begin{bmatrix} -2,886,159 & 0 & -3,135,683 & 0 \\ 0 & -2,813,685 & 0 & -3,119,594 \end{bmatrix}$

In Case 1, $\Delta \underline{v} = 0$ and $m = 0$ (no uncertainty); tracking error is essentially zero. In Case 2, $\Delta \underline{v} = 0$ but $m = 1$; tracking error is noticeable. In Case 3, $m = 1$, but the control from Table 3 associated with $\delta = \infty$ is used; tracking error is greatly reduced. Notice that, according to theory, δ actual may be very large with this set of gains, yet for this maneuver δ actual < 0.01, again indicating the conservatism in our design technique. Results for the three cases are given in Table 4.

Table 4 Simulation Results (Two Link)

Case	$m(\ell bm)$	outerloop	$\|q - q^d\|_{max}$ (rad)	$\|u_1\|_{max}$ $(ft - \ell bf)$	$\|u_2\|_{max}$ $(ft - \ell bf)$
1	0	$\Delta v = 0$	0.000,006,7	39.4	7.89
2	1	$\Delta v = 0$	0.11	43.0	9.50
3	1	$\delta = \infty$	0.004,2	42.5	9.29

DISCUSSION

Both examples demonstrated the conservatism of the design technique in that the actual tracking radius was much less than the guaranteed tracking radius for a given feedback matrix in the outer loop. Therefore the following revised procedure is recommended.

1) Calculate $\bar{E}$ only (not ρ).
2) With $\rho = 0$ and $\gamma = 1$, calculate $-\gamma B^T P$ (i.e. the $\delta = \infty$ case). This feedback guarantees δ-stability for some undetermined δ.
3) Simulate the desired maneuver with a variety of uncertainties, increasing γ until the desired δ is achieved (increase γ approximately as $\frac{1}{\delta^2}$ according to (17)).

Advantages of this revised method are:

1) Avoids the conservatism of using a large γ.
2) Avoids calculation of ρ (which is very complicated since it is a function of $\dot{q}$).

The Riccati method of [3], though not used completely in the revised method, allows us to conclude the system is δ - stable for some δ which is not exactly known, but which can be accurately estimated by simulation.

CONCLUSIONS

We have investigated a robotic manipulator with unknown but bounded parameters using an inverse dynamics control scheme. The nonlinear inner loop control is based on a nominal system model. The linear outer loop is designed to be robust against the uncertainties. The combined system achieves practical tracking. Calculations are straightforward, however we recommend simulation to avoid using conservatively large gains.

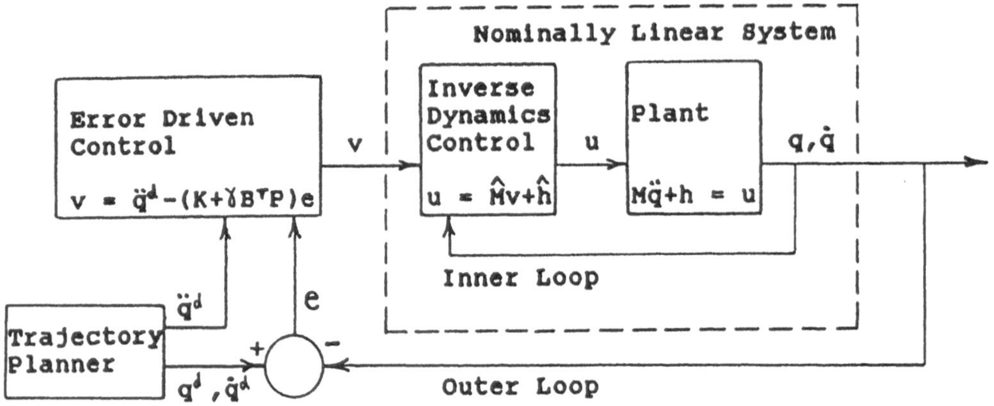

Figure 1 Robust Trajectory Control: Inner/Outer Loop Configuration

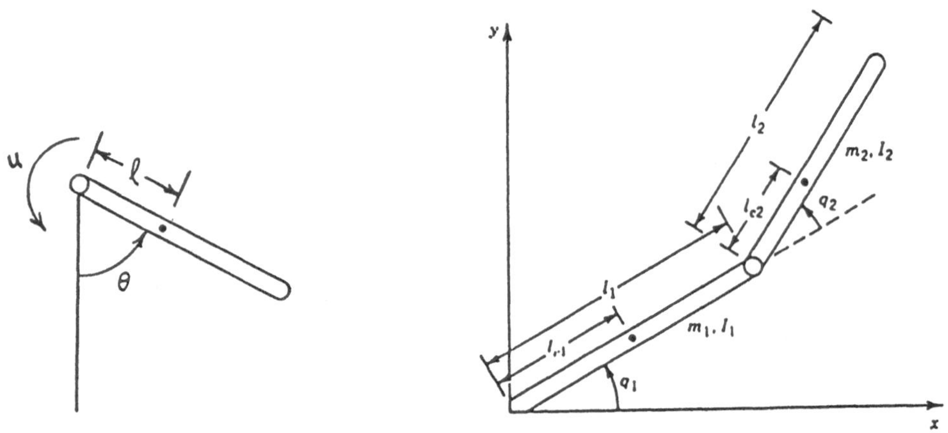

Figure 2 One Link and Two Link Robots

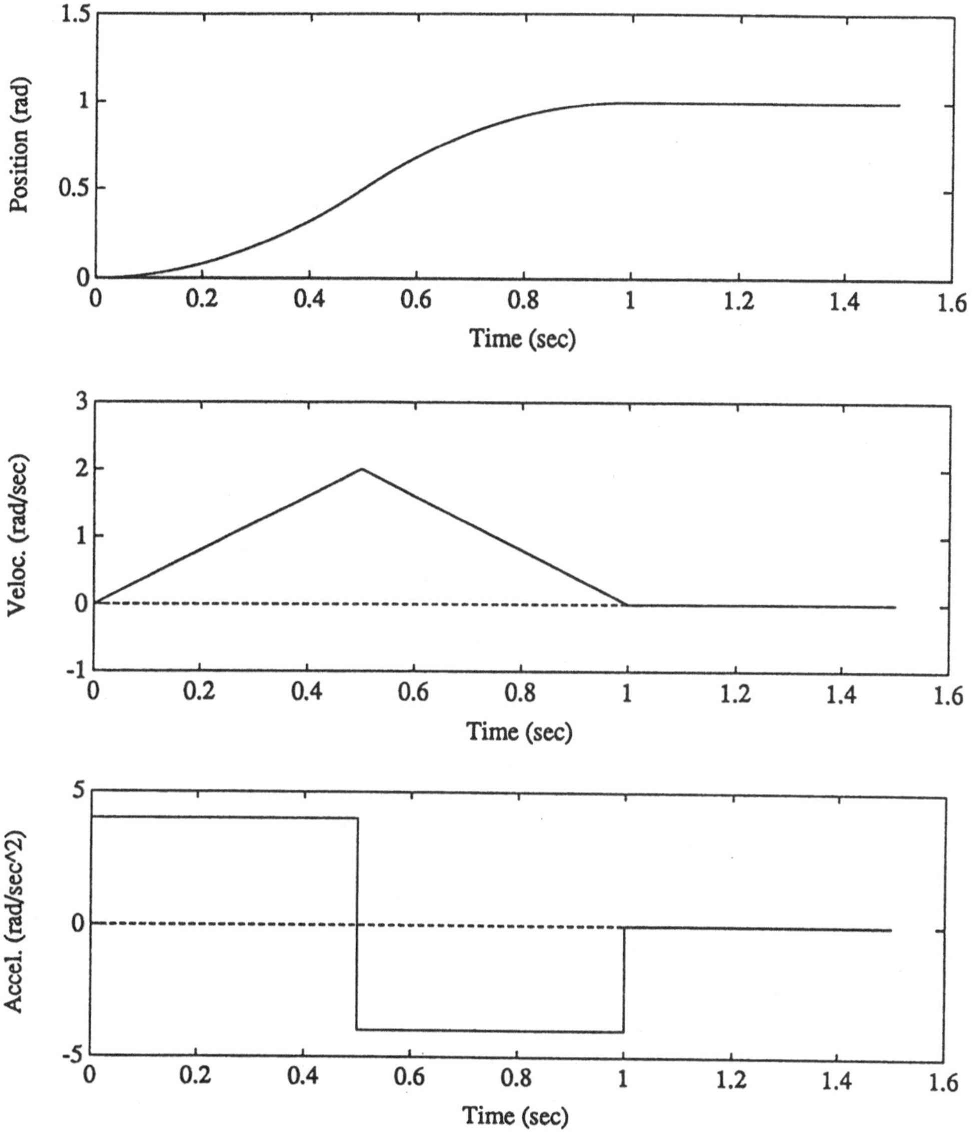

$$q^d = \begin{cases} 2t^2 & , \ 0 \le t < 0.5 \\ 1 - 2(1-t)^2 & , \ 0.5 \le t < 1 \\ 1 & , \ 1 \le t \end{cases}$$

Figure 3 Desired Trajectory

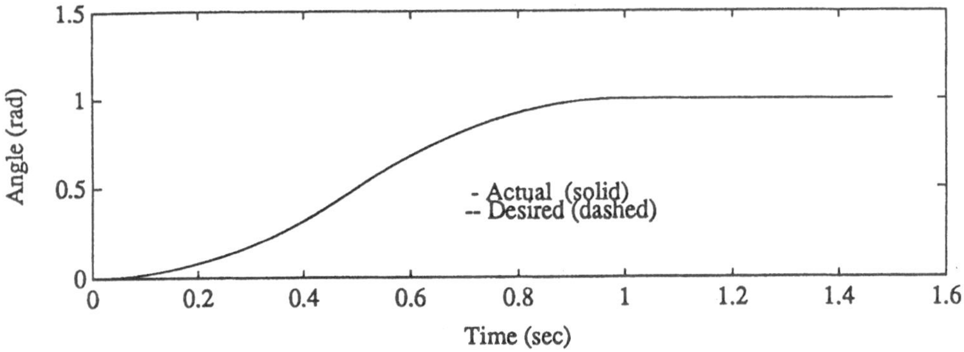

Figure 4(a) One Link, Case 1, Position vs. Time

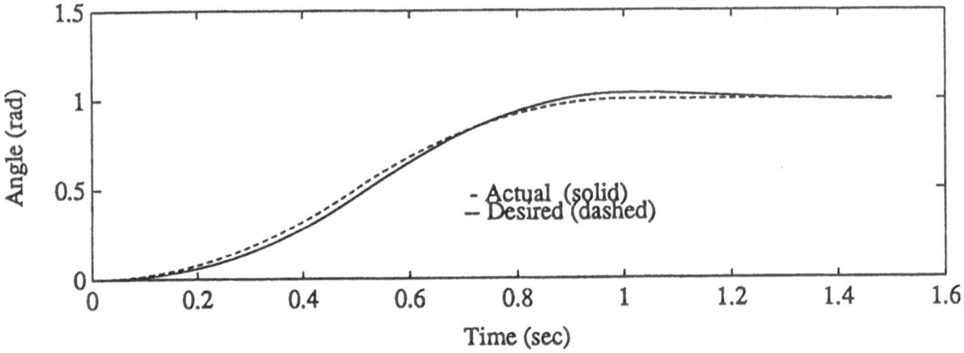

Figure 4(b) One Link, Case 2, Position vs. Time

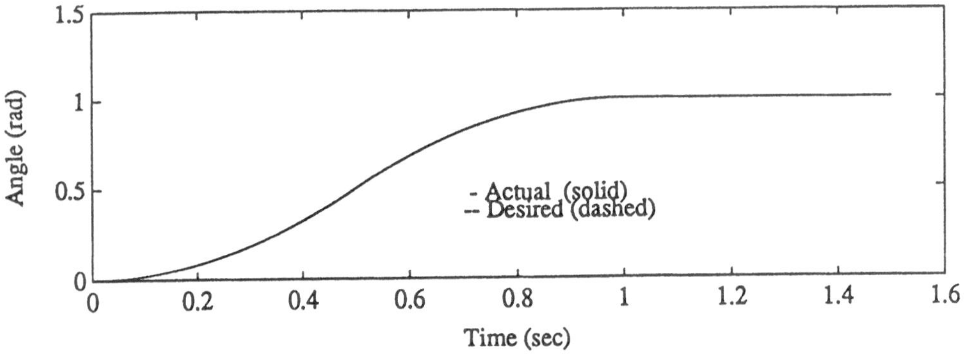

Figure 4(c) One Link, Case 3, Position vs. Time

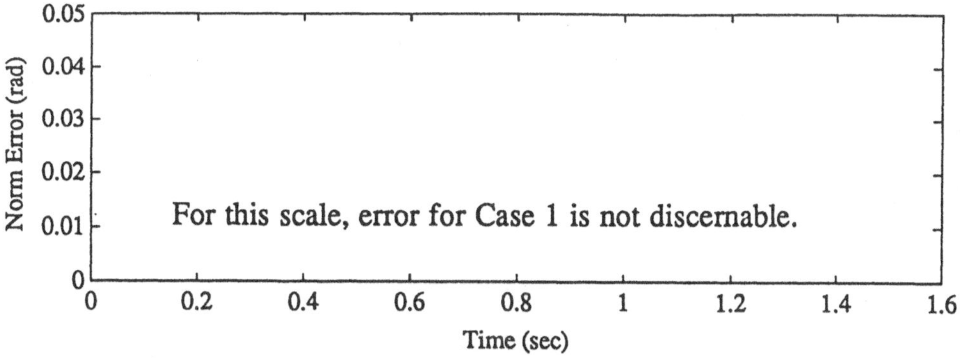

Figure 5(a) One Link, Case 1, Position Error vs. Time

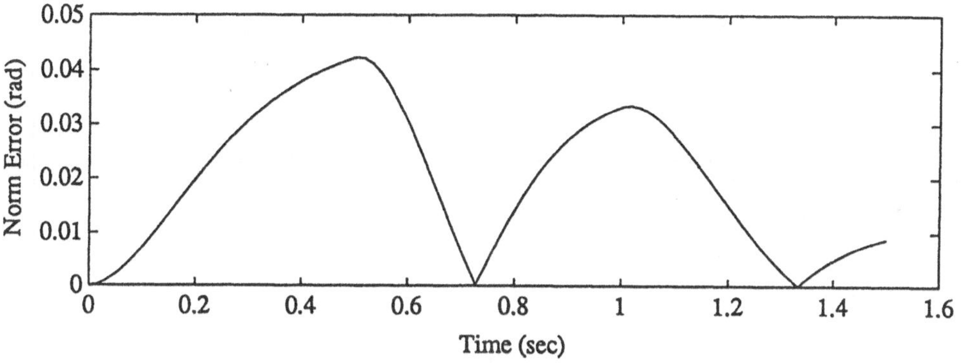

Figure 5(b) One Link, Case 2, Position Error vs. Time

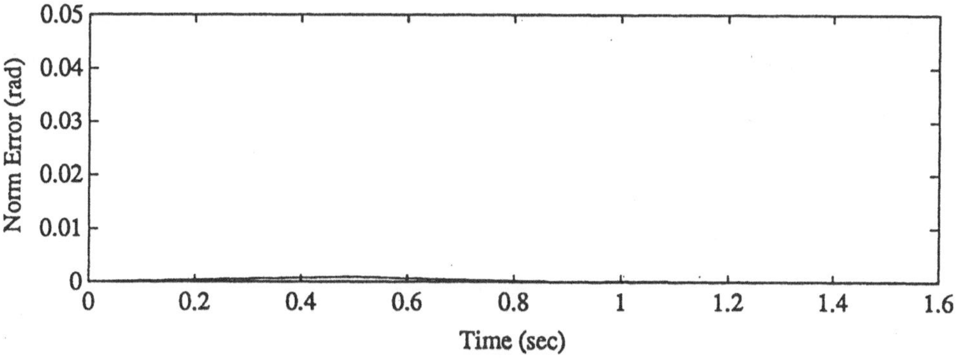

Figure 5 (c) One Link, Case 3, Position Error vs. Time

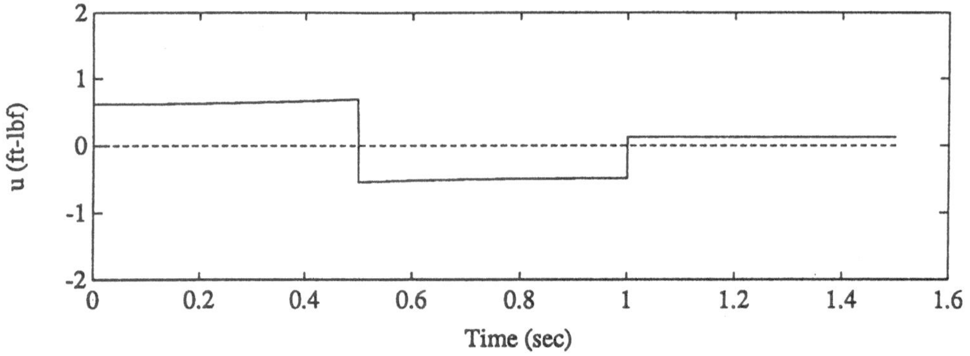

Figure 6(a) One Link, Case 1, Control Torque vs. Time

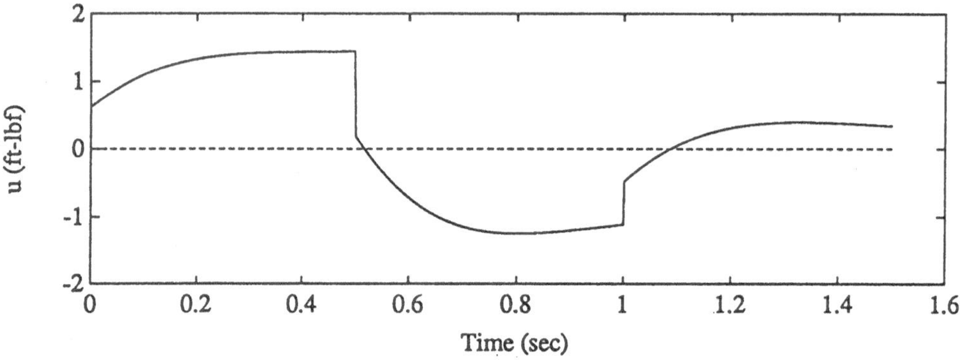

Figure 6(b) One Link, Case 2, Control Torque vs. Time

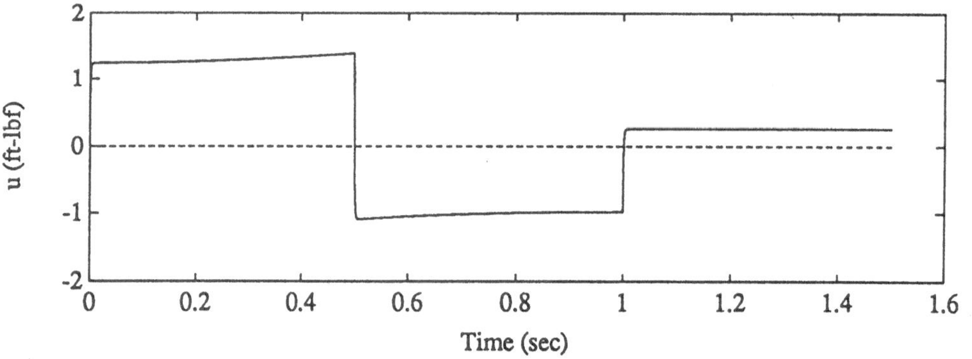

Figure 6(c) One Link, Case 3, Control Torque vs. Time

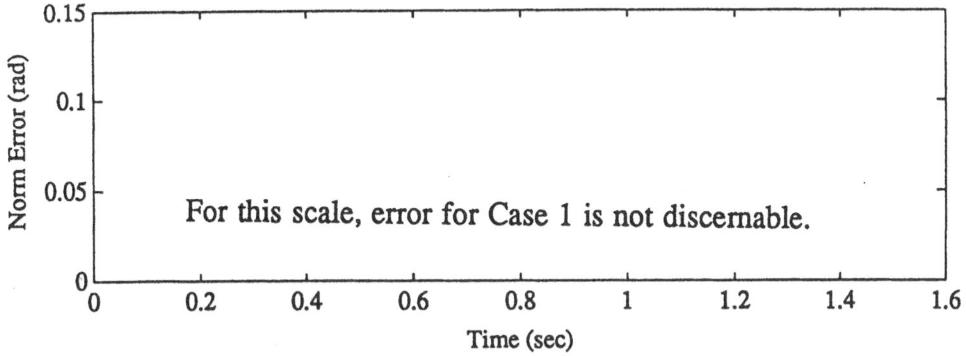

Figure 7(a) Two Link, Case 1, Position Error vs. Time

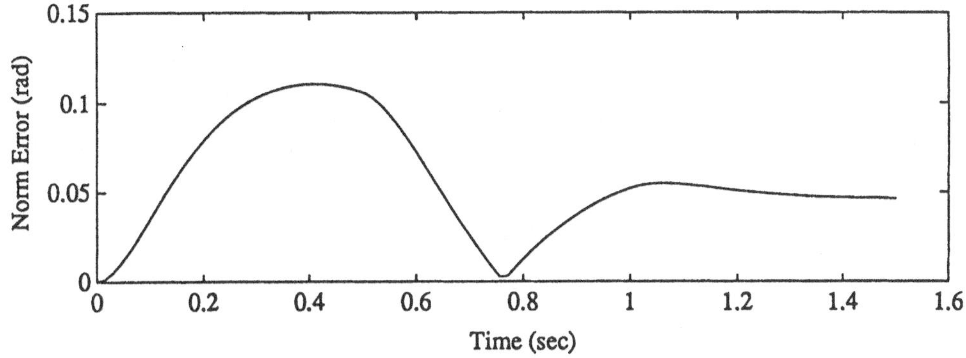

Figure 7(b) Two Link, Case 2, Position Error vs. Time

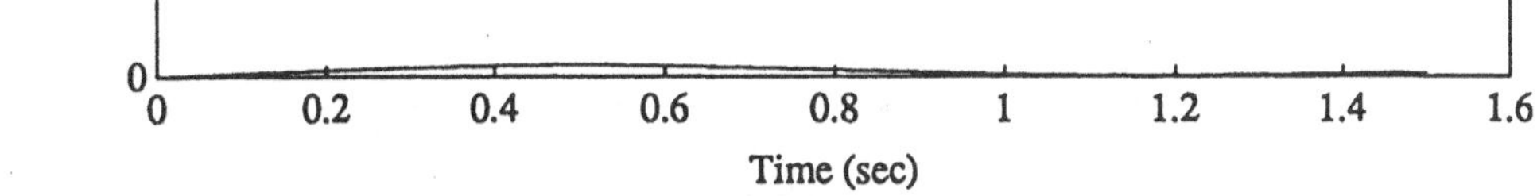

Figure 7(c) Two Link, Case 3, Position Error vs. Time

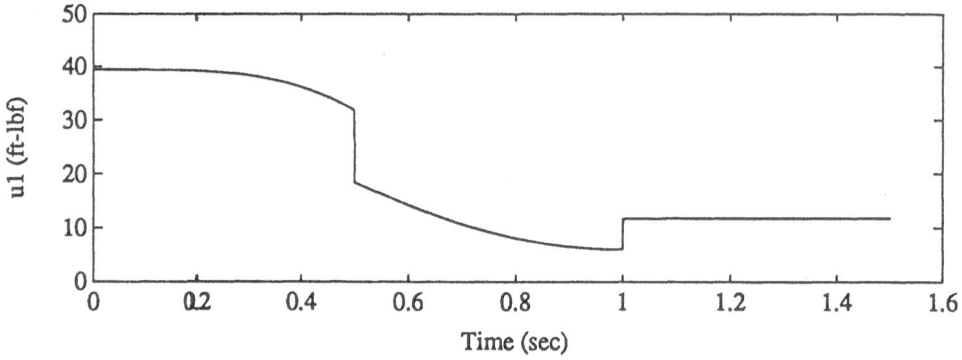

Figure 8(a) Two Link, Case 1, Control Torque 1 vs. Time

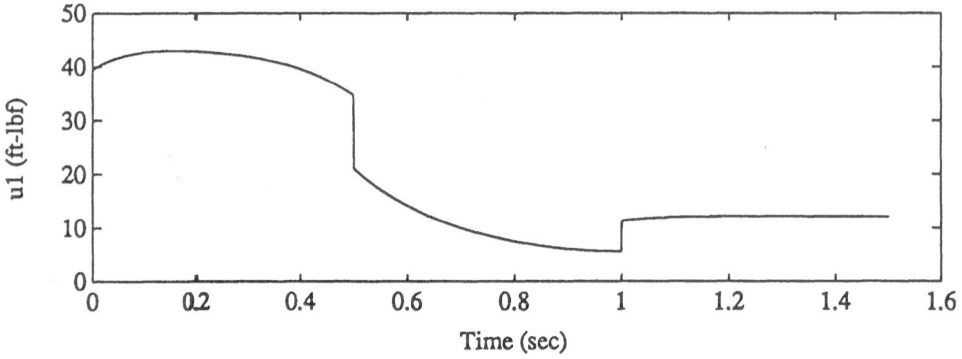

Figure 8(b) Two Link, Case 2, Control Torque 1 vs. Time

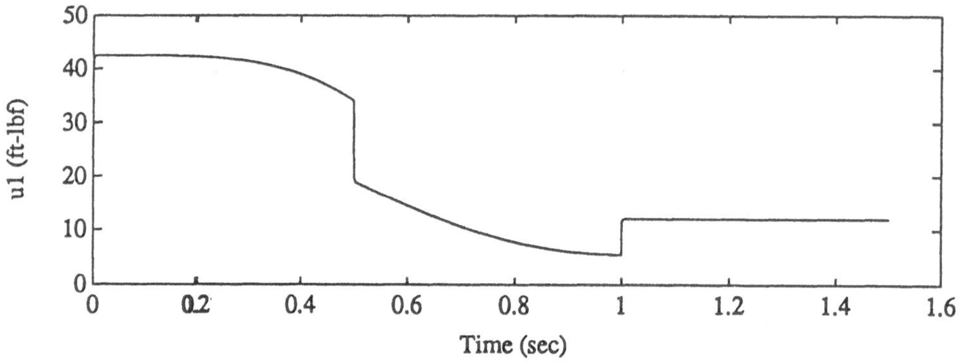

Figure 8(c) Two Link, Case 3, Control Torque 1 vs. Time

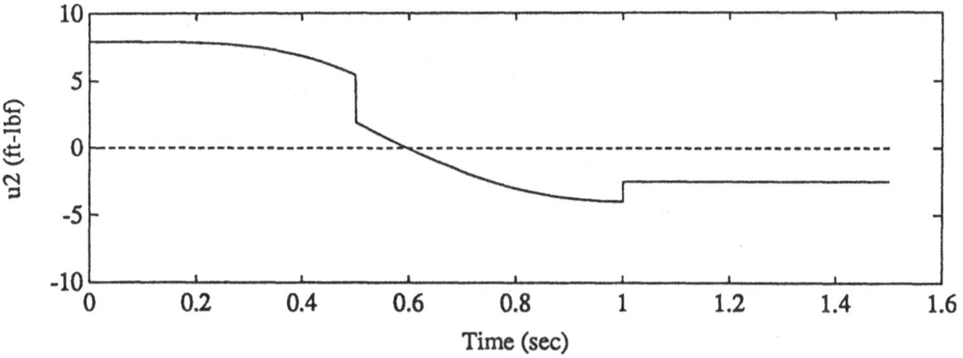

Figure 9(a) Two Link, Case 1, Control Torque 2 vs. Time

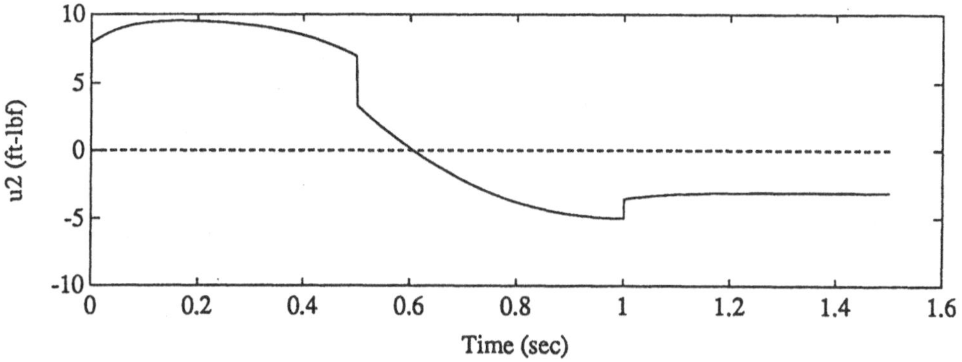

Figure 9(b) Two Link, Case 2, Control Torque 2 vs. Time

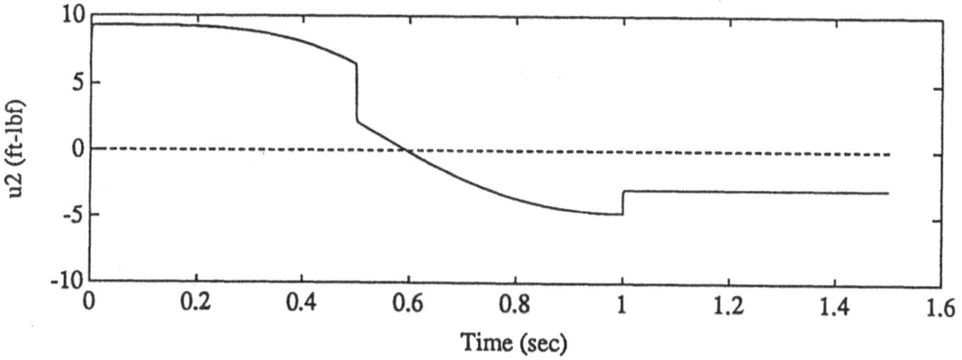

Figure 9(c) Two Link, Case 3, Control Torque 2 vs. Time

REFERENCES

[1] Spong, M.W. & Vidyasagar, M., Robot Dynamics and Control, Wiley, 1989.

[2] Asada, H. & Slotine, J.J.E., Robot Analysis and Control, Wiley, 1986.

[3] Dolphus, R.M. & Schmitendorf, W.E., "A Non-Iterative Riccati Approach to Robust Control Design," Proceedings American Control Conference, San Diego, May 1990.

[4] Leitmann, G., "On the Efficacy of Nonlinear Control in Uncertain Linear Systems," Journal of Dynamic Systems, Measurement, and Control, Vol. 103, 1981.

[5] Spong, M.W. & Vidyasagar, M., "Robust Linear Compensator Design for Nonlinear Robotic Control," IEEE Journal of Robotics and Automation, Vol. RA-3, No. 4, Aug. 1987.

APPENDIX A

A Non-Iterative Riccati Approach to Robust Control Design

The following summary is from [3]. Consider a system described by

$$\dot{x} = [A + \Delta A(r)]x + [B + \Delta B(s)]u + Hv \tag{A1}$$

where $x \varepsilon \Re^n$ and $u \varepsilon \Re^m$. We have uncertainties $r(t) \varepsilon R \subset \Re^{q_1}$, $s(t) \varepsilon\ s \subset \Re^{q_2}$, and $v(t) \varepsilon V \subset \Re^{q_3}$ where R, S, and V are compact sets and r(·), s(·), and v(·) are assumed to be Lebesgue measurable. We further assume that (A, B) is a controllable pair, the matrix functions ΔA(·) and ΔB(·) are continuous, and the following matching conditions are satisfied.

$$\Delta A(r) = BD(r) \tag{A2a}$$

$$\Delta B(s) = BE(s)\,, \qquad \bar{E} \triangleq \max\ \{\|E(s)\|: s\varepsilon\ S\} < 1 \tag{A2b}$$

$$H = BF. \tag{A2c}$$

We wish to find a linear δ - stabilizing control. Consider the choice

$$u = -\gamma B^T P x \tag{A3}$$

where γ and P are to be determined. Combining (A1), (A2), and (A3) gives

$$\dot{x} = (A + BD)x - (B + BE)\gamma B^T Px + BFv\ . \tag{A4}$$

We choose a candidate Lyapunov function $V = x^T Px$ and take its derivative

$$\begin{aligned}\dot{V} &= x^T\left(PA + A^TP - 2\gamma PBB^TP + PBD + D^TB^TP\right)x \\ &+ \gamma x^TPB\left(-E - E^T\right)B^TPx + x^TPBFv + v^TF^TB^TPx\ .\end{aligned} \tag{A5}$$

After various manipulations we obtain

$$\begin{aligned}\dot{V} &\le x^T\left(PA + A^TP - PBB^TP\left[2\gamma(1-\bar{E}) - (1-\bar{E})^2\right] + \frac{D^TD}{(1-\bar{E})^2}\right)x \\ &+ x^TPBFv + v^TF^TB^TPx\ .\end{aligned} \tag{A6}$$

If we solve the following Riccati equation for P (Q > 0)

$$PA + A^TP - PBB^TP\left(1-\bar{E}^2\right) + \frac{Qd}{(1-\bar{E})^2} + Q = 0 \tag{A7}$$

where $Q_d \ge D^TD$ for all rεR, then we can simplify (A6) further and get

$$\dot{V} \le -x^TQx + \frac{\rho^2}{2(\gamma-1)(1-\bar{E})} \tag{A8}$$

where $\rho \triangleq \max \{\|Fv\|: v \in V\}$. It can be shown using (A8) that the trajectory x(t) will eventually reach and stay inside a sphere of radius δ where

$$\delta > \sqrt{\left(\frac{\bar{\lambda}_P}{\lambda_P}\right)\frac{\rho^2}{\lambda_Q 2(\gamma-1)(1-\bar{E})}} \tag{A9}$$

Thus we have the following design procedure:

1) Determine D, E, F, $\bar{E}$, ρ, Q_d.
2) Choose $Q > 0$, solve Riccati (A7) for P.
3) For a given δ, choose

$$\gamma > \left(\frac{\bar{\lambda}_P}{\lambda_P}\right)\frac{\rho^2}{2\lambda_Q \delta^2(1-\bar{E})} + 1 \tag{A10}$$

4) The control (A3) is a δ-stabilizing control law.

NONLINEAR ADAPTIVE CONTROL OF ROBOTIC MANIPULATORS

H. Flashner and J. M. Skowronski
University of Southern California
Department of Mechanical Engineering
Los Angeles, California 90089-1453

ABSTRACT

We consider n-DOF manipulator arm with untruncated nonlinearity (Coriolis forces, nonreduced coupling) subject to bounded uncertainty in inertia, payload and gravity. The latter is compensated adaptively by varying the feedback controller, whose objective is to track a generally nonlinear reference model. The classical linear MRAC technique is augmented to serve the case concerned, under the condition that the system can be represented in the damped and forced (controlled) Hamiltonian format. The tracking controllers are designed using Liapunov formalism with a specified Liapunov function. A case study of a simple DOF manipulator illustrates the results. Key Words: Adaptive tracking, Hamiltonian modeling, Liapunov design.

INTRODUCTION

High speed of work, large angles of articulation (several equilibria), flexibility of links and increased demand for precision of robotic manipulators make the designer unable to avoid nonlinear modeling and adaptive control. Thus, one of the basic problems in robotics, namely tracking, requires adjustments done to the, so far commonly used, linear techniques. In particular, the presently popular MRAC technique subtracts the plant and reference-model equations to obtain the so called error equation, and then makes its zero-error solution asymptotically stable, see [1]. For nonlinear systems, with untruncated nonlinearitiy, such subtraction is in general not possible. Thus, a few years ago a technique was introduced avoiding the error equation by investigating the stability of a diagonal set in the product of the state spaces of plant and model. See [2], [3]. This technique expanded the dimensions of the system and required reduced observers in order to keep the computation time acceptable. It occurs however, that on the expense of turning the widely used Lagrangian representation of manipulator dynamics into the equivalent Hamiltonian format we may still obtain the error equation for highly nonlinear systems and moreover, have the added option of automatic inertial decoupling. The nonadaptive version of such approach had been published in [4]. We propose now to extend the study by making the controller adaptive and thus robust to bounded uncertainty in inertia, payload and gravity.

THE MOTION EQUATIONS

The equations of motion of the manipulator are given by

$$M(\overline{q},\overline{w})\ddot{\overline{q}}+\overline{Q}^{C}(\overline{q},\dot{\overline{q}})+\overline{Q}^{D}(\overline{q},\dot{\overline{q}},\overline{\lambda})+\overline{Q}^{P}(\overline{q},\overline{\lambda},\overline{w},)=\overline{Q}^{F}(\overline{q},\dot{\overline{q}},\overline{u}) \tag{1}$$

where $\overline{q}(t),\dot{\overline{q}}(t)$ are n-vectors of the lagrangian joint displacement and velocity variables, ranging in bounded work regions $\Delta_q,\Delta_{\dot{q}}$ respectively for $t \geq 0$ with $\Delta=\Delta_q\times\Delta_{\dot{q}}$ being the work envelope in the phase space $\Re^N, N=2n$. Moreover $\overline{w}(\ell)\in\dot{W}\subset\Re^{s}$ is an unknown uncertainty vector within the known band W and $\overline{\lambda}(t)\in\Lambda\subset\Re^{\ell}$ is a vector of adjustable parameters in a bounded Λ. The control variables form the vector $\overline{u}(t)\in U\subset\Re^{n}$ within bounded set of constraints U, all defined for $t\geq 0$. The matrix $M\in\Re^{n\times n}$ represents inertia and $\overline{Q}^C,\overline{Q}^D,\overline{Q}^P$ are the lagrangian forces of the Coriolis-gyro type, damping and potential, respectively. The vector $\overline{Q}^F$ represents the gearbox characteristics of the actuators collocated with joints.

Equation (1) can be written in the Hamiltonian state format by introducing the momentum vector $\overline{p}(t)$ via the Legendre's transformation:

$$\dot{q}=\frac{\partial H(\overline{q},\overline{p},\overline{w},\overline{\lambda})}{\partial p_i} \tag{2}$$

$$\dot{p}_i=-\frac{\partial H(\overline{q},\overline{p},\overline{w},\overline{\lambda})}{\partial q_i}+Q_i^D(\overline{q},\overline{p},\overline{w},\overline{\lambda})+Q^F(\overline{q},\overline{p},\overline{u})$$

with the stationary Hamiltonian H being the total energy of the system.

The vectors $\overline{\lambda}(t),\overline{u}(t)$ are generated successively by the specifically designed later adaptive law

$$\dot{\overline{\lambda}}=\overline{g}(\overline{q},\overline{p},\overline{\lambda}) \tag{3}$$

and the set valued feedback control program $P(\cdot):\Delta\times\Lambda\rightarrow$ subsets of U, defined by

$$\bar{u}(t) \in P(\bar{q}(t), \bar{p}(t), \bar{\lambda}(t)), t \geq 0 \tag{4}$$

accommodating possible discontinuities of $\bar{u}(\cdot)$.

Let $\Delta_L \subset \Delta$ enclose all equilibria and define the set in-the-large $C\Delta_L = \Delta - \Delta_L$. It will be assumed that given $\bar{w}, \bar{\lambda}$,

$$grad\, H(\bar{q}, \bar{p})^T \cdot \bar{q} > 0, \quad \forall \bar{q}, \bar{p} \in C\Delta_L \tag{5}$$

The known reference model to be tracked is given as

$$\dot{q}_{mi} = \frac{\partial H_m(\bar{q}_m, \bar{p}_m, \bar{\lambda}_m)}{\partial p_{mi}}$$

$$\dot{p}_{mi} = -\frac{\partial H_m(\bar{q}_m, \bar{p}_m, \bar{\lambda}_m)}{\partial p_{mi}} + Q^D_{mi}(\bar{q}_m, \bar{p}_m, \bar{\lambda}_m) \tag{6}$$

with $\bar{q}_m(t), \bar{p}_m(t)$ being the model displacement and momentum vectors ranging in Δ. The Hamiltonian equals to the model total energy and $\bar{Q}^D_m$ is the model damping force. We define the misalignments in displacements and moments as $\bar{e}_q(t) = \bar{q}(t) - \bar{q}_m(t), \bar{e}_p(t) = \bar{p}(t) - \bar{p}_m(t)$ and let $\bar{e}_\lambda(t) = \lambda(t) - \lambda_m$, with $\lambda_m \in \Lambda$, a known constant. Then we form the sum of the Hamiltonians concerned

$$H_e(\bar{e}_q, \bar{e}_p, \bar{q}_m, \bar{p}_m \bar{w}, \bar{e}_\lambda) = H(\bar{q}, \bar{p}, \bar{w}, \bar{\lambda}) + H_m(\bar{q}_m, \bar{p}_m \bar{\lambda}_m) \tag{7}$$

which is a function of $\bar{e}_q, \bar{e}_p$ through $\bar{q} = \bar{e}_q + \bar{q}_m, \bar{p} = \bar{e}_p + \bar{p}_m$, with parameters $\bar{w}, \bar{e}_\lambda$, and $\bar{q}_m, \bar{p}_m, \bar{\lambda}_m$, given by design. We denote

$$H^-(\bar{q}, \bar{p}) = \inf_{(\bar{w}, \bar{e}_\lambda)} + H(\bar{q}, \bar{p}, \bar{w}, \bar{e}_\lambda), \tag{8}$$

$$H_e^-(\bar{e}_q, \bar{e}_p) = \inf_{(\bar{w}, \bar{e}_\lambda)} H(\bar{e}_q, \bar{e}_p, \bar{w}, \bar{e}_\lambda). \tag{9}$$

Subtracting (6) from (1) we have

$$\left.\begin{aligned} \dot{e}_{qi} &= \frac{\partial H(\bar{q}, \bar{p}, \bar{\lambda}, \bar{w})}{\partial p_i} - \frac{\partial H_m(\bar{q}_m, \bar{p}_m, \bar{\lambda}_m)}{\partial p_{mi}} \\ \dot{e}_{pi} &= \frac{\partial H(\bar{q}, \bar{p}, \bar{\lambda}, \bar{w})}{\partial q_i} + Q^D_{mi}(\bar{q}, \bar{p}, \bar{\lambda}, \bar{w}) + Q^F_{mi}(\bar{q}, \bar{p}, \bar{u}) + \frac{\partial H_m(\bar{q}_m, \bar{p}_m, \bar{\lambda}_m)}{\partial q_{mi}} - Q^D_{mi}(\bar{q}_m, \bar{p}_m, \bar{\lambda}_m) \end{aligned}\right\} \tag{10}$$

Considering now that $q_i = e_{qi} + q_{mi}$, $p_i + e_{pi} + p_{mi}$ and $q_{mi} = q_i + e_{qi}$, $p_i - e_{pi}$ we have

$$\frac{\partial H_e}{\partial e_{qi}} = \frac{\partial H}{\partial q_i} - \frac{\partial H_m}{\partial q_{mi}}, \quad \frac{\partial H_e}{\partial e_{pi}} = \frac{\partial H}{\partial p_i} - \frac{\partial H_m}{\partial p_{mi}},$$

which makes (10) into the error equations

$$\left.\begin{aligned} \dot{e}_{qi} &= \frac{\partial H_e}{\partial e_{pi}} \\ \dot{e}_{pi} &= -\frac{\partial H_e}{\partial e_{qi}} + \left(Q_i^D - Q_{mi}^D\right) + Q_i^F \end{aligned}\right\} \qquad (11)$$

$$i = 1, \ldots, n$$

where all the functions have arguments $\overline{e}_q, \overline{e}_p$ and parameters $\overline{w}, \overline{e}_\lambda, \overline{q}_m, \overline{p}_m$ the latter two known.

THE TRACKING OBJECTIVE

Introducing the error state vector $\overline{e}(t) = \left(e_q(t), e_p(t), e_\lambda(t)\right)^T \in \Delta \times \Lambda \subset \Re^{N+\ell}$ and the vector

$$\overline{f} = \left(f_1, \ldots, f_{N+\ell}\right)^T, \text{with} f_i = \frac{\partial H_e}{\partial e_{pi}}, f_{n+i} = \frac{\partial H_e}{\partial e_{qi}} + \left(Q_i^D + Q_{mi}^D\right) + Q_i^F, f_{2n+j} = g_j, \subset n = 1, \ldots, n, j = 1, \ldots,$$

we can write (11) and (3) jointly in the contingent form

$$\dot{\overline{e}} \in \left\{ f(e, \overline{u}, \overline{w}) \middle| \overline{u} \in P(e), \overline{w} \in W \right\} \qquad (12)$$

Following Filippov [5], for suitable $f(\cdot), P(\cdot)$ through each $\overline{e}^0 = \overline{e}(0)$ the equation (12) has absolutely continuous solutions $\phi(\overline{e}^0, \cdot)$: $\Re \to \Delta \times \Lambda$ and there is a choice of $\overline{u}(\cdot), \overline{w}(\cdot)$ such that the corresponding solution satisfies the set of selector equations (11), (3). Given $\overline{e}^0$ the curves $\phi(t)$ are recognized as product trajectories of (11), (3) in $\Delta \times \Lambda$, forming a class $K(\overline{e}^0)$. Given $P(\cdot)$, the set $\Phi(\overline{e}^0, P, t) = \left\{ \phi(\overline{e}^0, t) \phi \in K \right\}$ is called the attainability set from $\overline{e}^0$ at t. We define now an open target set about {0}:

$$M_\mu = \left\{ \overline{e} \in \Delta \times \Lambda \middle| \|\overline{e}\| < \mu \right\} \qquad (13)$$

where $\mu < 0$ is a stipulated estimate for tracking precision, and $\|\cdot\|$ is any norm in $\Re^{N+\ell}$.

Definition

Given some $\Delta_0 \subset \Delta$ and μ, the system (2) tracks the model (6) on Δ_0 with stipulated precision μ, if and only if there is a controller $P(\cdot)$ and a law (3) such that

(a) The system (12) is positively Lagrange stable in $\Delta_0 \times \Lambda$:

$$\bar{e}^0 \in \Delta \times \Lambda \Rightarrow \Phi(\bar{e}^0, P, t) \subset \Delta_0 \times \Lambda, \forall t \geq 0$$

(b) There is a constant $T < \infty$ such that $\bar{e}^0 \in \Delta \times \Lambda \Rightarrow \Phi(\bar{e}^0, P, t) \subset M_\mu, \forall t \geq T$.

The time interval T after which the tracking is effective may be stipulated. Then we qualify the objective as "tracking after T".

Define now $CM_\mu = (\Delta_0 \times \Lambda) - M_\mu$ and the envelope $S \supset CM_\mu, S \cap \{0\} = \phi$.. Then let $V(\cdot): S \to \Re$ be a C^1 - function and let $V_0 = \inf V(\bar{e}) | \bar{e} \in \partial(\Delta_0 \times \Lambda)$.. Specifying $\|\cdot\|$ in (13) by $V(\cdot): M_\mu = \{\bar{e} | V(\bar{e}) < \mu\}$, we have

Tracking Conditions Definition holds, if given Δ_0, μ,, there are $P(\cdot), V(\cdot)$ such that for all $\bar{e} \in S$,

(i) $0 \leq V(\bar{e}) \leq v_0, \forall \bar{e} \in \overline{CM}_\mu$;

(ii) $V(\bar{e}) \leq \mu, \forall \bar{e} \in S \cap \overline{M}_\mu$;

(iii) $\forall \bar{u} \in P(\bar{e})$ there is $T > 0$ such that $\text{grad } V(\bar{e})^T \cdot \bar{f}(\bar{e}, \bar{u}, w) \leq -\frac{v_0 - \mu}{T}$, $\forall \bar{w} \in W$

The conditions follow by the same argument as the conditions for capture proved in [3]. Note that T may be stipulated.

THE CONTROLLERS AND ADAPTIVE LAWS

Let us now choose $V(\bar{e}) = H_e^-(\bar{e}_q, \bar{e}_p) + \bar{a}\bar{e}_\lambda$ where $\bar{a} = (\text{sgn } e_{\lambda 1}, \ldots, \text{sgn } e_{\lambda e})^T, \bar{e}_\lambda \neq 0$. Obviously the case $\bar{e}_\lambda = 0$ does not require adaptation. Then we define $\partial(\Delta_0 \times \Lambda): V(e) = v_0 = h_e^+ + e_\lambda^+$, where h_e^+, e_λ^+ are the upper bounds on H_e, e_λ respectively. With the assumed $V(\cdot)$, conditions (i),(ii) are obviously satisfied. Since $\bar{q}_m(t), \bar{p}_m(t), \bar{\lambda}_m$ and the extremizing $\bar{\lambda}^*, \bar{w}^*$ in $H_e^-(\cdot)$ are known values, the formal differentiation gives

$$\dot{V}(\bar{e}) = \sum_{i=1}^{n}\left[\left(\bar{Q}_i^F + \bar{Q}_i^D - \bar{Q}_{mi}^D\right)\dot{e}_{qi}\right] + F + \bar{a}\bar{e}_\lambda$$

where

$$F(\bar{e}_q, \bar{e}_q) = \sum_{i=1}^{n}\left(\frac{\partial H^-}{\partial q_{mi}}\frac{\partial H_m}{\partial p_{mi}} - \frac{\partial H^-}{\partial p_{mi}}\frac{\partial H_m}{\partial q_{mi}} + \frac{\partial H_e^-}{\partial p_{mi}}Q_{mi}^D\right)$$

is the known perturbation due to influence of known parameters $\bar{q}_m(t), \bar{p}_m(t)$. In order to satisfy condition (iii) we need

$$\min_{\bar{u}} \max_{\bar{w}} \sum_i \left[\left(Q_i^F + Q_i^D - Q_{mi}^D\right)\dot{e}_{qi}\right] + F + \bar{a}\dot{\bar{e}}_\lambda \le -\frac{h_e^t + e_\lambda^t - \mu}{T}$$

This is implied by the following conditions

(I) Control Condition

$$\min_{\bar{u}} \max_{\bar{w}}\left(\bar{Q}^F + \bar{Q}^D\right) = \bar{Q}_m^D, \forall \dot{e}_{qi} \ne 0$$

(II) Adaptation Condition

$$\bar{a}\dot{\bar{e}}_\lambda \le -\left(\frac{h_e^+ + e_\lambda^+ - \mu}{T} + |F|\right)|\bar{e}_\lambda|, \forall \bar{e}_\lambda \ne 0$$

The first is implied by the control program

$$\min_{\bar{u}} \max_{\bar{w}} \bar{Q}^F(\bar{q}, \bar{p}, \bar{u}, \bar{w}) = \begin{cases} \bar{Q}^F(\bar{q}_m, \bar{p}_m, \lambda_m) - \max_{\bar{w}} \bar{Q}^D(\bar{q}, \bar{p}, \bar{\lambda}, \bar{w}), \forall |\dot{e}_q| \ge \beta \\ 0, |\dot{\bar{e}}_q| < \beta \end{cases}$$

with $\beta > 0$ sufficiently small in order to secure smooth instantaneous crossing of the surface $\dot{\bar{e}}_q = 0$. Note that $\bar{a}\dot{\bar{e}}_\lambda \le -(c + |F|)|e_\lambda|$ implies $\bar{a}\dot{\bar{e}}_\lambda \le -(c + |F|)|e_\lambda|$ which implies $\bar{a}\dot{\bar{e}}_\lambda \le -(c + F)$, since $(c + |F|)|\bar{e}_\lambda| > c + F$, where $c = (h_e^+ + e_\lambda^+ - \mu)/T$. Hence the adaptation condition (II) is implied by the adaptive laws

$$\dot{e}_{\lambda i} = e_\lambda\left[|F| + \frac{h_e^+ + e_\lambda^+ - \mu}{T}\right], e_{\lambda i}^0 \ne 0$$
$$i = 1, \ldots, \ell$$

EXAMPLE

Consider the single DOF manipulator $m\ddot{q} + w\dot{q} + \lambda(q - 2q^3 + q^5) = u$ with $\lambda, w \in [1,2]$ and let the reference model be $\ddot{q}_m + 5\dot{q}_m + 2(q_m - 2q_m{}^3 + q_m{}^5) = 0$. The Hamiltonian format of these equations becomes

$$\dot{q} = p/m$$
$$\dot{p} = -\lambda(q - 2q^3 + q^5) + (wp/m) + u$$

with

$$H = (p^2/2m) + \lambda\left[\frac{1}{2}q^2 - \frac{1}{2}q^4 + \left(\frac{1}{6}\right)q^6\right],$$

and

$$\dot{q}_m = p_m$$
$$\dot{p}_m = -2(q_m - 2q_m{}^3 + q_m{}^5) + 5p_m,$$

with

$$H_m = \frac{1}{2}p_m{}^2 + 2\left[\frac{1}{2}q_m{}^2 - \frac{1}{2}q_m{}^4 + \left(\frac{1}{6}\right)q_m{}^6\right]$$

Considering the potential energy in H positive definite, we have the extremizing $\lambda^* = 2$ and

$$H^- = \frac{(e_p + p_m)^2}{2m} + (e_q + q_m)^2 - (e_q + p_m)^4 + \frac{1}{3}(e_q + q_m)^6 + H_m$$

Note that with $\lambda = \lambda^*$, $H(\cdot)$ and $H_m(\cdot)$ have the same extrema thus plant and model have the same equilibria. Our control program becomes

$$u = \begin{Bmatrix} 5p_m - \max_w(wp/m), \forall|p| \geq \beta \\ 0, \forall|p| < \beta \end{Bmatrix}$$

For adaptive law we calculate

$$F = 2e_q p_m(1 + 2e_q^2 + e_q^4) + 2q_m p_m(1 + 6e_q^2 + 5e_q^4)$$

$$+4e_q p_m q_m^2(3 + 20e_q^2) + 4q_m^3 p_m(1 + 5e_q^2)$$

$$-[2(e_p + p_m)(q_m - 2q_m^3 + q_m^5)/m] + 5p_m[e_p + (1+m)p_m]/m$$

Substituting data m = 1.5kg, we obtain the motions indicated in figure below. The definite convergence between the manipulator and the model occurs after about 4 sec.

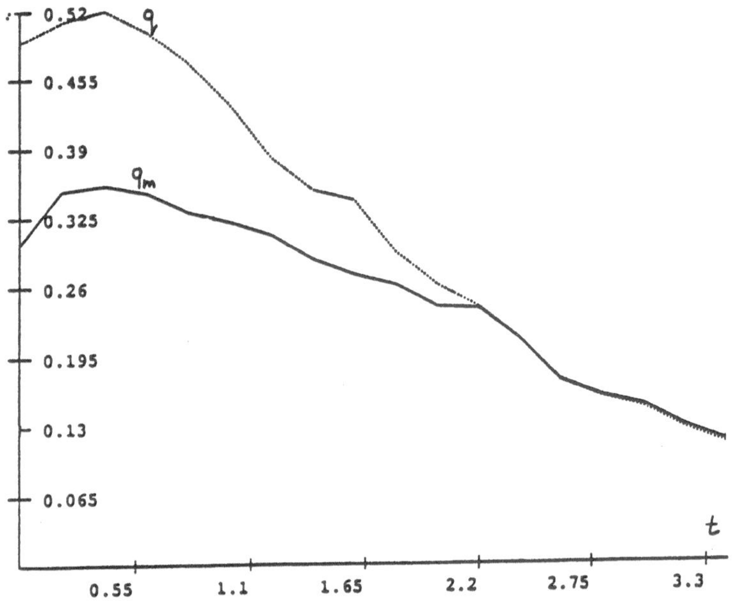

REFERENCES

1. Landau, I. D., Adaptive Control, Dekker, N. Y. 1979.

2. Skowronski, J. M., Nonlinear model reference adaptive control, J. Australian Math. Soc. Section 3 (Appl. Math), Vol 28, 1985, 23-35.

3. Skowronski, J. M., Control Dynamics of Robotic Manipulators, Acad. Press, 1986.

4. Flashner H., Skowronski, J. M., Model tracking by Hamiltonian Systems, Trans, ASME, J. Dyn. Syst. Mes. Control, Vol. 111, 1989, #4, 656-660.

5. Filippov, A. F., Existence of solutions of generalized differential equations, Math Notes (Russian) Vol. 10, 1971, 608-611.

DELTA ALGORITHM FOR GLOBAL OPTIMAL CONTROL

Efim A. Galperin

Département de mathématiques et d'informatique
Université du Québec à Montréal
C.P. 8888, Succ. A, Montréal, Qué., Canada H3C 3P8

Extended Summary

Consider the problem of steering a controlled system

$$\frac{dx}{dt} = f(x, t, u), t \in [t_0, T], x \in X \subset R^n, u \in R \tag{1}$$

from a given initial position x_0 in $X_0 \subseteq X$ to a given terminal position $x_1 \in X_1 \subseteq X$, $x_1 \neq x_0$ in some optimal way that will be specified later. Here X, X_0, X_1 are some domains (open connected sets) in R^n and $u = u(x(t), t)$ is a scalar control function with values in R which may represent an open-loop control or a feedback.

To guarantee the existence, uniqueness and continuation of solution over the entire segment $[t_0, T]$, whe have to impose some generally accepted hypotheses.

1) The function $f: R^n \times R \times R \longrightarrow R^n$ in (1) is Lipschitzian in x, u:

$$\| f(x, t, u) - f(\tilde{x}, t, \tilde{u}) \| \leq L (\| x - \tilde{x} \| + | u - \tilde{u} |), L = \text{const} > 0,$$
$$\forall t \in [t_0, T], x \in X, \tilde{x} \in X, | u | \leq M_0, | \tilde{u} | \leq M_0, \tag{2}$$

where $\| . \|$ is Euclidian norm.

2) For each $u(x, t)$, the function $f(x, t, u (x, t))$ is piecewise continuous in t and Lipschitzian in x:

$$\| f(x, t, u(x, t)) - f(\tilde{x}, t, u, (\tilde{x}, t)) \| \leq L_u \| x - \tilde{x} \|, L_u = \text{const} > 0,$$
$$\forall t \in [t_0, T], x \in X, \tilde{x} \in X, \tag{3}$$

where the constant L_u may vary for different choices of u(.). This hypothesis assures [1] that in a neighborhood of every point (x_0, t_0), $x_0 \in X_0 \subseteq X$ there is one and only one solution for every u(x, t) and every $x_0 \in X_0$ given.

3) For every $x_0 \in X_0$ and every u(.) $\in$ U belonging to some class U of controls specified later, the solution

$$x[x_0, t_0, t] = x(x_0, t_0, t, u(.)) \tag{4}$$

can be continued over the whole segment $[t_0, T]$. To guarantee the extendability of solutions over $[t_0, T]$, it is sufficient, see, e.g., [2], that f(x, t, u) satisfy the condition

$$x f(x,t,u) \leq k(t)(1+\|x\|^2), \int_{t_0}^{T} k(t)dt < \infty \tag{5}$$

If f does not satisfy (5), then the system may still be integrable over the whole $[t_0, T]$; example : $dx/dt = 1 + x^2$, $x(0) = 0$, over [0, 1]. In fact, we do not need the verification of (5) since the extendability of solutions is being checked by the algorithm automatically. Almost all real-life systems satisfy these hypotheses.

It is worthwhile to briefly mention some fundamental differencies between the delta algorithm and classical optimal control methods based on the calculus of variations (the Euler equation, the maximum principle, the Hamilton-Jacobi-Bellman equation).

1. Classical optimal control methods are bases on certain variational *optimality conditions* and are, thus, indirect. If a control function u(x, t) meets those conditions, then it presents the optimal control if the conditions are sufficient, or a candidate optimal control if the conditions are necessary. The delta algorithm is a direct method which does not need any optimality conditions. It minimizes the cost functional itself without fitting some conditions. Having an analitically solvable optimality condition presents an important advantage if it allows us to avoid the iteration process. However, for nontrivial problems the optimality conditions usually do not have an explicit closed form solution $u^0 = u^0(x, t)$ or $u^0 = u^0 (x_0, t_0, x_1, t_1, t)$. In this case it may be more appropriate to apply an iteration process aimed at minimizing the cost functional directly via delta algorithm than at fitting some optimality conditions.

2. Variational optimality conditions usually impose strong restrictions (cf. conditions A-B-C-D-E-F [3, pp. 268-270] of "regular synthesis" introduced in [4]) that are not related to the problem itself. They are called regularity conditions. The delta algorithm does not require any

regularity conditions except for representability of a control function u(x[t], t) by its Fourier series on some orthogonal basis.

3. The notion of variation is local. Therefore, optimal control methods based on variational optimality conditions are also local and solve the problem of local optimal control (i.e. they find $u^0(.)$ optimal with respect to *neighboring* trajectories, not with respect to all admissible trajectories). To obtain global solutions, certain global convexity requirements are imposed which essentially restrict the classes of problems that can be dealt with by a variational method. The delta algorithm delivers, irrespective of convexity considerations, the full global optimal solution, that is, the global minimum value and the set of all globally minimizing controls.

4. If an iteration process is used in the framework of a variational optimal control method, then it is a point-to-point process with "points" meant as elements of an appropriate functional space of controls. Such process represents a descent in some feasible direction. In contrast with such methods, the full version of the delta algorithm represents a set-to-set global descent onto the set of all global optimal solutions. It is a non-directional monotonic space sweep method with systematic deletions of whole sets not containing global minimizers.

Consider the Hilbert space $L_2(t_0, T)$ of square integrable functions and an orthonormal basis $\{\varphi_i\} \in L_2(t_0, T)$ of piecewise continuous functions. Consider controls $u(\cdot) \in L_2(t_0, T)$ representable by the Fourier series

$$u(.) = \sum_{i=1}^{\infty} c_i\varphi_i(.) = \lim_{N\to\infty} \sum_{i=1}^{N} c_i\varphi_i(.), \quad \sum_{i=1}^{\infty} c_i^2 < \infty \tag{6}$$

or its partial sum

$$u_N(t) = \sum_{i=1}^{N} c_i\varphi_i(t), \quad t \in [t_0, T], \quad N = 1,2,\ldots \tag{7}$$

Definition 1. The set U of admissible controls consists of functions u(t) such that

(a) u(t) is representable by (6) or (7);

(b) u(t) is bounded:

$$|u(t)| \leq M, \quad \forall t \in [t_0, T] \tag{8}$$

where a constant M can be chosen in advance;

(c) if u(t) is represented by (6), then it admits a left derivative and a right derivative everywhere in $[t_0, T]$, to satisfy the Dini condition of convergence, and is at most piecewise continuous.

Definition 2. The set $U_0 \subset U$ of exact terminal controls is defined as follows:

$$U_0 = \{u(.) \in U \mid x(x_0, t_0, t_1, u(.)) = x_1, t_0 < t_1 \le T\} \tag{9}$$

Problem 1. Find a control function $u(.) \in U_0$.
Problem 2. Find the entire set U_0 of (9).

These two problems are limited and full exact terminal control problems respectively. They can be formulated as optimal control problems in the following way.

Consider the functional

$$G(u) = \| x(x_0, t_0, t_1, u(.)) - x_1 \| \ge 0 \tag{10}$$

where x_0, t_0, x_1 are given and $t_1 \in (t_0, T]$ may be fixed or free. Then Problems 1, 2 can be regarded as *global* optimization problems of the form:

$$\min_{u \in U} G(u) = \lim_{N\to\infty} S_N^0 = \lim_{N\to\infty} \min_{c\in K_N} S_N(c), \tag{11}$$

where

$$S_N(c) = S_N(c_1,\ldots,c_N) = \| x(x_0,t_0,t_1, \sum_{i=1}^{N} c_i\varphi_i(.)) - x_1\|, \quad N = 1,2,\ldots \tag{12}$$

and K_N are octahedral sets induced by (8).

Definition 3. Given $\eta > 0$, the set U_η of η-approximate terminal controls is defined as follows:

$$U_\eta = \{u(.) \mid u(.) \in U, \| x(x_0, t_0, t_1, u(.)) - x_1 \| \le \eta\} \tag{13}$$

If $U_\eta \ne \emptyset$, then system (1) is said to be η-controllable with respect to x_0, x_1. If $U_\eta \ne \emptyset$ for any $x_0 \in X$, $x_1 \in X$ (or any $x_0 \in X_0 \subset X$ and any $x_1 \in X_1 \subset X$), then system (1) is said to be η-controllable on X (or with respect to X_0, X_1).

Suppose that $U_\eta \ne \emptyset$ for given x_0, x_1.

Problem 3. Find a control function $u(.) \in U_\eta$.
Problem 4. Find the entire set U_η of (13).

These two problems are limited and full approximate terminal control problems respectively. Obviously, $U_0 \subseteq U_\eta$, $\forall\eta \geq 0$.

Lemma 1. If $U_0 \neq \emptyset$, then for each M, every $\eta > 0$ and any fixed $\{\varphi_i(.)\}$ there exist $N = N(\eta, M, \{\varphi_i\})$ and $c \in K_N$, i = 1,..., N, such that for

$$\tilde{u}(.) = \sum_{i=1}^{N} c_i\varphi_i \ , \ |\tilde{u}(t)| \leq M \tag{14}$$

we have

$$\| x(x_0, t_0, t_1, \tilde{u}(.)) - x_1 \| \leq \eta \qquad \blacksquare \tag{15}$$

We see that for any $u_0(.) \in U_0$ one can always cut off the tail of a series in (6), or find another control $\tilde{u}(.)$, (14), equivalent to $u_0(.)$ in the sense (15), i.e., at the expense of a minor terminal imprecision. Due to informational and technological imprecisions as well as to elasticity of materials and media, exact values are not realizable in practice, therefore, for any realistic goal, one can consider problems defined up to certain finite precision $\eta > 0$, hence, in a finite dimensional space R^N corresponding to (15).

It is clear that with the help of a full global optimization method, such as cubic or beta algorithms [5], we can always find the entire set $B_\eta^N \subset K_N$ of coefficients, that is, the set U_η of η-approximate controls (14)-(15). However, if $\eta \longrightarrow 0$, then usually $N \longrightarrow \infty$ and, due to different norms in which convergence is considered, we have a complication in finding the exact solution U_0 of (9).

Let $u_m(.) \longrightarrow u_0(.)$ as $m \longrightarrow \infty$ with respect to the L_2-norm, that is,

$$\|u_m(.) - u_0(.)\|_{L_2}^2 = \int_{t_0}^{T} [u_m(t) - u_0(t)]^2 dt \rightarrow 0, \text{as } m \rightarrow \infty. \tag{16}$$

Lemma 2. If the hypotheses 1 to 3 are satisfied, then the solutions $x^m = x(x_0, t_0, t, u_m(.))$ converge to the solution $x^0 = x(x_0, t_0, t, u_0(.))$ *uniformly* on $[t_0, T]$. ■

Now we are prepared to formulate the final results.

Theorem. If $U_0 \neq \emptyset$ and a *full global* optimization method is employed in the delta algorithm, then

$$\lim_{N\rightarrow\infty} S_N^0 = 0. \qquad \blacksquare \tag{17}$$

We see that if $U_0 \neq \emptyset$, then the delta algorithm equipped with a full global optimization method delivers in the limit the entire sets U_0 of all exact terminal controls.

Suppose that a cost functional is given on trajectories of system (1). Then for some $\eta \geq 0$ we consider the *global* optimal control problem.

$$\inf_{u \in U_\eta} J(u) = \alpha h(x[t_1]) + \beta \int_{t_0}^{t_1} f_0(x[t], t, u(t))\, dt, \tag{18}$$

where $x[t_1]$, $x[t]$ are meant in the sense of (4), functions h, f are Lipschitz continuous and $\alpha \geq 0$, $\beta \geq 0$ are constants.

If $\eta = 0$ and U_0 is a singleton, then the unique terminal control $u^0(t) \in U_0$ is optimal for any J(u). Usually, U_0 is not a singleton and U_η, $\eta > 0$, is always not a singleton, in which cases the 2nd optimization by (18) can be applied after solving the exact ($\eta = 0$) terminal control problem by (11) or approximate ($\eta > 0$, open end $x_1 \in \mathcal{N}_\eta(x_0)$) terminal control problem by (12), (14), (15). Clearly, the 2nd optimization with respect to (18) can be performed in the same way, simply by continuing the application of the cubic or beta algorithms over the set B_η^N (where $N < \infty$ if $\eta > 0$, and $N \leq \infty$ if $\eta = 0$). Alternatively, the two criteria, (11)-(12) and (18) can be combined in several different ways and applied simultaneously to speed up the convergence, see [5], chapters 8, 10.

References

1. E.A. Coddington and N. Levinson. Theory of Ordinary Differential Equations, McGraw Hill, New York, 1955.

2. A. Friedman. Differential Games, Wiley-Interscience, New York, 1971.

3. I.N. Roitenberg. Théorie du contrôle automatique, Mir, Moscou, 1974.

4. V.G. Boltyanskii. Sufficient conditions of optimality and justification of the dynamic programming method, Izvestija Akad. Nauk SSSR, Mathem. series, Vol.28 (1964) No.3, pp.481-514 (in Russian).

5. E.A. Galperin. The Cubic Algorithm for Optimization and Control, NP Research Publ., 1990, P.O. Box 1691, Station B, Montreal, Que., Canada H3B 3L3.

DEVELOPMENT OF GUIDANCE LAWS FOR ACCELERATING MISSILE

R. Gazit[1] and S. Gutman[2]

Abstract

The most widely-used guidance law for short range homing missiles is Proportional Navigation (Pro. Nav.) . In Pro. Nav. the acceleration command is proportional to the line of sight angular velocity (L.O.S rate). Indeed, if a missile and a target move on collision course with constant speeds the L.O.S rate is zero.

The speed of a highly maneuverable modern missile varies cnsiderably during flight. The performance using Pro. Nav. is far from being satisfactory .

In this work we analyze the collision course for a variable speed missile and define a guidance law that turns the heading of the missile towards a collision course. We develop guidance laws based on optimal control and differential games, and note that the optimal laws coincide with the 'Guidance to Collision' law at the moment of impact.

We demonstrate the improvement in the missile performance using the new guidance law, relative to Pro. Nav. .

[1] Graduate student, Faculty of Aerospace Engineering, Technion - Israel Institute of Technology, Haifa, Israel.

[2] Associate Professor, Faculty of Mechanical Engineering, Technion - Israel Institute of Technology, Haifa, Israel.

1. Introduction

Proportional Navigation (Pro. Nav.) has been used for almost half a century for guidance of short range homing missiles.
The origin of Pro. Nav. is based on the following kinematic observation :if two vehicles maintain a constant relative bearing while closing in range, then collision is ensured. Thus Pro. Nav. seeks to null the Line Of Sight (L.O.S) rate in order to acheive a collision course.
In addition, Bryson and Ho [1] used Optimal Control Theory to show that Pro. Nav. is the optimal guidance law that minimizes final miss distance, for a constant speed missile pursuing a non-maneuvering target. Ho et. al.[2] and Gutman and Leitmann [3] used Differential Game Theory and proved that Pro. Nav. also ensures minimum miss distance against any permissible target maneuver.

The most critical assumption made in deriving Pro. Nav. is that the missile speed is constant. When variations in missile speed are small, Pro. Nav. performs quite well. However, when the missile is subject to large acceleration or deceleration, Pro. Nav. can not point to the collision course direction, and is far from optimal. This is a severe problem for a modern missile, maneuvering at high angles of attack, and consequently subject to large variations in speed.

A common solution to the variable speed guidance problem, is adding an 'Acceleration Compensation' term to the L.O.S rate in the guidance law [4]. The acceleration term helps the missile to null the L.O.S rate. We will show that, for a variable speed missile, nulling the L.O.S rate does not lead to a collision course.
Several theoretical ways to attack the variable speed guidance problem appear in the literature :
Refs. [5-8] accounted for speed variation by estimating time-to-go more accurately, and substituting in optimal guidance law obtained in the linear model. Even here, the missile speed variation is not a part of the linear model.

When the complete nonlinear equations of motion are considered, solutions can be obtained numerically [9-12]. In that case, direct formulation of guidance laws can not be done. The analytical approach [13,14] however, enables formulation of open loop optimal strategies. These guidance laws depend on the final states, and can not be directly implemented. Near optimal, closed loop solutions can be obtained by applying singular perturbation technique [15-17] for model order reduction. These guidance laws are not suitable for short range scenarios, where the singular perturbation technique model is not valid.

The purpose of this work is the development of an implementable guidance law for short range homing missiles. The law should take into account missile speed variations throughout the engagement.

The development of the new law is carried out in a similar way to the development procedure of Pro. Nav. :

First, we recall the kinematic basis of Pro. Nav. and define a collision course for a variable speed missile. We find some function of the state variables that vanishes whwn the missile is heading towards collision (like L.O.S rate, for a constant speed missile), and define a "Guidance to Collision" law, where the missile's turn rate is directly proportional to the value of that function.

Next, we define a linear model based on small deviations from collision course. The missile axial acceleration is part of the linear model. In this model, we find the optimal guidance law that minimizes final miss distance against a non-maneuvering target. Furthermore, we solve a simple linear differential game, to find the strategy ensuring minimal miss distance against any permissible target maneuver.

Finally, we notice that the optimal guidance laws coincide at the moment of impact with the kinematics based guidance to collision law. The terminal structure of those laws, is proposed as a new guidance law for short range homing missiles. A short performance study, comparing the new law with Pro. Nav. and Acceleration Compensated Pro. Nav., is performed using simulation of an ideal missile.

2. Guidance to Collision

Consider the collision triangle depicted in Fig. 1. Let M be the missile and T be the target, both moving along straight lines toward collision point C.

On the collision triangle, we have :

$$\frac{S_M}{\sin \phi_T} = \frac{S_T}{\sin \phi_M} = \frac{R}{\sin (\phi_T - \phi_M)} \tag{1}$$

Where :

S_M , S_T are the distances from M and T to the collision point.

ϕ_M , ϕ_T are the angles between the velocity vectors and the line of sight.

R is the range.

Assume that speeds V_M, V_T are constant :

$$S_M = V_M \cdot t_{go} \quad , \quad S_T = V_T \cdot t_{go} \tag{2}$$

where t_{go} is the time to collision.

Using (2) in (1), we obtain :

$$V_T \sin \phi_T - V_M \sin \phi_M = 0 \tag{3}$$

This is known as the relative velocity normal to the line of sight:

$$R\dot{\sigma} = V_T \sin \phi_T - V_M \sin \phi_M \tag{4}$$

where σ is the line of sight orientation in some inertial reference frame.

It follows that the line of sight rate $\dot{\sigma}$ is zero along a collision course :

$$\dot{\sigma} = 0 \qquad ; \quad R \neq 0 \tag{5}$$

when the missile and target speeds are constant.

To reach a collision course, Pro. Nav. states that the missile turn rate should be proportional to rate of change of the L.O.S :

$$\dot{\gamma}_M = N\,\dot{\sigma} \quad . \tag{6}$$

In a similar way, we define a guidance to collision law for a variable speed missile. First, we look for a function of the state variables that vanishes on collision course. Toward this end, assume that the target velocity is constant :

$$S_T = V_T \; t_{go} \quad .$$

Substituting in (1), we obtain the following equations :

$$t_{go} = \frac{R \sin \phi_M}{V_T \sin (\phi_T - \phi_M)} \quad ; \tag{7}$$

$$R \sin \phi_T - S_M(t_{go}) \sin (\phi_T - \phi_M) = 0 \quad . \tag{8}$$

$S_M(t_{go})$ is the distance M travels to collision, where t_{go} given by (7), is the time to collision, under the assumption that M is pointed toward collision. The distance M will travel during time t_{go}, depends on his current velocity V_M, and on his thrust and drag profile.

To get free of T's state variables V_T ,ϕ_T ,we use the line of sight equations of motion :

$$\dot{R} = V_T \cos \phi_T - V_M \cos \phi_M \tag{9}$$

$$R\dot{\sigma} = V_T \sin \phi_T - V_M \sin \phi_M \tag{10}$$

Substituting in (7),(8), we get :

$$t_{go} = \frac{R \sin \phi_M}{R\dot{\sigma} \cos \phi_M - \dot{R} \sin \phi_M} \quad ; \tag{11}$$

$$R \dot{\sigma} + (V_M - \frac{S_M(t_{go})}{t_{go}}) \sin \phi_M = 0 \quad . \tag{12}$$

Equations (11),(12) define a function that vanishes, when a variable speed missile is heading to collision with a constant speed target. In the same way Pro. Nav. was defined, we formulate a guidance to collision law, for a variable speed missile :

$$\dot{\gamma}_M = N [\dot{\sigma} + \frac{1}{R} (V_M - \bar{V}_M) \sin \phi_M] \quad ; \quad R \neq 0 \tag{13}$$

where :

$$\bar{V}_M \overset{\Delta}{=} \frac{S_M(t_{go})}{t_{go}}$$

and t_{go} is given in (11).

The first term in the guidance law is σ ,the line of sight rate. This term vanishes on collision course for a constant speed missile. The second term comes from missile speed variation. This term is composed of a variable gain, that is multiplied by the sine of lead angle ϕ_M .
When missile speed is constant,

$$\bar{V}_M = V_M$$

and we are left with Pro. Nav. :

$$\dot{\gamma}_M = N \dot{\sigma}$$

When the missile accelerates, its mean velocity in time t_{go} is greater than it's current velocity :

$$\bar{V}_M > V_M$$

Therefore, the speed variation term will have negative sign. When the missile slows down, the sign will change to positive.

This guidance law can be implemented by measuring range, range rate and missile velocity, and computing the future distance S_M using some pre-defined function. However, we wish to check whether this guidance to collision law is also the optimal guidance law that minimizes final miss distance.
We derive the optimal guidance law on a linear model describing the motion close to a collision course.

3. Linear State Model

The linear model is formulated by analyzing the final stage of homing. We assume, that at this stage the missile has already completed the initial turn towards a collision course, and its motion is composed of small deviations from that course.
At $t = t_o$, a collision course is defined by R_o ,V_{To} ,V_{Mo} ,γ_{To} ,γ_{Mo} .
We fix an inertial reference frame along the initial line of sight LOS_o . Let y_M and y_T be the locations of M and T normal to LOS_o , respectively.

Define :

$$y \overset{\Delta}{=} y_T - y_M \tag{14}$$

then :

$$\dot{y} = \dot{y}_T - \dot{y}_M = V_T \sin \gamma_T - V_M \sin \gamma_M \tag{15}$$

$$\ddot{y} = \dot{V}_T \sin \gamma_T + V_T \dot{\gamma}_T \cos \gamma_T - \dot{V}_M \sin \gamma_M - V_M \dot{\gamma}_M \cos \gamma_M \tag{16}$$

Assume constant speed target :

$$\dot{V}_T = 0 \tag{17}$$

Define the state vector $\underline{x} \in R^3$:

$$\begin{aligned} x_1 &\overset{\Delta}{=} y \\ x_2 &\overset{\Delta}{=} \dot{y} \\ x_3 &\overset{\Delta}{=} \sin \gamma_M \end{aligned} \tag{18}$$

The control variables are M and T's lateral accelerations normal to LOS_o :

$$\begin{aligned} u &\overset{\Delta}{=} V_M \dot{\gamma}_M \cos \gamma_M \\ v &\overset{\Delta}{=} V_T \dot{\gamma}_T \cos \gamma_T \end{aligned} \tag{19}$$

The equations of motion, become :

$$\begin{aligned} \dot{x}_1 &= \dot{y} = x_2 \\ \dot{x}_2 &= \ddot{y} = v - u - \dot{V}_M \; x_3 \\ \dot{x}_3 &= \dot{\gamma}_M \cos \gamma_M = \frac{u}{V_M} \end{aligned} \tag{20}$$

These equations must be joined by another expression for the missile speed change V_M. The complete system is nonlinear, because of the dependence on V_M. To allow formulation of the equations of motion as a linear system, we assume that the target axial acceleration is constant :

$$\dot{V}_M = a_x \quad . \tag{21}$$

Then,

$$V_M = V_{Mf} - a_x \; t_{go} \quad , \tag{22}$$

where V_{Mf} is the final value of V_M.
Thus, we obtain a linear, time dependent model describing the final homing phase of the engagement :

$$\dot{x} = \begin{bmatrix} 0 & 1 & 0 \\ 0 & 0 & -a_x \\ 0 & 0 & 0 \end{bmatrix} x + \begin{bmatrix} 0 \\ -1 \\ \dfrac{1}{V_{Mf}-a_x \; t_{go}} \end{bmatrix} u + \begin{bmatrix} 0 \\ 1 \\ 0 \end{bmatrix} v \tag{23}$$

$$x^T = [y \; , \dot{y} \; , \sin \gamma_M]$$

4. The Optimal Control Law, for a Non-Maneuvering Target

We use the linear model, assuming that the target does not maneuver,

$$v = 0 \quad . \tag{24}$$

We seek the optimal control law, that minimizes the cost function :

$$J = \frac{1}{2} m^2(t_f) + \frac{1}{2} \int_{t_o}^{t_f} k\, u^2(\tau)\, d\tau \tag{25}$$

m(t) is the "Zero Effort Miss" :

$$m(t) \overset{\Delta}{=} x_1(t_f) = d^T \Phi(t_f, t)\, x(t) \tag{26}$$

where $\Phi(t_f, t)$ is the transition matrix of (23), and $d^T=[1,0,0]$. This is a standard optimal linear quadratic control problem. The optimal control, obtained by letting $k \to 0$, can be formulated in a well known form of Augmented Pro. Nav. :

$$u^*(t) = \frac{N'(t_{go})}{t_{go}^2}\, m(t_{go}) \tag{27}$$

where the zero effort miss is given by :

$$m(t_{go}) = x_1 + x_2 t_{go} - x_3 a_x \frac{t_{go}^2}{2} \tag{28}$$

and the navigation gain is :

$$N'(t_{go}) = 3\,\frac{4 - 2\varepsilon}{4 - \varepsilon} \tag{29}$$

where ε is defined as the velocities ratio :

$$\varepsilon \overset{\Delta}{=} \frac{a_x\, t_{go}}{V_{Mf}} \tag{30}$$

The navigation gain N' is time dependent, and is a function of axial acceleration and final velocity. However, N' can also be formulated in a feedback form of the current velocity V_M :

$$N'(t_{go}) = 3\,\frac{4V_M + 2a_x\, t_{go}}{4V_M + 3a_x\, t_{go}} \tag{31}$$

In case of constant speed,

$a_x = 0$,

we get the original linear Pro. Nav., as obtained by Bryson and Ho [1] :

$$u^*\Big|_{a_x=0} = \frac{3}{t_{go}^{\,2}}\,(x_1 + x_2\, t_{go}) \tag{32}$$

5. The Optimal Guidance Law for a Maneuvering Target

In the previous section we applied optimal control theory assuming a non-maneuvering target. The targets future maneuver can be substituted into the linear law [18], but it must be completely defined, either in open loop (pre defined) or in closed loop form (measurement). The difficulty in target manuever estimation leads us to the formulation of a differential game between the missile and the target. This approach makes no assumption on future target maneuvers, but rather takes into consideration the targets maneuver capability. The optimal guidance law then guides the missile so as to minimize final miss distance against any permissible target maneuver. The solution of the guidance problem by formulating a simple differential game between the missile and the target, was introduced by Gutman [3] for the case of constant speed. We extend the analysis to the case of constant axial acceleration.

The final homing loop is decribed by the linear system (23). The control variables are bounded :

$$| u(t) | \leq u_m$$
$$| v(t) | \leq v_m \qquad (33)$$

Define the cost function :

$$J = |m(t_f)| \qquad (34)$$

Pursuer u wishes to minimize final miss distance, and evader v - to maximize it. The zero effort (of both players) miss is equal to the previous definition (28) .

Based on "simple class of differential games" introduced in [19], the optimal solution is obtained in the following way: The state space (m, t_{go}) is divided into two regions :

$$D_1 \overset{\Delta}{=} \{(m,t_{go}) : |m| > v_m \frac{t_{go}^2}{2} [\mu - 1 - \mu f(\varepsilon)] \}$$

$$D_2 = D_1^c \tag{35}$$

Where :

$$f(\varepsilon) = \frac{1}{\varepsilon^2} \left[\varepsilon + \frac{\varepsilon^2}{2} + \ln (1-\varepsilon) \right] \tag{36}$$

μ is the ratio of the control bounds :

$$\mu \overset{\Delta}{=} \frac{u_m}{v_m} \qquad ,\mu > 1 \tag{37}$$

and ε is the speeds ratio (30).
The state space partition is depicted in Fig. 2.
The strategies in region D_1, are :

$$u^* = u_m \text{ sign } [m(t)] \quad ,$$
$$v^* = v_m \text{ sign } [m(t)] \quad . \tag{38}$$

The trajectories here are parallel to the boundary between D_1 and D_2, and the controls don't change their sign.
The intersection of a trajectory with the $(|m|, t_{go} = 0)$ axis, is the optimal cost in this region.
In region D_2 the strategies are arbitrary, i.e one may choose any permissible control. The trajectories will reach the boundary and slide to the origin. Therefore the optimal cost in this region is zero.

In particular, we chose in D_2 a guidance law in the form of A.P.N (27):

$$u^* = \frac{N'(t_{go})}{t_{go}^2} m(t_{go})$$

To match the guidance laws in D_1 and D_2, we calculate the navigation gain N', such that :

$$u^* = u_m \operatorname{sign}[m(t_{go})]$$

on the boundary . The gain should therefore be :

$$N'(t_{go}) = \frac{2\mu}{\mu - 1 - \mu f(\varepsilon)} \tag{39}$$

The optimal law, which is now valid in the entire state space, is a saturation function of $m(t_{go})$,

$$u^* = u_m \operatorname{sat}\left[\frac{N'(t_{go})}{t_{go}^2} m(t_{go})\right] , \tag{40}$$

where $m(t_{go})$ is given by (28).

The meaning of the state space partition is now clear .

In region D_2 the missile maneuver is not saturated, and therefore it can force zero cost (miss) for any permissible target maneuver.

In region D_1, the missile maneuver is saturated, and the miss distance depends on initial conditions and target maneuver.

For a constant speed missile, one may use L'Hospital's rule twice, to obtain :

$$\lim_{\varepsilon \to 0} f(\varepsilon) = 0 \tag{41}$$

In that case, we get the navigation gain presented by Gutman and Leitmann [3] :

$$N'\Big|_{a_x \to 0} = \frac{2\mu}{\mu - 1} \tag{42}$$

6. Implementation

The optimal guidance laws are written in terms of linear system state variables (18) ,

$$[x_1 , x_2 , x_3] = [y , \dot{y} , \sin \gamma_M]$$

In order to implement these laws, they should be transformed to the terms of the original non linear system. The detailed procedure may be found in the appendix.

The optimal control law (28) takes now the form :

$$u = N'(t_{go}) \left[V_c \dot{\sigma} - \frac{a_x}{2} \sin \phi_M + \frac{a_x}{2} \cos \gamma_M \dot{\sigma} t_{go} \right] \tag{43}$$

$$N'(t_{go}) = 3 \frac{4 - 2\varepsilon}{4 - \varepsilon}$$

where u is the required acceleration normal to the L.O.S .

The optimal guidance law based on the solution of the differential game (40) is the saturation function of the same argument as in (43), with a different navigation gain :

$$u = u_m \operatorname{sat} \left[N'(t_{go}) \left[V_c \dot{\sigma} - \frac{a_x}{2} \sin \phi_M + \frac{a_x}{2} \cos \gamma_M \dot{\sigma} t_{go} \right] \right] \tag{44}$$

$$N'(t_{go}) = \frac{2\mu}{\mu - 1 - \mu f(\varepsilon)}$$

The time dependent navigation gains (through ε) are depicted in fig. 3 .

We now wish to compare these optimal laws, to the Guidance to Collision law (13), developed in section 2. The optimal laws are based on the assumption that the missile axial acceleration is constant :

$$\dot{V}_M = a_x$$

therefore, we have to substitute in (13) :

$$S_M(t_{go}) = V_M\, t_{go} + a_x \frac{t_{go}^2}{2} \qquad (45)$$

Furthermore, the guidance to collision law was expressed in a rate command form. We transform it to the acceleration command form, using the definition of the normal acceleration (19) :

$$u \overset{\Delta}{=} V_M\, \dot{\gamma}_M \cos \gamma_M$$

and the well known relation [20] :

$$N' \overset{\Delta}{=} N \frac{V_M \cos \gamma_{Mo}}{V_c} \qquad (46)$$

Finally we obtain the guidance to collision law, for a constant axial acceleration missile :

$$u = N'\, V_c \left[\dot{\sigma} - \frac{a_x}{2\left[\dfrac{R\,\sigma}{\mathrm{tg}\, \phi_M} - \dot{R} \right]} \sin \phi_M \right] \qquad (47)$$

Now one can clearly notice that at the moment of impact , the guidance to collision law coincides with the optimal law.

Substitute in (43), $t_{go} = 0$, or in (47), $R = 0$, we obtain :

$$u = N' \left[V_c\, \dot{\sigma} - \frac{a_x}{2} \sin \phi_M \right] \qquad (48)$$

$$N' = 3$$

or in the nonlinear form (44) :

$$u = u_m \text{ sat} \left[N' \left[V_c \dot{\sigma} - \frac{a_x}{2} \sin \phi_M \right] \right] \tag{49}$$

$$N' = \frac{2\mu}{\mu - 1}$$

where $V_c \stackrel{\Delta}{=} - \dot{R}$.

Equation (48) defines a new guidance law which we choose to call the "Terminal Guidance Law". This law joins the guidance laws derived in the previous sections, as a candidate for variable speed missile guidance .

From the nature of its definition, the terminal law coincides at the end of pursuit with the guidance to collision law and the optimal guidance laws.

The terminal law does not require range measurment or time-to-go estimation, and can be easily implemented in a missile, guided previously by Pro. Nav. . The additional required information is only missile axial accleration and off-boresight angle.

7. Acceleration Compensation and Performance Study

The terminal guidance law, resembles another law, known as the "Acceleration Compensated Pro. Nav." law.

The latter is based on the acceleration command of True Pro. Nav. , normal to the L.O.S :

$$u = N' V_c \dot{\sigma} \tag{50}$$

When there is some lead angle ϕ_M, the required acceleration is composed of two components (see fig. 5) : the axial acceleration a_x and the lateral accleration a_M :

$$u = a_x \sin \phi_M + a_M \cos \phi_M \tag{51}$$

Since a_x is usually not controlable, the lateral acceleration required to null the L.O.S rate should be :

$$a_M = N' \frac{V_c \dot{\sigma}}{\cos \phi_M} - a_x \tan \phi_M \tag{52}$$

This is the "Acceleration Compensated Pro. Nav." law.

We divide the terminal guidance law (48) by $\cos \phi_M$ to transform it from u - acceleration normal to the L.O.S, to a_M - lateral acceleration. The result is :

$$a_M = N' \frac{V_c \dot{\sigma}}{\cos \phi_M} - \frac{N'}{2} a_x \tan \phi_M \tag{53}$$

The difference between these two laws is demonstrated by a simple simulation of a planar pursuit. An ideal ,variable speed ,short range homing missile is fired against a crossing, non-maneuvering target. The missile is guided alternatively by :

(1) Pro. Nav. (The left term of (52))

(2) Acceleration Compensated Pro. Nav. (52)

(3) Terminal Guidance Law (53)

The missile and target trajectories are depicted in fig. 5. The L.O.S rates are depicted in fig. 6.

Due to variations in missile speed , the missile guided by Pro. Nav. can not null the L.O.S rate, and therefore is far from collision course.

Adding the accleration term helps the missile to null the L.O.S rate, but still can not lead to the desired collision course. This is because the L.O.S rate is not zero on the collision course of an accelerating missile as was shown in section 2 (12).

However, the terminal law, based on the true guidance to collision law, does lead the missile to collision course. The initial maneuver required by the terminal law is shorter, and the missile captures the target more rapidly.
The look angle required by missiles (1) and (2) during the initial maneuver is large, and can exceed the mechanical look angle limit. If this occurs, the missile will lose the target.

Fig. 7 depicts the inner launch boundary for the same ideal missile guided alternatively by laws (1)-(3), now fired against a maneuvering target. The target is initially located at the origin, moving to the right and turning counter-clockwise. The missile is launched from each point on the plane towards the target. The miss region is clearly sperated into two sub-regions . In region 1 the missile could not reach the target because of look angle limit. This limit was arbitrarily taken as 1 radian. In region 2 the reason for miss is seeker rate saturation, arbitrarily taken as 1 radian/sec.
The large region of miss due to look angle limit for the Pro. Nav. guided missile was almost completely eliminated by the terminal guiudance law. The seeker rate saturation miss region was not affected.

8. Summary

A new guidance law for a short range homing missile was derived, based on the kinematic analysis of a collision course and on Optimal Control Theory. The missile speed variation was taken into account in the engagememnt model, and not as a secondary effect that can be treated by some compensation scheme.
The new guidance law is easily implementable, and reduces substantially the minimum launch range of a Pro. Nav. guided missile.

Appendix : Transformation of linear guidance law to original system

The linear system state variables describe the motion normal to initial line of sight LOS_o. In order to get back to the original system, we should also consider the motion along LOS_o.

As in (14), define :

$$z \overset{\Delta}{=} z_T - z_M \tag{A1}$$

then :

$$\dot{z} = V_T \cos \gamma_T - V_M \cos \gamma_M \tag{A2}$$

We assume small deviations from collision course :

$$\cos \gamma_T \cong \cos \gamma_{To}$$
$$\cos \gamma_M \cong \cos \gamma_{Mo} \tag{A3}$$

As we assumed before, the missile axial acceleration a_x is constant :

$$V_M = V_{Mo} + a_x t \tag{A4}$$

and target speed is constant.

Substitute in (A2) :

$$\dot{z} = V_{To} \cos \gamma_{To} - (V_{Mo} + a_x t) \cos \gamma_{Mo} \tag{A5}$$

Integrate $\dot{z}$, from the boundary value :

$$z(t = t_f) = 0 \tag{A6}$$

We obtain :

$$z = V_c\, t_{go} + \frac{a_x}{2} \cos \gamma_{Mo}\, t_{go}^2 \tag{A7}$$

Where :

$$V_c \overset{\Delta}{=} -\dot{z} = - [V_T \cos \gamma_{To} - (V_{Mo} + a_x\, t) \cos \gamma_{Mo}] \tag{A8}$$

$$t_{go} \overset{\Delta}{=} t_f - t$$

From geometry (see fig. 2) :

$$\mathrm{tg}\, \sigma = \frac{y}{z} \tag{A9}$$

thus :

$$\frac{\dot{\sigma}}{\cos^2 \sigma} = \frac{1}{z^2} (\dot{y}\, z - \dot{z}\, y) \tag{A10}$$

Define :

$$a_R \overset{\Delta}{=} a_x \cos \gamma_{Mo} \tag{A11}$$

Substitute (A7) in (A10), we obtain :

$$\frac{\dot{\sigma}}{\cos^2\sigma} = \frac{1}{\left[V_c\, t_{go} + \frac{a_R}{2} t_{go}^2\right]^2} \left[\dot{y}\,(V_c\, t_{go} + \frac{a_R}{2} t_{go}^2) + V_c\, y\right] =$$

(A12)

$$= \frac{1}{t_{go}^2}\,(y + \dot{y}\, t_{go}) \frac{1}{\left[V_c + \frac{a_R}{2} t_{go}\right]} - \frac{y}{t_{go}} \frac{\frac{a_R}{2}}{\left[V_c + \frac{a_R}{2} t_{go}\right]^2}$$

Rearranging terms, one can obtain :

(A13)

$$\frac{1}{t_{go}^2}\,(y + \dot{y}\, t_{go}) = \frac{\dot{\sigma}}{\cos^2\sigma}\left[V_c + \frac{a_R}{2} t_{go}\right] + \frac{a_R}{2}\, tg\, \sigma$$

Recall the optimal control law (27) :

(A14)

$$u^*(t) = \frac{N'(t_{go})}{t_{go}^2}\, m(t_{go}) = \frac{1}{t_{go}^2}\,(y + \dot{y}\, t_{go}) - \frac{a_x}{2} \sin \gamma_M$$

Substitute (A13), and after some algebraic and trigonometric operations, we obtain :

(A15)

$$u^* = N'(t_{go})\left[V_c \frac{\dot{\sigma}}{\cos^2\sigma} - \frac{a_x}{2\cos\sigma} \sin \phi_M + \frac{a_x}{2} \cos \gamma_M \frac{\dot{\sigma}}{\cos^2\sigma}\, t_{go}\right]$$

where lead angle ϕ_M is defined as :

$$\phi_M \overset{\Delta}{=} \gamma_M - \sigma \qquad \text{(A16)}$$

We assume now that σ is small. This is possible because the analysis is done near collision course, and the reference line was chosen as the initial line of sight LOS_o, such that :

$$\sigma_o = 0 \qquad (A17)$$

Substitute in (A15), we finally obtain :

$$u = N'(t_{go}) \left[V_c \dot{\sigma} - \frac{a_x}{2} \sin \phi_M + \frac{a_x}{2} \cos \gamma_M \dot{\sigma} t_{go} \right] \qquad (A18)$$

where u is the required acceleration normal to the L.O.S .

References

[1] Bryson, A.E., and Yu - Chi Ho, Applied Optimal Control; Hemisphere Publishing Corp., pp. 154-155, 1975.

[2] Ho, Y.C., Bryson, A.E., and Baron, S., "Differential games and optimal pursuit-evasion strategies", IEEE Trans. of Automatic Control, Vol. AC-10, 4, Oct. 1965, pp. 385-389.

[3] Gutman, S., Leitmann, G., "Optimal Strategies in the Neighborhood of a Collision Course", AIAA Journal, Vol. 14, No. 9, pp. 1210-1212, Sept. 1976.

[4] Garnell, P., Guided Weapon Control Systems, 2nd ed.; Pergamon Press, 1980.

[5] Riggs, T., "Linear Optimal Guidance for Short Range Air to Air Missile", Proc. of National Aerospace and Electronics Conf., May 1979, Vol. II, pp. 757-764.

[6] Lee, G.K.F, "Estimation of the Time-to-Go Parameter for Air to Air Missiles", J. Guidance, Vol. 8, No. 2, pp. 262-266, Mar.-Apr. 1985.

[7] Vergez, P.L., "Linear Optimal Guidance for an AIM-9L Missile", J. Guidance and Control, Vol. 4, No. 6, pp. 662-663, Nov.-Dec. 1981.

[8] Hull , D.G., Mack, R.E., "Prediction of Time-To-Go for A Homing Missile Using Bang-Bang Control", AIAA paper No. 88-4065-CP.

[9] Jarmark B.S.A, Merz, A.W., Breakwell, J.V., "The Variable Speed Tail Chase Aerial Combat Problem", J. Guidance and Control, Vol. 4, No. 3, May-June 1981, pp. 323-328.

[10] Hillberg, C., Jarmark, B.S.A., "Pursuit-Evasion Between Two Realistic Aircraft", J. Guidance, Vol. 7, No. 6, Nov.-Dec. 1984, pp. 690-694.

[11] Yavin, Y., De Villiers, R., "Proportional Navigation and The Game of Two Cars : The Case of Pursuer with Variable Speed", Computers Math. Applic., Vol. 18, No. 1-3, pp. 69-76, 1989.

[12] Yavin, Y., De Villiers, R., "Game of Two Cars : Case of Variable Speed", Journal of Optimization Theory and Applications, Vol. 60, No. 2, Feb. 1989, pp. 327-339.

[13] Prasad, U.R., Rajan, N., Rao, N.J., "Planar Pursuit-Evasion with Variable Speeds, Part 1, Extremal Trajectory Maps", and "... Part 2, Barrier Sections", Journal of Optimization Theory and Applications, Vol. 33, No. 3, pp. 401-432, Mar. 1981.

[14] Green, A., Shinar, J., Guelman, M., "Guidance Law Synthesis Based on A Planar Pursuit-Evasion Game Solution",<u>Differential Games and Applications</u> ;Lecture Notes in Control and Information Sciences, no. 119, T.S Basar, P. Bernhard (Eds.), Springer Verlag.

[15] Sridhar, B., Gupta, N.K, "Missile Guidance Laws Based on Singular Perturbation Methodology", J. Guidance and Control, Vol. 3, No. 2, Mar.-Apr. 1980.

[16] Cheng, V.H.L, Gupta, N.K, "Advanced Midcourse Guidance for Air to Air Missiles", Journal of Guidance, Control and Dynamics, vol. 9, Mar.-Apr. 1986, pp. 135-142.

[17] Goldstein, F., Calise, A., "Non Linear State Feedback Control for Near Optimal Intercept in Three Dimensions", AIAA Paper no. 79-1673.

[18] Garber, V., "Optimum Interpcept Laws for Accelerating Targets", AIAA Jouranl, Vol. 6, No. 7, pp. 2196-2198, Nov. 1968.

[19] Gutman, S., "On Optimal Guidance for Homing Missiles", J. Guidance and Control, Vol. 2, No. 4, pp. 296-300, Jul.-Aug. 1979.

[20] Jerger, J., Systems Preliminary Design; D. Van Nostrand Company Inc., 1960, eq. 5-38.

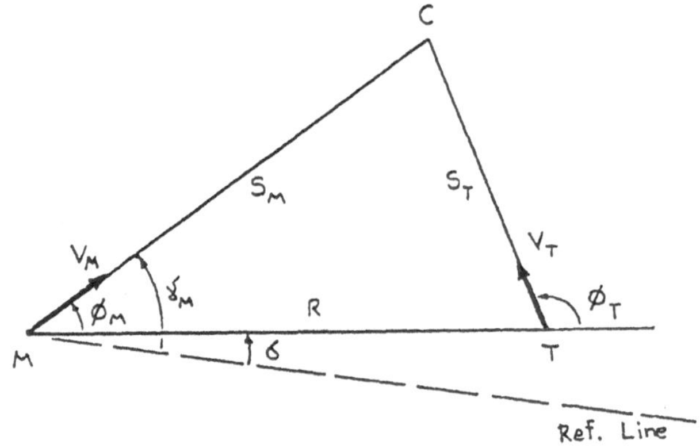

Fig. 1 : Collision course geometry.

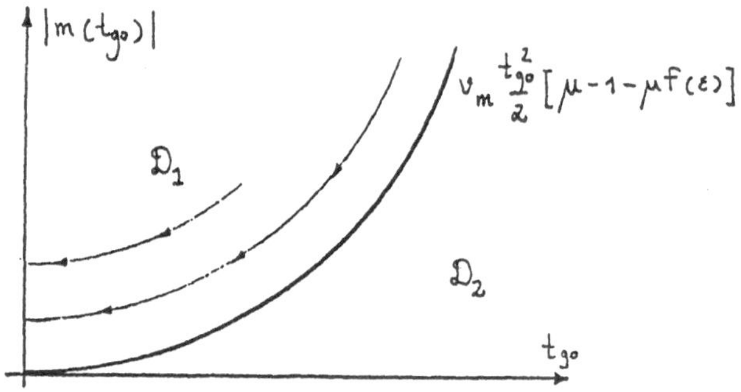

Fig. 2 : State space partition.

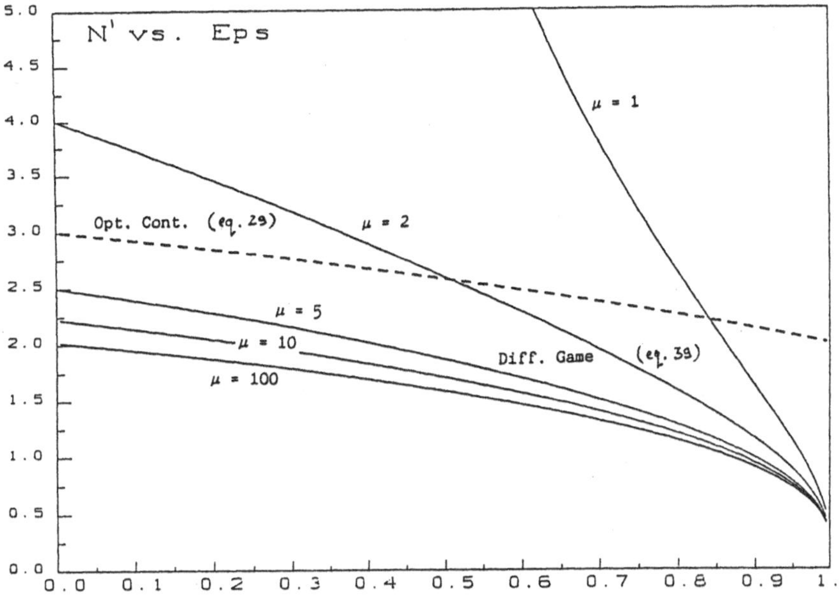

Fig. 3 : Navigation gains.

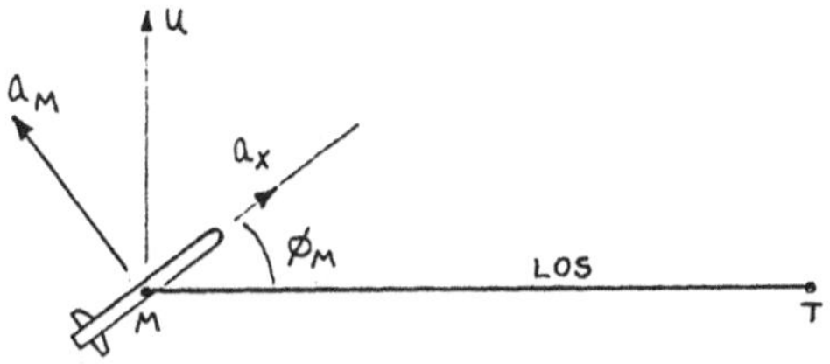

Fig. 4 : Missile accelerations.

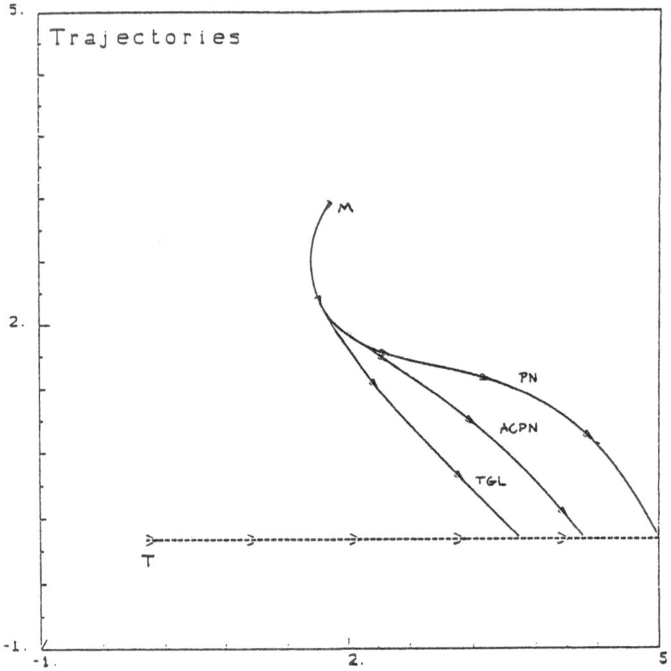

Fig. 5 : Trajectories

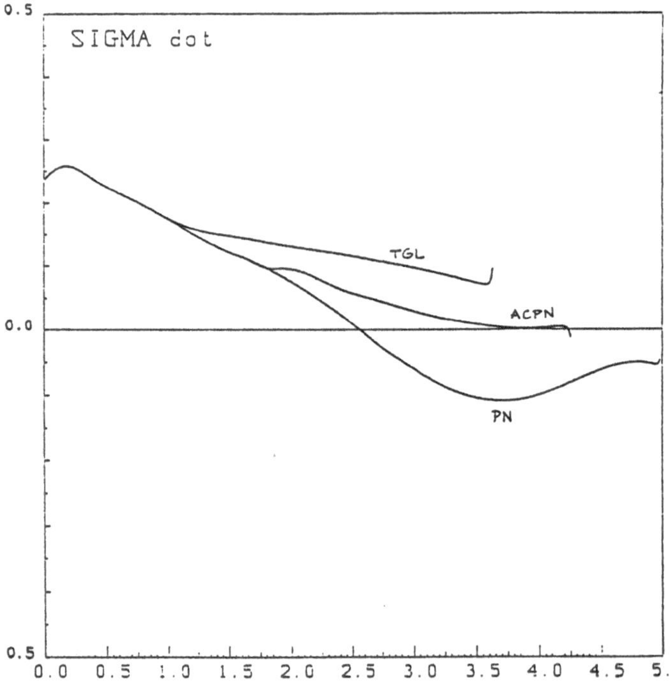

Fig. 6 : L.O.S rates.

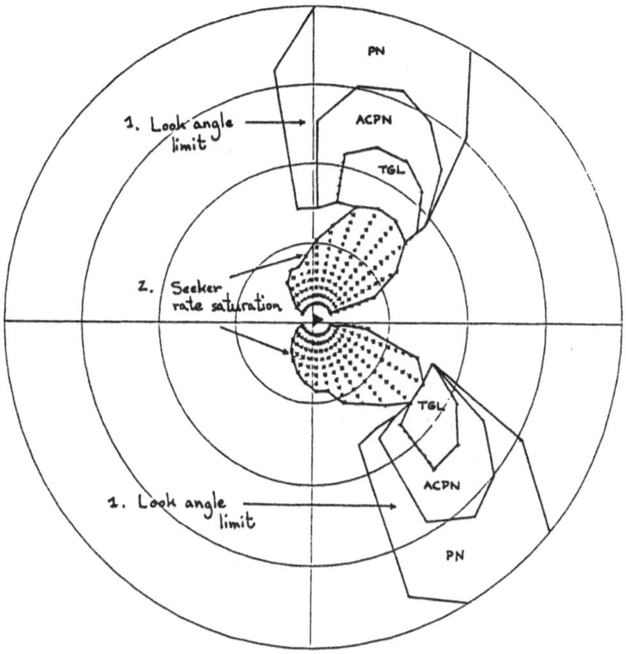

Fig. 7 : Inner launch boundaries.

ON THE EXISTENCE OF A SOLUTION OF RICCATI EQUATION AND THE MISMATCH THRESHOLD IN PRACTICAL STABILIZATION OF UNCERTAIN DYNAMICAL SYSTEMS

F. Hamano

Department of Electrical Engineering
California State University, Long Beach
1250 Bellflower Boulevard
Long Beach, California 90840-8303

Abstract *This paper presents a sufficient condition for the existence of a symmetric positive definite solution of a Riccati equation that appears in the design of a practically stabilizing controller for mismatched uncertain dynamical systems. It is used to compare two existing methods for stabilizing mismatched systems; one that involves measure of mismatch and the other that uses the Riccati equation. The result indicates that, if the measure of mismatch is within the threshold of mismatch (which is a sufficient condition for the first method to be applicable), then a symmetric positive definite solution exists for the Riccati equation (which is a sufficient condition for the second method to be applicable). The type of Riccati equation under consideration also appears in Game Theory and H^∞ Optimal Control.*

1. INTRODUCTION

There have been many studies in the control literature which are concerned with designing state feedback controllers to stabilize uncertain dynamical systems for which uncertainties are characterized by certain upper or lower bounds. The majority of the earlier articles (e.g., [2], [3], [4]) assumed the matching condition (i.e., the uncertainties must enter through the same channels as the controls). But more and more attention has been paid in the recent literature to the mismatched situations where the above condition is not met or partially met. See [5]-[11]. This paper compares the two important approaches in the mismatched cases which require different sufficient conditions for the existence of stabilizing controllers. The essences of the results for these schemes are as follows: One states that a sufficient condition for the existence of a suitable controller is that "the measure of mismatch" is less than the "threshold of mismatch". See [5] and [11]. The other states that a sufficient condition for the existence of a suitable controller is the existence of a symmetric positive definite solution of a certain Riccati equation. See for instance [7]-[10]. Therefore a natural question is whether the conditions are equivalent, if not, whether one is stronger than the other. In this paper a new sufficient condition for the existence of a solution of the Riccati equation is presented and a constructive proof is given based on "scaling". The result is applied to the problem of practical stabilization to a neighbourhood for a class of nonlinear uncertain dynamical systems to show that, if the measure of mismatch is less than the threshold of mismatch in the first approach, then the Riccati equation in the second approach has a symmetric positive definite solution.

Notation: Throughout the paper the small letters n, m and m' denote non-negative integers and all the matrices appearing in this article are real matrices and therefore the word "real"will be omitted when we refer to the matrices. In particular, the capital letters A, B and D denote nxn, nxm and nxm' real matrices and the capital letters P and Q possibly with subscripts or " ~ " denote nxn real matrices. Moreover, if P, P_1 and P_2 are symmetric matrices, then $P > 0$ (or $P \geq 0$) means that P is positive (or, respectively, positive semi-) definite, and $P_1 > P_2$ (or $P_1 \geq P_2$) iff $P_1 - P_2 > 0$ (or, respectively, $P_1 - P_2 \geq 0$). The small Greek letters α, β, μ, ν, etc. will indicate real numbers. If x is a k-dimensional column vector, $\|x\|$ denotes the Euclidean norm of x, i.e., $\|x\| := \sqrt{x'x}$, and the (induced) norm $\|M\|$ of an rxk matrix M is defined by $\|M\| := \sup_{\|x\|\neq 0}\|Mx\|/\|x\|$ where k and r are integers.

2. EXISTENCE CONDITIONS FOR SOLUTION OF RICCATI EQUATION

The following theorem and corollary will be used to compare the conditions for the existence of stabilizing feedback controller.

Theorem 2.1: Let Q_o and Q_1 be symmetric positive definite and positive semi-definite matrices, respectively, and also let μ_o and $\nu \geq 0$. If there is a symmetric positive definite matrix P_o satisfying

$$A'P_o + P_oA - \mu_oP_oBB'P_o + Q_o + Q_1 = 0 \tag{2.1}$$

and

$$\sqrt{\nu}\|D\| < \frac{\lambda_{min}(Q_o)}{\lambda_{max}(P_o)}, \tag{2.2}$$

then the equation

$$A'P + PA - P(\mu BB' - \nu DD')P + Q = 0 \tag{2.3}$$

has a symmetric positive definite solution P for some $\mu \geq 0$ and for some symmetric matrix $Q > 0$.

The equations of the forms (2.1) and (2.3) will be referred to as Riccati equations in this article regardless of the values of μ_o, μ and ν though they are normally called Lyapunov equations in the literature if $\mu_o = 0$ for (2.1) and if $\mu = \nu = 0$ for (2.3).

Corollary 2.1: Under the conditions given in Theorem 2.1, (2.3) holds for $\mu := \mu_o\lambda_{min}(Q_o)$ and for some $Q > Q_1/\lambda_{min}(Q_o)$.

Remark: The Riccati equation (2.3) appears in many areas of control theory such as game theory (e.g., [12]), H^∞ optimization (e.g.,[13]) as well as stabilizing controller design for uncertain systems (e.g., [7]-[10]) which is the subject of this paper.

To prove the above theorem we need the following lemmas. We first observe

Lemma 2.1: Let Q_1 be symmetric matrix satisfying $Q_1 \geq 0$, and let $\mu \geq 0$ and $\nu \geq 0$. Then, a Riccati equation described by

$$A'P + PA - P(\mu BB' - \nu DD')P + Q = 0 \tag{2.3}$$

has a symmetric positive definite solution P for some symmetric $Q > Q_1$ if and only if, for some symmetric $Q_o > 0$, a Riccati equation

$$A'P + PA - \mu PBB'P + Q_o + Q_1 = 0 \tag{2.4}$$

has a symmetric positive definite solution P satisfying

$$\nu PDD'P < Q_o. \tag{2.5}$$

Proof: For sufficiency, simply note

$$\begin{aligned} &A'P + PA - \mu PBB'P + Q_o + Q_1 \\ &= A'P + PA - \mu PBB'P + (Q_o - \nu PDD'P) + \nu PDD'P + Q_1 \\ &= A'P + PA - \mu PBB'P + \nu PDD'P + Q \end{aligned}$$

where $Q := Q_o - \nu PDD'P + Q_1$. By (2.5), clearly $Q > Q_1$. To prove necessity, note

$$\begin{aligned} &A'P + PA - P(\mu BB' - \nu DD')P + Q \\ &= A'P + PA - \mu PBB'P + Q_o + Q_1 \end{aligned}$$

where $Q_o := Q - Q_1 + \nu PDD'P$. Clearly, $Q_o - \nu PDD'P = Q - Q_1 > 0$. Q.E.D.

Remark: If a symmetric positive definite matrix P is a solution of (2.4) satisfying (2.5), then P is a solution for (2.3), and vice versa.

Remark: It is well-known that (2.4) has a unique symmetric positive definite solution if the pair (A, B) is stabilizable for $\mu > 0$, and if A is stable (i.e., Hurwitz) for $\mu = 0$.

Lemma 2.2: Let P_o and Q_o be symmetric and satisfy $P_o > 0$ and $Q_o > 0$ and let $\nu \geq 0$. Assume $\lambda_{min}(Q_o) \leq 1$. Then, (2.2) implies

$$\nu P_o DD'P_o < Q_o. \tag{2.6}$$

Proof: First note that $\sqrt{\nu}\|D'P_o\| = \sqrt{\nu}\|P_o D\| \leq \sqrt{\nu}\|D\|\,\|P_o\| \leq \sqrt{\nu}\|D\|\lambda_{max}(P_o)$. Therefore, (2.2) implies $\sqrt{\nu}\|D'P_o\| < \lambda_{min}(Q_o)$. But, by assumption $\lambda_{min}(Q_o) \leq 1$, and so $\lambda_{min}(Q_o) \leq \lambda_{min}(Q_o^{1/2})$. Therefore, $\sqrt{\nu}\|D'P_o\| < \lambda_{min}(Q_o^{1/2})$, which implies $\nu\|D'P_o\|^2 < \lambda_{min}(Q_o^{1/2})^2 = \lambda_{min}(Q_o)$. Thus, for any nonzero $x \in R^n$, $x'\nu P_o DD'P_o x \leq \nu\|D'P_o\|^2\|x\|^2 < \lambda_{min}(Q_o)\|x\|^2 \leq x'Q_o x$, which proves (2.6). Q.E.D.

If $\lambda_{min}(Q_o) > 1$, (2.6) is satisfied by scaling down P_o and Q_o as follows.

Lemma 2.3: Let $P_o > 0$ and assume $\lambda_{min}(Q_o) > 1$. Choose a number q so that $\lambda_{min}(Q_o) \leq q$. Then, (2.2) implies

$$\nu(P_o/q)DD'(P_o/q) < (Q_o/q). \tag{2.7}$$

Proof: From (2.2)

$$\sqrt{\nu}\|D\| < \frac{\lambda_{min}(Q_o)}{\lambda_{max}(P_o)} = \frac{\lambda_{min}(Q_o)/q}{\lambda_{max}(P_o)/q} = \frac{\lambda_{min}(Q_o/q)}{\lambda_{max}(P_o/q)}.$$

Let $\tilde{P}_o := P_o/q$ and $\tilde{Q}_o := Q_o/q$. Then,

$$\sqrt{\nu}\|D\| < \frac{\lambda_{min}(\tilde{Q}_o)}{\lambda_{max}(\tilde{P}_o)}.$$

Moreover, $\lambda_{min}(\tilde{Q}_o) = (1/q)\lambda_{min}(Q_o) \leq 1$. Therefore, by Lemma 2.2, we obtain $\nu\tilde{P}_oDD'\tilde{P}_o < \tilde{Q}_o$, which is clearly equivalent to (2.7). Q.E.D.

We are now ready to prove Theorem 2.1.

Proof of Theorem 2.1: Case 1) Assume $\lambda_{min}(Q_o) \leq 1$. Then, by Lemma 2.2, (2.6) holds. Therefore, by Lemma 2.1, (2.3) follows where $Q := Q_o - \nu P_oDD'P_o + Q_1$, $P := P_o$ and $\mu = \mu_o$.

Case 2) Assume $\lambda_{min}(Q_o) > 1$. Choose a number q so that $\lambda_{min}(Q_o) \leq q$ as in Lemma 2.3. Define $\tilde{Q}_o := Q_o/q$, $\tilde{Q}_1 := Q_1/q$ and $\tilde{P}_o := P_o/q$. Then, dividing the both sides of (2.1) by q, we obtain

$$A'(P_o/q) + (P_o/q)A - q\mu_o(P_o/q)BB'(P_o/q) + (Q_o/q) + (Q_1/q) = 0,$$

i.e.,

$$A'\tilde{P}_o + \tilde{P}_oA - q\mu_o\tilde{P}_oBB'\tilde{P}_o + \tilde{Q}_o + \tilde{Q}_1 = 0. \tag{2.8}$$

where $\tilde{Q}_o > o$ and $\tilde{Q}_1 \geq 0$. By (2.2) and by Lemma 2.3, $\nu\tilde{P}_oDD'\tilde{P}_o < \tilde{Q}_o$. It follows by Lemma 2.1 that (2.3) holds for the symmetric positive definite $P := \tilde{P}_o$, for $Q := \tilde{Q}_o - \nu\tilde{P}_oDD'\tilde{P}_o + \tilde{Q}_1 > \tilde{Q}_1 \geq 0$ and for $\mu := q\mu_o$. Q.E.D.

Proof of Corollary 2.1: In the proof of Theorem 2.1, Case 2, choose $q := q_m := \lambda_{min}(Q_o)$. The remaining part of the proof is the same as the proof of Theorem 2.1, Case 2.

3. SYSTEM DESCRIPTION AND ASSUMPTIONS

The system under consideration is described by the equation

$$\dot{x}(t) = Ax(t) + B\phi(u(t); x(t), t) + Dw(x(t), t) \tag{3.1}$$

where $x(t) \in R^n$ and $u(t) \in R^m$ are, respectively, the state and the input at time $t \in [0, \infty)$, A, B and D are, respectively, nxn, nxm and nxm' matrices with real elements, and $\phi(u; x, t)$ and $w(x, t)$ are respectively m and m' vectors whose entries are continuous functions of $u \in R^m$, $x \in R^n$ and $t \in [0, \infty)$. The initial time and the initial state of the system will be denoted as t_o and $x(t_o) = x_o$, respectively.

We assume that the system satisfies the following conditions.

Assumption 1: The solution of (3.1) exists and is unique for each initial state x_o when the state feedback law of the form $u(t) = f(x(t))$ is applied where $f: R^n \rightarrow R^m$ is a continuous function. For this it is sufficient, for instance, that Lipschitz conditions are satisfied for $\phi(f(x); x, t)$ and $w(x, t)$ with respect to $x \in R^n$ for each $t \in [0, \infty)$.

Assumption 2: $\phi(u; x, t)$ satisfies the sector type nonlinearity with respect to u, i.e., there is $\gamma > 0$ such that $\gamma u'u \leq u\phi(u; x, t)$ for all $u \in R^m$, $x \in R^n$ and $t \in [0, \infty)$.

Assumption 3: The nonlinearity $w(x, t)$ has an affine norm bound, i.e., there are nonnegative constants δ and β such that $\|w(x,t)\| \leq \beta(\delta + \|x\|)$ for all $x \in R^n$, and $t \in [0, \infty)$.

Application of the state feedback law $u(t) = f(x(t))$ yields the closed loop system Σ_f described by

$$\dot{x}(t) = Ax(t) + B\phi(f(x(t)); x(t), t) + Dw(x(t), t). \tag{3.2}$$

4. PRACTICAL STABILITY

We adopt the following definition for practical stability. (See [9] and [2]-[4].)

Definition 4.1: Let N be a region around the origin of the state space. Then the system Σ_f is said to be practically stable with respect to N if it satisfies the following conditions:

(i) State trajectories are uniformly bounded, i.e., given any $r > 0$ and any state trajectory $x(.) : [t_o, \infty) \rightarrow R^n$ of Σ_f with $x(t_o) = x_o$, there is a constant $\delta(r) > 0$ such that $\|x_o\| \leq r$ implies $\|x(t)\| \leq \delta(r)$ for all $t \in [t_o, \infty)$.

(ii) Any state trajectory of Σ_f is equiuniformly ultimately bounded with respect to N, i.e., for any $r > 0$ and any state trajectory $x(.)$ satisfying $\|x_o\| \leq r$ there is a number $T_N(r) > 0$ such that $x(t) \in N$ for all $t \geq t_o + T_N(r)$.

5. EXISTENCE OF STABILIZING STATE FEEDBACK CONTROLLER

The following theorem follows immediately from [11].

Theorem 5.1: Suppose a symmetric positive definite matrix P_o satisfies the Riccati equation

$$A'P_o + P_oA - 2\rho_o\gamma P_oBB'P_o + 2Q_o = 0 \tag{5.1}$$

and the condition

$$\beta\|D\| < \frac{\lambda_{min}(Q_o)}{\lambda_{max}(P_o)} \tag{5.2}$$

for some positive constant ρ_o and for some symmetric positive definite matrix Q_o. Then, there is a continuous state feedback law $f(x(t))$ such that the system Σ_f is practically stable with respect to some neighbourhood of the zero state.

Remark. $\beta\|D\|$ is called the measure of mismatch and $\dfrac{\lambda_{min}(Q)}{\lambda_{max}(P)}$ is called the (critical) mismatch threshold in [5] and [11].

The following theorem is analogous to the one in [10]. Also see [7]-[9].

Theorem 5.2: Suppose a symmetric positive definite solution P exists for the Riccati equation

$$A'P + PA - P(2\rho\gamma BB' - \beta^2 DD')P + Q = 0 \tag{5.3}$$

for a positive constant ρ, a symmetric matrix $Q > I$. Then, the system Σ_f is practically stable with respect to some neighbourhood of the origin for the linear state feedback law given by

$$f(x(t)) = -\rho B'Px(t). \tag{5.4}$$

Proof: Given in Appendix.

Minor modifications to the proof of Theorem 5.2 prove

Corollary 5.1: Suppose $\delta = 0$. Suppose also that $\phi(0, 0) = 0$. Then, Σ_f in Theorem 5.2 is globally asymptotically stable.

6. RELATION BETWEEN MISMATCH THRESHOLD AND EXISTENCE OF RICCATI EQUATION SOLUTION

As an immediate consequence of Corollary 2.1, we will now see that, if the measure of mismatch is smaller than the mismatch threshold, then the Game Theoretic Riccati equation has a solution and therefore there is a state feedback control law such that the closed loop system is practically stable.

Theorem 6.1: Suppose the Riccati equation (5.1) has a real symmetric positive definite solution P_o satisfying (5.2) for some positive constants ρ_o and γ and some symmetric positive semi-definite matrix Q_o. Then, the Riccati equation (5.3) has a symmetric positive definite solution for some positive constant ρ and for some $Q > I$.

Proof: In (2.1) let $Q_1 = Q_o$, $\mu_o = 2\rho\gamma$ and in (2.2) let $\nu = \beta^2$. Note that $Q_1/\lambda_{min}(Q_o) = Q_o/\lambda_{min}(Q_o) \geq I$. Then the result follows immediately from Corollary 2.1. Note that the choice of Q is given by $Q := (Q_o/q_m) - \nu(P_o/q_m)DD'(P_o/q_m) + (Q_o/q_m)$ where $q_m := \lambda_{min}(Q_o)$.

7. CONCLUSION

A new sufficient condition has been introduced for the existence of a symmetric positive definite solution for the Riccati equation which originally appeared in Game Theory. It has been used to compare two methods of designing practically stabilizing controllers for a class of uncertain dynamical systems.

8. ACKNOWLEDGEMENT

The author is grateful to Professor G. Leitmann, University of California at Berkeley for his valuable advice.

REFERENCES

[1] R. E. Kalman and J. E. Bertram, "Control system analysis and design via the "Second Method" of Lyapunov I, Continuous-time systems", J. Basic Engineering, Trans. ASME, vol. 82, No. 2, pp. 371-393, 1960.
[2] M. Corless and G. Leitmann, "Continuous state feedback guaranteeing uniform ultimate boundedness for uncertain dynamic systems", IEEE Trans. Automat. Contr., vol.AC-26, No.5, pp.1139-1144, 1981.
[3] G. Leitmann, "On the efficacy of nonlinear control in uncertain linear systems", ASME Journal of Dynamic Systems, Measurement and Control, vol. 102, pp.95-102, June 1981.
[4] B. R. Barmish, M. Corless and G. Leitmann, "A new class of stabilizing controllers for uncertain dynamical systems", SIAM J. Control and optimization,Vol. 21, No2, pp.246 - 255, March 1983.
[5] B. R. Barmish and G. Leitmann, "On ultimate boundedness control of uncertain systems in the absence of matching assumptions", IEEE Trans. Automat. Contr., vol. AC-27, No. 1, pp.153-158, 1982.
[6] B. R. Barmish, "Necessary and sufficient conditions for quadratic stabilizability of an uncertain system", J. Optimization Theory and Applications, vol. 46, No. 4, pp. 399-408, Aug. 1985.
[7] I. R. Petersen and C. V. Hollot, "A Riccati equation approach to the stabilization of uncertain linear systems", Automatica, vol. 22, No. 4, pp. 397-411, 1986.
[8] W. E. Schmitendorf, "Designing stabilizing controllers for uncertain systems using the Riccati equation approach", IEEE Trans. Automat. Contr., vol. 33, No. 4, pp. 376-379, April 1988.
[9] F. Hamano, "Practical stabilization of uncertain dynamical systems by continuous state feedback based on Riccati equation and a sufficient condition for robust practical stability", Proc. of the 28th Conference on Decision and Control, Tampa, Dec.89.
[10] F. Hamano, "Exponential stabilization to a neighborhood of zero state for nonlinear uncertain dynamical systems by linear feedback", Proc. of 28th SICE (Society of Instrumentation and Control Engineers) Annual Conference, Matsuyama, Japan, July, 1989
[11] G. Leitmann and S. Pandey, "A controller for a class of mismatched uncertain systems", Proc. of the 28th Conference on Decision and Control, Tampa, Dec. 1989.
[12] E. F. Mageirou, "Iterative techniques for Riccati game equations", J. Optimization Theory and Applications, vol. 22, No. 1, pp. 51, May 1977.
[13] P. P. Khargonekar, I. R. Petersen and M. A. Rotea, "H^∞-optimal control with state-feedback", IEEE Trans. Automat. Contr., vol. 33, No. 8, pp. 786-788, 1988.
[14] M. Wonham, Linear Multivariable Control, 3rd Edition, Springer, 1985.

APPENDIX

Proof of Theorem 5.1: Proof of uniform equi-ultimate boundedness: It will be shown that for some N to be defined later $T_N(r)$ will be upper bounded by some number a(r).

For this define a candidate Lyapunov function $V(x)$ by

$$V(x) := x'Px \tag{A1}$$

where $x \in R^n$. Then,

$$\dot{V}(x(t)) = \dot{x}(t)'Px(t)x(t)(t) + x(t)'Px(t)\dot{x}(t)$$
$$= \{Ax(t)+B\phi(Fx(t);\, x(t),\, t) + Dw(x(t),\, t)\}'Px(t) + x'(t)P\{Ax(t)+B\phi(Fx(t);\, x(t),\, t) + Dw(x(t),\, t)\}$$
$$= x'(t)(A'P + PA)x(t) + 2x'(t)PB\phi(Fx(t);\, x(t),\, t) + 2x'(t)PDw(x(t),\, t)$$

By using the Riccati equation (5.3), we have

$$\dot{V}(x(t)) = x'(t)\{P(2\rho\gamma BB' - \beta^2 DD')P - Q\}x(t) + 2x'(t)PB\phi(-\rho B'Px(t);\, x(t),\, t) + 2x'(t)PDw(x(t),\, t). \tag{A2}$$

By Assumption 3 it is easy to see that

$$2x'(t)PB\phi(-\rho B'Px(t);\, x(t),\, t) \le -2\gamma\rho x'(t)(B'P)'B'Px(t), \tag{A3}$$

and by Assumption 4

$$2x'(t)PDw(x(t),\, t) \le 2\|D'Px(t)\|\,\|w(x(t),t)\| \le 2\|D'Px(t)\|(\beta(\delta + \|x\|))$$
$$\le \beta^2\|D'Px(t)\|^2 + (\delta + \|x\|)^2.$$
$$= \beta^2 x'(t)PDD'Px(t) + \delta^2 + 2\delta\|x(t)\| + \|x(t)\|^2. \tag{A4}$$

Using the relations (A2)-(A4), we get

$$\dot{V}(x(t)) \le x'(t)\{P(2\rho\gamma BB' - \beta^2 DD')P - Q\}x(t) - 2\gamma\rho x'(t)(B'P)'B'Px(t) + \beta^2 x'(t)PDD'Px(t)$$
$$+ \delta^2 + 2\delta\|x(t)\| + \|x(t)\|^2$$
$$= -x'(t)(Q - I)x(t) + \delta^2 + 2\delta\|x(t)\|$$
$$\le -\lambda_{min}(Q - I)\|x(t)\|^2 + 2\delta\|x(t)\| + \delta^2. \tag{A5}$$

Note that $\lambda_{min}(Q - I) > 0$ since $Q > 0$ by assumption. It follows that $\dot{V}(x(t)) < 0$ for $\|x(t)\| > \varepsilon_o$ where

$$\varepsilon_o := \frac{\delta(1 + \sqrt{1+\lambda_{min}(Q-I)})}{\lambda_{min}(Q-I)}$$

Let $\varepsilon := \varepsilon_o + \tilde{\varepsilon}$. where $\tilde{\varepsilon}$ is an arbitrary (small) positive number and define $N := \{x \in R^n \mid x'Px \le \lambda_{max}(P)\varepsilon^2\}$. Then, by (A5)

$$\dot{V}(x(t)) \le -2\tilde{\varepsilon}\delta\sqrt{1+\lambda_{min}(Q-I)} \tag{A6}$$

for $\|x(t)\| \ge \varepsilon$ and, therefore, for $x \notin N$.

Since the maximum value of $V(x)$ for $x \in R^n$ satisfying $\|x\| = r$ is given by

$$\max_{\|x\|=r} x'Px = \lambda_{max}(P)r^2, \tag{A7}$$

we obtain from (A6) - (A7)

$$T_N(r) \le (\lambda_{max}(P)r^2 - \lambda_{max}(P)\varepsilon^2)/(2\tilde{\varepsilon}\delta\sqrt{1+\lambda_{min}(Q-I)}). \tag{A8}$$

<u>Proof of uniform boundedness</u>: Suppose $\varepsilon_o \le \|x_o\| \le r$. Since $\dot{V}(x(t)) \le 0$ for $\|x(t)\| \ge \varepsilon_o$,

$$\lambda_{min}(P)\|x(t)\|^2 \le V(x(t)) \le V(x_o) \le \lambda_{max}(P)\|x_o\|^2 \le \lambda_{max}(P)r^2$$

which implies

$$\|x(t)\| \le \sqrt{\frac{\lambda_{max}(P)}{\lambda_{min}(P)}}\; r.$$

Now suppose $\|x_o\| \leq \varepsilon_o$. Then by a similar argument,

$$\|x(t)\| \leq \sqrt{\frac{\lambda_{max}(P)}{\lambda_{min}(P)}}\ \varepsilon_o. \qquad \text{Q. E. D.}$$

VECTOR CHANNEL LATTICE FILTERS FOR THE DELTA OPERATOR: COMPLETE DERIVATION

Faryar Jabbari
Department of Mechanical Engineering
University of California, Irvine, CA 92717

Abstract- In this paper, the complete derivation of the vector-channel lattice filter algorithm for least-squares estimation of an input/output model based on the delta operator is presented. Algorithms for residual errors and the auto regressive coefficients are presented also.

I. INTRODUCTION

In this paper, the complete derivation of the vector-channel for the delta operator based least-squares adaptive estimation is presented. The main motivation for delta operator lattices is to combine superior numerical characteristics of the lattices with the advantages of the delta operator approach. Further, lattices allow order-recursive implementation which provides the opportunity for on-line order identification and/or change of order. The derivation here concerns the vector-channel measurements in which there could be p measurement channels, each with m independent measurements. The vector-channel structure forces the sensors (or measurements) in each channel 'see' the same dynamics (see [[JG1] and [W1]) for details). Note that the vector-channel lattice is a generalization of standard least-squares lattices. Indeed, by setting $m = 1$ in the derivation, the standard least-squares lattice is obtained.

In [J1], the derivation for the single (and scalar) channel lattice is presented. Though the derivation in [J1] captures the essence of the delta lattice, it is not suited for control and identification applications, since even a SISO system requires, through imbedding, a two channel structure. This paper presents the complete derivation for the vector-channel lattice filter, including algorithms for the residual filters and the AR coefficients. Since the main contribution of this paper is the derivation itself, we leave much of the discussion on advantages of the lattices or, for that matter, delta operator approach to references such as [JG1], [J1], [MG1], etc. In the following, however, we briefly outline the main motivation behind this work.

Consider the discrete-time representation of a linear system is based often on the backward shift operator; i.e., an Nth model of the form

$$y(k) = \sum_{i=1}^{N} \hat{a}_i y(k-i) + \hat{b}_i u(k-i) = (\sum_{i=1}^{N} \hat{a}_i q^{-i}) y(k) + (\sum_{i=1}^{N} \hat{b}_i q^{-i}) u(k) \tag{1}$$

where $y(k)$ and $u(k)$ are the output(s) and input(s), respectively, at the kth sampling instant. In this formulation q^{-1} is the backward shift operator (i.e., $q^{-1}x(k) = x(k-1)$). Similarly, q is the forward shift operator ($qx(k) = x(k+1)$). It is well known that if this model is obtained by discretization of a continuous-time plant, certain problems arise as the sampling rate is increased. The poles of the system tend to cluster around $z = 1$ and the MA coefficients (the $\hat{b}_i$'s in the equation above) tend to zero. Both can cause problems due to the finite word-length rounding error. Further, in adaptive control applications additional difficulties will be encountered. For example, in adaptive pole placement, the resultant matrix that is inverted (see [GS1] page 482 for details) becomes close to singular as $\hat{b}_i$'s go to zero, causing severe problems.

Goodwin, et al ([MG1], [GSM], etc.) have suggested the use of an alternative approach for modeling the system. This new approach is based on the delta operator, as defined by

$$\delta = \frac{q-1}{T} \tag{2}$$

where q is the forward shift operator, and T is the sampling period. The corresponding ARMA model has the following form

$$\delta^n y(k) = \sum_{i=1}^{N} a_i \delta^{n-i} y(k) + b_i \delta^{n-i} u(k) \tag{3}$$

where as T goes to zero (i.e., δ becomes closer to d/dt), a_i's and b_i's tend to the denominator and numerator coefficients of the continuous-time transfer function. Advantages of this method have been presented in the references listed above, with a complete treatment in [MG2].

The derivation here combines the general approach used for the standard (shift) version of the vector-channel lattice in [JG1], and the techniques used in [J1]. The (infinite) history vectors that are used for forming the underlying subspaces onto which all projections are performed, differ in the two formulation. In [JG1], these vectors were shifted version of one another, while for the delta formulation they have a different structure that creates certain complications. For lattices, the model is imbedded into a larger dimension AR model, of the form

$$\delta^n y(k) = \sum_{i=1}^{N} A_i^N \delta^{n-i} y(k) \tag{4}$$

where A_i^N denotes the ith coefficient for an Nth order model. The derivation of Section II (subsections A, B, and C) solves the least-squares identification problem for (4). For discussion on imbedding, and recovery of the ARMA model, see [JG1]. In Section II.D complete algorithms, with appropriate initializations, are presented. All other technical issues are identical to the standard (shift) lattice and are omitted here for the sake of brevity. The reader is referred to [JG1] for further details.

II. DERIVATION

II.A DEFINITIONS AND ORDER UPDATE EQUATIONS

In this derivation, it is assumed that there are p measurement channels, each of which has m scalar measurements. The measurements from channel i at the kth sampling time, therefore, are

$$Y^i(k) = [y_1^i(k)\ y_2^i(k)\ \cdots y_m^i(k)]^T \quad i = 1, 2, \cdots, p. \tag{5}$$

For this channel, the infinite history vector is defined by

$$\psi^i(k) = [(Y^i(k))^T\ (Y^i(k-1))^T\ \cdots\ \cdots]^T. \tag{6}$$

In the following, the history vectors of the form in (6) are used to derive the lattice equations for filters of any order. These history vectors are assumed to be in the following Hilbert space

$$l_2(R^m, \lambda) = \{\psi = [x_1^T, x_2^T, \ldots]^T\ : <\psi, \psi> = \|\psi\|^2 < \infty\}, \tag{7}$$

where each x_j is a real m-vector. The inner product is defined by

$$<\psi, \hat{\psi}> = \sum_{j=1}^{\infty} \lambda^{j-1} x_j \hat{x}_j, \tag{8}$$

where $\lambda \leq 1$ is the forgetting factor. We will need the following definitions

$$S(k-n) = span\{\psi^1(k-n), \cdots, \psi^p(k-n)\} \tag{9}$$

$$H_n(k) = S(k-n) \oplus \delta S(k-n) \oplus \cdots \oplus \delta^{n-1} S(k-n) \ , \ H_0(k) = \{0\}, \tag{10}$$

$$P_n(k) = Orthogonal\ Projection\ onto\ H_n(k) \tag{11}$$

$$E_n^f(k) = [I - P_n(k)]\ \delta^n S(k-n) \quad forward\ error\ space, \tag{12}$$

$$E_n^b(k) = [I - P_n(k+1)]\ S(k-n) \quad backward\ error\ space, \tag{13}$$

Note that in [JG1], the vectors spanning $H_n(k)$ were shifted versions of $S(k)$; i.e., $S(k-1)$, $S(k-2)$, *etc.* From the definition of the delta operator in (1), it is clear that $\delta\psi^i(k) = \frac{1}{T}\psi^i(k+1) - \frac{1}{T}\psi^i(k)$. As a result,

$$S(k-n) \oplus \delta S(k-n) \oplus \cdots \oplus \delta^{n-1} S(k-n) =$$
$$S(k-n) \oplus S(k-n+1) \oplus \delta S(k-n+1) \cdots \oplus \delta^{n-2} S(k-n+1)$$

which can be used in (10) to obtain

$$\begin{aligned} H_{n+1}(k) &= S(k-n-1) \oplus \delta S(k-n-1) \oplus \cdots \oplus \delta^n S(k-n-1) \\ &= S(k-n-1) \oplus H_n(k) = E_n^b(k-1) \overset{\perp}{\oplus} H_n(k) \end{aligned} \tag{14}$$

where the notation $\overset{\perp}{\oplus}$ indicates the direct sum of orthogonal subspaces of $l_2(R^m, \lambda)$. Therefore, from elementary properties of orthogonal projections, we have

$$P_{n+1}(k) = P_n(k) + P_n^b(k-1), \tag{15}$$

where $P_n^b(k-1)$ is the orthogonal projection onto $E_n^b(k-1)$. From (15), we can write

$$[I - P_{n+1}(k)] = [I - P_n^b(k-1)]\ [I - P_n(k)]. \tag{16}$$

Similarly, for the forward error we can write

$$\begin{aligned} H_{n+1}(k+1) &= S(k-n) \oplus \delta S(k-n) \oplus \cdots \oplus \delta^n S(k-n) \\ &= \delta^n S(k-n) \oplus H_n(k) = E_n^f(k) \overset{\perp}{\oplus} H_n(k). \end{aligned} \tag{17}$$

Again, we have

$$[I - P_{n+1}(k+1)] = [I - P_n^f(k)]\ [I - P_n(k)], \tag{18}$$

where $P_n^f(k)$ is the orthogonal projection onto $E_n^f(k)$. Similar to (12) and (13), we define the forward and backward residual error vectors

$$f_0^i(k) = \psi^i(k), \quad f_n^i(k) = [I - P_n(k)]\ \delta^n \psi^i(k-n), \quad i = 1, 2, \cdots p, \quad forward\ error, \tag{19}$$

$$b_0^i(k) = \psi^i(k), \quad b_n^i(k) = [I - P_n(k+1)]\ \psi^i(k-n), \quad i = 1, 2, \cdots p \ \textit{backward error.} \tag{20}$$

Then,

$$E_n^f(k) = span\{f_n^i(k)\} \ \textit{and} \ E_n^b(k) = span\{b_n^i(k)\}, \quad i = 1, 2, \cdots, p. \tag{21}$$

To obtain the order update equations for the forward error we use (16) and (19)

$$\begin{aligned} f_{n+1}^i(k) &= [I - P_{n+1}(k)]\ \delta^{n+1}\psi^i(k-n-1) \\ &= [I - P_{n+1}(k)]\ \{\frac{1}{T}\delta^n\psi^i(k-n) - \frac{1}{T}\delta^n\psi^i(k-n-1)\} \\ &= \frac{1}{T}[I - P_{n+1}(k)]\ \delta^n\psi^i(k-n) \\ &= \frac{1}{T}[I - P_n^b(k-1)]\ [I - P_n(k)]\ \delta^n\psi^i(k-n) \\ &= \frac{1}{T}f_n^i(k) - \frac{1}{T}P_n^b(k-1)\ f_n^i(k) \end{aligned} \tag{22}$$

for $i = 1, 2, \cdots p$. To simplify equation (22) we introduce the following definitions

$$K_{n+1}(k) \triangleq p \times p \ \textit{matrix with} \ (i,j) \ \textit{element} \ < f_n^i(k), b_n^j(k-1) >, \tag{23}$$

$$R_n^e(k) \triangleq p \times p \ \textit{matrix with} \ (i,j) \ \textit{element} \ < f_n^i(k), f_n^j(k) >, \tag{24}$$

$$R_n^r(k) \triangleq p \times p \ \textit{matrix with} \ (i,j) \ \textit{element} \ < b_n^i(k), b_n^j(k) >, \tag{25}$$

and finally, for notational simplicity, we use the following notation

$$[f_n(k)] = [f_n^1(k) \ \cdots f_n^p(k)], \quad [b_n(k)] = [b_n^1(k) \ \cdots b_n^p(k)], \tag{26}$$

e.g., $[f_n(k)]$ is the matrix that contains the p infinitely long columns of $f_n^i(k)$. Following standard results on orthogonal projections, (22) can be written as

$$[f_{n+1}(k)] = \frac{1}{T}\ [\ f_n(k)] - \frac{1}{T}[b_n(k-1)]R_n^{-r}(k-1)K_{n+1}^T(k). \tag{27}$$

where $R_n^{-r}(k-1)K_{n+1}^T(k)$ means any matrix α such that $R_n^r(k-1)\alpha = K_{n+1}^T(k)$. If R_n^r is nonsingular, R_n^{-r} is its inverse. For more details, concerning existence of such a matrix and computational issues, see Lemma 2.1, the discussion following equation (2.27) and Section III.D in [JG1].

For the backward error, similar manipulations yield

$$\begin{aligned} b_{n+1}^i(k) &= [I - P_{n+1}(k+1)]\ \psi^i(k-n-1) \\ &= [I - P_n^f(k)]\ [I - P_n(k)]\ \psi^i(k-n-1) \\ &= [I - P_n^f(k)]\ b_n^i(k-1) \\ &= b_n^i(k-1) - P_n^f(k)\ b_n^i(k-1) \end{aligned}$$

which, using the notation of (26) and similar to (27), yields

$$[b_{n+1}(k)] = [b_n(k-1)] - [f_n(k)]R_n^{-e}(k)K_{n+1}(k). \tag{28}$$

As in the standard (shift operator) case, only the top rows of (27) and (28) are used in calculations; i.e., the following equations are used in the algorithm

$$e_{n+1}(k) = \frac{1}{T}\ \{\ e_n(k) - r_n(k-1)R_n^{-r}(k-1)K_{n+1}^T(k)\ \} \tag{29}$$

$$r_{n+1}(k) = r_n(k-1) - e_n(k)R_n^{-e}(k)K_{n+1}(k) \tag{30}$$

where $e_n(k)$ is the top m rows of $[f_n(k)]$ and $r_n(k)$ is the top m rows of $[b_n(k)]$. Matrices R_n^r and R_n^e are also updated in the lattice algorithm. Using (27) in (24) and (28) in (25) result in

$$R_{n+1}^e(k) = \frac{1}{T^2}\ \{\ R_n^e(k) - K_{n+1}(k)R_n^{-r}(k-1)K_{n+1}^T(k)\ \}, \tag{31}$$

$$R_{n+1}^r(k) = R_n^r(k-1) - K_{n+1}^T(k)R_n^{-e}(k)K_{n+1}(k). \tag{32}$$

Note that in the four equations above (i.e., 29,30, 31 and 32), two equations are similar to the ones for the shift operator while the other two differ by a factor of T (or T^2) at each stage of the lattice algorithm. As shown in Section II.C, this difference is more pronounced in the equations used to obtain the AR coefficients.

II.B TIME UPDATE EQUATIONS

The following vectors are needed for the time-update equations

$$\phi^i \triangleq (0\ \ 0\ \ \cdots\ \ \underset{\uparrow\, i^{th}\text{ position}}{1}\ \cdots\ 0 \cdots)^T,\quad i = 1,2,\cdots m, \tag{33}$$

and their projections onto $H_n(k+1)$,

$$\hat{\phi}_n^i(k) \triangleq [I - P_n(k+1)]\ \phi^i\ ,\ i = 1,2,\cdots m. \tag{34}$$

where, consistent with the notations used above, $P_n(k+1)$ is the projection operator onto $H_n(k+1)$. We also introduce the following subspace

$$\hat{H}_{n+1}(k) = span\{\phi^i,\ \cdots\phi^m\} \oplus S(k-n+1) \oplus \delta S(k-n+1) \cdots \oplus \delta^{n-1}S(k-n+1) \tag{35}$$

$$= span\{\phi^i,\ \cdots\phi^m\} \oplus H_n(k+1) \tag{36}$$

or alternatively from (6) and (9),

$$\begin{aligned}\hat{H}_n(k+1) &= span\{\phi^i,\cdots \phi^m\} \oplus span\left(\begin{bmatrix} 0^{(m)} \\ \psi^1(k-n)\end{bmatrix},\cdots\begin{bmatrix} 0^{(m)} \\ \psi^p(k-n)\end{bmatrix}\right)\\ &\oplus\; span\left(\begin{bmatrix} 0^{(m)} \\ \delta\psi^1(k-n)\end{bmatrix},\cdots\begin{bmatrix} 0^{(m)} \\ \delta\psi^p(k-n)\end{bmatrix}\right)\oplus\cdots\\ &\oplus\; span\left(\begin{bmatrix} 0^{(m)} \\ \delta^{n-1}\psi^1(k-n)\end{bmatrix},\cdots\begin{bmatrix} 0^{(m)} \\ \delta^{n-1}\psi^p(k-n)\end{bmatrix}\right)\end{aligned}\tag{37}$$

where $0^{(m)}$ is the zero matrix of appropriate dimension (e.g., $m \times 1$ in equation (37)). Also, note that from (36)

$$\hat{H}_n(k+1) = H_n(k+1) \overset{\perp}{\oplus} span\{\hat{\phi}_n^1(k),\cdots \hat{\phi}_n^m(k). \tag{38}$$

Now if we define

$$\hat{P}_n(k+1) = projection\ onto\ \hat{H}_n(k+1) \tag{39}$$

and

$$Q_n(k) = projection\ onto\ span\ \{\hat{\phi}_n^1(k),\cdots,\hat{\phi}_n^m(k)\} \tag{40}$$

it follows from (38) that

$$[I - P_n(k+1)] = [I - \hat{P}_n(k+1)] + Q_n(k). \tag{41}$$

Applying (41) to (20), we have

$$b_n^j(k) = [I - P_n(k+1)]\ \psi^j(k-n) = [I - \hat{P}_n(k+1) + Q_n(k)]\ \psi^j(k-n). \tag{42}$$

We can also exploit the structure of $\psi^j(k-n)$, from (6), through the following

$$b_n^j(k) = [I - \hat{P}_n(k+1)]\ \left\{\sum_{h=1}^{m} y_h^j(k-n)\phi^h + \begin{pmatrix} 0^{(m)} \\ \psi^j(k-n-1)\end{pmatrix}\right\} + Q_n(k)\ \psi^j(k-n) \tag{43}$$

which, in light of (37) and (20), simplifies to

$$b_n^j(k) = \begin{pmatrix} 0^{(m)} \\ b_n^j(k-1)\end{pmatrix} + Q_n(k)\ \psi^j(k-n). \tag{44}$$

Using the same approach used in (27), and the discussion following it, we have

$$Q_n(k)\ \psi^j(k-n) = [\hat{\phi}_n(k)]\ G_n^{-1}(k)\ d \tag{45}$$

where $[\hat{\phi}_n(k)] = [\hat{\phi}_n^1(k)\cdots\hat{\phi}_n^m(k)]$,

$$G_n(k) \triangleq m\times m\ matrix\ with\ (i,j)\ element\ <\hat{\phi}_n^i(k),\hat{\phi}_n^j(k)>, \tag{46}$$

and

$$d = m - vector\ with\ h^{th}\ element\ < \hat{\phi}_n^h(k), \psi^j(k-n) > . \quad (47)$$

Since $[I - P_n(k+1)]$ is self adjoint (being an orthogonal projection), considering (34) and (20), we have

$$h\ element\ of\ d = < \hat{\phi}_n^h(k), \psi^j(k-n) > = < \phi^h, b_n^j(k) > = r_n^{h,j}(k). \quad (48)$$

i.e., the vector d in (45) is the jth column of the matrix $r_n(k)$, defined in (30). To obtain the time update equation for $K_n(k)$ we will need (44-45) and (48)

$$\begin{aligned} K_{n+1}^{i,j}(k) &= < f_n^i(k), b_n^j(k-1) > = < [I - P_n(k)]\ \delta^n \psi^i(k-n), [I - P_n(k)]\psi^j(k-n-1) > \\ &= < \delta^n \psi^i(k-n), [I - P_n(k)]\psi^j(k-n-1) > . \end{aligned}$$

Therefore,

$$\begin{aligned} K_{n+1}^{i,j}(k+1) &= \langle\ \delta^n \psi^i(k-n+1), b_n^j(k)\rangle \\ &= \langle \delta^n \psi^i(k-n+1), \begin{pmatrix} 0^{(m)} \\ b_n^j(k-1) \end{pmatrix} \rangle + \langle \delta^n \psi^i(k-n+1), Q_n(k)\psi^j(k-n)\rangle \\ &= \lambda \langle \delta^n \psi^i(k-n), b_n^j(k-1)\rangle + \langle \delta^n \psi^i(k-n+1), [\hat{\phi}_n(k)]\ G_n^{-1}(k)\ d\ \rangle. \end{aligned} \quad (49)$$

Since

$$\begin{aligned} \langle \delta^n \psi^i(k-n+1), \hat{\phi}_n^h(k)\rangle &= \langle [I - P_n(k+1)]\delta^n \psi^i(k-n+1), \phi^h\rangle \\ &= \langle f_n^i(k+1), \phi^h\rangle = e_n^{h,i}(k+1), \end{aligned} \quad (50)$$

equation (49), in conjunction with (45) and (46), results in

$$\langle \delta^n \psi^i(k-n+1), Q_n(k)\psi^j(k-n)\rangle = [i^{th} column\ of\ e_n(k+1)]^T\ G_n^{-1}(k)\ [j^{th}\ column\ of\ r_n(k)]. \quad (51)$$

Finally, with (49) and (51) we have

$$K_{n+1}(k+1) = \lambda K_{n+1}(k) + e_n^T(k+1)\ G_n^{-1}(k) r_n(k),$$

or

$$K_{n+1}(k) = \lambda K_{n+1}(k-1) + e_n^T(k) G_n^{-1}(k-1) r_n(k-1). \quad (52)$$

The last equation needed to complete the residual filter algorithm is the update equation for $G_n(k)$. Using (46) and (34)

$$\begin{aligned} G_n^{i,j}(k) &= \langle \hat{\phi}_n^i(k), \hat{\phi}_n^j(k) \rangle = \langle \hat{\phi}_n^i(k), \phi^j \rangle \\ &= \langle \phi^i, \phi^j \rangle - \langle P_n(k+1)\, \phi^i, \phi^j \rangle. \end{aligned} \tag{53}$$

Since the backward error vectors are orthogonal to one another, the subspace $H_n(k+1)$ can be written as

$$H_n(k+1) = E_{n-1}^b(k) \overset{\perp}{\oplus} E_{n-2}^b(k) \cdots \overset{\perp}{\oplus} E_0^b(k) \tag{54}$$

so that

$$P_n(k+1) = P_{n-1}^b(k) + P_{n-2}^b(k) + \cdots + P_0^b(k). \tag{55}$$

According to (13) and Lemma 2.1 of [JG1] and recalling the structure of ϕ^i vectors, for each h we have

$$\langle P_h^b(k)\phi^i, \phi^j \rangle = [r_h^{i,1}(k) \cdots r_h^{i,p}(k)]\ R_h^{-r} \begin{bmatrix} r_h^{j,1}(k) \\ \vdots \\ r_h^{j,p}(k) \end{bmatrix}. \tag{56}$$

In view of (53) and (56)

$$G_n(k) = I - \sum_{h=0}^{n-1} r_h(k) R_h^{-r}(k) r_h^T(k), \tag{57}$$

and therefore

$$G_{n+1}(k) = G_n(k) - r_n(k) R_n^{-r}(k) r_n^T(k). \tag{58}$$

II.C AR COEFFICIENTS

It is clear that $P_n(k)\delta^n\psi^i(k-n)$ is a linear combinations of vectors spanning $H_n(k)$ and $P_n(k+1)\psi^i(k-n)$ is a linear combination of vectors spanning $H_n(k+1)$. Consequently, considering the definitions (19), (20) and (34), we can write

$$[f_n(k)] = [\delta^n\psi(k-n)] - \sum_{j=1}^{n}[\delta^{n-j}\psi(k-n)]A_j^n(k) \tag{59}$$

$$[b_n(k)] = [\psi(k-n)] - \sum_{j=1}^{n}[\delta^{n-j}\psi(k-n+1)]B_j^n(k) \tag{60}$$

$$[\hat{\phi}_n(k)] = [\phi] - \sum_{j=1}^{n}[\delta^{n-j}\psi(k-n+1)]C_j^n(k). \tag{61}$$

where $A_j^n(k)$ and $B_j^n(k)$ are $p \times p$ matrices, $C_j^n(k)$ is $p \times m$, and $[f_n(k)]$, and $[b_n(k)]$ are in accordance with the notation defined in (26), while $[\phi] = [\phi^1 \cdots \phi^m]$. Similarly,

$$[\delta^{n-j}\psi(k-n)] = [\delta^{n-j}\psi^1(k-n)\cdots\delta^{n-j}\psi^p(k-n)], \quad for \;\; j = 0,1,2,\cdots.$$

Recall the order update equation for f_n from equation (27):

$$[f_{n+1}(k)] = \frac{1}{T}[f_n(k)] - \frac{1}{T}[b_n(k-1)]R_n^{-r}(k-1)K_{n+1}^T(k).$$

Substituting (59) in the above equation, left and right hand sides become

$$\begin{aligned} R.H.S. \;\; &= \;\; \frac{1}{T}[\delta^n\psi(k-n)] - \sum_{j=1}^{n}\frac{1}{T}[\delta^{n-j}\psi(k-n)]A_j^n(k) \\ &\quad -\frac{1}{T}\;\{\;[\psi(k-n-1)] - \sum_{j=1}^{n}[\delta^{n-j}\psi(k-n)]B_j^n(k-1)\;\}\;R_n^{-r}(k-1)K_{n+1}^T(k) \\ &= \;\; \frac{1}{T}[\delta^n\psi(k-n)] - \sum_{j=1}^{n}\;\frac{1}{T}[\delta^{n-j}\psi(k-n)]\;\{\;A_j^n(k) - B_j^n(k-1)R_n^{-r}(k-1)K_{n+1}^T(k)\;\} \\ &\quad -\frac{1}{T}[\psi(k-n-1)]R_n^{-r}(k-1)K_{n+1}^T(k) \end{aligned} \tag{62}$$

$$L.H.S. \;\; = \;\; [\delta^{n+1}\psi(k-n-1)] - [\delta^n\psi(k-n-1)]A_1^{n+1}(k) - \cdots - [\psi(k-n-1)]A_{n+1}^{n+1}(k)$$

$$
\begin{aligned}
&= \frac{1}{T}[\delta^n\psi(k-n)] - \frac{1}{T}[\delta^n\psi(k-n-1)] - [\delta^n\psi(k-n-1)]A_1^{n+1}(k) - \cdots \\
&= \frac{1}{T}[\delta^n\psi(k-n)] - [\delta^n\psi(k-n-1)](A_1^{n+1}(k)+\frac{1}{T}) - [\delta^{n-1}\psi(k-n-1)]A_2^{n+1}(k)\cdots - \\
&= \frac{1}{T}[\delta^n\psi(k-n)] - \frac{1}{T}[\delta^n\psi(k-n)](A_1^{n+1}+\frac{1}{T}) - \\
&\quad [\delta^{n-1}\psi(k-n-1)]\{A_2^{n+1}(k) - \frac{1}{T}(A_1^{n+1}(k)+\frac{1}{T})\}\cdots
\end{aligned}
$$

Continuing this substitution and dropping momentarily the time index on A

$$
\begin{aligned}
L.H.S. \;=\; & \frac{1}{T}[\delta^n\psi(k-n)] - \frac{1}{T}[\delta^n\psi(k-n)]\{A_1^{n+1} - \frac{1}{T}(-I)\} - \\
& \frac{1}{T}[\delta^{n-1}\psi(k-n)]\{A_2^{n+1} - \frac{1}{T}(A_1^{n+1} - \frac{1}{T}(-I))\} \\
& \cdots - \frac{1}{T}[\psi(k-n)]\{A_n^{n+1} - \frac{1}{T}(A_{n-1}^{n+1} - \frac{1}{T}(\cdots))\} \\
& -[\psi(k-n-1)]\;\{A_{n+1}^{n+1} - \frac{1}{T}(A_n^{n+1} - \frac{1}{T}(\cdots))\}.
\end{aligned} \tag{63}
$$

Matching powers of δ in (63) and (62) yields the following

$$\tilde{A}_i^{n+1}(k) = A_i^n(k) - B_i^n(k-1)R_n^{-r}(k-1)K_{n+1}^T(k) \tag{64}$$

$$A_i^{n+1}(k) = \tilde{A}_i^{n+1}(k) + \frac{1}{T}\tilde{A}_{i-1}^{n+1}(k) \tag{65}$$

with the following end conditions

$$\tilde{A}_0^{n+1}(k) = -I \quad , \quad \tilde{A}_{n+1}^{n+1}(k) = \frac{1}{T}R_n^{-r}(k-1)K_{n+1}^T(k). \tag{66}$$

Similarly, consider the backward error equation of (28)

$$[b_{n+1}(k)] = [b_n(k-1)] - [f_n(k)]R_n^{-e}(k)K_{n+1}(k).$$

Using (59) and (60) in this equation results in

$$L.H.S. = [\psi(k-n-1)] - \sum_{j=1}^{n+1}[\delta^{n+1-j}\psi(k-n)]B_j^{n+1}(k)$$

$$
\begin{aligned}
R.H.S. \;=\; & [\psi(k-n-1)] - \sum_{j=1}^{n}[\delta^{n-j}\psi(k-n)]B_j^n(k-1) \\
& -\{\;[\delta^n\psi(k-n)] - \sum_{j=1}^{n}[\delta^{n-j}\psi(k-n)]A_j^n(k)\;\}R_n^{-e}(k)K_{n+1}(k).
\end{aligned}
$$

Matching powers of δ,

$$B_{i+1}^{n+1}(k) = B_i^n(k-1) - A_i^n(k) R_n^{-e}(k) K_{n+1}(k) \quad i = 1, 2, \ldots n, \tag{67}$$

$$with \;\; B_1^{n+1}(k) = R_n^{-e}(k) K_{n+1}(k). \tag{68}$$

Equations (64) and (67) require $B^n(k-1)$. If the AR algorithm is used at every time step, this does not create any difficulties. However, to reduce computational cost and to provide the ability to increase or decrease the order of the model from one time step to another, it is preferable to calculate $B^n(k-1)$ at time k directly. This can be achieved by adding two extra equations. First, we start by rewriting (44)-(45) as

$$[b_n(k)] = \left[\begin{pmatrix} 0^{(m)} \\ b_n^1(k-1) \end{pmatrix} \cdots \begin{pmatrix} 0^{(m)} \\ b_n^p(k-1) \end{pmatrix} \right] + [\hat{\phi}_n(k)] G_n^{-1}(k) r_n(k)$$

and using (60) and (61) in the above equation to obtain

$$\begin{aligned} [\psi(k-n)] \;\; &- \;\; \sum_{j=1}^{n} [\delta^{n-j} \psi(k-n+1)] B_j^n(k) = \left[\begin{pmatrix} 0^{(m)} \\ \psi^1(k-n-1) \end{pmatrix} \cdots \begin{pmatrix} 0^{(m)} \\ \psi^p(k-n-1) \end{pmatrix} \right] \\ &- \sum_{j=1}^{n} \left[\begin{pmatrix} 0 \\ \delta^{n-j} \psi^1(k-n) \end{pmatrix} \cdots \begin{pmatrix} 0 \\ \delta^{n-j} \psi^p(k-n) \end{pmatrix} \right] B_j^n(k-1) \\ &+ \{ \, [\phi] - \sum_{j=1}^{n} [\delta^{n-j} \psi(k-n+1)] C_j^n(k) \, \} G_n^{-1}(k) r_n(k). \end{aligned} \tag{69}$$

Considering row $(m+1)$ down only, and matching powers of δ results in

$$B_i^n(k-1) = B_i^n(k) - C_i^n(k) G_n^{-1}(k) r_n(k). \tag{70}$$

The last equation needed is the recursion on $C_i^n(k)$. From the development outlined in (53)-(55)

$$\begin{aligned} \hat{\phi}_{n+1}^j(k) = \phi^j - \sum_{i=0}^{n} P_i^b(k) \, \phi^j \;\; &= \;\; \phi^j - \sum_{i=0}^{n-1} P_i^b(k) \, \phi^j - P_n^b(k) \phi^j \\ &= \;\; \hat{\phi}_n^j(k) - [b_n(k)] \; R_n^{-r} \; [r_n^{j,1}(k) \cdots r_n^{j,p}(k)]^T. \end{aligned} \tag{71}$$

Putting all m of these equations next to one another yields

$$[\hat{\phi}_{n+1}(k)] = [\hat{\phi}_n(k)] - [b_n(k)] R_n^{-r}(k) r_n^T(k). \tag{72}$$

Substituting (60) and (61) in (72) will result in

$$
\begin{aligned}
R.H.S. &= [\phi] - \sum_{j=1}^{n}[\delta^{n-j}\psi(k-n+1)]C_j^n(k) \\
&\quad -\{\ \psi(k-n) - \sum_{j=1}^{n}[\delta^{n-j}\psi(k-n+1)]B_j^n(k)\ \}\ R_n^{-r}(k)r_n^T(k) \\
&= [\phi] - \sum_{j=1}^{n}[\ \delta^{n-j}\psi(k-n+1)]\ \{\ C_j^n(k) - B_j^n(k)R_n^{-r}(k)r_n^T(k)\ \} - \psi(k-n)R_n^{-r}(t)r_n^T(k)
\end{aligned}
$$

$$
\begin{aligned}
L.H.S. &= [\phi] - \sum_{j=1}^{n+1}[\delta^{n+1-j}\psi(k-n)]C_j^{n+1}(k) \\
&= [\phi] - \frac{1}{T}[\delta^{n-1}\psi(k-n+1)]C_1^{n+1} + \frac{1}{T}[\delta^{n-1}\psi(k-n)]C_1^{n+1} - [\delta^{n-1}\psi(k-n)]C_2^{n+1} - \cdots \\
&= [\phi] - \frac{1}{T}[\delta^{n-1}\psi(k-n+1)]C_2^{n+1} - \frac{1}{T}[\delta^{n-2}\psi(k-n+1)]\{\ C_2^{n+1} - \frac{1}{T}C_1^{n+1}\ \} \\
&\quad -\frac{1}{T}[\delta^{n-3}\psi(k-n+1)]\{\ C_3^{n+1} - \frac{1}{T}(C_2^{n+1} - \frac{1}{T}C_1^{n+1})\ \}\ - \cdots .
\end{aligned}
$$

Matching the powers of δ

$$\tilde{C}_i^{n+1}(k) = T\ \{\ C_i^n(k) - B_i^n(k)R_n^{-r}(t)r_n^T(k)\ \} \tag{73}$$

$$C_i^{n+1}(k) = \tilde{C}_i^{n+1}(k) + \frac{1}{T}\tilde{C}_{i-1}^{n+1}(k), \tag{74}$$

$$with\ \ \tilde{C}_0^{n+1}(k) = 0\ \ ,\ \ \tilde{C}_{n+1}^{n+1}(k) = R_n^{-r}(k)r_n^T(k). \tag{75}$$

Equations (70), (73), (64) and (467, with the corresponding end conditions, form the complete AR coefficient algorithm. This algorithm can be used whenever a new set of coefficients are needed and for any order desired.

II.D. COMPLETE ALGORITHMS WITH INITIAL CONDITIONS

The Residual Error Lattice Algorithm

Let $Y(k) = [Y^i(k)\ Y^2(k) \cdots Y^p(k)]$ be the $m \times p$ measurement matrix.

Initialize:

$$R_0^e(-1) = R_0^r(-1) = \sigma I,\ \ , K_{n+1}(k) = 0,\ \ for\ n+1 \geq 0$$

where σ is a small number (e.g., $\sigma = 10^{-9}$)

For each $k \geq 0$:

$$e_0(k) = r_0(k) = Y(k), \; G_0(k) = I$$

$$R_0^e(k) = R_0^r(k) = Y^T(k)Y(k) + \lambda R_0^e(k-1)$$

For each $k \geq 1$, for $n = 0$ to $k-1$:

$$K_{n+1}(k) = \lambda K_{n+1}(k-1) + e_n^T(k)G_n^{-1}(k-1)r_n(k-1)$$

$$G_{n+1}(k) = G_n(k) - r_n(k)R_n^{-r}(k)r_n^T(k)$$

$$e_{n+1}(k) = \frac{1}{T}\;\{\; e_n(k) - r_n(k-1)R_n^{-r}(k-1)K_{n+1}^T(k) \;\}$$

$$R_{n+1}^e(k) = \frac{1}{T^2}\{\; R_n^e(k) - K_{n+1}(k)R_n^{-r}(k-1)K_{n+1}^T(k)]$$

$$r_{n+1}(k) = r_n(k-1) - e_n(k)R_n^{-e}(k)K_{n+1}(k)$$

$$R_{n+1}^r(k) = R_n^r(k-1) - K_{n+1}^T(k)R_n^{-e}(k)K_{n+1}(k).$$

<u>The AR Coefficients Algorithm</u>

For $n = 1\;,\; k-1$ and for $i = 1, N$:

$$\tilde{C}_i^{n+1}(k) = T\;\{\; C_i^n(k) - B_i^n(k)R_n^{-r}(k)r_n^T(k) \;\}$$

$$C_i^{n+1}(k) = \tilde{C}_i^{n+1}(k) + \frac{1}{T}\tilde{C}_{i-1}^{n+1}(k),$$

$$B_i^n(k-1) = B_i^n(k) - C_i^n(k)G_n^{-1}(k)r_n(t)$$

$$B_{i+1}^{n+1}(k) = B_i^n(k-1) - A_i^n(k)R_n^{-e}(k)K_{n+1}(k) \quad i = 1, 2, \ldots n,$$

$$\tilde{A}_i^{n+1}(k) = A_i^n(k) - B_i^n(k-1)R_n^{-r}(k-1)K_{n+1}^T(k)$$

$$A_i^{n+1}(k) = \tilde{A}_i^{n+1}(k) + \frac{1}{T}\tilde{A}_{i-1}^{n+1}(k)$$

with

$$\tilde{A}_0^{n+1}(k) = -I \quad , \quad \tilde{A}_{n+1}^{n+1}(k) = \frac{1}{T}R_n^{-r}(k-1)K_{n+1}^T(k),$$

$$B_1^{n+1}(k) = R_n^{-e}(k)K_{n+1}(k),$$

$$\tilde{C}_0^{n+1}(k) = 0 \quad , \quad \tilde{C}_{n+1}^{n+1}(k) = R_n^{-r}(k) r_n^T(k).$$

REFERENCES

CG1 Chen, H-f, Guo, L., "Consistent Estimation of the Order of Stochastic Control Systems", *IEEE Transaction on Automatic Control*, Vol 32, No 6, June 87, pp 531-535.

GSM Goodwin. G.C., Salgado, M.E., Middleton, R.H., "Indirect Adaptive Control: An Integrated Approach", Proceeding of 1988 ACC, Atlanta, Georgia, pp 2440-2445.

J1 Jabbari, F., "Lattice Filters for RLS Estimation of a Delta Operator Based Model," to appear in *IEEE Trans. on AC.*

JG1 Jabbari, F., Gibson, J.S., "Vector Channel Lattice Filters and Identification of Flexible Structures", *IEEE Transactions on Automatic Control*, Vol 33, No 5, May 88, pp 448-456.

MG1 Middleton, R.H., Goodwin. G.C., "Improved Finite Word Length Characteristics in Digital Control Using Delta Operators", *IEEE Transaction on Automatic Control*, Vol 31, No 11, November 86, pp 1015-1021.

MG2 Middleton, R.H., Goodwin. G.C., *Digital Estimation and Control: A Unified Approach*, Prentice-Hall, New Jersey, 1989.

W1 Wiberg, D.M., "Frequencies of vibration Estimated by Lattices", *J. Astronautical Sciences*, Vol 33, 85, pp 35-47.

APPLICATION OF ROBUST DETERMINISTIC CONTROL TO ROBOTIC MANIPULATORS

Chul Goo Kang, *Ph.D.*
Roberto Horowitz, *Associate Professor*
George Leitmann, *Professor*

Department of Mechanical Engineering
University of California at Berkeley
Berkeley, CA 94720, U.S.A.

Abstract

This paper presents an application of the robust deterministic control scheme proposed in [6] to the tracking control of robotic manipulators. Inertia variations due to kinematic changes, unknown and time-varying payload, Coriolis and centrifugal accelerations, and friction and gravitational disturbances are considered as uncertainties in the formulation. "Parasitic" actuator and sensor dynamics of the manipulator system are considered in the controller design and analysis by means of a singular perturbation technique. Simulation and experimental studies have been conducted for a two degree of freedom, direct drive SCARA manipulator to evaluate the performance of the control scheme proposed in [6] and the method of application presented in this paper.

1 Introduction

In recent years, considerable attention has been given to the control of robotic manipulators with the increasing number of robotic manipulators used in industrial applications. Since the manipulator dynamics are highly nonlinear and time-varying, the control of robotic manipulators is not an easy problem but a challenging one. In robotic manipulator systems, decoupling control by means of the exact cancellation of all dynamic disturbances such as varying inertia, Coriolis and centrifugal accelerations, and gravitational and friction disturbances, may not be computationally feasible. Furthermore, there may exist significant "parasitic" actuator and sensor dynamics. One way of solving these difficulties is to use a robust deterministic control scheme such as the one in [6] which can account for "parasitic" (or unmodelled) dynamics, and nonlinearities and time-varying properties of control systems.

The application of deterministic control schemes to robotic manipulators has been addressed by several authors (see e.g. [2], [3], [5], [10]). In all of the above mentioned implementation papers, "parasitic" sensor and actuator dynamics are neglected in the controller design and in the stability analysis. However, some robotic systems, such as the one considered in this paper, have significant "parasitic" dynamics which may cause instability if their effect is neglected in the controller design.

In this paper, the robust deterministic control scheme developed in [6] is applied to the tracking control of n degree of freedom (DOF) direct drive robotic manipulators. The "parasitic" actuator and sensor dynamics of the manipulator system are considered in the controller design and analysis by means of a singular perturbation technique. Inertia variations due to kinematic changes, unknown and time-varying payload, Coriolis and centrifugal accelerations, and friction torques are considered as uncertainties in the formulation. The performance of this deterministic control is evaluated by computer simulations and experiments for a *two* degree of freedom direct drive SCARA manipulator. The detailed description of the application to the *two* degree of freedom direct drive SCARA manipulator is given in a forthcoming paper [7].

This paper is organized as follows. Section 2 describes the model of n degree of freedom manipulator systems. Section 3 presents a method of application of the robust deterministic control scheme to n degree of freedom manipulator systems, and shows the control scheme. As an example of application to a real robotic manipulator, Section 4 considers a *two* degree of freedom direct drive SCARA manipulator, and shows simulation and experimental results. Conclusions are given in Section 5.

2 Dynamic model of n DOF manipulator systems

We consider the dynamics of n degree of freedom rigid link, direct drive robotic manipulators with some "parasitic" actuator and sensor dynamics. The structure of the input/output dynamics of these systems can be described in terms of the block diagram shown in Fig. 1. The dynamics in Fig. 1 are comprised of: single input single output (SISO) blocks which describe the "parasitic" actuator dynamics, SISO blocks which describe the "parasitic" velocity and position sensor dynamics, a multi-input multi-output (MIMO) block which describes the manipulator nonlinear rigid body dynamics, and pure integrator blocks which describe the input-output relations between joint velocities and positions.

The equations of motion for the manipulator nonlinear dynamics block are well known (cf. [1], [4], [11]). The equations of motion in the joint space can be expressed in the following form:

$$\mathbf{M}(\theta(t))\dot{\omega}(t) + \mathbf{v}(\theta(t),\omega(t)) = \tau(t) - \mathbf{f}(\omega(t),\tau(t)) - \mathbf{g}(\theta(t)) \tag{2.1}$$

where $\theta(t)$ is the joint angular position vector, $\omega(t)$ is the joint angular velocity vector, $\tau(t)$ is the input torque vector supplied by the actuators, $\mathbf{M}(\theta(t))$ is the symmetric and positive definite generalized inertia matrix, $\mathbf{v}(\theta(t),\omega(t))$ is the vector due to Coriolis and centrifugal forces, $\mathbf{f}(\omega(t),\tau(t))$ is the friction torque vector and $\mathbf{g}(\theta(t))$ is the gravitational torque vector.

In some direct drive robotic manipulator systems, such as the one considered in this paper, the "parasitic" actuator and sensor dynamics can be considered as linear. Then, the dynamics of the actuator and sensor blocks can be represented by transfer functions with high frequency poles and zeros. The actuator and sensor dynamics can be identified from the experimental frequency response method by way of curve fittings of magnitude and phase plots.

3 Robust deterministic control of n DOF manipulator systems

This section presents a method of application of the robust deterministic control scheme proposed in [6] to the tracking control of n DOF direct drive robotic manipulators. For the summary and terminology of the robust deterministic control scheme, the reader is referred to [9]. The control scheme proposed in [6] is the one for regulation. For the tracking control of robotic manipulators, we define the slow state vector $\mathbf{x}(t)$ and the output vector $\mathbf{z}(t)$ in the full-order system of [6] as follows:

$$\mathbf{x}(t) \triangleq \begin{bmatrix} \theta_d(t) - \theta(t) \\ \omega_d(t) - \omega(t) \end{bmatrix} \triangleq \begin{bmatrix} \mathbf{x}_\theta(t) \\ \mathbf{x}_\omega(t) \end{bmatrix} \tag{3.2}$$

$$\mathbf{z}(t) \triangleq \begin{bmatrix} \hat{\theta}_d(t) - \hat{\theta}(t) \\ \hat{\omega}_d(t) - \hat{\omega}(t) \end{bmatrix} \triangleq \begin{bmatrix} \mathbf{z}_\theta(t) \\ \mathbf{z}_\omega(t) \end{bmatrix}. \tag{3.3}$$

Here, $\theta_d(t)$ is the desired joint position vector and $\omega_d(t)$ is the desired joint velocity vector. $\theta(t)$ is the joint position vector and $\omega(t)$ is the joint velocity vector. $\hat{\theta}(t)$ is the measured joint position vector, i.e., the output of position sensors, $\hat{\omega}(t)$ is the measured joint velocity vector, i.e., the output of velocity sensors. $\hat{\theta}_d(t)$ is the filtered desired position vector, i.e., the prefilter output of $\theta_d(t)$, and $\hat{\omega}_d(t)$ is the filtered desired velocity vector, i.e., the prefilter output of $\omega_d(t)$. For convenience, $\mathbf{x}(t)$ is divided into the two sub-vectors $\mathbf{x}_\theta(t)$ and $\mathbf{x}_\omega(t)$, and $\mathbf{z}(t)$ into the two sub-vectors $\mathbf{z}_\theta(t)$ and $\mathbf{z}_\omega(t)$.

Fig. 2 shows the block diagram of an overall control system. Now we will show how the control system in Fig. 2 can be formulated in terms of the singularly perturbed uncertain system (called the full-order system) described in [6].

3.1 Formulation of the full-order system

Let the transfer function matrices of the position and velocity sensors be stable and be denoted by $G_\theta(s)$ and $G_\omega(s)$, respectively, where s represents the Laplace variable. Then, the measured position and velocity vectors can be expressed by

$$\hat{\theta}(s) = G_\theta(s)\theta(s)$$
$$\hat{\omega}(s) = G_\omega(s)\omega(s).$$

Define the filters of the desired positions and velocities, denoted by $G_{\theta_d}(s)$ and $G_{\omega_d}(s)$, as some convenient factor of $G_\theta(s)$ and $G_\omega(s)$, respectively. Then, the filtered desired position and velocity vectors can be expressed by

$$\hat{\theta}_d(s) = G_{\theta_d}(s)\theta_d(s)$$
$$\hat{\omega}_d(s) = G_{\omega_d}(s)\omega_d(s).$$

In the Laplace transform domain, the two output sub-vectors $\mathbf{z}_\theta(t)$ and $\mathbf{z}_\omega(t)$ can be written as follows:

$$\begin{aligned} \mathbf{z}_\theta(s) &= G_\theta(s)[\theta_d(s) - \theta(s)] + [G_{\theta_d}(s) - G_\theta(s)]\theta_d(s) \\ &= \mathbf{z}_1(s) + \mathbf{z}_2(s) \end{aligned}$$

where

$$\mathbf{z}_1(s) \triangleq G_\theta(s)[\theta_d(s) - \theta(s)] = G_\theta(s)\mathbf{x}_\theta(s), \tag{3.4}$$

$$\mathbf{z}_2(s) \triangleq [G_{\theta_d}(s) - G_\theta(s)]\theta_d(s) \triangleq \Delta G_\theta(s)\theta_d(s), \tag{3.5}$$

and

$$\begin{aligned} \mathbf{z}_\omega(s) &= G_\omega(s)[\omega_d(s) - \omega(s)] + [G_{\omega_d}(s) - G_\omega(s)]\omega_d(s) \\ &= \mathbf{z}_3(s) + \mathbf{z}_4(s) \end{aligned}$$

where

$$\mathbf{z}_3(s) \triangleq G_\omega(s)[\omega_d(s) - \omega(s)] = G_\omega(s)\mathbf{x}_\omega(s), \tag{3.6}$$

$$\mathbf{z}_4(s) \triangleq [G_{\omega_d}(s) - G_\omega(s)]\omega_d(s) \triangleq \Delta G_\omega(s)\omega_d(s). \tag{3.7}$$

The following state space realizations can be defined for the transfer function matrices in Eqs. (3.4) and (3.6):

$$\begin{aligned} \mathbf{z}_1(t) &= \mathbf{C}_\theta \mathbf{y}_\theta(t) \\ \dot{\mathbf{y}}_\theta(t) &= \mathbf{F}_\theta \mathbf{y}_\theta(t) + \mathbf{B}_\theta \mathbf{x}_\theta(t), \end{aligned} \tag{3.8}$$

$$\begin{aligned} \mathbf{z}_3(t) &= \mathbf{C}_\omega \mathbf{y}_\omega(t) \\ \dot{\mathbf{y}}_\omega(t) &= \mathbf{F}_\omega \mathbf{y}_\omega(t) + \mathbf{B}_\omega \mathbf{x}_\omega(t), \end{aligned} \tag{3.9}$$

where $\mathbf{F}_\theta$ and $\mathbf{F}_\omega$ are chosen to be real block-diagonal matrices, and $\mathbf{B}_\theta, \mathbf{B}_\omega, \mathbf{C}_\theta, \mathbf{C}_\omega$ are selected as real matrices of appropriate dimensions.

Similar to the "parasitic" sensor dynamics, the input-output relations for the "parasitic" actuator dynamics can be expressed by

$$\tau(s) = G_u(s)\mathbf{u}(s). \tag{3.10}$$

The state space realization for the transfer function matrix in Eq. (3.10) can be defined by

$$\dot{\tau}(t) = \mathbf{F}_u\tau(t) + \mathbf{B}_u\mathbf{u}(t), \tag{3.11}$$

where $\mathbf{F}_u$ is chosen as a real block-diagonal matrix, and $\mathbf{B}_u$ is given as an real matrix with appropriate dimension.

The singular perturbation parameter μ is now defined by

$$\mu \triangleq \frac{1}{|\,\sigma\,|_{min}} \tag{3.12}$$

where $|\,\sigma\,|_{min}$ is the smallest value among the absolute values of the real parts in the eigenvalues of $\mathbf{F}_\theta, \mathbf{F}_\omega$ and $\mathbf{F}_u$ in Eqs. (3.8), (3.9) and (3.11).

Equation (3.5) can be written as

$$\mathbf{z}_2(s) = \Delta G_\theta(s)\theta_d(s) \triangleq \Delta G_{1\theta}(s)\Delta G_{2\theta}(s,\mu)\theta_d(s), \tag{3.13}$$

where $\Delta G_{2\theta}(s,\mu)$ is selected such that $\Delta G_{2\theta}(s,0) = 0$. This condition may be satisfied by selecting $\Delta G_{1\theta}(s) = G_{\theta_d}(s)$ and, as previously mentioned, selecting $G_{\theta_d}(s)$ as a factor of $G_\theta(s)$. A state space realization for Eq. (3.13) can be defined by

$$\begin{aligned} \mathbf{z}_2(t) &= \mathbf{C}_{\tilde{\theta}}\mathbf{y}_{\tilde{\theta}}(t) \\ \dot{\mathbf{y}}_{\tilde{\theta}} &= \mathbf{F}_{\tilde{\theta}}\mathbf{y}_{\tilde{\theta}}(t) + \mathbf{B}_{\tilde{\theta}}\mathbf{g}_{21}(t,\mu) \end{aligned} \tag{3.14}$$

where $\mathbf{g}_{21}(t,\mu) = \mathcal{L}^{-1}\{\Delta G_{2\theta}(s,\mu)\theta_d(s)\}$ and $\mathcal{L}^{-1}$ denotes the inverse Laplace transform operator. As in previous cases, $\mathbf{F}_{\tilde{\theta}}$ is selected to be a block-diagonal matrix, and $\mathbf{B}_{\tilde{\theta}}$ and $\mathbf{C}_{\tilde{\theta}}$ are selected as real matrices of appropriate dimensions.

Similarly, Eq. (3.7) can be expressed as

$$\mathbf{z}_4(s) = \Delta G_\omega(s)\omega_d(s) \triangleq \Delta G_{1\omega}(s)\Delta G_{2\omega}(s,\mu)\omega_d(s), \tag{3.15}$$

where $\Delta G_{2\omega}(s,\mu)$ is chosen such that $\Delta G_{2\omega}(s,0) = 0$. A state space realization for Eq. (3.15) can be defined by

$$\begin{aligned} \mathbf{z}_4(t) &= \mathbf{C}_{\tilde{\omega}}\mathbf{y}_{\tilde{\omega}}(t) \\ \dot{\mathbf{y}}_{\tilde{\omega}}(t) &= \mathbf{F}_{\tilde{\omega}}\mathbf{y}_{\tilde{\omega}}(t) + \mathbf{B}_{\tilde{\omega}}\mathbf{g}_{22}(t,\mu), \end{aligned} \tag{3.16}$$

where $\mathbf{g}_{22}(t,\mu) = \mathcal{L}^{-1}\{\Delta G_{2\omega}(s,\mu)\omega_d(s)\}$. Note that $\mathbf{g}_{2i}(t,0) = 0$, $i = 1,2$, because $\mathcal{L}^{-1}$ is a linear operator. Since $\Delta G_{2\theta}(s,\mu)$ and $\Delta G_{2\omega}(s,\mu)$ are stable transfer functions, and $\theta_d(t)$ and $\omega_d(t)$ are assumed to be bounded, $\mathbf{g}_{2i}(t,\mu)$ is bounded for given $\mu \in (0,\infty)$.

Defining the state vector $\mathbf{y}(t)$ of the fast subsystem by

$$\mathbf{y}^T(t) \triangleq [\mathbf{y}_\theta^T(t), \mathbf{y}_{\tilde{\theta}}^T(t), \mathbf{y}_\omega^T(t), \mathbf{y}_{\tilde{\omega}}^T(t), \tau^T(t)], \tag{3.17}$$

and using the definitions of the slow state vector $\mathbf{x}(t)$, and output vector $\mathbf{z}(t)$ in Eqs. (3.2) and (3.3), we can formulate the following single state equation from Eqs. (3.8), (3.9), (3.11), (3.14) and (3.16); this state equation corresponds to the fast subsystem in [6] :

$$\mu\dot{\mathbf{y}}(t) = \mathbf{F}[\mathbf{A}_{21}\mathbf{x}(t) + \mathbf{y}(t) + \mathbf{B}_2\mathbf{u}(t)] + \mathbf{g}_2(t,\mathbf{x}(t),\mathbf{y}(t),\mu) \tag{3.18}$$

$$\mathbf{z}(t) = \mathbf{C}\mathbf{y}(t)$$

where

$$\mathbf{F} = \frac{1}{|\sigma|_{min}} diag(\mathbf{F}_\theta, \mathbf{F}_{\tilde{\theta}}, \mathbf{F}_\omega, \mathbf{F}_{\tilde{\omega}}, \mathbf{F}_u),$$

$$\mathbf{A}_{21} = \begin{bmatrix} \mathbf{F}_\theta{}^{-1}\mathbf{B}_\theta & 0 \\ 0 & 0 \\ 0 & \mathbf{F}_\omega{}^{-1}\mathbf{B}_\omega \\ 0 & 0 \\ 0 & 0 \end{bmatrix}, \quad \mathbf{B}_2 = \begin{bmatrix} 0 \\ 0 \\ 0 \\ 0 \\ -\mathbf{I} \end{bmatrix},$$

$$\mathbf{C} = \begin{bmatrix} \mathbf{C}_\theta & \mathbf{C}_{\tilde{\theta}} & 0 & 0 & 0 \\ 0 & 0 & \mathbf{C}_\omega & \mathbf{C}_{\tilde{\omega}} & 0 \end{bmatrix}, \quad \mathbf{S} = 0,$$

and

$$g_2(t,\mu) = \frac{1}{|\sigma|_{min}} \begin{bmatrix} 0 & 0 \\ B_{\tilde{\theta}} & 0 \\ 0 & 0 \\ 0 & B_{\tilde{\omega}} \\ 0 & 0 \end{bmatrix} \begin{bmatrix} g_{21}(t,\mu) \\ g_{22}(t,\mu) \end{bmatrix}.$$

We will now derive the corresponding state equation for the slow subsystem considered in [6], using the nonlinear manipulator equations of motion in Eq. (2.1). Eq. (2.1) can be written as

$$\dot{\omega}(t) = \mathbf{B}(\theta(t)) \, [\tau(t) - \mathbf{v}(\theta(t),\omega(t)) - \mathbf{f}(\omega(t),\tau(t)) - \mathbf{g}(\theta(t))], \tag{3.19}$$

where $\mathbf{B}(\theta) \triangleq \mathbf{M}^{-1}(\theta) \triangleq \mathbf{B}_o[\mathbf{I} + \mathbf{E}(\theta)]$ with

$$\mathbf{B}_o \triangleq \frac{\max_\theta\{\lambda_{max}(\mathbf{B}(\theta))\} + \min_\theta\{\lambda_{min}(\mathbf{B}(\theta))\}}{2}. \tag{3.20}$$

$\lambda_{max}(\mathbf{B}(\theta))$ and $\lambda_{min}(\mathbf{B}(\theta))$ are the maximum and minimum eigenvalues of $\mathbf{B}(\theta)$ for given θ. The definition (3.20) guarantees that $\beta \triangleq max\|\mathbf{E}(\theta)\| < 1$, which is a required condition in Assumption 3 (iv) of [6].

Equation (3.19) can be rewritten as

$$\begin{aligned} \dot{\omega}(t) = &\mathbf{B}_o\tau(t) + \mathbf{B}_o\mathbf{E}(\theta(t))\tau(t) \\ &-\mathbf{B}(\theta(t))[\mathbf{v}(\theta(t),\omega(t)) + \mathbf{f}(\omega(t),\tau(t)) + \mathbf{g}(\theta(t))]. \end{aligned} \tag{3.21}$$

Using the definitions of $\mathbf{x}(t)$ and $\mathbf{y}(t)$ given in Eq. (3.2) and Eq. (3.17), the following state equation is derived from Eq. (3.21):

$$\dot{\mathbf{x}}(t) = \mathbf{A}_{11}\mathbf{x}(t) + \mathbf{A}_{12}\mathbf{y}(t) + \mathbf{g}_1(t,\mathbf{x}(t),\mathbf{y}(t)) \tag{3.22}$$

where

$$\mathbf{A}_{11} = \begin{bmatrix} 0 & \mathbf{I} \\ 0 & 0 \end{bmatrix},$$

$$\mathbf{A}_{12} = \begin{bmatrix} 0 & 0 & 0 & 0 & 0 \\ 0 & 0 & 0 & 0 & -\mathbf{B}_o \end{bmatrix},$$

$$\mathbf{B}_1 = 0,$$

$$\mathbf{g}_1(t,\mathbf{x}(t),\mathbf{y}(t)) = \begin{bmatrix} 0 \\ \mathbf{g}_{12} \end{bmatrix}, \tag{3.23}$$

$$\begin{aligned} \mathbf{g}_{12} = &\dot{\omega}_d(t) - \mathbf{B}_o\mathbf{E}(\theta(t))\tau(t) \\ &+ \mathbf{B}(\theta(t))[\mathbf{v}(\theta(t),\omega(t)) + \mathbf{f}(\omega(t),\tau(t)) + \mathbf{g}(\theta(t))]. \end{aligned}$$

The system described by equations (3.18) and (3.22) is included in the class of the full-order systems considered in [6].

3.2 Robust deterministic controller

The robust deterministic controller of [6] which assures a desired stability property (specifically called "global uniform attractivity") of system (3.18) and (3.22) [1], is of the form

$$\mathbf{u}(t) = \mathbf{p}_l(\mathbf{z}(t)) + \mathbf{p}_n(\mathbf{z}(t)) \tag{3.24}$$

with

$$\mathbf{p}_l(\mathbf{z}(t)) \triangleq -\gamma_1 \bar{\mathbf{B}}^T \mathbf{K}\mathbf{z}(t), \tag{3.25}$$

$$\mathbf{p}_n(\mathbf{z}(t)) \triangleq -\frac{\bar{\mathbf{B}}^T \mathbf{K}\mathbf{z}(t)}{\|\bar{\mathbf{B}}^T \mathbf{K}\mathbf{z}(t)\| + \epsilon}\rho, \tag{3.26}$$

where $\epsilon > 0$ is a design parameter, and the intervals of γ_1 and ρ for the stable contol system are given in [6]. In Eqs. (3.25) and (3.26), $\mathbf{K}$ is the positive definite solution of the Riccati equation

$$\mathbf{K}\bar{\mathbf{A}} + \bar{\mathbf{A}}^T\mathbf{K} + \mathbf{Q} - 2\gamma_0 \mathbf{K}\bar{\mathbf{B}}\bar{\mathbf{B}}^T\mathbf{K} = 0, \tag{3.27}$$

where $\mathbf{Q}$ is a symmetric and positive definite matrix and γ_0 is a non-negative constant. The stability of system (3.18) and (3.22) implies that the real positions and velocities of the robotic manipulator track the desired positions and velocities.

4 Simulation and experiment

The problem formulation in this paper was applied to a *two* degree of freedom direct drive SCARA manipulator (called Berkeley/NSK manipulator) which was designed by the Department of Mechanical Engineering of the University of California at Berkeley and the Nippon Seiko K.K. (NSK) Corporation. Fig. 3 shows the schematic drawing of the Berkeley/NSK manipulator. The manipulator consists of two NSK Megatorque motors, two NSK motor drivers, two aluminum links, and two resolvers. The link dimensions and important model parameters of the Berkeley/NSK manipulator are given in Table 1. Note that gravitational torques are negligible in this manipulator. The details of the computer simulation model and the physical system of the Berkeley/NSK manipulator can be found in [8]. The detailed expressions for the transfer functions and the matrices in the state eqations for the Berkeley/NSK manipulator are given in [7].

Computer simulation and experimental studies for the Berkeley/NSK manipulator were conducted to test and evaluate the proposed control scheme. The desired trajectory for both axes is given by a seventh order polynomial, simulating a pick-and-place operation. Fig. 4 shows the desired trajectory and payload changes. The design parameters used in the simulations and the experiments were $\mathbf{Q} = diag(16000, 5, 170, 0.1)$, $\gamma_0 = 0.13$, $\gamma_1 = 5.8$, $\epsilon = 40$, $\rho = 50$.

The existence of μ^*, which is the upper limit of the singular perturbation parameter μ that assures a stable control system, was determined in successive numerical simulations. Simulation tests were performed by artificially increasing the values of μ until the feedback system became unstable. These results are shown in Fig. 5 in which the system is stable for $\mu = 1.25 \times 5.16 \times 10^{-3}$, but becomes unstable for $\mu = 2 \times 5.16 \times 10^{-3}$. Note that the actual singular perturbation parameter of the Berkeley/NSK manipulator is $\mu = 5.16 \times 10^{-3}$.

[1] All assumptions of [6], except one, are satisfied by system (3.18) and (3.22); the assumption which is not fully met is satisfied approximately; see [7].

To investigate the possibility of implementing the control law with a digital computer, the continuous time control was discretized and approximated by a zero-order hold. In the simulations, nonlinearities such as torque saturations of the actuators, voltage saturations of the D/A converters and quantization effects, as well as "parasitic" actuator and sensor dynamics, were considered.

Fig. 6, Fig. 7 and Fig. 8 show the simulation results for the position tracking, position tracking errors and control inputs of both axes when the discrete control is used with sampling time = 8 ms, payload = 0 and computational delay = 3.8 ms. The 3.8 ms is the actual time required to execute all the calculations of the control algorithm with an IBM-AT (80286-8 MHz, 80287-8 MHz).

When the task is executed with picking up an 6.8 Kg payload, the tracking errors increase. The simulation results in this case are shown in Fig. 9 and Fig. 10. Notice that, when the payload is between zero and 6.8 Kg, the control system is always stable. In other words, the proposed control scheme is robust with respect to bounded payload variations.

Fig. 11 shows the experimental results for the position tracking errors when the manipulator is operated without payload. In the experiments, the same design parameters were used as the ones in the simulations. Fig. 7 and Fig. 11 show that the tracking error trends are consistent in the simulation and the experiment.

Furthermore, with many simulations and/or experiments which are not shown in this paper, the following additional facts for this control scheme were found: (i) the tracking errors can be made smaller by reducing ϵ in the control and adjusting other design parameters, provided that chattering of input signals can be endured, and (ii) the effect of discretization (by a zero order hold) of the continuous time control is not significant, but including the computational delay in the discretized control has a significant effect. For example, computational delays led to chattering problems in the control inputs. This is one motivation for deriving a digital version of this control scheme instead of the one obtained by discretization of the continuous time control

5 Conclusions

This paper presents an application of the robust deterministic control scheme proposed in [6] to the tracking control of robotic manipulators. Inertia variations due to kinematic changes, unknown and time-varying payload, Coriolis and centrifugal accelerations, and friction and gravitational disturbances have been considered as uncertainties in the formulation. "Parasitic" actuator and sensor dynamics of the manipulator system have been considered in the controller design and analysis by means of a singular perturbation technique.

Simulation and experimental studies were conducted for a two degree of freedom, direct drive SCARA manipulator to evaluate the performance of the control scheme proposed in this paper. The simulation and experimental results showed that: (i) the deterministic control scheme proposed in this paper is effective for the tracking control of a robotic manipulator even in the presence of "parasitic" actuator and sensor dynamics, (ii) the controller is robust with respect to inertia variations due to kinematic changes and payload variations, and (iii) the experimental results are consistent with the simulation results.

The method of application of the control scheme presented in this paper is specific to the control of an n degree of freedom direct drive robotic manipulator. However, it can be used to control any general n degree of freedom robotic manipulator or other mechanical system where "parasitic" dynamics are present and the measurement or calculation of the varying system parameters is not practical. Work is in progress to modify the method so that all the assumptions, including Assumption 3, are satisfied. The development of the digital version of the deterministic control scheme is another topic for further study.

Acknowledgements

This work was supported by the National Science Foundation under grant MSM-8657520 and grant ECS-8602524.

References

[1] H. Asada and J. Slotine. *Robot Analysis and Control.* Wiley-Interscience, 1986.

[2] Y. H. Chen. Robust control of mechanical manipulators. *ASME J. Dynam. Syst. Meas. Contr.*, submitted.

[3] M. Corless and G. Leitmann. Deterministic control of uncertain systems. In *Proc. American Control Conf.*, pages 2019–2025, Atlanta, Georgia, June 1988.

[4] John J. Craig. *Introduction to Robotics: Mechanics and Control.* Addison-Wesley Publ. Co., 1986.

[5] R. Horowitz, H. Stephens, and G. Leitmann. Experimental verification of a deterministic controller for a d.c. motor with uncertain dynamics. In *Proc. American Control Conf.*, pages 1900–1907, Minneapolis, MN, June 1987.

[6] C. G. Kang, R. Horowitz, and G. Leitmann. Robust deterministic control for robotic manipulators - Part I: Control theory. *ASME J. Dynam. Syst. Meas. Contr.*, submitted.

[7] C. G. Kang, R. Horowitz, and G. Leitmann. Robust deterministic control for robotic manipulators - Part II: Application to robot. *ASME J. Dynam. Syst. Meas. Contr.*, submitted.

[8] C. G. Kang, W. W. Kao, M. Boals, and R. Horowitz. Modeling, identification and simulation of a two link Scara manipulator. In *Proc. Winter Annual Meeting of ASME: Symposium on Robotics*, pages 393–407, Chicago, IL, Nov. 1988.

[9] C. G. Kang, G. Leitmann, and R. Horowitz. Robust deterministic controller design of a two degree of freedom SCARA manipulator. In *Proc. American Control Conf.*, pages 1457–1462, Pittsburgh, PA, 1989.

[10] R. Shoureshi, M. Corless, and M. D. Roesler. Control of industrial manipulator with bounded uncertainties. *ASME J. Dynam. Syst. Meas. Contr.*, 109:53–59, 1987.

[11] M. W. Spong and M. Vidyasagar. *Robot Dynamics and Control.* John Wiley & Sons, 1989.

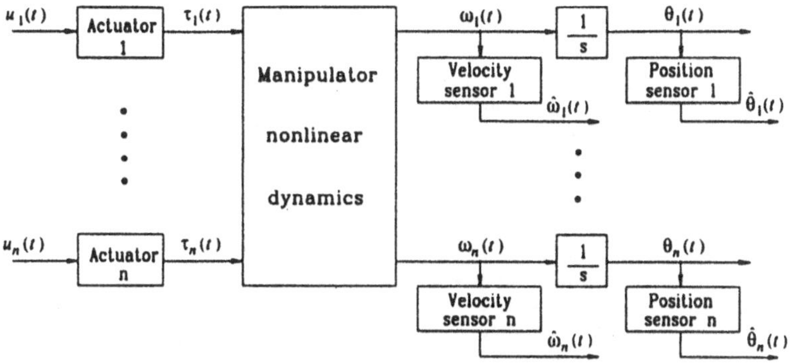

Fig. 1 Modelling structure of an n degree of freedom direct drive manipulator system

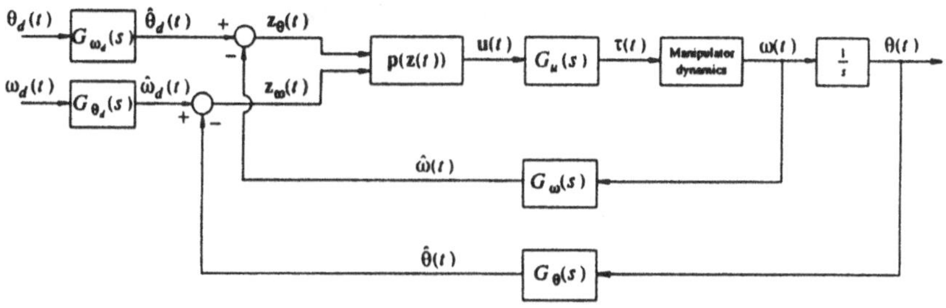

Fig. 2 Block diagram of the overall control system

Table 1 Physical parameters of the Berkeley/NSK manipulator

Definition	Value
Lower link inertia (about c.g.)	0.360 $Kg \cdot m^2$
Upper link inertia (about c.g.)	0.051 $Kg \cdot m^2$
Rotor inertia of the lower motor	0.2675 $Kg \cdot m^2$
Rotor inertia of the upper motor	0.0077 $Kg \cdot m^2$
Stator and housing inertia of the upper motor	0.040 $Kg \cdot m^2$
Payload inertia (about c.g.)	0.046 $Kg \cdot m^2$
Lower link mass	10.6 Kg
Upper link mass	4.85 Kg
Lower motor mass	73.0 Kg
Upper motor mass	12.0 Kg
Payload mass	6.81 Kg
Lower link length	0.36 m
Upper link length	0.24 m
Distance from c.g. of the lower link to the lower motor axis	0.139 m
Distance from c.g. of the upper link to the upper motor axis	0.099 m
Lower axis Coulomb friction	5.7 $N \cdot m$
Upper axis Coulomb friction	0.9 $N \cdot m$

Note: c.g. implies the center of gravity.

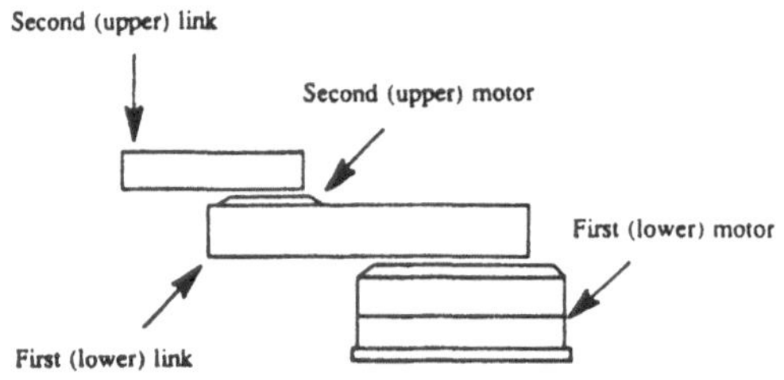

Fig. 3 Schematic drawing of the Berkeley/NSK manipulator

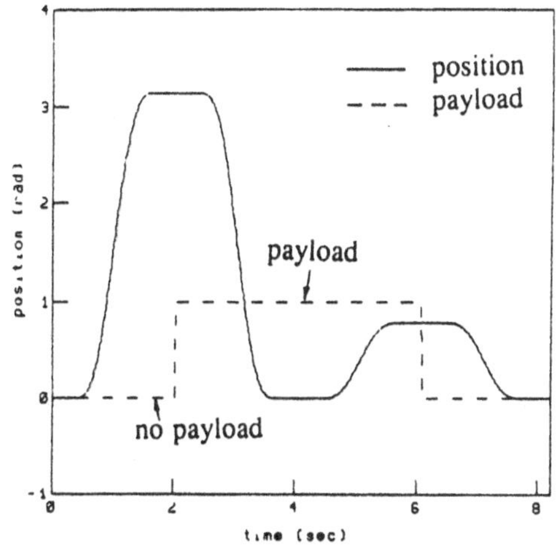

Fig. 4 Desired trajectory and payload changes

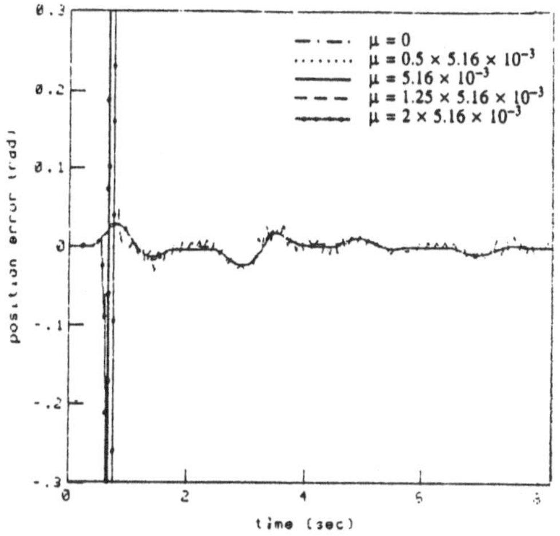

Fig. 5 Position tracking errors $(\theta_d - \theta)$ of the first (lower) axis when continuous time control is used without payload; simulation

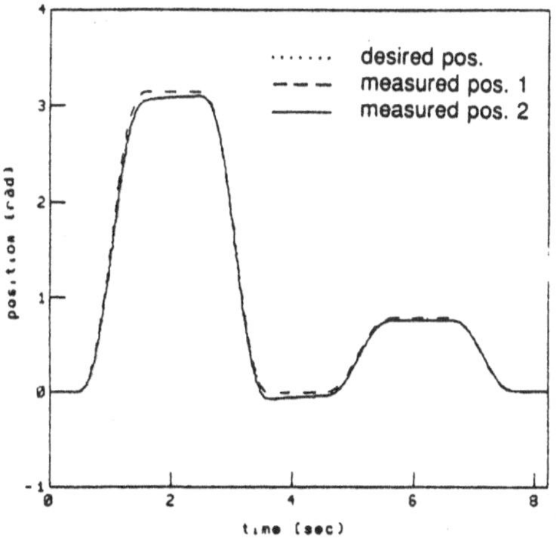

Fig. 6 Position tracking when discrete control is used without payload; simulation

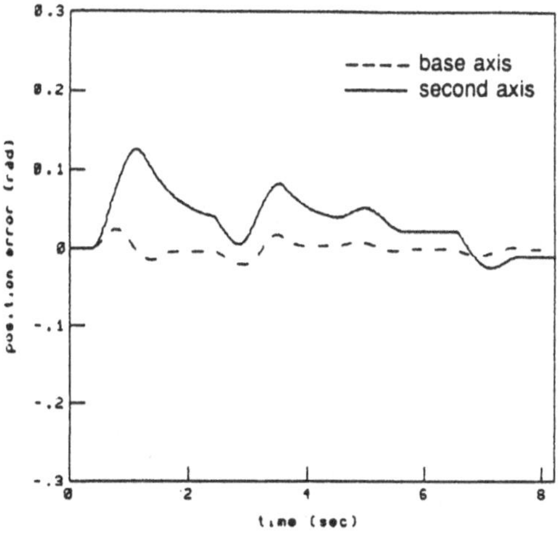

Fig. 7 Position tracking errors $(\theta_d - \hat{\theta})$ when discrete control is used without payload; simulation

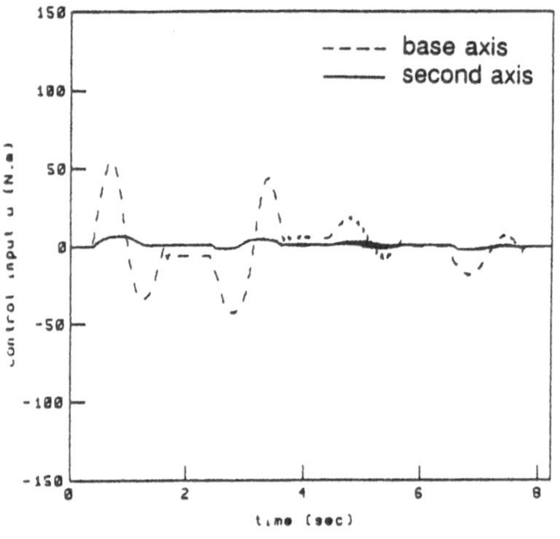

Fig. 8 Control input u when discrete control is used without payload; simulation

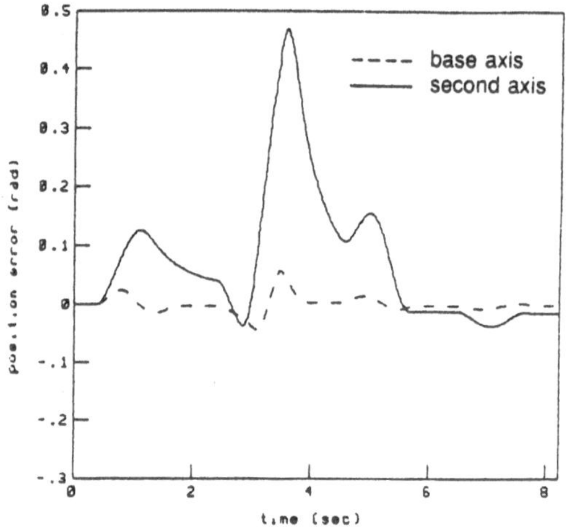

Fig. 9 Position tracking errors $(\theta_d - \hat{\theta})$ when discrete control is used with payload; simulation

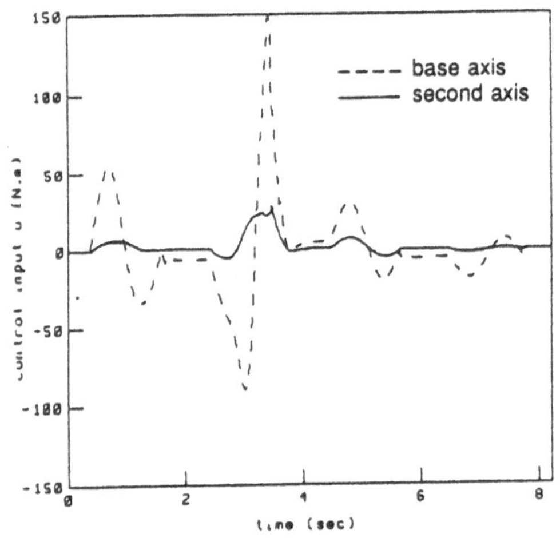

Fig. 10 Control input u when discrete control is used with payload; simulation

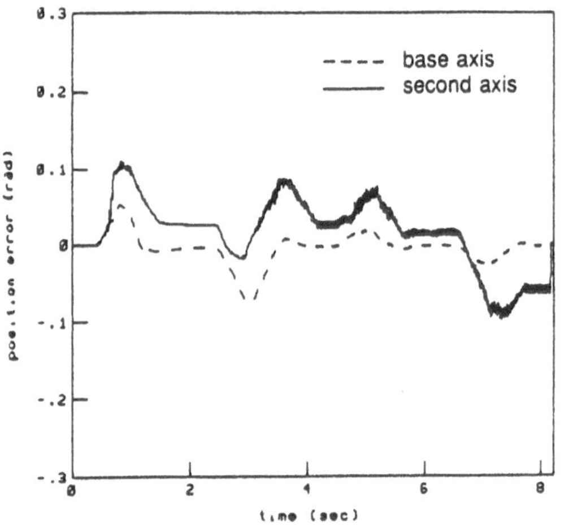

Fig. 11 Position tracking errors $(\theta_d - \hat{\theta})$ when discrete control is used without payload; experiment

ESTIMATION OF REGIONS OF ASYMPTOTIC STABILITY WITH SLIDING FOR RELAY-CONTROL SYSTEMS

S. Mehdi Madani-Esfahani*
School of Electrical Engineering
Purdue University
West Lafayette, IN 47907

Stefen Hui**
Department of Mathematical Sciences
San Diego State University
San Diego, CA 92182-0314

and

Stanislaw H. Żak*
School of Electrical Engineering
Purdue University
West Lafayette, IN 47907

Abstract

The object of this paper is the estimation of stability boundaries and regions of asymptotic stability with sliding for a class of relay-control systems. The direct method of Lyapunov is used to obtain these estimates. A coordinate transformation which brings the system into a special canonical form is utilized to facilitate the stability analysis. The proposed approach to stability regions estimation is applied to a class of second-order systems and analytical expressions for stability regions are derived.

*The work of these authors was supported by the School of Electrical Engineering of Purdue University, West Lafayette, IN 47907.

**The work of this author was partially supported by the National Science Foundation.

Key Words:

Direct method of Lyapunov, Variable structure control, State transformation, State feedback, Stability.

AMS Subject Classification:

93D05, 93D20, 34D20, 93B17.

1. Introduction

Relay-control systems and variable structure controls, the control of dynamical systems with discontinuous feedback controllers, have been studied over the last twenty five years. (See for example [1]-[6]; [8]-[11]; see [2], [4] for surveys). This type of control rests on the concept of changing the structure of the controller in response to the changing states of the system to obtain a desire response. This is accomplished by using high speed switching which forces the trajectories of the system onto the switching surface where they are maintained thereafter. The main advantage of this type of control is its robustness against parameter variations and disturbances while the trajectories are on the switching surface. The motion of the trajectory while on the switching surface is referred to as sliding. To properly describe sliding and the regions where this occurs, we need

Definition 1.1 ([2]) A domain Δ contained in the switching surface is a sliding mode domain if for each $\epsilon > 0$ there exists a $\delta > 0$ such that any motion starting in n-dimensional δ-neighborhood of Δ may leave the n-dimensional ϵ-neighborhood of Δ only through the n-dimensional ϵ-neighborhood of the boundary of Δ.

Note that trajectory starting in Δ can leave Δ only through its boundary in the switching surface.

In this paper we study linear time-invariant single-input relay-control systems. The main goal is to obtain regions of asymptotic stability (RAS) and RAS with sliding for this class of systems. Section 2 gives some background materials and notation to facilitate our analysis. We utilize a special canonical form described in Section 3. The main tool we use is a class of Luré-type Lyapunov functions ([12]) which is given in Section 5. In Section 6 we analyze in detail a class of second order relay-control systems and give explicit expressions for regions of asymptotic stability. We also include a numerical example to illustrate the obtained results. Concluding remarks are given in Section 7.

2. System Description and Definitions

In this paper we consider a class of dynamical systems modeled by

$$\dot{x} = Ax + bu \tag{2.1}$$

where $x \in \mathbb{R}^n$, $A \in \mathbb{R}^{n \times n}$, $b \in \mathbb{R}^{n \times 1}$, $u \in \mathbb{R}$. We assume that the pair (A,b) is completely controllable and hence (2.1) is equivalent ot the controller canonical form

$$\dot{x} = \begin{bmatrix} 0 & 1 & 0\ 0\ 0 & & 0 \\ 0 & 0 & 1\ 0\ 0 & & 0 \\ \vdots & \vdots & & \vdots & \vdots \\ 0 & 0 & \cdots & 0 & 1 \\ \alpha_1 & \alpha_2 & \cdots & \alpha_{n-1} & \alpha_n \end{bmatrix} x + \begin{bmatrix} 0 \\ 0 \\ \vdots \\ 0 \\ 1 \end{bmatrix} u \, . \tag{2.2}$$

An important concept in variable structure control is that of an attractive manifold on which certain desired dynamical behaviors are guaranteed. Trajectories of the system should be steered towards the manifold and subsequently constrained to remain on it. The manifold we use has the form

$$\Omega = \{x \mid sx = 0\} \, , \ s \in \mathbb{R}^{1 \times n} \, .$$

We can assume that $s = [s_1 \dots, s_{n-1}, 1]$. Observe that if the system is in the controller canonical form, then $sb = 1$.

Definition 2.1. The solution of the algebraic equation in u of

$$s\dot{x} = sAx + sbu = 0$$

is called the equivalent control and denoted by u_{eq}. If $sb \neq 0$, then

$$u_{eq} = -(sb)^{-1} sAx \, .$$

In the controller canonical form, we have

$$u_{eq} = -sAx = -\sum_{i=1}^{n} (\alpha_i + s_{i-1}) \, x_i \, ,$$

where $s_0 \triangleq 0$

Definition 2.2. The equivalent system is the system that is obtained when $u = u_{eq}$ in the system equation. If sb is nonsingular, then the equivalent system is

$$\dot{x} = [I_n - b(sb)^{-1}s]Ax \,. \tag{2.3}$$

When the system dynamics is given by the controller canonical form, then the equivalent system is

$$\dot{x} = \begin{bmatrix} 0 & 1 & 0 & \dots & 0 & 0 \\ 0 & 0 & 1 & \dots & 0 & 0 \\ \vdots & & & & & \\ 0 & 0 & \dots & \dots & 0 & 1 \\ 0 & -s_1 & -s_2 & \dots & -s_{n-2} & -s_{n-1} \end{bmatrix} x \,. \tag{2.4}$$

The controller on which we will concentrate is

$$u = -\mu \mathrm{sgn}(sx) \,, \tag{2.5}$$

where μ is a positive real number. Note that this controller is bounded by μ.

3. Design of the Switching Surface and the State-Variable Transformation

There are two basic steps in the design of a variable structure controller.

(1) The design of the switching surface so that the system is asymptotically stable on the switching surface.

(2) The design of the control strategy to steer the system to the switching surface and to maintain it there.

We choose the switching surface so that the system restricted to the surface has prescribed distinct negative eigenvalues $-\lambda_1,\dots,-\lambda_{n-1}$, $\lambda_i > 0$, $i = 1,\dots,n-1$. In sliding mode the system is governed by (2.3) and $sx = 0$. If the system is in controller canonical form, then in sliding mode the system is described by (2.4) and $sx = 0$. The order of the system in sliding is n-1 and its characteristic equation is given by

$$\lambda^{n-1} + s_{n-1}\lambda^{n-2} + \dots + s_1 = 0 \tag{3.1}$$

The prescribed eigenvalues $-\lambda_1,\dots,-\lambda_{n-r}$ must satisfy (3.1) and hence we have the linear equations

$$\begin{bmatrix} 1 & -\lambda_1 & (-\lambda_1)^2 & \dots & (-\lambda_1)^{n-1} \\ \vdots & & & & \\ 1 & -\lambda_{n-1} & (-\lambda_{n-1})^2 & \dots & (-\lambda_{n-1})^{n-1} \end{bmatrix} \begin{bmatrix} s_1 \\ \vdots \\ s_{n-1} \\ 1 \end{bmatrix} = 0 \, . \tag{3.2}$$

Since $\lambda_1, \dots \lambda_{n-1}$ are distinct, the coefficient matrix has full rank and $s_1, \dots, s_{n-1}$ are uniquely determined. This completes the design of the switching surface $\Omega = \{x \mid sx = 0\}$.

To facilitate the analysis we introduce a state-variable transformation. For ℓ,k positive integers, we let

$$V_\ell(\beta_1, \dots, \beta_k) = \begin{bmatrix} 1 & 1 & \dots & 1 \\ \beta_1 & \beta_2 & & \beta_k \\ \vdots & \vdots & & \vdots \\ \beta_1^\ell & \beta_2^\ell & & \beta_k^\ell \end{bmatrix} \in \mathbb{R}^{(\ell+1)\times k} \, .$$

Note that that (3.2) can be written as $sV_{n-1}(-\lambda_1, \dots, -\lambda_{n-1}) = 0$. Let

$$W = \left[V_{n-1}(-\lambda_1, \dots, -\lambda_{n-1}) \text{ diag } (p_1, \dots, p_{n-1}) \right] \in \mathbb{R}^{n\times(n-1)}$$

and

$$W^g = [\{V_{n-2}(-\lambda_1, \dots, -\lambda_{n-1}) \text{ diag } (p_1, \dots, p_{n-1})\}^{-1} \mid 0] \in \mathbb{R}^{(n-1)\times n} \, .$$

Note that $W^g W = I_{n-1}$. The p_i's are to be chosen so that the system matrix will have a desired form to be given.

Let

$$M = \begin{bmatrix} W^g \\ s \end{bmatrix} .$$

Observe that $M^{-1} = [W \;\; b]$. We introduce the new coordinates

$$\begin{bmatrix} z \\ y \end{bmatrix} = Mx = \begin{bmatrix} W^g x \\ sx \end{bmatrix}, \tag{3.3}$$

where $z \in \mathbb{R}^{n-1}$, $y \in \mathbb{R}$. In these coordinates, the system (2.2) has the form

$$\begin{aligned} \dot{z} &= A_{11} z + A_{12} y \\ \dot{y} &= A_{21} z + A_{22} y + u \, , \end{aligned} \tag{3.4}$$

where $A_{11} = \text{diag } (-\lambda_1, \dots, -\lambda_{n-1})$. If we use the controller

$$u = -\mu\, \mathrm{sgn}(sx) = -\mu\, \mathrm{sgn}\, y\ ,$$

then the system is described by

$$\begin{aligned} \dot{z} &= A_{11} z + A_{12} y \\ \dot{y} &= A_{21} z + A_{22} y - \mu\, \mathrm{sgn}\, y\ . \end{aligned} \tag{3.5}$$

We can observe further form (3.2) that

$$A_{21} = sAW = [d_1 \ldots d_{n-1}]\ , \tag{3.6}$$

where

$$d_i = p_i \left\{ \alpha_1 + \sum_{j=1}^{n-1} (-\lambda_i)^j (s_j + \alpha_{j+1}) \right\} = -p_i \{ (-\lambda_i)^n - \sum_{j=1}^{n} (-\lambda_i)^{j-1} \alpha_j \}\ ,$$

and

$$A_{22} = sAb = \alpha_n + \sum_{i=1}^{n-1} \lambda_i\ . \tag{3.7}$$

If we let

$$p_i = (-1)^n \prod_{\substack{j=1 \\ j \neq i}}^{n-1} \frac{1}{\lambda_i - \lambda_j}\ , \quad n > 2\ , \quad i = 1, \ldots, n-1,$$

then from the form of the inverses of the Vandermonde matrices we have

$$A_{12} = W^g Ab = \begin{bmatrix} 1 \\ 1 \\ \vdots \\ 1 \end{bmatrix} \in \mathbb{R}^{n-1}\ . \tag{3.8}$$

The above relation can be obtained if one notices that the last column of $V_{n-2}^{-1} \triangleq \left[V_{n-2}(-\lambda_1, \ldots, -\lambda_{n-1}) \right]^{-1}$ has the form

$$\frac{1}{\det V_{n-2}} \begin{bmatrix} (-1)^{1+(n-1)} \det V_{n-3}(-\lambda_2, -\lambda_3, \ldots, -\lambda_{n-1}) \\ (-1)^{2+(n-1)} \det V_{n-3}(-\lambda_1, -\lambda_3, \ldots, -\lambda_{n-1}) \\ \vdots \\ (-1)^{(n-1)+(n-1)} \det V_{n-3}(-\lambda_1, -\lambda_2, \ldots, -\lambda_{n-2}) \end{bmatrix},$$

where

$$\det V_{n-2} = \Big[(-\lambda_{n-1}) - (-\lambda_{n-2})\Big]\Big[(-\lambda_{n-1}) - (-\lambda_{n-3})\Big] \cdots \Big[(-\lambda_{n-1}) - (-\lambda_1)\Big]$$
$$\vdots$$
$$\times \Big[(-\lambda_3) - (-\lambda_2)\Big]\Big[(-\lambda_3) - (-\lambda_1)\Big]$$
$$\times \Big[(-\lambda_2) - (-\lambda_1)\Big].$$

If we let the entries of the last column of $\mathrm{diag}(p_i^{-1},\ldots,p_{n-1}^{-1})V_{n-2}^{-1}$ be all equal to one, then p_i's and (3.8) are easily obtained.

Example 1: Consider the system given by

$$\begin{bmatrix}\dot{x}_1\\ \dot{x}_2\end{bmatrix} = \begin{bmatrix}0 & 1\\ \alpha_1 & \alpha_2\end{bmatrix}\begin{bmatrix}x_1\\ x_2\end{bmatrix} + \begin{bmatrix}0\\ 1\end{bmatrix} u\,. \tag{3.9}$$

We choose the switching surface so that the system restricted to the switching surface has eigenvalue $-\lambda_1$. From (3.1) and (3.2) the switching surface is

$$s_1 x_1 + x_2 = \lambda_1 x_1 + x_2 = 0\,. \tag{3.10}$$

We use the controller

$$u = -\,\mu\,\mathrm{sgn}(\lambda_1 x_1 + x_2)\,. \tag{3.11}$$

A block diagram representation of the closed-loop system is given in Fig. 1.

Using the method described above we have

$$M = \begin{bmatrix}W^g\\ s\end{bmatrix} = \begin{bmatrix}1 & 0\\ \lambda_1 & 1\end{bmatrix},$$

and the system in the new coordinates is

$$\begin{aligned}\dot{z} &= -\,\lambda_1 z + y\\ \dot{y} &= -\,(\lambda_1^2 + \lambda_1\alpha_2 - \alpha_1)z + (\lambda_1 + \alpha_2)y - \mu\,\mathrm{sgn}\,y\,.\end{aligned} \tag{3.12}$$

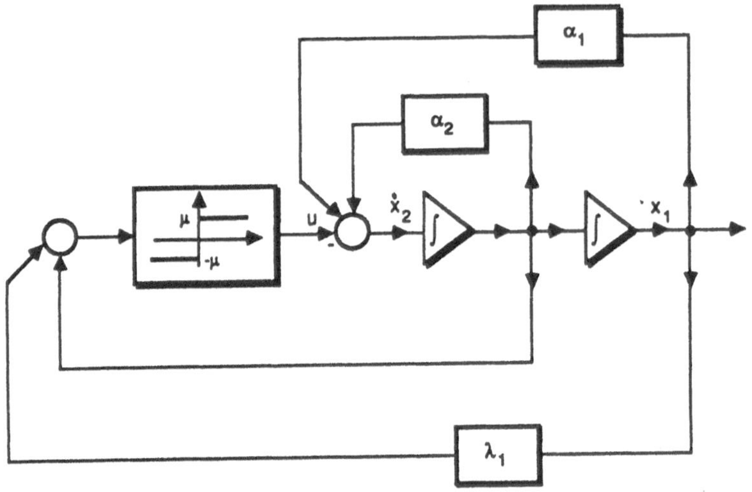

Fig. 1. Block diagram of the closed-loop system in Example 1.

4. A First Approximation of the Region of Asymptotic Stability with Sliding

In this Section we give a first approximation of the region of asymptotic stability (RAS) with sliding. A better approximation will be given in the next Section. We start with the following lemma.

Lemma 4.1: For $0 < \epsilon < \mu$, the region

$$\Delta = \{(z,0) \mid |A_{21}z| < \mu-\epsilon\} \subset \Omega$$

is a sliding mode domain.

Proof: Let $\Delta_\epsilon = \{(z,y) \mid |A_{21}z| < \mu-\epsilon\,,\ |y| < \frac{\epsilon}{|A_{22}|}\}$. Then in $\Delta_\epsilon \backslash \{(z,y) \mid y = 0\}$

$$\frac{1}{2}\frac{d}{dt}(y^2) = y\,\dot{y} = y(A_{21}z + A_{22}y) - \mu\,|y|$$

$$< |y|\,(\mu - \epsilon + \epsilon - \mu) = 0\,.$$

Therefore a trajectory starting in Δ_ϵ can leave Δ_ϵ only through the $\epsilon/|A_{22}|$-neighborhood of the boundary of Δ in Ω.

□

Observe that if the initial point is in Δ, then the system will be in sliding for some positive time. However, there is no guarantee that we stay in Δ for subsequent times. From the fact that $A_{11} = \text{diag}(-\lambda_1, ..., -\lambda_{n-1})$ we have the following. Let B_r denote the ball centered at 0 with radius r.

Proposition 4.2: Let

$$R = \sup \{ r \mid B_r \cap \Omega \subset \Delta \},$$

and

$$\Sigma = B_R \cap \Omega .$$

Then Σ is a region of asymptotic stability (RAS) with sliding.

Proof: While in sliding the system is governed by $\dot{z} = \text{diag}(-\lambda_1, ..., -\lambda_{n-1})z$. Hence $z_i = z_i(0)e^{-\lambda_i t}$ for $i = 1, ..., n-1$. Thus if $\sum_{i=1}^{n-1} z_i^2(0) < R^2$, then $\sum_{i=1}^{n-1} z_i^2(t) < R^2$ and by Lemma 4.1 $z(t) \in \Sigma$ for $t \geq 0$.

□

Note that Σ is the largest circular region that is contained in Δ. We can easily see that $R = \mu / a_{21}$, where $a_{21} = \|A_{21}\|$, the Euclidean norm of A_{21}.

From the explicit form of the solutions of $\dot{z} = A_{11} z$, we can see that the largest RAS on Δ is given by

$$\{(z, y) \mid y = 0 , \sup_{t \geq 0} \mid \sum_{i=1}^{n-1} d_i z_i e^{-\lambda_i t} \mid < \mu \},$$

where the d_i's are the entries of A_{21} given in (3.6).

5. Improved Estimates of RAS with Sliding

In the previous Section we obtained a RAS with sliding contained in the switching surface. We now use this information to obtain RAS's that are not constrained to the switching surface. The method we use is that of finding RAS's whose restriction to the switching surface is contained in Σ, and hence will be a RAS with sliding. Our main tool is a Luré-like Lyapunov function candidate

$$V(z,y;\beta,\eta,h) = (a_{21}\|z\|)^2 + 2\beta(A_{21}z)y + hy^2 + \mu\eta\,|y|\,, \tag{5.1}$$

where β, h, and η are positive constants. When there is no ambiguity, we will write V(z,y) for V(z,y;β,η,h). Observe that

$$V(z,y) \geq (a_{21}\|z\| - \beta\,|y|)^2 + (h - \beta^2)y^2 + \mu\eta\,|y|\,. \tag{5.2}$$

Hence if $h - \beta^2 \geq 0$, then V is positive definite in $\mathbb{R}^n \backslash 0$. If $\beta^2 - h > 0$ then V is positive in $\{(z,y) \mid |y| < \frac{\mu\eta}{\beta^2 - h}\} \setminus 0$. Since V contains a multiple of $|y|$, the Lyapunov derivative $\dot{V}$ may not exist on a trajectory which intersects $\Omega = \{(z,y) \mid y = 0\}$. However, when restricted to Ω the system takes the form $\dot{z} = A_{11}z = \text{diag}\,(-\lambda_1,...,-\lambda_{n-1})z$. Therefore if the trajectory (z(t),y(t)) is in Ω for $t_1 \leq t \leq t_2$, then we must have $\|z(t_2)\| \leq \|z(t_1)\|$. Since the restriction of level sets of V to Ω are circular regions, the trajectories of the system cannot leave a sublevel set $\{(z,y) \mid V(z,y) < a^2\}$ of V through Ω. Therefore if Γ is a region such that

(i) V is positive in Γ,

(ii) $\dot{V}$ is negative in $\Gamma \backslash \Omega$,

then the largest sublevel set of V contained in Γ is a RAS. If in addition we have

(iii) the restriction of Γ to Ω is contained in Σ,

then the largest sublevel set of V contained in Γ is a RAS with sliding. Note that we do not need to consider $\dot{V}$ on Ω.

Theorem 5.1. Suppose $\eta \geq 2\beta$, $a_{21} \neq 0$. Then there exist positive constants $C_1 = C_1(A,\mu,\beta,\eta)$ and $C_2 = C_2(A,\mu,\beta,\eta)$ such that for $\delta \in (0,\mu)$ and $h \geq \max\left\{\frac{C_1}{\delta^2} + \beta^2\,,\ \frac{C_2}{\delta}\right\}$ the region

$$\{(z,y)\,|V(z,y;\eta,\beta,h) < (\mu - \delta)^2\}$$

is a RAS with sliding.

Proof: Suppose $0 < \delta < \mu$. Let

$$\Gamma = \{(z,y)\,|V(z,y) < (\mu - \delta)^2\}\,.$$

From the discussion before the theorem we need to show that

(i) $V > 0$ in Γ,

(ii) $\dot{V} < 0$ in $\Gamma \backslash \Omega$,

(iii) $\Gamma \cap \Omega \subset \Sigma$.

Condition (i) holds since by assumption $h \geq \max\left\{\frac{C_1}{\delta_2} + \beta^2 , \frac{C_2}{\delta}\right\}$, and thus $h > \beta^2$ which implies that V is positive definite.

Condition (ii) holds by definition. Indeed $V(z,0) = (a_{21}||z||)^2 < (\mu - \delta)^2$ implies $a_{21}||z|| < \mu - \delta$. Hence by Lemma 4.1 $\Gamma \cap \Omega \subset \Sigma$.

We need to prove (iii). After some manipulations one obtains

$$\begin{aligned}\dot{V} &= 2a_{21}^2 z^T A_{11} z + \left[2\beta(A_{21} z)^2 - \eta\mu^2\right] \\ &+ 2\,h\,|y|\left[A_{21} z \operatorname{sgn} y - \mu\right] + \mu A_{21} z\left[\eta - 2\beta\right]\operatorname{sgn} y \\ &+ A_{22}\ |y|\ \left[2h\,|y| + 2\beta(A_{21} z)\operatorname{sgn} y + \mu\eta\right] \\ &+ 2y\left[\beta A_{21}' A_{11} z + \beta y A_{21} A_{12} + a_{21}^2 z^T A_{12}\right].\end{aligned}$$

Choose $\tau > 0$ so that $\beta\tau < \frac{1}{4}$ and $a_{22}\tau < \frac{1}{4}$. Let $(z,y) \in \Gamma$. If $h \geq \left(\frac{\mu-\delta}{\tau\delta}\right)^2 + \beta^2$ we sec from (5.2) that $|y| < \tau\delta$. Indeed $(z,y) \in \Gamma$ implies

$$(a_{21}||z|| - \beta\,|y|)^2 + (h - \beta^2)y^2 + \mu\eta\,|y| < (\mu - \delta)^2 .$$

If $h \geq \left(\frac{\mu - \delta}{\tau\delta}\right)^2 + \beta^2$ then

$$\left(\frac{\mu - \delta}{\tau\delta}\right)^2 y^2 < (a_{21}||z|| - \beta\,|y|)^2 + \left(\frac{\mu - \delta}{\tau\delta}\right)^2 y^2 + \mu\eta\,|y| < (\mu - \delta)^2 .$$

Therefore

$$y < \tau\delta .$$

If in addition $\beta\tau < \frac{1}{4}$ then

$$(a_{21}||z|| - \beta\,|y|)^2 < (a_{21}||z|| - \beta\,|y|)^2 + \left(\frac{\mu - \delta}{\tau\delta}\right)^2 y^2 + \mu\eta\,|y| < (\mu - \delta)^2$$

and hence

$$a_{21}\|z\| - \beta|y| < \mu - \delta$$

implies

$$a_{21}\|z\| < \mu - \frac{3}{4}\delta .$$

Since $\eta \geq 2\beta$ and $|A_{21}z| < \mu$ (see Lemma 4.1), we have

$$[2\beta(A_{21}z)^2 - \eta\mu^2] + \mu A_{21}z[\eta - 2\beta] \operatorname{sgn} y$$

$$\leq 2\beta(A_{21}z)^2 - \eta\mu^2 + \mu|A_{21}z|[\eta - 2\beta]$$

$$= (2\beta|A_{21}z| + \eta\mu)(|A_{21}z| - \mu) < 0 .$$

This combined with the fact that $A_{11} = \operatorname{diag}(-\lambda_1,\dots,-\lambda_{n-1})$, $-\lambda_i < 0$, and $\|z\| < \left(\mu - \frac{3}{4}\delta\right)/a_{21} \leq \mu/a_{21}$ shows that one can find a constant $K = K(A,\mu,\eta,\beta)$ such that

$$\dot{V} < 2h|y|(|A_{21}z| - \mu + a_{22}|y|) + K(|y| + |y|^2) .$$

Since $|y| < \tau\delta$, $|A_{21}z| - \mu < -\frac{3}{4}\delta$, and $a_{22}\tau < \frac{1}{4}$, we have

$$\dot{V} < |y|\left[2h\left(-\frac{3}{4}\delta + a_{22}\tau\delta\right) + K(1 + |y|)\right].$$

Hence

$$\dot{V} < |y|(-\delta h + K(1 + \tau\delta)) .$$

Therefore if $h > K(1 + \tau\delta)/\delta$, in addition to the previous requirement that $h \geq \left(\frac{\mu - \delta}{\tau\delta}\right)^2 + \beta^2$, we have $\dot{V} < 0$ in $\Gamma\backslash\Omega$.

Let

$$C_1 = \frac{\mu^2}{\tau^2} , \quad C_2 = K(1 + \tau\mu) .$$

Note that τ depends only on β and a_{22}. Then $\dot{V} < 0$ in $\Gamma\backslash\Omega$ if $h \geq \max\left\{\frac{C_1}{\delta^2} + \beta^2 , \frac{C_2}{\delta}\right\}$ and the proof is complete.

□

The importance of this theorem lies not in the form of the RAS that it gives but rather in the fact that the parameteric Lyapunov functions given can be used to obtain RAS. In the next Section we will utilize these Lyapunov functions to obtain stability boundary approximations.

6. Estimation of RAS for Second Order Relay-Control System

In this Section we use the Lyapunov functions (5.1) to estimate RAS for second order relay-control systems modeled by (3.9), (3.10), (3.11). We perform a detailed case by case analysis based on the coefficients of the characteristic equation of the open-loop system. We will be working with the system given by (3.12) and for notational convenience we represent it as

$$\begin{cases} \dot{z} = -\lambda z + y \\ \dot{y} = dz + ey - \mu \operatorname{sgn} y \,, \end{cases} \tag{6.1}$$

where $\lambda = \lambda_1$, $d = -(\lambda_1^2 + \lambda_1 \alpha_2 - \alpha_1)$, $e = \lambda_1 + \alpha_2$.

Let $\eta = 2\beta$. The Lyapunov function (5.1) for the system (6.1) takes the form

$$V(z,y) = d^2 z^2 + 2\beta dzy + hy^2 + 2\beta\mu\,|y|\,. \tag{6.2}$$

Let $\mathcal{R}$ be the connected component of $\{(z,y)\,|V(z,y) < \mu^2\alpha^2,\ \alpha^2 > 1\}$ which contains the origin. We need the following lemma in our analysis.

Lemma 6.1: If $\beta^2 - h > 0$ and $\alpha^2 \leq \dfrac{\beta^2}{\beta^2 - h}$, then the region $\mathcal{R}$ exists and is contained into the region $\{(z,y)\,|\ |y| < \dfrac{\mu\beta}{\beta^2 - h}\}$ and hence $V(z,y)$ is positive definite in $\mathcal{R}$.

Proof: The area of the closed region

$$(dz)^2 + 2\beta y(dz) + hy^2 + 2\beta\mu\,|y| - \mu^2\alpha^2 = 0$$

is non-zero and hence $\mathcal{R}$ exists if $(\beta^2 - h)y^2 - 2\beta\mu\,|y| + \mu^2\alpha^2 > 0$. Therefore for all $(z,y)\in\mathcal{R}$, we have

$$|y| < \frac{\mu\beta - \sqrt{\mu^2\beta^2 - (\beta^2 - h)\mu^2\alpha^2}}{\beta^2 - h} \le \frac{\mu\beta}{\beta^2 - h}.$$

Using the argument following (5.2) with $\eta = 2\beta$ we conclude that V(z,y) is positive definite in $\mathcal{R}$.

In what follows we let $\eta = 2\beta$ and $\alpha^2 = \frac{\beta}{\beta - \lambda}$.

Case 1. $\alpha_1 \ge 0$ and $\alpha_2 \le \frac{2\alpha_1}{\lambda}$. Denote the roots of the characteristic polynomial by s_1, s_2, $s_1 \ge s_2$. It is easy to see that $s_1 > 0 \ge s_2$. Note that $e = \lambda + s_1 + s_2$, $d = -(\lambda + s_1)(\lambda + s_2)$, $\alpha_1 = -s_1 s_2$, and $\alpha_2 = s_1 + s_2$. Note also that we need only to consider the case $y \ne 0$.

Theorem 6.2: Let $\beta = \frac{(\lambda + s_1)^2}{\lambda + 2s_1}$ and $h = \lambda\beta$. Then $\mathcal{R}_1$, the connected component of

$$\{(z,y) \mid 0 \le V(z,y) < \mu^2\alpha^2\}$$

which contains the origin, is a RAS for the system (6.1).

Proof: Note that $\beta^2 - h = \beta(\beta - \lambda) = \beta\,\frac{s_1^2}{\lambda + 2s_1} > 0$. We have

$$\alpha^2 = \frac{\beta}{\beta - \lambda} \cdot \frac{\beta}{\beta} = \frac{\beta^2}{\beta^2 - h}.$$

Therefore by Lemma 6.1, V(z,y) is positive definite in $\mathcal{R}_1$. Observe that

$$V(z,y) - \mu^2\alpha^2 = (dz\,\mathrm{sgn}\,y)^2 + 2\frac{(\lambda+s_1)^2}{\lambda+2s_1}dzy + \lambda\frac{(\lambda+s_1)^2}{\lambda+2s_1}|y|^2 + 2\frac{(\lambda+s_1)^2}{\lambda+2s_1}\mu|y| - \mu^2\alpha^2$$

$$= \{dz\,\mathrm{sgn}\,y + (\lambda+s_1)|y| - \mu\alpha\}\{dz\,\mathrm{sgn}\,y + \frac{\lambda(\lambda+s_1)}{\lambda+2s_1}|y| + \mu\alpha\}. \tag{6.3}$$

Hence the boundary of $\mathcal{R}_1$ forms a parallelogram (See Fig. 2).

We next show that $\dot{V} < 0$ in $\mathcal{R}_1$ and hence $\mathcal{R}_1$ is a RAS for the system (6.1). First note that if $y \ne 0$, then

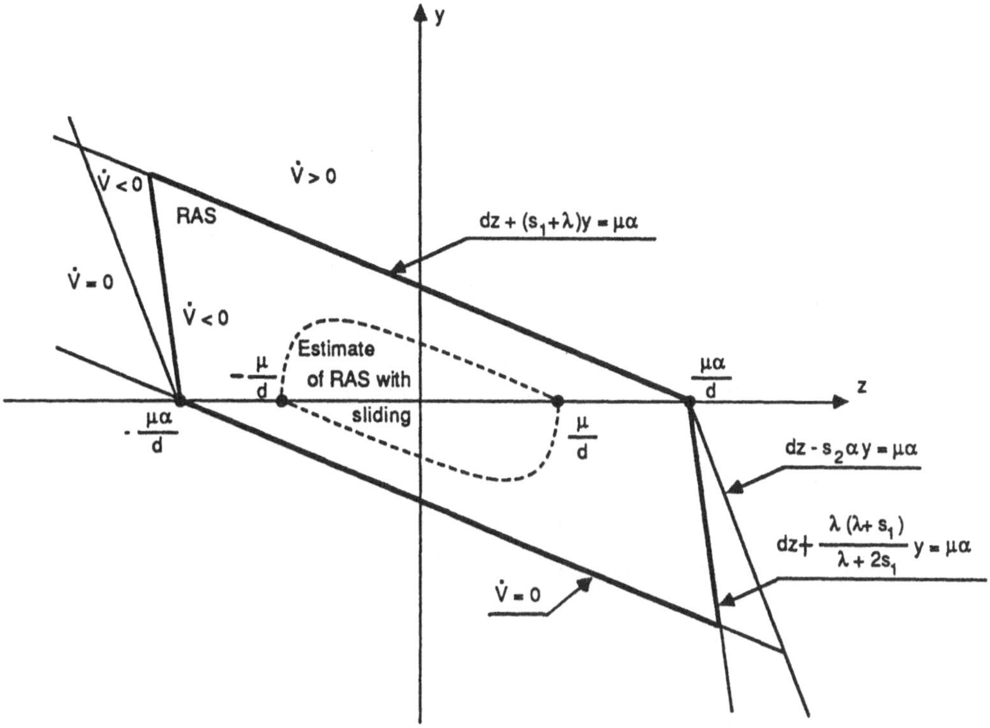

Fig. 2. Illustration of the proof of Theorem 6.2 for d > 0.

$$\frac{d}{dt}\,|y(t)| = \begin{cases} \dot{y}(t) & \text{if } y(t) > 0 \\ -\dot{y}(t) & \text{if } y(t) < 0 \end{cases} = (\text{sgn} y)\,\dot{y}\,.$$

The Lyapunov derivative of V is

$$\dot{V} = 2d^2 z\dot{z} + 2\beta dy\dot{z} + 2\beta dz\dot{y} + 2hy\dot{y} + 2\mu\beta(\text{sgn} y)\dot{y}$$

$$= 2d^2 z(-\lambda z + y) + 2\beta dy(-\lambda z + y)$$

$$+ 2\beta dz(dz + ey - \mu\,\text{sgn}\,y)$$

$$+ 2\lambda\beta y(dz + ey - \mu \operatorname{sgn} y)$$

$$+ 2\mu\beta \operatorname{sgn} y(dz + ey - \mu \operatorname{sgn} y)$$

We can represent $\dot{V}$ as

$$\dot{V} = 2(\beta - \lambda)d^2z^2 + 2(d + e\beta)dzy + 2(d + e\lambda)\beta y^2 + 2\mu\beta\ |y|\ (e - \lambda) - 2\mu^2\beta$$

$$= 2(\beta - \lambda)\{dz \operatorname{sgn} y + (\lambda + s_1)\ |y| - \mu\alpha\}\ \{dz \operatorname{sgn} y - \frac{s_2}{s_1}\ (\lambda + s_1)\ |y| + \mu\alpha\}\ . \quad (6.4)$$

Comparing (6.3) and (6.4), we see that if $2\alpha_1 \geq \lambda\alpha_2$, then the region where $\dot{V} < 0$ forms a parallelogram which contains $\mathscr{R}_1$ (See Fig. 2).

□

Corollary 6.3: If $2\alpha_1 = \lambda\alpha_2$, then $\mathscr{R}_1$ is the true RAS.

Note that the condition $2\alpha_1 = \lambda\alpha_2$ is equivalent to the condition $-2s_1s_2 = \lambda(s_1+s_2)$.

Proof: From (6.3) and (6.4), we see that if $2\alpha_1 = \lambda\alpha_2$, then $\{(z,y)\ |\ \dot{V} < 0\} = \mathscr{R}_1$.

□

Case 2. $\alpha_2 > 0$, $\alpha_2 \geq \dfrac{2\alpha_1}{\lambda}$, and $\alpha_2^2 + 4\alpha_1 > 0$. Note that in this case we have $e = \lambda + \alpha_2 > 0$ and $d = -\lambda e + \alpha_1 \leq -\dfrac{\lambda(2e - \alpha_2)}{2} = -\dfrac{\lambda(e+\lambda)}{2} < 0$.

Theorem 6.4: There is a constant $M_2 < \dfrac{\alpha_2}{2}$ so that if $0 \leq M \leq M_2$ and if $\beta = \dfrac{e + \lambda}{2} - M = \lambda + \left(\dfrac{\alpha_2}{2} - M\right)$ and $h = -d - 2\beta M$, then the connected component of

$$\{(z,y)\ |\ V(z,y) < \mu^2\alpha^2\}$$

which contains the origin is a RAS for the system (6.1).

Proof: Let $\mathscr{R}_2$ be the connected component of

$$\{(z,y) \mid V(z,y) < \mu^2\alpha^2\}$$

that contains the origin. We perform the following computation

$$h-\lambda\beta = -d - 2\beta M - \lambda\beta$$

$$= -d - 2M\left(\frac{e+\lambda}{2} - M\right) - \lambda\left(\frac{e+\lambda}{2} - M\right)$$

$$= \lambda^2 + \lambda\alpha_2 - \alpha_1 - 2M\left(\frac{\lambda+\alpha_2+\lambda}{2} - M\right) - \lambda\left(\frac{\lambda+\alpha_2+\lambda}{2} - M\right)$$

$$+ 2M^2 - 2M^2 - \alpha_2 M + \alpha_2 M + \lambda M - \lambda M$$

$$= 4M^2 - (2\alpha_2 + 3\lambda)M + \frac{\lambda\alpha_2 - 2\alpha_1}{2} + 2\beta M\,, \tag{6.5}$$

and

$$\beta^2 - h = \left(\frac{e+\lambda}{2} - M\right)^2 + d + 2\left(\frac{e+\lambda}{2} - M\right)M = -M^2 + \frac{\alpha_2^2 + 4\alpha_1}{4}\,. \tag{6.6}$$

We also have

$$\dot{V} = 2d^2 z(-\lambda z + y) + 2\beta dy(-\lambda z + y)$$

$$+ (2\beta dz + 2hy + 2\mu\beta \operatorname{sgn} y)(dz + ey - \mu \operatorname{sgn} y)$$

$$= 2(\beta - \lambda)(d^2 z^2 + 2\beta dzy + hy^2 + 2\mu\beta\,|y| - \mu^2\alpha^2)$$

$$+ 2[h + d - 2\beta(\beta - \frac{e+\lambda}{2})]dzy + 2[h(e - \beta + \lambda) + \beta d]y^2$$

$$+ 2\mu(-h - 2\beta^2 + 2\beta\lambda + e\beta)\,|y| \tag{6.7}$$

Using the definitions of β and h, we can represent (6.7) as follows

$$\dot{V} = 2\left(\frac{\alpha_2}{2} - M\right)\left[V(z,y) - \mu^2\alpha^2\right] - 4M\left(-M^2 + \frac{\alpha_2^2 + 4\alpha_1}{4}\right)y^2$$

$$- 2\mu\left[4M^2 - (2\alpha_2 + 3\lambda)M + \frac{\lambda\alpha_2 - 2\alpha_1}{2}\right]|y|\,,$$

and from (6.5) and (6.6) we have

$$\dot{V} = 2\left(\frac{\alpha_2}{2} - M\right)\left[V(z,y) - \mu^2\alpha^2\right] - 4M(\beta^2 - h)y^2$$

$$- 2\mu(h - \lambda\beta - 2\beta M)\ |y|\ . \qquad (6.8)$$

Choose M_2 such that $M_2 < \frac{\alpha_2}{2}$ and for $0 \le M \le M_2$ the inequalities $\beta^2 - h \ge 0$ and $h - \lambda\beta - 2\beta M \ge 0$ are satisfied. Note that this is possible since if $M = 0$, from (6.5) and (6.6) we have $\beta^2 - h = \frac{\alpha_2^2 + 4\alpha_1}{4} > 0$ and $h - \lambda\beta - 2\beta M = \frac{\lambda\alpha_2 - 2\alpha_1}{2} \ge 0$. Therefore for $0 \le M \le M_2$, $\dot{V} < 0$ in $\mathcal{R}_2$.

Since we choose M_2 so that $\beta^2 - h \ge 0$ and $h - \lambda\beta \ge 2\beta M \ge 0$, therefore we have

$$\alpha^2 = \frac{\beta}{\beta - \lambda} = \frac{\beta^2}{\beta^2 - \lambda\beta} \le \frac{\beta^2}{\beta^2 - h}\ .$$

Hence by (5.2) and Lemma 6.1, $V(z,y)$ is positive definite in $\mathcal{R}_2$.

□

Corollary 6.5: If there exists $0 \le M < \frac{\alpha_2}{2}$ so that $\beta^2 - h = h - \lambda\beta - 2\beta M = 0$, then $\mathcal{R}_2$ is the true RAS.

Proof: This follows from (6.8).

□

Since $\beta^2 - h$ and $h - \beta - 2\beta M$ are quadratics in M, and since they are nonnegative when $M = 0$, the optimal M_2 in Theorem 6.4 is given by the smallest nonnegative root of $\beta^2 - h = 0$ and $h - \lambda\beta - 2\beta M = 0$. The value of $M = M_2$ gives the best RAS obtainable by the technique given above. Note that if $\alpha_1 < 0$, then from (6.6) the positive root of $\beta^2 - h$ is less than $\frac{\alpha_2}{2}$. Otherwise, if $\alpha_1 \ge 0$, then from (6.5) the value of $h - \lambda\beta - 2\beta M$ is negative for $M = \frac{\alpha_2}{6}$ and hence $h - \lambda\beta - 2\beta M$ has a positive root less than $\frac{\alpha_2}{6}$.

Case 3. $\alpha_2 > 0$, $\alpha_1 < 0$, and $\alpha_2^2 + 4\alpha_1 < 0$. We define β and h as before. In this case we see that

$$\beta^2 - h = -M^2 + \frac{\alpha_2^2 + 4\alpha_1}{4} < 0 .$$

Using (5.2) we conclude that V(z,y) is positive definite everywhere in the state space.

Theorem 6.6: There is a constant $M_2 < \frac{\alpha_2}{2}$ so that if $0 < M \leq M_2$ then

$$\mathcal{R}_3 = \{(z,y) \mid V(z,y) < \mu^2 \alpha^2\}$$

is a RAS for the system (6.1).

Proof: From (6.8) we have

$$\dot{V}(z,y) = 2\left(\frac{\alpha_2}{2} - M\right)\left[V(z,y) - \mu^2\alpha^2\right]$$

$$+ 4M(h - \beta^2)\ |y|\left[|y| - \frac{h - \lambda\beta - 2\beta M}{2M(h - \beta^2)}\ \mu\right]. \tag{6.9}$$

From (6.5) and (6.6) we have

$$\lim_{M \to 0}\ (h - \lambda\beta - 2\beta M) = \frac{\lambda\alpha_2 - 2\alpha_1}{2} > 0 , \tag{6.10}$$

$$\lim_{M \to 0}\ (h - \beta^2) = -\frac{\alpha_2^2 + 4\alpha_1}{4} > 0 , \tag{6.11}$$

and

$$\lim_{M \to 0}\ \alpha^2 = \frac{\alpha_2 + 2\lambda}{\alpha_2} \tag{6.12}$$

Hence for sufficiently small M, $|y|$ is uniformly bounded in $\mathcal{R}_3$ independent of M. We conclude that $\dot{V} < 0$ in $\mathcal{R}_3$ for $M < \frac{\alpha_2}{2}$ sufficiently small, and hence M_2 exists. In order to calculate M_2 we need to show that $|y| - \frac{h - \lambda\beta - 2\beta M}{2M(h - \beta^2)}\ \mu < 0$ in $\mathcal{R}_3$.

One can check that

$$\sup\{\,|y|\;\;|\;(z,y)\in\mathcal{R}_3\} = \frac{\mu\beta}{h-\beta^2}\left[\sqrt{\frac{h-\lambda\beta}{\beta^2-\lambda\beta}}-1\right]. \tag{6.13}$$

Indeed, the boundary of $\mathcal{R}_3$ is described by the equation

$$V(z,y) - \mu^2\alpha^2 = d^2z^2 + 2\beta dzy + hy^2 + 2\beta\mu\,|y\,| - \mu^2\alpha^2 = 0\,.$$

The discriminant of the above quadratic in z is

$$\beta^2 d^2 y^2 - d^2[hy^2 + 2\beta\mu\,|y\,| - \mu^2\alpha^2]$$

$$= d^2[(\beta^2-h)\,|y\,|^2 - 2\beta\mu\,|y\,| + \mu^2\alpha^2]\,.$$

Equating the discriminant to zero and computing $|y\,|$, we arrive at (6.13). To obtain M_2 we find $M \le M_2$ so that

$$\frac{\mu\beta}{h-\beta^2}\left[\sqrt{\frac{h-\lambda\beta}{\beta^2-\lambda\beta}}-1\right] \le \frac{h-\lambda\beta-2\beta M}{2M(h-\beta^2)}\,\mu\,. \tag{6.14}$$

Using the definition of β and (6.6), this is equivalent to solving

$$L(M) = 2M^3 - 3\lambda M^2 - \frac{\alpha_2^2+2\lambda\alpha_2-2\alpha_1}{2}M + \frac{\alpha_2(\lambda\alpha_2-2\alpha_1)}{4} \ge 0\,. \tag{6.15}$$

Indeed, (6.14) is equivalent to

$$\beta\left[\sqrt{\frac{h-\lambda\beta}{\beta^2-\lambda\beta}}-1\right] \le \frac{h-\lambda\beta-2\beta M}{2M}\,.$$

The above inequality, on the other hand, can be represented as

$$\beta(h-\lambda\beta)[(\beta-\lambda)(h-\lambda\beta) - 4M^2\beta] \ge 0\,. \tag{6.16}$$

Note that $h-\lambda\beta = (h-\beta^2)+\beta(\beta-\lambda) > 0$. Utilizing now $\beta = \lambda + \frac{\alpha_2}{2} - M$ and (6.6) we arrive at (6.15). One can check that if $\alpha_2 > 0$, then

$$L(0) = \frac{\alpha_2(\lambda\alpha_2-2\alpha_1)}{4} > 0\,,$$

and

$$L\left(\frac{\alpha_2}{6}\right) = \frac{\alpha_2^2}{9} - \frac{\alpha_2^2 + 4\alpha_1}{4} > 0\,,$$

and

$$L\left(\frac{\alpha_2}{2}\right) = -\,4\lambda\alpha_2 < 0\,.$$

Thus there exists a unique $M \in \left(\frac{\alpha_2}{6}, \frac{\alpha_2}{2}\right)$ which gives the optimal RAS and for which $L(M)=0$.

Note that

$$h - \lambda\beta - 2\beta M = (h - \beta^2) + \beta(\beta - \lambda) - 2\beta M = (h - \beta^2) + 3\beta\left(\frac{\alpha_2}{6} - M\right)$$

As $h - \beta^2 > 0$ in this case, therefore $h - \lambda\beta - 2\beta M$ in (6.14) is positive at least for $M \leq \frac{\alpha_2}{6}$. In what follows we show that it is also positive for all M which satisfy $L(M) \geq 0$ and hence (6.14) is valid. If $L(M) \geq 0$, then from (6.16) we have

$$h - \lambda\beta - 2\beta M \geq \frac{4M^2\beta}{\beta - \lambda} - 2\beta M\,.$$

It follows that

$$h - \lambda\beta - 2\beta M \geq 6M\beta\,\frac{\left(M - \frac{\alpha_2}{6}\right)}{\left(\frac{\alpha_2}{2} - M\right)}\,.$$

The right hand side of the above inequality is positive in the open interval $\left(\frac{\alpha_2}{6}, \frac{\alpha_2}{2}\right)$. Therefore M_2 is equal to the root of $L(M) = 0$ in $\left(\frac{\alpha_2}{6}, \frac{\alpha_2}{2}\right)$ and satisifies (6.14). The value of $M = M_2$ gives the best RAS obtained by the technique given above.

□

Remark: Suppose that $\alpha_2 = 0$. If $\alpha_1 \geq 0$, then we apply Theorem 6.2 and Corollary 6.3. Otherwise, if $\alpha_1 < 0$, then conditions of Case 3 are satisified. In such a case $M = 0$, $\beta = \lambda = e$, $h = -d = \lambda^2 - \alpha_1$, $\alpha^2 = \infty$, and from (6.7) we have

$$\dot{V} = -2\mu^2\lambda + 2\mu\alpha_1 |y| < 0 .$$

Hence, the whole state space is the attractivity region for the system (6.1).

In this Section we obtained estimates of RAS which were regions of the form

$$\{(z,y) \mid V(z,y) < \mu^2\alpha^2\}$$

where $\alpha^2 > 1$. Since $V(z,0) = d^2z^2$, we see that with the same parameters β, η, h the restriction of

$$\{(z,y) \mid V(z,y) < \mu^2\}$$

to the switching line $y = 0$ is contained in the sliding mode domain Σ given in Proposition 4.2. Thus

$$\{(z,y) \mid V(z,y) < \mu^2\}$$

is a RAS with sliding.

Example 2: Consider a dynamical system modeled by the following equations

$$\begin{cases} \dot{x}_1 = x_1 + 2x_2 + u \\ \dot{x}_2 = x_1 + x_2 + 2u \ , \quad |u| \leq 10 . \end{cases}$$

The system representation in the controllable canonical form is given by

$$\begin{cases} \dot{x}_1^* = x_2^* \\ \dot{x}_2^* = x_1^* + 2x_2^* + u \ , \quad |u| \leq 10 , \end{cases}$$

where $x_1^* = \frac{1}{7}[2x_1 - x_2]$ and $x_2^* = \frac{1}{7}[x_1 + 3x_2]$.

It is required that the eigenvalue of the reduced order system be equal to -1. Hence, $\alpha_2 = 2$, $\alpha_1 = 1$, $\lambda = 1$, and $u = -10\,\mathrm{sgn}(x_1^* + x_2^*)$. Note that $\lambda\alpha_2 = 2\alpha_1$, $\beta = 2$, and $h = 2$. In this case, by Corollary 6.3 the optimal stability domain estimate is also the true stability region and is given by

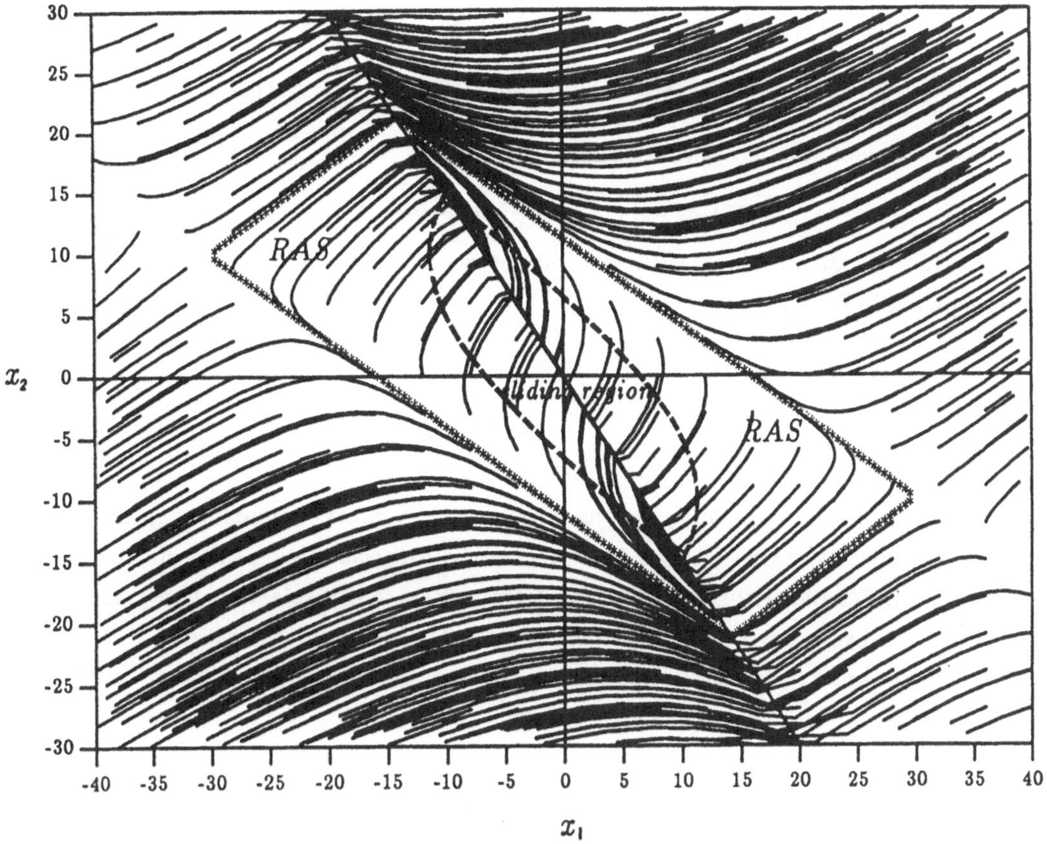

Fig. 3. True stability domain $\mathscr{R}_1$, and the RAS with sliding Γ (dashed line) for Example 2.

$$\mathscr{R}_1 = \{(x_1^*,x_2^*) \mid x_2^{*2} - x_1^{*2} - 2x_1^*x_2^* + 20\,|x_1^* + x_2^*| < 100\}$$

or

$$\mathscr{R}_1 = \{(x_1^*,x_2^*) \mid -\sqrt{2}\ x_1^*\mathrm{sgn}(x_1^* + x_2^*) + (\sqrt{2}+1)\,|x_1^* + x_2^*| < 10\,,$$
$$-\sqrt{2}\ x_1^*\mathrm{sgn}(x_1^* + x_2^*) + (\sqrt{2}-1)\,|x_1^* + x_2^*| > -10\}\,.$$

In the (x_1,x_2) coordinates, $\mathscr{R}_1$ has the form

$$\mathscr{R}_1 = \{(x_1,x_2) \mid -\sqrt{2}(2x_1 - x_2)\mathrm{sgn}(3x_1 + 2x_2) + (\sqrt{2} + 1)\,|3x_1 + 2x_2| < 70\,,$$
$$-\sqrt{2}\,(2x_1 - x_2)\mathrm{sgn}(3x_1 + 3x_2) + (\sqrt{2} - 1)\,|3x_1 + 2x_2| > -70\}\,.$$

The region of stability with sliding is

$$\Gamma = \{(x_1,x_2) \mid \ 2x_2^2 - x_1^2 + 20\,|3x_1 + 2x_2| < 350\}\,.$$

Regions $\mathscr{R}_1$ and Γ along with the computer generated trajectories for this system are depicted in Figure 3.

7. Conclusions

This paper is devoted to the problem of stability boundary approximation for relay-control systems. A tool that we used is a special canonical form (3.5) in which some essential structural properties of the system are displayed. Using the canonical form and Lyapunov functions of the form (5.1) we obtained estimates of RAS for a class of single input systems. An interesting problem would be to extend the results of this paper to multi-input relay-control systems.

References

[1] V. I. Utkin, "Sliding modes and their application in variable structure systems," Mir Publishers, Moscow 1978.

[2] V. I. Utkin, "Variable structure systems with sliding modes," IEEE Trans. Automat. Contr., Vol. AC-22, No. 2, pp. 212-222, 1977.

[3] B. Draženović, "The invariance conditions in variable structure systems," Automatica, Vol. 5, No. 3, pp. 287-295, 1969.

[4] R. A. DeCarlo, S. H. Żak, and G. P. Matthews, "Variable structure control of nonlinear multivariable systems: A tutorial," Proc. of the IEEE, Vol. 76, No. 3, pp. 212-232, 1988.

[5] S. Weissenberger, "Stability-boundary approximations for relay-control systems via a steepest-ascent construction of Lyapunov functions," J. of Basic Engineering, Trans. of the ASME Ser. D, Vol. 38, No. 2, pp. 419-428, 1966.

[6] O. M. E. El-Ghezawi, A. S. I. Zinober, and S. A. Billings, "Analysis and design of variable structure systems using a geometric approach," Int. J. Control, Vol. 38, No. 3, pp. 657-671, 1983.

[7] H. H. Rosenbrock, "A method of investigating stability," IFAC Proc., Basel, Switzerland, pp. 590-594, 1963.

[8] J. R. Hewit, and C. Storey, "Comparison of numerical methods in stability analysis," Int. J. Control, Vol. 10, No. 6, pp. 687-701, 1969.

[9] E. J. Davison, and E. M. Kurak, "A computational method for determining quadratic Lyapunov functions for non-linear systems," Automatica, Vol. 7, No. 5, pp. 627-636, 1971.

[10] V. G. Borisov, and S. N. Diligenskii, "Numerical method of analysis of nonlinear systems," Automation and Remote Control, Vol. 46, No. 11, Part 1, pp. 1373-1380, Nov. 1985.

[11] D. N. Shields, and C. Storey, "The behavior of optimal Lyapunov functions," Int. J. Control, Vol. 21, No. 4, pp. 561-573, 1975.

[12] R. E. Kalman, "Lyapunov function for the problem of Luŕe in automatic control," Proc. Nat. Acad. Sci. U.S.A., Vol. 49, pp. 201-205, Feb. 1963.

THEORY OF RESIDENCE TIME CONTROL BY OUTPUT FEEDBACK

S. M. Meerkov

Department of Electrical Engineering
and Computer Science
The University of Michigan
Ann Arbor, MI 48109-2122

and

T. Runolfsson
Department of Electrical and
Computer Engineering
The Johns Hopkins University
Baltimore, MD 21218

Abstract. The problem of residence time control by the observer based output feedback is formulated and solved for the case of linear systems with small additive input noise. Both noiseless and noisy measurements are considered. In the noiseless measurements case, it is shown that the fundamental bounds on the achievable residence time depend on the nonminimum phase zeros of the system. In the noisy measurements case, the achievable residence time is shown to be always bounded, and an estimate of this bound is given. Controller design techniques are presented. The development is based on the asymptotic large deviations theory.

1. INTRODUCTION

Consider the following Ito stochastic system

$$\begin{aligned} dx &= (Ax + Bu)dt + \epsilon C dw \\ y &= Dx \ , \end{aligned} \tag{1.1}$$

where $x \in R^n, u \in R^m, y \in R^p, w(t)$ is a standard r-dimensional Brownian motion, A, B, C, D are matrices of appropriate dimensionality, and $0 < \epsilon \ll 1$. For a given u, the behavior of (1.1) in a bounded domain $\Psi \subset R^p$ can be characterized by the first passage time [1],

$$\tau^\epsilon(u) = \inf\{t \geq 0 : \ y(t,u) \in \partial\Psi | y(0,u) \in \Psi\} \ ,$$

($\partial\Psi$ is the boundary of Ψ), or by its average value

$$\bar{\tau}^\epsilon(u) = E[\tau^\epsilon(u)] \ .$$

The $\bar{\tau}^{\epsilon}(u)$ is referred to as the (average) residence time of (1.1) in Ψ.

Assume that control specifications for (1.1) are given in the form of an aiming (pointing) problem: maintain $y(t)$ in a given domain $\Psi \subset R^p$ during a specified time interval $[0, T]$, $T < \infty$. In terms of the average residence time this problem has the form

$$\bar{\tau}^{\epsilon}(u) \geq T \ . \tag{1.2}$$

Technical examples of this problem can be found in [2].

To accomplish (1.2), the feedback control approach can be utilized. Papers [2] and [3] address this problem under the assumption that all states x are available for control and u is chosen as

$$u = Kx \ . \tag{1.3}$$

In [2] it was assumed that $D = I$, i.e. the pointing of states has been considered, and in [3] the general case of output aiming has been analyzed. It has been shown that from the point of view of satisfying (1.2), all systems (1.1) can be partitioned into two groups: weakly and strongly residence time controllable. Roughly speaking, (1.1) is weakly residence time controllable (*wrt*-controllable) if there exists $T^* < \infty$ such that the closed loop system (1.1), (1.3) satisfies (1.2) for $T < T^*$ and some K and does not satisfy (1.2) for $T > T^*$ and any K. System (1.1) is strongly residence time controllable (*srt*-controllable) if $T^* = \infty$. It has been shown in [2] that (1.1) with $D = I$ is *wrt*-controllable if and only if (A, B) is stabilizable and *srt*-controllable if and only if $Im\ C \subseteq Im\ B$. It has been shown in [3] that system (1.1) *wrt*-controllable in states can, in fact, be *srt*-controllable in outputs $y \neq x$. In particular, it was shown that a SISO system (1.1) is *srt*-controllable if and only if all non-minimum phase zeros of $G_s(s) \triangleq D(sI - A)^{-1}B$ coincide with non-minimum phase zeros of $G_n(s) \triangleq D(sI - A)^{-1}C$. This means, of course, that minimum phase plants are pointable with any precision whereas non-minimum phase ones may or may not be, depending on the location of the right half plane zeros of $G_n(s)$.

In the present paper we address problem (1.2) under the assumption that only (measured) outputs are available for control and, therefore, the output feedback has to be utilized. To simplify the problem, we consider here the observer based output feedback, i.e. controllers of the form:

$$\begin{aligned} u &= K\hat{x} \\ \dot{\hat{x}} &= A\hat{x} + Bu + L(z - E\hat{x}) \end{aligned} \tag{1.4}$$

if the measured output,

$$z = Ex \ , \quad z \in R^q \ , \quad E \in R^{q \times n} \ ,$$

is noise free, or

$$\begin{aligned} u &= K\hat{x} \\ d\hat{x} &= (A\hat{x} + Bu)dt + L(dz - E\hat{x}dt) \end{aligned} \tag{1.5}$$

if the measured output,

$$dz = Exdt + \epsilon\, Fdw_1 \quad ,$$

is noisy. Here $w_1(t)$ is a q-dimensional standard Brownian motion and $0 < \epsilon \ll 1$. In each case, (1.4) and (1.5), the problem is to choose the pair (K, L) so that (1.2) is satisfied.

To this end, in this paper we derive the following results:

1. System (1.1) with feedback (1.4) is *srt*-controllable if and only if the system is invertible and minimum phase in an appropriate sense.

2. If this is not the case, the maximal achievable residence time T^* for system (1.1), (1.4) coincides with that for system (1.1), (1.3) if and only if $G_{n_1}(s) \triangleq E(sI - A)^{-1}C$ is left invertible and minimum phase; otherwise the output controllers lead to a smaller residence time.

3. System (1.1) with feedback (1.5) is never *srt*-controllable. Thus, the measurement noise has a much more severe effect on the residence time than the input noise.

4. The observer gain L that ensures the largest possible residence time in system (1.1), (1.5) coincides with that of the corresponding Kalman filter. Thus, Kalman filter is optimal not only with respect to the standard performance measure, i.e., the mean square estimation error, but also from the point of view of the residence time.

5. The feedback gain K that ensures the largest possible residence time in system (1.1), (1.5) is dependent on the optimal value of L mentioned above. Thus, although the separation principle does not take place, the situation here can be characterized as semi-separation: the optimal observations do not depend on optimal control but the optimal control does depend on optimal observations. As a result, the maximal achievable residence time for controllers derived in this paper is larger than that for LQG-designed systems.

The remainder of this paper is organized as follows: In Section 2 some mathematical preliminaries are discussed. In Sections 3 - 5 system (1.1) with controllers (1.4) and (1.5), respectively, is considered and in Section 6 an illustrative example is given. In Section 7 the conclusions are formulated. The proofs are given in the Appendix.

2. PRELIMINARIES

In this section, the notion of logarithmic residence time, i.e., the main tool of asymptotic analysis of (1.1) with (1.3)-(1.5), is introduced and utilized for a precise formulation of problem (1.2).

Consider the linear Ito system

$$\begin{aligned} dx &= Axdt + \epsilon\, Cdw \\ y &= Dx \end{aligned} \qquad (2.1)$$

where, as before, $x \in R^n, y \in R^p, w(t)$ is a standard r-dimensional Brownian motion and $0 < \epsilon \ll 1$. Let $\Psi \subset R^p$ be again a bounded domain with the origin in its interior and a smooth boundary $\partial\Psi$. Define

$$\Omega_0 \triangleq \{x \in R^n : \ y = Dx \in \Psi\} \ , \tag{2.2}$$

$$\Omega \triangleq \{x \in R^n : \ De^{At}x \in \Psi, \ t \geq 0\} \ . \tag{2.3}$$

Assume that $x(0) = x_0 \in \Omega_0$ and introduce the first passage time as

$$\tau^\epsilon(x_0) \triangleq \inf\{t \geq 0 : \ y(t, x_0) \in \partial\Psi\} \ , \tag{2.4}$$

where $y(t, x_0)$ is the solution of (2.1). The following theorem was proved in [3].

Theorem 2.1: Suppose A is Hurwitz and (A, C) is disturbable, i.e. rank $[C \ AC \ldots A^{n-1}C] = n$. Then uniformly for all x_0 belonging to compact subsets of Ω we have:

$$\lim_{\epsilon \to 0} \epsilon^2 \ln \bar{\tau}^\epsilon(x_0) = \hat{\mu} \ , \tag{2.5}$$

where, as before, $\bar{\tau}^\epsilon(x_0) = E_{x_0}\tau^\epsilon(x_0)$ and

$$\begin{gathered} \hat{\mu} = \min_{y \in \partial\Psi} \frac{1}{2} y^T N y \ , \\ N = (DXD^T)^{-1}, \ \ AX + XA^T + CC^T = 0 \ . \end{gathered} \tag{2.6}$$

Constant $\hat{\mu}$ is referred to as the logarithmic residence time of (2.1) in Ψ.

Let $\bar{y}(t, x_0, \hat{x}_0, K, L)$ be the solution of the deterministic system

$$\begin{aligned} \begin{bmatrix} \dot{x} \\ \dot{\hat{x}} \end{bmatrix} &= \begin{bmatrix} A & BK \\ LE & A + BK - LE \end{bmatrix} \begin{bmatrix} x \\ \hat{x} \end{bmatrix}, \ \begin{bmatrix} x(0) \\ \hat{x}(0) \end{bmatrix} = \begin{bmatrix} x_0 \\ \hat{x}_0 \end{bmatrix} , \\ y &= Dx \end{aligned} \tag{2.7}$$

and define

$$\Omega(K, L) = \left\{ \begin{bmatrix} x_0 \\ \hat{x}_0 \end{bmatrix} \in R^{2n} \middle| \bar{y}(t, x_0, \hat{x}_0, K, L) \in \Psi, t \geq 0 \right\} \ . \tag{2.8}$$

Then, with regard to control system (1.1) and controllers (1.4) or (1.5), Theorem 2.1 allows us to conclude that for sufficiently small ϵ and $\binom{x_0}{\hat{x}_0} \in \Omega(K, L)$, problem (1.2) can be replaced by an alternative problem of selecting the pair (K, L) such that

$$\hat{\mu}(\Psi; K, L) > \mu \tag{2.9}$$

where $\hat{\mu}(\Psi; K, L)$ is the logarithmic residence time of the closed loop system (1.1), (1.4) or (1.1), (1.5) and $\mu = \epsilon^2 \ln T$. This is the problem solved in this paper.

As it was pointed out in the Introduction, the solution of this problem is given in terms of the weak and strong residence time controllability defined precisely below. In order to simplify the notations, we drop argument Ψ in (2.9).

Definition 2.1: (i) System (1.1) is called weakly residence time controllable if for any bounded domain $\Psi \subset R^p (0 \in \Psi)$ there exists controller (1.4) (or (1.5)) such that $\hat{\mu}(K, L) > 0$;

(ii) System (1.1) is said to be strongly residence time controllable if for any bounded $\Psi \subset R^p (0 \in \Psi)$ and $\mu > 0$ there exists controller (1.4) (or (1.5)) such that $\hat{\mu}(K, L) > \mu$.

In what follows, we will be assuming that:

A1: (A, C) is disturbable,

A2: (D, A) is detectable,

A3: $FF^T > 0$, and $w(t)$ and $w_1(t)$ are independent Brownian motions,

A4: Transfer matrices $G_s(s) = D(sI - A)^{-1}B, G_n(s) = D(sI - A)^{-1}C$ and $G_{n1}(s) = E(sI - A)^{-1}C$ have full normal rank.

3. NOISELESS MEASUREMENTS CASE

Let $\mathcal{K} \triangleq \{K \in R^{m \times n} : A + BK \text{ is Hurwitz }\}$, $\mathcal{L} \triangleq \{L \in R^{n \times P} : A - LE \text{ is Hurwitz }\}$ and define the maximal logarithmic residence time of (1.1), (1.4) or (1.1), (1.5) in Ψ as

$$\hat{\mu}^* = \sup_{\substack{K \in \mathcal{K} \\ L \in \mathcal{L}}} \hat{\mu}(K, L) \ . \tag{3.1}$$

Introduce the following hypotheses:

H1: $G_s(s)$ is right invertible and minimum phase.

H2: $G_{n1}(s)$ is left invertible and minimum phase.

H3: There exists an $m \times r$ rational matrix $U(s)$ with no poles in Re $s > 0$ such that $G_n(s) + G_s(s)U(s) = 0$.

H4: There exists a $p \times q$ rational matrix $V(s)$ with no poles in Re $s > 0$ such that $G_n(s) + V(s)G_{n1}(s) = 0$.

Theorem 3.1: System (1.1) is

(i) weakly residence time controllable by controller (1.4) if and only if (A, B) is stabilizable and (E, A) is detectable,

(ii) strongly residence time controllable by controller (1.4) if and only if (A, B) is stabilizable, (E, A) is detectable and either H1 and H4 or H2 and H3 are satisfied.

Proof: See the Appendix.

Remark 3.1: As it was shown in [3], H3 is the condition for strong residence time controllability with respect to the state space feedback $u = Kx$. Furthermore, H1 is a stronger condition than H3.

Thus, either H4 or H2 are the additional condition that has to be satisfied when the state space feedback is replaced by the output feedback.

Remark 3.2: In SISO case with $D = E$, Theorem 3.1 implies that for strong residence time controllability $G_s(s)$ should be minimum phase.

A comparison of the fundamental bounds on the residence time achievable by state space (1.3) and output (1.4) feedback can be given as follows:

Consider the closed loop system (1.1), (1.3), i.e.

$$dx = (A + BK)xdt + \epsilon\, Cdw \quad , \tag{3.2}$$

and define as

$$\mu^* = \sup \mu(\Psi; K) \tag{3.3}$$

its maximal logarithmic residence time in Ψ.

Theorem 3.2: Equality $\hat{\mu}^* = \mu^*$ takes place if and only if $G_{n1}(s)$ has a left inverse with no poles in Re $s > 0$.

Proof: See the Appendix.

4. NOISY MEASUREMENTS CASE

Theorem 4.1: Let P be the unique positive definite solution of the (Kalman filter) Riccati equation:

$$AP + PA^T + CC^T - PE^T(FF^T)^{-1}EP = 0 \ . \tag{4.1}$$

Then the maximal logarithmic residence time of the closed loop system (1.1), (1.5) in Ψ satisfies the bound

$$\hat{\mu}^* \leq \min_{y \in \partial\Psi} \frac{1}{2}\, y^T(DPD^T)^{-1}y \ . \tag{4.2}$$

Proof: See the Appendix.

Remark 4.1: It follows, in particular, from Theorem 4.1 that since the upper bound in (4.2) is always finite, system (1.1) with control (1.5) is never strongly residence time controllable. Therefore, the measurement noise in (1.5) has a greater limiting effect on the achievable residence time than the input noise in (1.1).

Theorem 4.2: The upper bound (4.2) is attained if and only if there exists a rational matrix $W(s)$ with no poles in Re $s > 0$ such that

$$G_l(s) + G_s(s)W(s) = 0 \quad , \tag{4.3}$$

where $G_s(s)$ is defined as previously and

$$G_l(s) = D(sI - A)^{-1}\hat{L} \ , \tag{4.4}$$

$$\hat{L} = PE^T(FF^T)^{-1} \ . \tag{4.5}$$

Proof: See the Appendix.

Remark 4.2: Theorem 4.2 illustrates that the upper bound in (4.2) is attainable. Therefore, it is the best possible upper bound.

5. DESIGN TECHNIQUES

In the two previous sections we have characterized the fundamental bounds on the achievable logarithmic residence time. In this section we develop the controller design techniques that achieve these bounds. First system (1.1) with control (1.5) is considered and then system (1.1) with control (1.4) is addressed. An example is given in Section 6.

To select the pair $\{K, L\}$ that maximizes $\hat{\mu}(K, L)$ assume, for simplicity, that domain Ψ is an ellipsoid

$$\Psi = \{y \in R^p : \ y^T S y \leq r^2, \ S = S^T > 0\} \ . \tag{5.1}$$

Let $W \in R^{p \times p}$ be a nonsingular matrix such that $S = W^T W$. Then by direct calculations we obtain:

$$\hat{\mu}(K, L) = \frac{r^2}{2\lambda_{\max}[WDX(K,L)D^T W^T]} \ , \tag{5.2}$$

where $X(K, L)$ is given by

$$\begin{bmatrix} A & BK \\ LE & A+BK-LE \end{bmatrix} \begin{bmatrix} X(K,L) & T(K,L) \\ T^T(K,L) & \widehat{X}(K,L) \end{bmatrix} \tag{5.3}$$

$$+ \begin{bmatrix} X(K,L) & T(K,L) \\ T^T(K,L) & \widehat{X}(K,L) \end{bmatrix} \begin{bmatrix} A & BK \\ LE & A+BK-LE \end{bmatrix}^T + \begin{bmatrix} CC^T & 0 \\ 0 & LFF^T L^T \end{bmatrix} = 0$$

Therefore, the pair $\{K, L\}$ is optimal if and only if it minimizes the largest eigenvalue of $\Gamma(K, L) \triangleq WDX(K, L)D^T W^T$. The $\lambda_{max}(\Gamma)$ can be characterized as follows:

Lemma 5.1: Let $\theta \geq 0$ be a scalar, $l \geq 1$ be an integer and select $K_l \in \mathcal{K}$ and $L_l \in \mathcal{L}$ such that

$$\text{Tr}\ \Gamma(K_l, L_l)^l \leq (1 + \theta) \inf\{\text{Tr}\ \Gamma(K, L)^l | K \in \mathcal{K}, L \in \mathcal{L}\} \ . \tag{5.4}$$

Then

$$\lim_{l \to \infty} \lambda_{\max}(\Gamma(K_l, L_l)) = \inf\{\lambda_{\max}(\Gamma(K, L)) | K \in \mathcal{K}, L \in \mathcal{L}\} \tag{5.5}$$

Proof: The proof of this lemma is similar to the proof of Theorem 2.1 in [7]. We omit the details here.

Thus, in order to minimize $\lambda_{\max}(\Gamma)$, we need only to minimize $\mathrm{Tr}\ \Gamma(K,L)^l$, $l = 1,2,3\ldots$. To accomplish this, introduce

$$J_\gamma^l(K,L) = \mathrm{Tr}\ \Gamma(K,L)^l + \gamma\ \mathrm{Tr}\ K\widehat{X}(K,L)K^T \quad , \tag{5.6}$$

where $\widehat{X}(K,L)$ is given by (5.3).

Lemma 5.2: Assume that $K_l^\gamma \in \mathcal{K}$ and $L_l^\gamma \in \mathcal{L}$ minimize $J_\gamma^l(K,L)$. Then

$$\lim_{\gamma\to 0} J_\gamma^l(K_l^\gamma, L_l^\gamma) = \inf_{\substack{K\in\mathcal{K}\\ L\in\mathcal{L}}} \mathrm{Tr}\ \Gamma(K,L)^l \ . \tag{5.7}$$

Proof: The proof of this lemma is similar to the first part of the proof of the Theorem in [4]. We omit the details here.

From Lemmas 5.1 and 5.2 follows:

Corollary 5.1: Assume that the pair (K_l^γ, L_l^γ) with $K_l^\gamma \in \mathcal{K}$ and $L_l^\gamma \in \mathcal{L}$ minimizes $J_\gamma^l(K,L)$. Then

$$\lim_{l\to\infty} \lim_{\gamma\to 0} \hat{\mu}(K_l^\gamma, L_l^\gamma) = \hat{\mu}^* \ . \tag{5.8}$$

Thus, K_l^γ and L_l^γ provide the solution to (3.1). A necessary condition for the optimality of (K_l^γ, L_l^γ) in the sense of functional (5.6) can be formulated as follows.

Theorem 5.1: Assume that $K_l^\gamma \in \mathcal{K}$ and $L_l^\gamma \in \mathcal{L}$. Then in order for (K_l^γ, L_l^γ) to minimize $J_\gamma^l(K,L)$ it is necessary that

$$L_l^\gamma = \hat{L} = PE^T(FF^T)^{-1} \ , \tag{5.9}$$

$$K_l^\gamma = -\frac{1}{\gamma}\, B^T Q_l^\gamma \ , \tag{5.10}$$

where P is given by (4.1) and

$$A^T Q_l^\gamma + Q_l^\gamma A + D^T W^T M_l^\gamma W D - \frac{1}{\gamma}\, Q_l^\gamma B B^T Q_l^\gamma = 0 \ , \tag{5.11}$$

$$M_l^\gamma = l(WD(\widehat{X}_l^\gamma + P)D^T W^T)^{l-1} \ , \tag{5.12}$$

$$(A + BK_l^\gamma)\widehat{X}_l^\gamma + \widehat{X}_l^\gamma(A + BK_l^\gamma)^T + \hat{L}\hat{L}^T = 0 \ . \tag{5.13}$$

Proof: See the Appendix.

Thus, in particular, the optimal observation gain is independent of optimal control while the optimal control gain is a function of optimal observations.

Since (4.1) has a positive definite solution, $L_l^\gamma = \hat{L} \in \mathcal{L}$, $\forall\ \gamma, l$. The following lemma gives a condition for $K_l^\gamma \in \mathcal{K}$.

Lemma 5.3: Assume that $M_l^\gamma > 0$. Then $K_l^\gamma \in \mathcal{K}$.

Proof: See the Appendix.

Remark 5.1: As it follows from Theorem 5.1, the optimal estimator gain $\widehat{L}$ given in (5.9) is the Kalman filter gain. Thus, the Kalman filter is optimal in optimization problem (3.1). Moreover, consider the equation for the estimation error $e \triangleq x - \hat{x}$:

$$de = (A - LE)dt + \epsilon\, C\, dw - \epsilon\, LF dw_1 \tag{5.14}$$

and define its logarithmic residence time in any domain $\Lambda \subset R^n (0 \in \Lambda)$ as $\hat{\mu}(\Lambda; L)$. Then

$$\hat{\mu}(\Lambda; L) = \min_{e \in \partial\Lambda} \frac{1}{2}\, e^T P^{-1}(L) e \ , \tag{5.15}$$

where $P(L)$ is the positive definite solution of

$$(A - LE)P(L) + P(L)(A - LE)^T + CC^T + LFF^T L^T = 0 \ . \tag{5.16}$$

Since P given by (4.1) satisfies the inequality

$$P \leq P(L), \ \ \forall L \in \mathcal{L} \ , \tag{5.17}$$

we conclude that

$$\hat{\mu}(\Lambda; \widehat{L}) = \min_{e \in \partial\Lambda} \frac{1}{2}\, e^T P^{-1} e \geq \hat{\mu}(\Lambda; L), \ \forall L \in \mathcal{L} \ . \tag{5.18}$$

Thus, the Kalman filter is optimal in the sense of optimization of the estimation error residence time in every bounded domain of R^n.

The optimal control law for system (1.1) with control (1.4) can be obtained from (5.9) - (5.10) by selecting $F = \alpha I$ and letting $\alpha \to 0$. Indeed, since the optimal estimator law for (1.1), (1.5) is the Kalman filter, we know from optimal filtering theory that the optimal (singular) filter for (1.1), (1.4) is obtained in the limit $\alpha \to 0$ (see, e.g., [4]). Therefore, the maximal logarithmic residence time for (1.1), (1.4) is given by

$$\hat{\mu}^* = \lim_{l \to \infty} \lim_{\gamma \to 0} \lim_{\alpha \to 0} \hat{\mu}(K_l^{\gamma,\alpha}, L_l^{\gamma,\alpha}) \ , \tag{5.19}$$

where $L_l^{\gamma,\alpha}$ and $K_l^{\gamma,\alpha}$ are given by (5.9) - (5.13) with $FF^T = \alpha^2 I$.

6. EXAMPLE

Consider the second order system

$$\begin{aligned} \dot{x} &= \begin{bmatrix} 0 & 1 \\ -1 & 0 \end{bmatrix} x + \begin{bmatrix} 0 \\ 1 \end{bmatrix} u + \epsilon \begin{bmatrix} 1 \\ 0 \end{bmatrix} \dot{w} \ , \\ y &= [0 \quad 1] x \ , \\ z &= [1 \quad 0] x + \epsilon\, F \dot{w}_1 \ . \end{aligned} \tag{6.1}$$

For this system

$$G_s(s) = \frac{s}{s^2+1} \;,\; G_n(s) = \frac{-1}{s^2+1} \;,\; G_{n1}(s) = \frac{s}{s^2+1} \;. \tag{6.2}$$

Thus, since $G_s(s) = G_{n1}(s)$ is minimum phase, this system is *srt*-controllable by controller (1.4) when $F \equiv 0$.

Assume that $F \neq 0$. Then, by Theorem 4.1, the logarithmic residence time in the interval $\Psi = (-a, b)$, $a, b > 0$, is bounded by

$$\min_{y \in \partial\Psi} \frac{1}{2}\, y^T (DPD^T)^{-1}\, y = \frac{(\min(a,b))^2}{2|F|} \;. \tag{6.3}$$

Furthermore, when $a = b$, the (sub)optimal controller can be calculated using (5.9) - (5.13) to be

$$\hat{L} = \begin{bmatrix} \frac{1}{|F|} \\ 0 \end{bmatrix} \;,\; K_l^\gamma = -[0 \quad K_2] \tag{6.4}$$

where $K_2 > 0$ satisfies the equation

$$\frac{K_2^2 \gamma}{l|F|^{l-1}} = \left(1 + \frac{|F|}{2K_2}\right)^{l-1} \;. \tag{6.5}$$

The logarithmic residence time with this control is

$$\hat{\mu}(K_l^\gamma, \hat{L}) = \frac{a^2}{2|F|} \cdot \frac{2K_2}{2K_2 + |F|} \;. \tag{6.6}$$

Note that $\hat{\mu}(K_e^\gamma, \hat{L})$ is the upper bound in (6.3) multiplied by the factor

$$\rho = \frac{2K_2}{2K_2 + |F|} \;. \tag{6.7}$$

Thus, in order to obtain logarithmic residence time as close as desired to the maximal value, (6.3), (6.7) can be used to calculate the necessary K_2 (for a given ρ) and l and γ can be determined from (6.5).

As $\gamma \to 0$ equation (6.5) simplifies considerably. Indeed, in this case $K_2 \to \infty$ and, thus, for small γ (6.5) becomes

$$\frac{K_2^2 \gamma}{l|F|^{l-1}} \simeq 1 \;. \tag{6.8}$$

Therefore,

$$K_2 \simeq \sqrt{\frac{l}{\gamma}}\, |F|^{\frac{l-1}{2}} \;. \tag{6.9}$$

7. CONCLUSIONS

It is shown in this paper that the observer based output feedback can be efficiently used for pointing of linear systems subject to both input and measurement noise. The fundamental bounds on the achievable precision of pointing depend on the locations of the right half plane zeros of the various transfer functions involved. Roughly speaking, the best precision of pointing is obtained for minimum phase systems. Any desired precision of aiming is attainable only if no measurement noise is present. Therefore, the effect of the measurement noise on the achievable precision of aiming is more detrimental than that of the input noise.

APPENDIX

Proof of Theorem 3.1: The proof of (i) parallels the proof of Theorem 3.1 in [3]. We omit the details here. In order to prove (ii) we first derive the inequality

$$\frac{r^2}{2\ \mathrm{Tr}\ DX(K,L)D^T} \leq \hat{\mu}(K,L) \leq \frac{pR^2}{2\ \mathrm{Tr}\ DX(K,L)D^T}\ , \tag{A.1}$$

where $K \in \mathcal{K},\ L \in \mathcal{L}$. To get the left inequality note that

$$\begin{aligned}
\hat{\mu}(K,L) &\geq \frac{1}{2}\ \lambda_{\min}[(DX(K,L)D^T)^{-1}] \min_{y \in \partial\Psi} y^T y \\
&= \frac{r^2}{2\lambda_{\max}[DX(K,L)D^T]} \\
&\geq \frac{r^2}{2\ \mathrm{Tr}\ DX(K,L)D^T}\ .
\end{aligned}$$

For the right inequality we have $(R^2 = \max_{y \in \partial\Psi} y^T y,\ B(0,R) = \{y | y^T y \leq R^2\})$

$$\begin{aligned}
\hat{\mu}(K,L) &= \min_{y \in \partial\Psi} \frac{1}{2}\ y^T (DX(K,L)D^T)^{-1} y \\
&\leq \min_{y \in \partial B(0,R)} \frac{1}{2}\ y^T (DX(K,L)D^T)^{-1} y \\
&= \frac{1}{2}\ \lambda_{\min}[(DX(K,L)D^T)^{-1}] R^2 \\
&= \frac{R^2}{2\lambda_{\max}[DX(K,L)D^T]} \\
&\leq \frac{pR^2}{2\ \mathrm{Tr}\ DX(K,L)D^T}\ .
\end{aligned}$$

It follows from (A.1) that $\hat{\mu}^* = \infty$ is equivalent to

$$\inf_{\substack{K \in \mathcal{K} \\ L \in \mathcal{L}}}\ \mathrm{Tr}\ DX(K,L)D^T = 0\ . \tag{A.2}$$

Next note that it follows from linear quadratic theory [4], [5] that

$$\inf_{\substack{K\in\mathcal{K}\\ L\in\mathcal{L}}} \text{Tr } DX(K,L)D^T = \lim_{\substack{\gamma\to 0\\ \alpha\to 0}} \text{Tr } DX(K^\gamma,L^\gamma)D^T \tag{A.3}$$

where

$$K^\gamma = -\frac{1}{\gamma} B^T Q^\gamma, \quad A^T Q^\gamma + Q^\gamma A + D^T D - \frac{1}{\gamma} Q^\gamma B B^T Q^\gamma = 0 \ , \tag{A.4}$$

$$L^\alpha = -\frac{1}{\alpha} P^\alpha E^T, \quad A P^\alpha + P^\alpha A + C C^T - \frac{1}{\alpha} P^\alpha E^T E P^\alpha = 0 \ . \tag{A.5}$$

Furthermore,

$$\begin{aligned} \lim_{\substack{\gamma\to 0\\ \alpha\to 0}} \text{Tr } DX(K^\gamma,L^\gamma)D^T &= \lim_{\substack{\gamma\to 0\\ \alpha\to 0}} \text{Tr } (DP^\alpha D^T + \alpha L^{\alpha^T} Q^\gamma L^\alpha) \\ &= \lim_{\substack{\gamma\to 0\\ \alpha\to 0}} \text{Tr } (C^T Q^\gamma C + \gamma\, K^\gamma P^\alpha K^{\gamma^T}) \ . \end{aligned} \tag{A.6}$$

Therefore, with $\tilde{C} = \lim_{\alpha\to 0} \sqrt{\alpha}\, L^\alpha$ and $\hat{D} = \lim_{\alpha\to 0} \sqrt{\gamma}\, K^\gamma$, we have

$$\begin{aligned} \inf_{\substack{K\in\mathcal{K}\\ L\in\mathcal{L}}} \text{Tr } DX(K,L)D^T &= \text{Tr } (DP^0 D^T + \tilde{C}^T Q^0 \tilde{C}) \\ &= \text{Tr } (\hat{D} P^0 \hat{D}^T + C^T Q^0 C) \ . \end{aligned} \tag{A.7}$$

Each of the terms Tr DP^0D^T, Tr $\tilde{C}^TQ^0\tilde{C}$, Tr $\hat{D}P^0\hat{D}^T$ and Tr C^TQ^0C is nonnegative. Thus system (1.1) with control (1.4) is strongly residence time controllable if and only if all four terms are zero.

It was shown in [3] that Tr $C^TQ^0C = 0$ if and only if there exists a rational matrix $U(s)$, with no poles in Re $s > 0$, such that

$$G_n(s) + G_s(s)U(s) = 0 \ . \tag{A.8}$$

Similarlily, Tr $\tilde{C}^TQ^0\tilde{C} = 0$, Tr $DP^0D^T = 0$ and Tr $\widehat{D}P^0\widehat{D}^T = 0$ if and only if there exist rational matrices $\tilde{U}(s)$, $V(s)$ and $\hat{V}(s)$, with no poles in Re $s > 0$, such that

$$\tilde{G}_n(s) + G_s(s)\tilde{U}(s) = 0 \ , \tag{A.9}$$

$$G_n(s) + V(s)G_{n1}(s) = 0 \ , \tag{A.10}$$

$$\hat{G}_n(s) + \hat{V}(s)G_{n1}(s) = 0 \tag{A.11}$$

where

$$\tilde{G}_n(s) = D(sI-A)^{-1}\,\tilde{C} \ , \tag{A.12}$$

$$\hat{G}_n(s) = \widehat{D}(sI-A)^{-1}\,C \ . \tag{A.13}$$

Now, if H1 is satisfied then $U(s) = -G_s^{-1}(s)G_n(s)$ and $\tilde{U}(s) = -G_s^{\prime -1}(s)\tilde{G}_n(s)$ ($G_s^{-1}(s)$ is the right inverse of $G_s(s)$) are both without poles in Re $s > 0$ and satisfy (A.8) and

(A.9). Therefore, Tr $C^T Q^0 C$ = Tr $\tilde{C}^T Q^0 \tilde{C} = 0$. Furthermore, in this case $D^T D = \widehat{D}^T \widehat{D}$ (see, e.g., [4]) and, thus, H4 implies that $0 = \text{Tr } DP^0D^T = \text{Tr } P^0D^TD = \text{Tr } P^0\widehat{D}^T\widehat{D} = \text{Tr } \widehat{D}P^0\widehat{D}^T$. Therefore, by (A.7) the system is strongly residence time controllable. Similarily, if H3 is satisfied, then $V(s) = -G_n(s)G_{n1}^{-1}(s)$ and $\hat{V}(s) = -G_n(s)G_{n1}^{-1}(s)$ are both without poles in Re $s > 0$ and, thus, Tr DP^0D^T = Tr $\widehat{D}P^0\widehat{D}^T = 0$. Furthermore, $CC^T = \tilde{C}\tilde{C}^T$ and, therefore, H3 implies that $0 = \text{Tr } C^TQ^0C = \text{Tr } \tilde{C}^TQ^0\tilde{C}$. This proves the sufficiency part of the theorem.

Assume now that (1.1), (1.4) is strongly residence time controllable. Then (A.8) - (A.11) are satisfied and, thus, H3 and H4 are true. Note that the existence of $U(s)$ such that (A.8) is satisfied and A4 imply that $m \geq \min(p, r)$. Similarly, the existence of $V(s)$ and A4 imply that $q \geq \min(p, r)$. Assume $p \leq r$. Then $m \geq p$ and, thus, $G_s(s)$ is right invertible. Similarly, if $p \geq r$, then $q \geq r$ and $G_{n1}(s)$ is left invertible. Next, it can be shown that (A.10) implies that $G_n(s)G_n^T(-s) = \tilde{G}_n(s)\tilde{G}_n^T(-s)$. Furthermore, $\tilde{G}_n(s)$ has no zeroes in Re $s > 0$ (see, e.g. [6]). Similarly, (A.8) implies that $G_n^T(-s)G_n(s) = \hat{G}_n^T(-s)\hat{G}_n(s)$ and $\hat{G}_n(s)$ has no zeroes in Re $s > 0$. Thus, if $p \leq r$, i.e. $G_s(s)$ is right invertible, then it follows from (A.10) and (A.9) that $G_s(s)$ has no zeroes in Re $s > 0$. Thus H1 is satisfied. Similarily, if $p \geq r$, then (A.8) and (A.11) imply that $G_{n1}(s)$ is left invertible and minimum phase, i.e., H2 is true. **Q.E.D.**

Proof of Theorem 3.2: Let $\mu(K)$ be the logarithmic residence time of (3.12). Then, obviously, for any $K \in \mathcal{K}$ and $L \in \mathcal{L}$ we have

$$\hat{\mu}(K, L) \leq \mu(K) \tag{A.14}$$

and, thus,

$$\sup_{L \in \mathcal{L}} \hat{\mu}(K, L) \leq \mu(K) \quad . \tag{A.15}$$

Furthermore, using a similar argument to the one in the proof of Theorem 4.1 (see below) we have

$$\sup_{L \in \mathcal{L}} \hat{\mu}(K, L) = \lim_{\alpha \to 0} \hat{\mu}(K, L^\alpha) \tag{A.16}$$

$$L^\alpha = P^\alpha E^T \ , \ AP^\alpha + P^\alpha A^T + CC^T - \frac{1}{\alpha} P^\alpha E^T E P^\alpha = 0 \quad . \tag{A.17}$$

Thus, we want to show that left invertibility and minimum phase of $G_{n1}(s)$ is necessary and sufficient for

$$\begin{aligned} \lim_{\alpha \to 0} \hat{\mu}(K, L^\alpha) &= \min_{y \in \partial \Psi} \frac{1}{2} y^T (D(\hat{X}(K) + P^0)D^T)^{-1} y \\ &= \mu(K) \end{aligned} \tag{A.18}$$

where

$$(A + BK)\hat{X}(K) + \hat{X}(K)(A + BK) + \tilde{C}\tilde{C}^T = 0 \tag{A.19}$$

for all $K \in \mathcal{K}$. However, since

$$\mu(K) = \min \frac{1}{2} y^T (DX(K)D^T)^{-1} y \quad , \tag{A.20}$$

$$(A+BK)X(K)+X(K)(A+BK)^T+CC^T=0 \ , \tag{A.21}$$

it follows that (A.18) is true if and only if $DP^0D^T = 0$ and $CC^T = \tilde{C}\tilde{C}^T$. These are exactly the necessary and sufficient conditions for $G_{n1}(s)$ to be left invertible and minimum phase. **Q.E.D.**

Proof of Theorem 4.1: It is straight forward to show that $X(K,L) \geq P$ (see, e.g. equation (A.25) below). Therefore, since

$$\hat{\mu}(K,L) = \min_{y \in \partial\Psi} \frac{1}{2}\, y^T(DX(K,L)D^T)^{-1}y \ , \tag{A.22}$$

inequality (4.2) follows. **Q.E.D.**

Proof of Theorem 4.2: The logarithmic residence time in a system with the optimal estimator gain $\hat{L} = PE^T(FF^T)^{-1}$ is

$$\hat{\mu}(K,\hat{L}) = \min_{y \in \partial\Psi} \frac{1}{2}\, y^T(D(\widehat{X}(K)+P)D^T)^{-1}y \tag{A.23}$$

where

$$(A+BK)\widehat{X}(K)+\widehat{X}(K)+\hat{L}\hat{L}^T=0 \ . \tag{A.24}$$

Thus, the upper bound (4.2) is attained if and only if $\inf_{K \in \mathcal{K}} \operatorname{Tr} D\widehat{X}(K)D^T = 0$. However, by the same argument as was used in the proof of Theorem 3.1, this happens if and only if (4.3) is satisfied. **Q.E.D.**

Proof of Theorem 5.1: Let $\hat{L}$ be the Kalman filter gain (5.9) and define

$$d\tilde{x} = (A\tilde{x}+BK\hat{x})dt+\hat{L}(dz-E\tilde{x}dt) \tag{A.25}$$

where $\hat{x}$ is the estimate (1.5) for an arbitrary L. Then [5]

$$X(K,L)=\widetilde{X}(K,L)+P \tag{A.26}$$

where P satisfies (4.1) and $\widetilde{X}$ is given by

$$\begin{pmatrix} A & BK \\ LE & A+BK-LE \end{pmatrix}\begin{pmatrix} \widetilde{X} & Z \\ Z^T & \widehat{X} \end{pmatrix}+\begin{pmatrix} \widetilde{X} & Z \\ Z^T & \widehat{X} \end{pmatrix}\begin{pmatrix} A & BK \\ LE & A+BK-LE \end{pmatrix}^T$$
$$+\begin{pmatrix} \hat{L}FF^T\hat{L}^T & \hat{L}FF^TL^T \\ LFF^T\hat{L}^T & LFF^TL^T \end{pmatrix}=0 \ . \tag{A.27}$$

Define

$$\begin{pmatrix} \widetilde{X} & X_1 \\ X_1^T & X_2 \end{pmatrix}=\begin{pmatrix} I & 0 \\ I & -I \end{pmatrix}\begin{pmatrix} \widetilde{X} & Z \\ Z^T & \widehat{X} \end{pmatrix}\begin{pmatrix} I & I \\ 0 & -I \end{pmatrix} \ . \tag{A.28}$$

Then

$$\begin{pmatrix} A+BK & -BK \\ 0 & A-LE \end{pmatrix}\begin{pmatrix} \widetilde{X} & X_1 \\ X_1 & X_2 \end{pmatrix}+\begin{pmatrix} \widetilde{X} & X_1 \\ X_1^T & X_2 \end{pmatrix}\begin{pmatrix} A+BK & -BK \\ 0 & A-LE \end{pmatrix}^T$$

$$+\begin{pmatrix} \hat{L}FF^T\hat{L}^T & \hat{L}FF^T(\hat{L}-L)^T \\ (\hat{L}-L)FF^T\hat{L} & (\hat{L}-L)FF^T(\hat{L}-L)^T \end{pmatrix} = 0 \ . \tag{A.29}$$

In order to show that $K_l^\gamma, \hat{L}$ satisfy the necessary conditions for minimizing $J_\gamma^l(K,L)$ we have to show that $\left(F = \binom{K}{L^T}\right)$

$$\frac{\partial J_\gamma^l(F)}{\partial F} = 0 \tag{A.30}$$

gives $F = F_l^\gamma = \binom{K_l^\gamma}{\hat{L}^T}$. (A.30) is equivalent to showing that

$$\frac{d}{d\epsilon}\, J_\gamma^l(F+\epsilon\Delta F)\bigg|_{\epsilon=0} = \left\langle \frac{\partial J_\gamma^l}{\partial F}\,(F),\ \Delta F \right\rangle = \mathrm{Tr}\ \Delta F^T \frac{\partial J_\gamma^l}{\partial F}(F) = 0 \tag{A.31}$$

for all $\Delta F = \binom{\Delta K}{\Delta L^T}$. In order to simplify notation we assume $WD = I$. Evaluating $\frac{d}{d\epsilon} J_\gamma^l(F+\epsilon\Delta F)\Big|_{\epsilon=0}$ gives

$$\begin{aligned} \frac{d}{d\epsilon}\, J_\gamma^l(F+\epsilon\Delta F)\bigg|_{\epsilon=0} &= l\ \mathrm{Tr}\ X^{l-1}X' + \gamma\ \mathrm{Tr}\ K\hat{X}'K^T \\ &\quad + \gamma\ \mathrm{Tr}\ \Delta K\widehat{X}K^T + \gamma\ \mathrm{Tr}\ K\widehat{X}\Delta K^T \ . \end{aligned} \tag{A.32}$$

where

$$X' = \frac{d}{d\epsilon}\, X(K+\epsilon\Delta K,\ L+\epsilon\Delta L)\bigg|_{\epsilon=0} \ , \tag{A.33}$$

$$\widehat{X}' = \frac{d}{d\epsilon}\, \widehat{X}(K+\epsilon\Delta K, L+\epsilon\Delta L)\bigg|_{\epsilon=0} \ . \tag{A.34}$$

From (A.26) and (A.28) we get $\widehat{X} = \widetilde{X} - X_1^T - X_1 + X_2$ and

$$\widehat{X}' = \widetilde{X}' - X_1'^T - X_1' + X_2' \ , \tag{A.35}$$

$$X' = \widetilde{X}' \ \ (\text{since } P = \text{const.}) \tag{A.36}$$

where

$$\widetilde{X}' = \frac{d}{d\epsilon}\ \widetilde{X}\ \Big|_{\epsilon=0} \ , \tag{A.37}$$

$$X_1' = \frac{d}{d\epsilon}\ X_1\Big|_{\epsilon=0} \ , \tag{A.38}$$

$$X_2' = \frac{d}{d\epsilon}\ X_2\Big|_{\epsilon=0} \ . \tag{A.39}$$

Using this in (A.32) gives

$$\begin{aligned} \frac{d}{d\epsilon}\, J_\gamma^l(F+\epsilon\Delta F)\bigg|_{\epsilon=0} &= l\ \mathrm{Tr}\ X^{l-1}\widetilde{X}' \\ + &\ \gamma\ \mathrm{Tr}\ K^TK(\widetilde{X}' - X_1'^T - X_1' + X_2') \\ + &\ \gamma\ \mathrm{Tr}\ \widehat{X}K^T\Delta K + \gamma\ \mathrm{Tr}\ \Delta K^TK\widehat{X} \ . \end{aligned} \tag{A.40}$$

From (A.29) we get the following equations for $\widetilde{X}, X_1$ and X_2

$$(A+BK)\widetilde{X}+\widetilde{X}(A+BK)^T - BKX_1 - X_1^T K^T B^T + \hat{L}FF^T\hat{L}^T = 0 \ , \tag{A.41}$$

$$(A+BK)X_1 + X_1(A-LE)^T - BKX_2 + \hat{L}FF^T(\hat{L}-L)^T = 0 \ , \tag{A.42}$$

$$(A-LE)X_2 + X_2(A-LE)^T + (\hat{L}-L)FF^T(\hat{L}-L)^T = 0 \ . \tag{A.43}$$

Thus, $\widetilde{X}', X_1'$ and X_2' satisfy

$$\begin{aligned}(A+BK)\widetilde{X}' + \widetilde{X}'(A+BK)^T + B\Delta K\widetilde{X} + \widetilde{X}\Delta K^T B^T \\ - BKX_1' - X_1'^T K^T B^T - B\Delta K X_1 - X_1^T \Delta K^T B^T = 0 \ ,\end{aligned} \tag{A.44}$$

$$\begin{aligned}(A+BK)X_1' + X_1'(A-LE)^T + B\Delta K X_1 - X_1 E^T \Delta L^T \\ - BKX_2' - B\Delta K X_2 - \hat{L}FF^T\Delta L^T = 0 \ ,\end{aligned} \tag{A.45}$$

$$\begin{aligned}(A-LE)X_2' + X_2'(A-LE)^T - \Delta L E X_2 - X_2 E^T \Delta L^T \\ - (\hat{L}-L)^T FF^T \Delta L^T - \Delta L FF^T(\hat{L}-L)^T = 0 \ .\end{aligned} \tag{A.46}$$

Next we rewrite (A.40) using (A.35) and the adjoint equation for (A.44). This gives

$$\begin{aligned}\frac{d}{d\epsilon} J_\gamma^l(F+\epsilon\Delta F)\Big|_{\epsilon=0} &= \text{Tr } (\widetilde{X}QB + \gamma\widehat{X}K^T - X_1QB)\Delta K \\ +\text{Tr } \Delta K^T(B^TQ\widetilde{X} + \gamma K\widehat{X} - B^TQX_1^T) &- \text{Tr}(QB+\gamma K^T)KX_1' \\ - \text{Tr } X_1'^T K^T(B^TQ+\gamma K) &+ \gamma \text{ Tr } K^TKX_2'\end{aligned} \tag{A.47}$$

where

$$(A+BK)^TQ + Q(A+BK) + lX^{l-1} + \gamma K^TK = 0 \ . \tag{A.48}$$

Now, it follows from (A.46) and the last term in (A.47) that in order for (A.47) to be zero for any ΔL it is necessary that $X_2' = 0$. Thus, $EX_2 + FF^T(\hat{L}-L)^T = 0$. Substituting $\hat{L}-L = -X_2E^T(FF^T)^{-1}$ into (A.43) gives $X_2 = 0$. Therefore $L = \hat{L}$. Furthermore, with $L = \hat{L}$ and $X_2 = 0$ it follows from (A.42) that for any $K \in \mathcal{K}$ we have $X_1 = 0$. Therefore $\widetilde{X} = \widehat{X}$ and the first two terms on the right hand side of (A.47) give $\gamma K + B^TQ = 0$. However, this makes the third and fourth terms in the right hand side of (A.47) also equal to zero. Therefore, in order for (A.47) to be identically zero for any ΔF we must have $L = L$ and $K = -\frac{1}{\gamma}B^TQ$.

Finally, substituting $K = K_l^\gamma = -\frac{1}{\gamma}B^TQ$ into (A.48) gives (5.11) and (A.26), (A.41) with $\widetilde{X} = \widehat{X}$ gives (5.13). **Q.E.D.**

Proof of Lemma 5.3: Note that if $M_l^\gamma > 0$ then $M_l^\gamma = N_l^{\gamma T} N_l^\gamma$ for some nonsingular N_l^γ. Furthermore, since (D, A) is detectable it follows that $(N_l^\gamma WD, A)$ is detectable. Thus, $Q_l^\gamma \geq 0$ and $K_l^\gamma \in \mathcal{K}$. **Q.E.D.**

REFERENCES

[1] M.I. Freidlin and A.D. Wentzell, *Random Perturbations of Dynamical Systems*, New York: Springer-Verlag, 1984.

[2] S.M. Meerkov and T. Runolfsson, "Residence Time Control", *IEEE Trans. Automat. Contr.*, Vol. AC-33, No. 4, pp. 323-332, April 1988.

[3] S.M. Meerkov and T. Runolfsson, "Output Residence Time Control," *IEEE Trans. Automat. Contr.*, to appear, 1989.

[4] H. Kwakernaak and R. Sivan, "The Maximally Achievable Accuracy of Linear Optimal Regulators and Linear Optimal Filters," *IEEE Trans. Automat. Contr.*, Vol. AC-17, 1972.

[5] D.L. Russell, *Mathematics of Finite Dimensional Control Systems*, New York: Marcel Dekker, 1979.

[6] U. Shaked and E. Soroka, "A Simple Solution to the Singular Linear Minimum-Variance Estimation Problem," *IEEE Trans. Automat. Contr.*, Vol. AC-32, pp. 81-84, January 1987.

[7] J.C. Allwright and J.Q. Mao, "Optimal Output Feedback by Minimizing $||K(F)||_2$," *IEEE Trans. Automat. Contr.*, Vol. AC-27, 1982.

MAXIMUM-MISS AIRCRAFT COLLISION AVOIDANCE

A. W. Merz

Staff Scientist

Advanced Systems Studies

Lockheed Missiles and Space Co.
Palo Alto, CA 94304

ABSTRACT

Aircraft densities in terminal areas increase each year, and the risk of collision grows proportionally. The maintenance of clearance between aircraft in this environment sometimes calls for evasive maneuvers, which depend on the relative position and relative velocity of two aircraft. Small amplitude maneuvers are found for both aircraft in near-miss configurations with short time-to-go. Using low-order dynamics, individual maneuvers are found which maximize the miss-distance. These evasive maneuvers are optimal with respect to the low-order dynamics, and they combine longitudinal (speed) and normal (lift) accelerations. The signs of the accelerations of both aircraft depend on their magnitudes. An evasive climb maneuver, for example, becomes a dive maneuver if the acceleration amplitude exceeds a certain value. The maximum-miss maneuvers appear to have practical potential, because they can be determined on-line from estimated position data for both aircraft.

INTRODUCTION

Optimal collision-avoidance maneuvers are defined as those for which the resulting miss-distance is maximized. In this paper, they are determined for the one-on-one encounter of two aircraft, moving arbitrarily in three dimensions. These maneuvers are appropriate when the uncontrolled miss and the time-to-go are small enough to cause anxiety to the air traffic controller and/or the pilots of the aircraft. Maneuver accelerations involve both normal (lift/turn) and tangential (thrust/flap) accelerations. The acceleration limits on both aircraft must be observed. These scalar limits, with the current position and velocity vectors, sum to 14 variables which determine the controls.

Refinements to the guidance logic relate to ground avoidance, other air traffic, more detailed acceleration limits, higher-order aircraft dynamics, navigation regulations, distance off-track, and control time-lags, among other things. Including any of these refinements will increase the number of parameters, but when a collision is imminent, the cited constraints seem to be most significant.

ANALYSIS

The optimal maneuvers are determined in a straightforward way, if it is assumed that the aircraft can accelerate equally in any direction. This assumption is valid when the acceleration level is small enough that longitudinal and normal components can be of equal magnitudes. The results of the study are *not* meant to apply to a "maximum possible miss" requirement, when limit values of the controls would be used, leading to wide departures from the linearized equations to be implemented.

The envelope of the available acceleration is a sphere of specified radius, and the vector equations of motion of either aircraft are:

$$\dot{\mathbf{r}}_i = \mathbf{v}_i, \quad \dot{\mathbf{v}}_i = \mathbf{a}_i, \quad |\mathbf{a}_i| \leq A_i, \tag{1}$$

where i = 1 or 2, and where $\mathbf{r}_i, \mathbf{v}_i$ and $\mathbf{a}_i$ are vectors. The homogeneous trajectory for each aircraft is a straight line, described at constant speed, such that lift and thrust balance drag and weight. The accelerations in Eq.(1) represent a departure from this flight condition, and are the sum of changes in thrust, lift and drag due to control changes from the nominal. These vectors are reasonable approximations to actual values during the brief time-interval of 10 or 20 seconds prior to minimum range, if the magnitude is low enough to be realizable in both tangential and normal directions. Typical trajectories to be discussed are generated by control accelerations of less than .3 g, or some 10 ft/sec^2, which can be shared between the two aircraft.

The position and velocity of Aircraft 2 relative to Aircraft 1 are given by the differences, $\mathbf{r} = \mathbf{r}_2 - \mathbf{r}_1$, and $\mathbf{v} = \mathbf{v}_2 - \mathbf{v}_1$, so that

$$\dot{\mathbf{r}} = \mathbf{v}, \tag{2}$$

$$\dot{\mathbf{v}} = \mathbf{a}. \tag{3}$$

Here $|\mathbf{a}| = A$, which equals $A_1 + A_2$ when the two aircraft accelerate in opposite directions. The time to minimum range with $\mathbf{a} = \mathbf{0}$ is

$$t_f = -\mathbf{r}_0 \cdot \mathbf{v}_0 / v_0^2 > 0. \tag{4}$$

The vector miss for no control accelerations is therefore,

$$\mathbf{r}_f = \mathbf{r}_0 + \mathbf{v}_0 t_f \tag{5}$$

The control criterion is the miss distance, to be maximized subject to the acceleration limits. Optimality requires that the relative acceleration be constant in time, and in the direction of the final range vector. This must be determined from the initial conditions and the magnitude of the acceleration.

The vector equations of this relative motion are:

$$\mathbf{r}_f = \mathbf{r}_0 + \mathbf{v}_0 t_f + \mathbf{a} t_f^2/2, \tag{6}$$

for the final position vector, and

$$\mathbf{v}_f = \mathbf{v}_0 + \mathbf{a} t_f \tag{7}$$

for the final velocity. The optimal acceleration is

$$\mathbf{a} = (A/|\mathbf{r}_f|)\mathbf{r}_f \tag{8}$$

The final time of minimum range is implied by $\mathbf{r}_f \cdot \mathbf{v}_f = 0$, and by using Eqs.(6) and (7), this becomes a cubic in t_f; i.e.,

$$a^2 t_f^3 + 3\mathbf{a} \cdot \mathbf{v}_0 t_f^2 + 2(v_0^2 + \mathbf{a} \cdot \mathbf{r}_0) t_f + 2\mathbf{r}_0 \cdot \mathbf{v}_0 = 0. \tag{9}$$

A parameter K apportions the control acceleration between the Aircraft 1 and 2. Real space paths can be shown in terms of this scalar, which takes a value between 0 and 1, so that the individual control vectors are

$$\mathbf{a}_1 = -K\mathbf{a}, \quad \mathbf{a}_2 = (1-K)\mathbf{a}. \tag{10}$$

For a known relative initial condition $(\mathbf{r}_0, \mathbf{v}_0)$, and control magnitude A, an iterative process finds the time and value of the minimum range vector. That is, an estimate of $\mathbf{r}_f$ with $\mathbf{a} = 0$ is given by Eq.(5). This defines the acceleration vector by Eq.(8), and the coefficients in Eq.(9). Factoring this cubic gives an improved estimate of t_f and of $\mathbf{r}_f = [x_f, y_f, z_f]'$, the location of Aircraft 2 relative to Aircraft 1 at minimum range.

The geometry for a coplanar encounter is shown in Fig. 1. Aircraft 1 is initially at the origin, with the axis system defined by its velocity direction. The "stability axes" are shown here as the horizontal and vertical, but the discussion to follow treats Aircraft 1 as "taking off," while Aircraft 2 is "landing." The crossing trajectories indicate only that a close encounter is to occur, and the paths are independent of the local vertical, since the uncontrolled velocities are constant. For intuitive purposes, it may be helpful to visulize the axes as being rotated counterclockwise by 10 or 15 deg.

The relative velocity $\mathbf{v}_0$ is the vector difference of the two velocities, as shown, and the varying position of 2 with respect to 1 is shown as the dashed line. The real-space paths are also shown, and the minimum-range condition occurs when $\mathbf{r} \cdot \mathbf{v} = d|\mathbf{r}|/dt = 0$. Because the control envelope is spherical, maximizing the magnitude of the final range is accomplished by applying accelerations in the direction of the final range vector. Notice that the control acceleration is in the plane of the relative position and velocity, $\mathbf{r}_0$ and $\mathbf{v}_0$. The relative trajectory is planar, whatever the initial geometry. And, if Aircraft 1 or 2 cannot or will not maneuver, then K=0 or 1, as shown in Eq.(10).

The notion developed here is that aircraft follow slowly varying flight paths over the brief interval before minimum range. The computations entirely ignore the dynamics of the control systems, orientational dynamics, aerodynamic force variations, etc. This is because no operational system could have these figures available, in general, and because adequate accuracy appears to be present in the model defined here, if the control envelope is small and spherical.

While attitude maneuver dynamics have been ignored, an approximation to this transient might be a dead-time in the pitch degree of freedom, and a larger dead-time in the lateral (heading) degree of freedom. The effect would be to include two more known parameters in the final time implied by Eq.(9).

A test of the validity of the algorithm described here would compare the miss with that resulting from the "exact" equations of motion, for which the controls are derived from the low-order dynamics of Eq.(2). Typically, studies such as these use intuitive or "textbook" maneuvers which depend on very few of the parameters; e.g., elevation and heading.

NUMERICAL RESULTS

Two sample problems illustrate the algorithm derived. These are meant to be representative of near-miss geometries in a close-quarter airport environment. The first example is a coplanar encounter in the vertical plane, modeling a slower aircraft taking off, as a faster aircraft moving in the same direction approaches in the landing configuration from above. The second is a crossing encounter, which shows turn and speed- change dynamics for two aircraft with different initial altitudes. In both cases, the reference x-axis is aligned with the initial velocity of Aircraft 1, with y in the local horizontal plane.

Results are found from the vector equations derived, as shown in Figs. 2-6. These illustrate three general features of the collision-avoidance problem:

1. It is sometimes not apparent how the aircraft should maneuver to maximize the miss, because of the need to predict (*i.e.*, to integrate) ahead to the end geometry. This is analogous to the Pedestrian's Problem: Should he return to the curb or continue across the street, to maximize the distance from the approaching traffic?

2. Optimal maneuvers take place in the plane defined by the initial relative position and relative velocity vectors. Only sub-optimal maneuvers can cause a change in orientation of this plane. This follows from the assumption of a spherical acceleration envelope. The first example is in the vertical plane, and none of the available control is "wasted" by applying it laterally, out of this plane. A general geometry is shown in Fig. 5. Here the relative position and velocity define an inclined plane in which both optimal accelerations are applied, parallel to the miss (minimum-range) vector. Use of the optimal control implies that subsequent relative positions will remain in this plane.

3. The direction or sign of a turn maneuver can depend on the magnitude of the acceleration control. For example, Fig. 6a shows Aircraft 2 initially above Aircraft 1. But minimum range for no control occurs when Aircraft 2 is below Aircraft 1. Small control magnitudes require that Aircraft 1 brake and climb, while 2 thrusts and dives, to the small terminal miss of Fig. 6b. Larger controls lead to sequentially larger misses, by having Aircraft 2 use thrust alone, Fig. 6c, by Aircraft 1 braking without turning while 2 thrusts and climbs, Fig. 6d, and by Aircraft 1 braking and diving, Fig. 6e.

The point made in Fig. 6 applies equally to general paths.* Notice that the term "climb," for example, does not mean that the motion is upward, but only that the control force is upward. This is a type 2 system, for which two integrations of the control acceleration precede the position output, which is the quantity being controlled.

A second, more obvious point apparent from Figs. 2 and 6 is that the miss can vary between the uncontrolled value of Fig. 6a, with $A = 0$, to the initial range r_0, with infinite A; Fig. 4. Acceleration levels of modest amplitude, when applied soon enough, will provide adequate miss when the direction is properly chosen.

---

* The terms "climb" and "dive" refer to the sign of the normal acceleration in the vertical plane. The terms would be "right" and "left" turns if the motion were in the horizontal plane.

APPLYING AND TESTING OPTIMAL MANEUVERS

Implementation of the maneuvers requires tolerable estimates of the relative position and velocity of the aircraft, and values for the acceleration limits. Because the dynamic model used here is of low order, the derived maneuvers must be tested by comparing miss distances to those derived from simulated pilot controller inputs. An unbiased test would involve noise in the raw data, and a range of initial geometries. Comparison of average miss distances for pilot-input controls and for algorithm-input controls, for these short-range initial geometries, would determine the areas of validity of the derived maneuvers. Assuming that such limits to the algorithm can be found by simulation, the maneuver-determination concept developed here appears to be practical for short-range air traffic control.

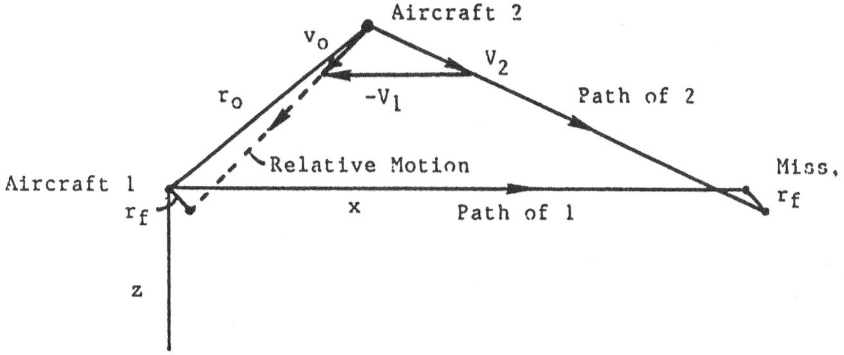

Fig. 1 - Landing Configuration and Trajectories

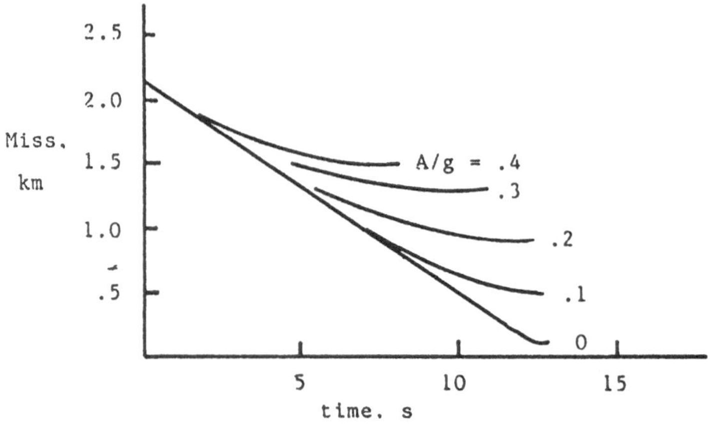

Fig. 2 Time Variation of Range with G-Level

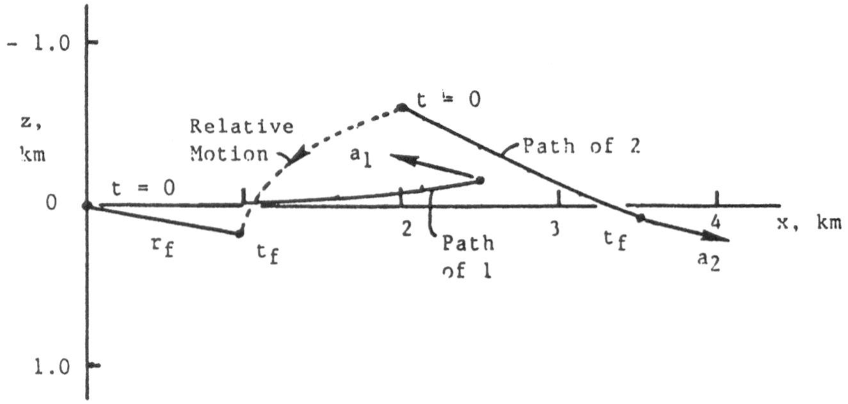

Fig. 3 Relative and Real Space Trajectories

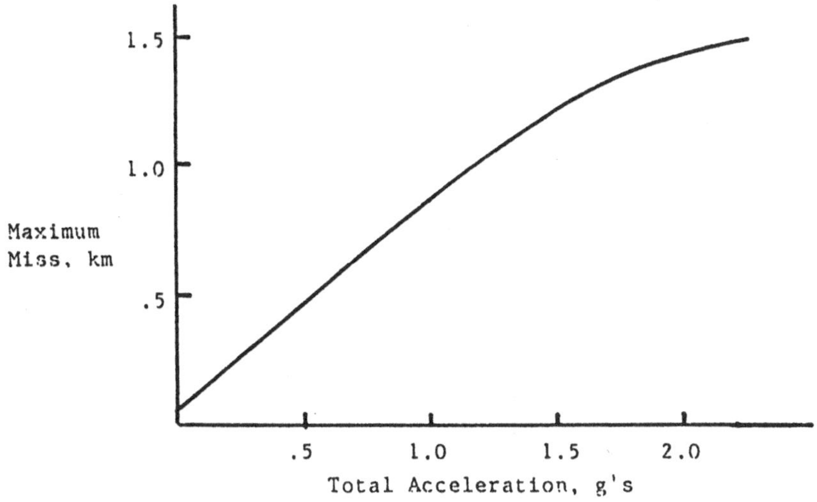

Fig. 4 - Typical Variation of Miss with G-Level

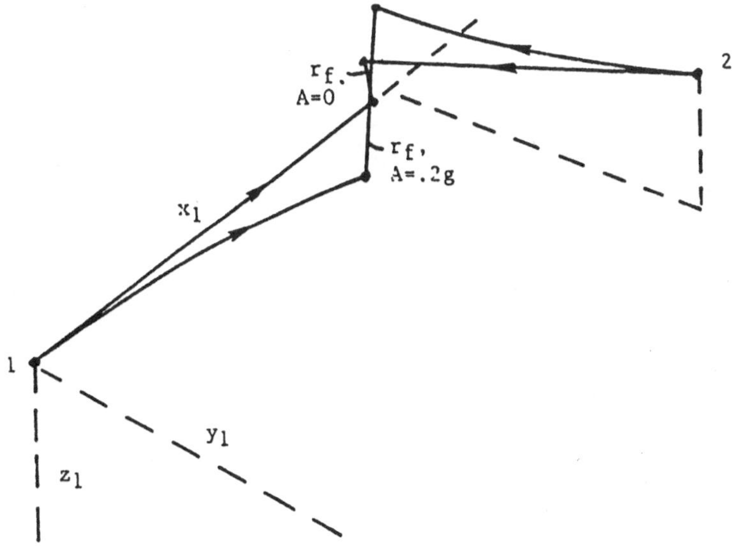

Fig. 5 - Crossing Configuration and Trajectories

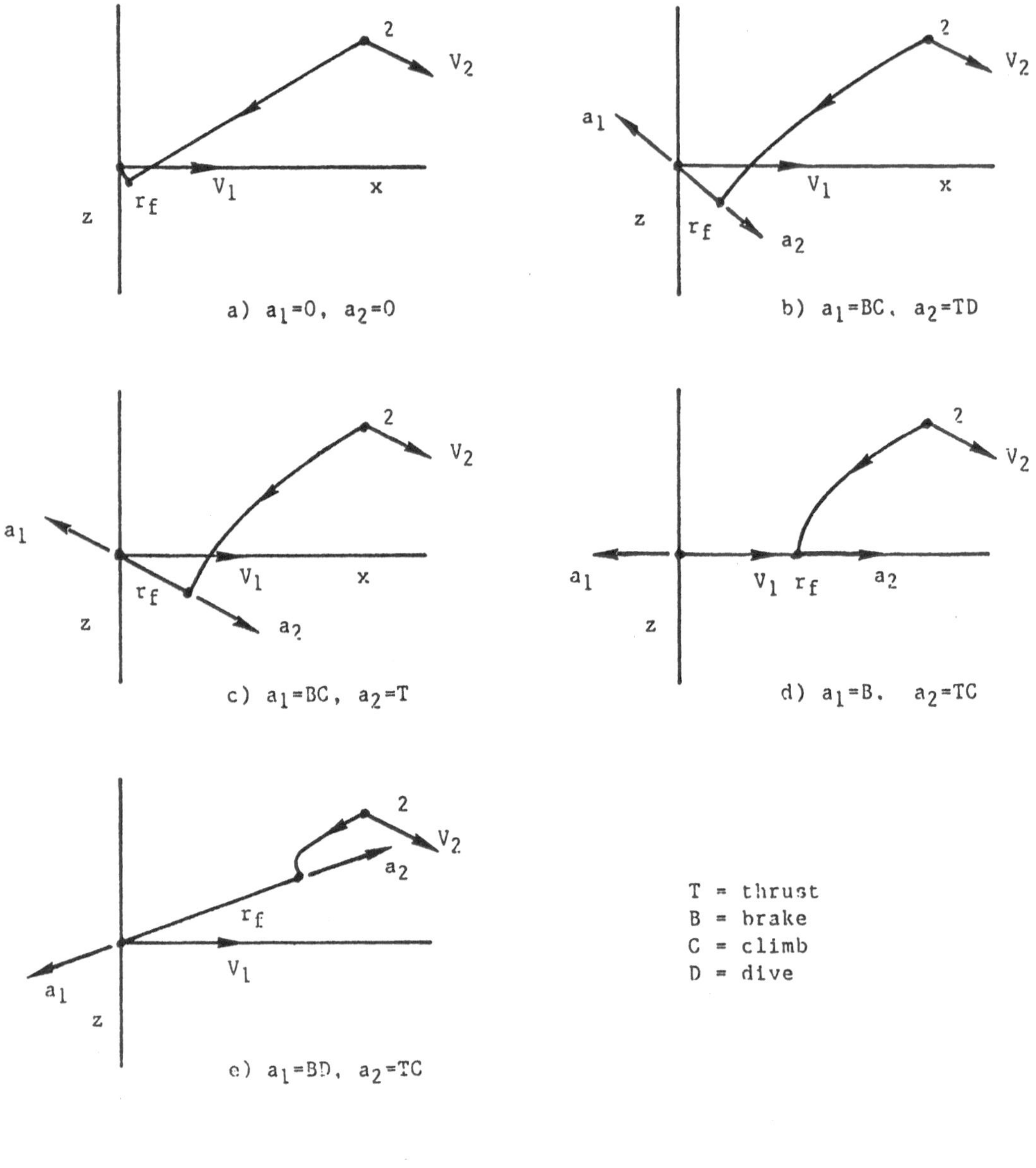

Fig. 6 - Variation of Evasive Trajectories with G-Level

AIRCRAFT SURVIVAL IN WINDSHEAR FLIGHT[1]

A. Miele
Aero-Astronautics Group
Rice University
Houston, Texas 77251-1892

1. INTRODUCTION

Low-altitude windshear is a threat to the safety of aircraft in take-off and landing. Over the past 25 years, at least 30 aircraft accidents have been attributed to windshear. The most notorious ones are the crashes of Eastern Airlines 066 at JFK International Airport, PANAM 759 at New Orleans International Airport, and Delta Airlines 191 at Dallas-Fort Worth International Airport. These crashes, which occurred in 1975, 1982, and 1985, respectively, involved considerable loss of human life and expensive insurance settlements (Refs. 1-2).

To offset the windshear threat, there are two basic systems: windshear warning systems and windshear recovery systems. A windshear warning system is designed to alert the pilot to the possible occurrence of a windshear encounter; here, the intent is to avoid a microburst. A windshear recovery system is designed to guide the pilot in the course of a windshear encounter; here, the intent is to safely traverse a microburst.

Examples of windshear warning systems are: ground-based mechanical systems (anemometers), ground-based radar systems (Doppler radar), and airborne systems (radar or lidar). Examples of windshear recovery systems are: maximum angle of attack guidance, constant pitch guidance, and variable pitch guidance (acceleration guidance and gamma guidance). At this time, some of

---

[1]This research was supported by NASA Langley Research Center, by Boeing Commercial Airplane Company, by Air Line Pilots Association, and by Texas Advanced Technology Program.

the above warning systems and recovery systems appear to be promising. Further research is both necessary and desirable prior to making large commitments of funds to one system or another. At any rate, it must be noted that windshear warning systems and windshear recovery systems are not mutually exclusive, but complementary to one another.

2. WINDSHEAR RECOVERY SYSTEMS

Research on windshear recovery systems was started by this writer at Rice University in 1983 at the urging of Captain W. W. Melvin of Delta Airlines and ALPA. The motivating concepts were as follows: as of 1983, considerable research had been already done on the meteorological, aerodynamic, instrumentation, and stability aspects of the windshear problem; relatively little had been done on the flight mechanics aspects; therefore, a fundamental study was needed in order to better understand the dynamic behavior of aircraft in windshear. Also, it seemed clear that the determination of good strategies for coping with windshear situations was essentially an optimal control problem;that the methods of optimal control theory were needed; and that, only after having found optimal control solutions, one could properly address the guidance problem.

In due time, the above concepts led to an extended research effort which was funded principally by NASA Langley Research Center, Boeing Commercial Airplane Company, Air Line Pilots Association, and Texas Advanced Technology Program. The main topics were the optimization and guidance of flight trajectories in the presence of windshear (Refs. 3-15). An overview of this research is presented in the following pages for three basic problems: take-off (Section 3), abort landing (Section 4), and penetration landing (Section 5).

3. TAKE-OFF PROBLEM

3.1. Optimal Trajectories. Optimal take-off trajectories were studied with the aid of the sequential gradient-restoration algorithm for optimal control problems (Refs. 16-17). Several performance indexes were considered: the most reliable was found to be the deviation of the absolute path inclination from the nominal value. Because the deviation has a peak value along the flight path, it was decided to minimize the peak deviation. The resulting optimization task is a minimax problem or Chebyshev problem of optimal control. After computing several hundred optimal trajectories for the Boeing B-727, B-737, and B-747 aircraft, certain general conclusions became apparent:

(i) the optimal trajectories achieve minimum velocity at the end of the shear;

(ii) the optimal trajectories require an initial decrease in the angle of attack, followed by a gradual increase; the maximum permissible angle of attack (stick-shaker angle of attack) is achieved at the end of the shear;

(iii) for weak-to-moderate windshears, the optimal trajectories are characterized by a continuous climb; the average path inclination decreases as the intensity of the shear increases;

(iv) for relatively severe windshears, the optimal trajectories are characterized by an initial climb, followed by nearly-horizontal flight, followed by renewed climbing after the aircraft has passed through the shear region;

(v) weak-to-moderate windshears and relatively severe windshears are survivable employing an optimized flight strategy; however, extremely severe windshears are not survivable, even employing an optimized flight strategy;

(vi) in relatively severe windshears, the optimal trajectories have better survival capability than maximum angle of attack trajectories and constant pitch trajectories.

3.2. Guidance Schemes. The computation of optimal trajectories requires global information on the wind flow field; that is, it requires the knowledge of the wind components at every point of the space in which the aircraft is flying. In practice, global information is not available; even if it were available, there would not be sufficient computing capability onboard and time to process it adequately. As a consequence, the optimal trajectories are merely a benchmark that it is desirable to approach in actual flight.

In the absence of global information, one is forced to employ local information on the wind acceleration and the downdraft. Therefore, the guidance problem must be posed as follows: Assuming that local information is available on the wind acceleration, the downdraft, as well as the state of the aircraft, we wish to guide an aircraft automatically or semi-automatically in such a way that the properties of the optimal trajectories are preserved.

Based on the idea of preserving the properties of the optimal trajectories, two guidance schemes were developed at Rice University: (a) acceleration guidance, based on the relative acceleration; and (b) gamma guidance, based on the absolute path inclination. For the Boeing B-727 aircraft, it was found that both the acceleration guidance and the gamma guidance produce trajectories which are close to the optimal trajectories. In addition, the resulting near-optimal trajectories have better survival capability than the trajectories arising from maximum angle of attack guidance and constant pitch guidance.

3.3. Simplified Guidance Schemes. The previous guidance schemes require local information on the wind flow field and the state of the aircraft. While this information will be available in future aircraft, it might not be available on current aircraft.

For current aircraft, one way to survive a windshear encounter is to use the constant pitch guidance. An alternative, simple technique is the quick transition to horizontal flight, based on the properties of the optimal trajectories.

The quick transition to horizontal flight requires an initial decrease in the angle of attack, so as to decrease the path inclination to nearly zero. Then, nearly-horizontal flight is maintained during the windshear encounter. Climbing flight is resumed after the shear is past.

For relatively severe windshears, the quick transition to horizontal flight yields trajectories which are competitive with those of the guidance schemes (a) and (b). In addition, for relatively severe windshears, the resulting near-optimal trajectories have better survival capability than the trajectories arising from maximum angle of attack guidance and constant pitch guidance.

4. ABORT LANDING PROBLEM

4.1. Optimal Trajectories. Optimal abort landing trajectories were studied with the aid of the sequential gradient-restoration algorithm for optimal control problems (Refs. 16-17). Several performance indexes were considered: the most reliable was found to be the altitude drop. Because the altitude drop has a peak value along the flight path, it was decided to minimize the peak altitude drop. Once more, the resulting optimization task is a minimax problem or Chebyshev problem of optimal control. A large number of optimal trajectories was generated for several combinations of windshear intensities, initial altitudes, and power setting rates. With reference to the Boeing B-727, B-737, and B-747 aircraft flying in strong-to-severe windshears, certain general conclusions became apparent:

(i) the optimal trajectory includes three branches: a descending flight branch, followed by a nearly-horizontal flight branch, followed by an ascending flight branch after the aircraft has passed through the shear region;

(ii) along an optimal trajectory, the point of minimum velocity is reached at the end of the shear;

(iii) the peak altitude drop depends on the windshear intensity, the initial altitude, and the power setting rate; it increases as the windshear intensity and the initial altitude increase; and it decreases as the power setting rate increases;

(iv) the peak altitude drop of the optimal abort landing trajectory is less than the peak altitude drop of the maximum angle of attack trajectory and the constant pitch trajectory;

(v) the survival capability of the optimal abort landing trajectory in a severe windshear is superior to that of the maximum angle of attack trajectory and the constant pitch trajectory.

4.2. Guidance Schemes. Based on the idea of preserving the properties of the optimal trajectories, while using local information on the wind and the state of the aircraft, five guidance schemes were developed at Rice University: (a) target altitude guidance, based on the initial altitude and the total wind velocity difference; (b) safe target altitude guidance, based on the initial altitude; (c) acceleration guidance, based on the relative acceleration; (d) gamma guidance, based on the absolute path inclination; and (e) modified constant pitch guidance, based on two target pitches, the first to be employed at the shear onset and the second to be employed at the end.

Numerical results for the Boeing B-727 aircraft indicate that, for the same horizontal shear and the same downdraft, the guidance trajectories (a)-(e) are close to the corresponding optimal trajectory. In addition, these guidance trajectories yield a smaller altitude drop than the maximum angle of attack trajectory and the constant pitch trajectory.

The survival capability of an aircraft in a severe windshear was examined, and the maximum permissible shear/downdraft combination that an aircraft can withstand without crashing was computed. Once more, the survival capability of the guidance trajectories (a)-(e) appears to be superior to that of the maximum angle of attack trajectory and the constant pitch trajectory.

5. PENETRATION LANDING PROBLEM

5.1. Optimal Trajectories. Optimal penetration landing trajectories were studied with the aid of the sequential gradient-restoration algorithm for optimal control problems (Refs. 16-17). Here, the optimal control task is formulated as a Bolza problem. The performance index being minimized measures the deviation of the flight trajectory from the nominal trajectory. In turn, the nominal trajectory includes two parts: the approach part, in which the slope is constant; and the flare part, in which the slope is a linear function of the horizontal distance. In the optimization process, the absolute path inclination at touchdown is to be -0.5 deg; the velocity at touchdown is to be within 30 knots of the nominal value; and the horizontal distance at touchdown is to be within 1000 ft of the nominal value.

Three power setting schemes were investigated: (S1) maximum power setting; (S2) constant power setting; and (S3) variable power setting. In Schemes (S1) and (S2), there is only one control: the angle of attack; in Scheme (S3), there are two controls: the angle of attack and the power setting. With reference to strong-to-severe windshears, the following conclusions were reached:

In Scheme (S1), the touchdown requirements can be satisfied for relatively low initial altitudes, while they cannot be satisfied for relatively high initial altitudes; the major inconvenient is excess of velocity at touchdown.

In Scheme (S2), the touchdown requirements cannot be satisfied, regardless of the initial altitude; the major inconvenient is defect of horizontal distance at touchdown.

In Scheme (S3), the touchdown requirements can be satisfied, and the optimal trajectory exhibits the following characteristics:

(i) the angle of attack has an initial decrease, followed by a gradual, sustained increase; the largest value of the angle of attack is attained at the end of the shear; in the aftershear region, the angle of attack decreases gradually;

(ii) initially, the power setting increases rapidly until maximum power setting is reached; then, maximum power setting is maintained in the shear region; in the aftershear region, the power setting decreases gradually;

(iii) the relative velocity decreases in the shear region and increases in the aftershear region; the point of minimum velocity occurs at the end of the shear;

(iv) depending on the windshear intensity and the initial altitude, the deviations of the flight trajectory from the nominal trajectory can be considerable in the shear region; however, these deviations become small in the aftershear region, and the optimal flight trajectory recovers the nominal trajectory.

A comparison was made between the optimal trajectory of Scheme (S3), the fixed control trajectory (fixed angle of attack, coupled with fixed power setting), and the autoland trajectory (angle of attack controlled via path inclination signals, coupled with power setting controlled via velocity signals). The superiority of the optimal trajectories of Scheme (S3) was shown in terms of the ability to meet the path inclination, velocity, and distance requirements at touchdown. Therefore, guidance schemes based on the properties of the optimal trajectories of Scheme (S3) should prove to be superior to the fixed control guidance scheme and the autoland guidance scheme.

5.2. Guidance Scheme. A guidance scheme for penetration landing in a windshear was studied under the following assumptions: the touchdown absolute path inclination is to be -0.5 deg; the touchdown velocity is to be within 30 knots of the nominal value; the touchdown distance is to be within 1000 ft of the nominal value; and the deviation of the flight trajectory from the nominal trajectory is to be relatively small.

First, a quasi-optimal trajectory was investigated in connection with the following power setting scheme: (S4) the power setting determination is based on the windshear detection and the velocity. It was shown that the quasi-optimal trajectory of Scheme (S4) is close to the optimal trajectory of Scheme (S3). This implies that no major coupling relation exists between the power setting and the angle of attack for the purposes of constructing a guidance scheme capable of approximating the behavior of the optimal trajectory.

In the resulting guidance scheme, the power setting and the angle of attack are decoupled controls. The power setting determination is based on the windshear intensity and the velocity. The angle of attack determination is based on the windshear intensity, the absolute path inclination, and the glide slope angle. Numerical results indicate that the guidance trajectory is close to the optimal trajectory. Also, in strong-to-severe windshears, the guidance trajectory has better survival capability than the autoland trajectory.

5.3. Simplified Guidance Scheme. From a practical point of view, penetration landing techniques should be employed only if the windshear encounter is initially detected at lower altitudes. If the windshear encounter is initially detected at higher altitudes, abort landing techniques should be preferred.

If the windshear encounter is initially detected at lower altitudes, the guidance scheme can be simplified as follows: (a) maximum power setting is to be applied; and (b) the angle of attack is determined only via absolute path inclination signals.

6. DISCUSSION AND CONCLUSIONS

In this paper, with particular regard to windshear recovery systems, we have reviewed the research performed at Rice University on two aspects of the windshear problem: determination of optimal trajectories and development of near-optimal guidance schemes. It now appears that, over the next few years, an advanced windshear control system can be developed, capable of functioning in different wind models and covering the entire spectrum of flight conditions, including take-off, abort landing, and penetration landing.

The advanced windshear control system must combine an advanced windshear warning system and an advanced windshear recovery system. The advanced windshear warning system requires the use of real-time identification techniques and must be characterized by small computational time, coupled with limited memory requirements. The advanced windshear recovery system must incorporate four basic properties explained below:

(i) completeness, which means that the system should be able to function in take-off, abort landing, and penetration landing;

(ii) continuation, which means that the system should cover a variety of situations, ranging from zero windshear to moderate windshear to strong-to-severe windshear; the switch from no-windshear operation to windshear operation should be smooth;

(iii) near-optimality, which means that the system should be constructed so as to supply a good approximation to the properties of the optimal trajectories;

(iv) simplicity, which means that the system should be as simple as possible and should emphasize the use of existing instrumentation, whenever possible.

With reference to the continuation property, it must be noted that any adverse wind gradient is both preceded and followed by a favorable wind gradient. The advanced windshear recovery system must not only react in a near-optimal way to adverse wind gradients, but must exploit to the best advantage of the aircraft favorable wind gradients. This means that, in an increasing headwind scenario, kinetic energy must be increased; conversely, in a decreasing tailwind scenario, potential energy must be increased. Clearly, this requires that not only the current windshear signals be measured, but that previous windshear signals be memorized, such that favorable wind gradients can be detected and utilized. Hence, some modification of the guidance schemes described in Sections 3-5 is in order.

To sum up, it is felt that an advanced windshear control system, endowed with the properties described above, should improve considerably the survival capability of aircraft in severe windshear.

REFERENCES

1. LEITMANN, G., and PANDEY, S., "Aircraft Control under Conditions of Windshear", Control and Dynamic Systems, Advances in Theory and Applications, Edited by C. T. Leondes, Academic Press, New York, New York, Vol. 34, pp. 1-79, 1990.

2. MIELE, A., WANG, T., WANG, H., and MELVIN, W. W., "Overview of Optimal Trajectories for Flight in a Windshear", Control and Dynamic Systems, Advances in Theory and Applications, Edited by C. T. Leondes, Vol. 34, pp. 81-123, 1990.

3. MIELE, A., WANG, T., and MELVIN, W. W., "Optimal Take-Off Trajectories in the Presence of Windshear", Journal of Optimization Theory and Applications, Vol. 49, No. 1, pp.1-45, 1986.

4. MIELE, A., WANG, T., and MELVIN, W. W.,"Guidance Strategies for Near-Optimum Take-Off Performance in a Windshear", Journal of Optimization Theory and Applications, Vol. 50, No. 1, pp. 1-47, 1986.

5. MIELE, A., WANG, T., MELVIN, W. W., and BOWLES, R. L., "Maximum Survival Capability of an Aircraft in a Severe Windshear", Journal of Optimization Theory and Applications, Vol. 53, No. 2, pp. 181-217, 1987.

6. MIELE, A., WANG, T., and MELVIN, W. W.,"Quasi-Steady Flight to Quasi-Steady Flight Transition in a Windshear: Trajectory Optimization and Guidance", Journal of Optimization Theory and Applications, Vol. 54, No. 2, pp. 203-240, 1987.

7. MIELE, A., WANG, T., TZENG, C. Y., and MELVIN, W. W., "Optimal Abort Landing Trajectories in the Presence of Windshear", Journal of Optimization Theory and Applications, Vol. 55, No. 2, pp. 165-202, 1987.

8. MIELE, A., WANG, T., and MELVIN, W. W.,"Optimization and Acceleration Guidance of Flight Trajectories in a Windshear", Journal of Guidance, Control, and Dynamics, Vol. 10, No. 4, pp. 368-377, 1987.

9. MIELE, A., WANG, T., WANG, H., and MELVIN, W. W., "Optimal Penetration Landing Trajectories in the Presence of Windshear", Journal of Optimization Theory and Applications, Vol. 57, No.1, pp. 1-40, 1988.

10. MIELE, A., WANG, T., MELVIN, W. W., and BOWLES, R. L.,"Gamma Guidance Schemes for Flight in a Windshear", Journal of Guidance, Control, and Dynamics, Vol. 11, No. 4, pp. 320-327, 1988.

11. MIELE, A., WANG, T., and MELVIN, W. W.,"Quasi-Steady Flight to Quasi-Steady Flight Transition for Abort Landing in a Windshear: Trajectory Optimization and Guidance", Journal of Optimization Theory and Applications, Vol. 58, No. 2, pp. 165-207, 1988.

12. MIELE, A., WANG, T., MELVIN, W. W., and BOWLES, R. L., "Acceleration, Gamma, and Theta Guidance for Abort Landing in a Windshear", Journal of Guidance, Control, and Dynamics, Vol. 12, No. 6, pp. 815-821, 1989.

13. MIELE, A., WANG, T., and MELVIN, W. W.,"Penetration Landing Guidance Trajectories in the Presence of Windshear", Journal of Guidance, Control, and Dynamics, Vol. 12, No. 6, pp. 806-814, 1989.

14. MIELE, A., WANG, T., TZENG, C. Y., and MELVIN, W. W.,"Abort Landing Guidance Trajectories in the Presence of Windshear", Journal of the Franklin Institute, Vol. 326, No. 2, pp. 185-220, 1989.

15. MIELE, A.,"Final Report on NASA Grant No. NAG-1-516, Optimization and Guidance of Flight Trajectories in the Presence of Windshear, 1984-89", Rice University, Aero-Astronautics Report No. 244, 1989.

16. MIELE, A., "Gradient Algorithms for the Optimization of Dynamic Systems", Control and Dynamic Systems, Advances in Theory and Applications, Edited by C. T. Leondes, Academic Press, New York, New York, Vol. 16, pp. 1-52, 1980.

17. MIELE, A.,"Optimal Trajectories of Aircraft and Spacecraft", Aircraft Trajectories: Computation, Prediction, and Control, Edited by A. Benoit, AGARDograph No. AG-301, AGARD/NATO, Paris, France, Vol. 1, pp. 201-256, 1990.

PROPERTIES OF THE RECURSIVE LYAPUNOV EQUATION

ANDRZEJ OLAS
Oregon State University
Department of Mechanical Engineering
Corvallis, OR 97331

1. Introduction

In recent years the Lyapunov equation has become an important tool in a significant number of applications. Besides its classical application in stability investigation, it is used for controller design for uncertain systems as by Leitmann in [1] and estimation of controller robustness, see Siljak review in [2]. Estimation of solution to Lyapunov equation is used to determine uncertainty threshold as by Barmish and Leitmann in [3]. Pole-assignment problem is addressed by utilizing Lyapunov equation as by Juang et al. in [4] and by Dib in [5], where also a relation between Lyapunov and Riccati equation was considered.

The present paper analyzes properties of a recursive Lyapunov equation, i.e. the equation

$$A^T P_{i+1} + P_{i+1} A = -P_i \quad , \quad i=1,2,3,...$$

Numerical solving of the equation involves the loop containing the same simple procedure as when solving the regular Lyapunov equation. Simplicity of this procedure stimulated the author's interest in analyzing properties of the equation. The equation was first considered by Jury in [6]; Olas in [7] found that the equation is related to recursive Lyapunov function.

To analyze the properties of the solution to recursive Lyapunov equation, the associated recursive equation has been introduced and its solution—the matrix $S^{(\infty)}$ has been found. It has been proved that after sufficiently large number of steps, the matrix P_i attains an interesting property, namely

$$P_{i+1} \approx \frac{1}{2\mu} P_i$$

where, denoting by μ_i, $i=1,...,n$, the real parts of eigenvalues of A, ordered so that

$$|\mu_1| < |\mu_2| < ... < |\mu_n|$$

we have

$$\mu = \mu_1$$

i.e., μ is equal to this real part of the eigenvalue, which is the closest to zero. Further investigation discloses that the solution of the associated recursive Lyapunov equation when used to design state-variable feedback provides the horizontal shift of the selected real pole or a pair of complex-conjugate poles, while leaving the remaining poles intact.

2. Problem Statement

The autonomous system

$$\dot{x} = f(x) \quad , \quad x \in R^n \tag{1}$$

defined on the set $Z = \{|x| < H > 0\}$ and such that $f(0) = 0$ is considered under the assumption that $f \in C^1(Z)$ satisfies the condition for existence and uniqueness of solutions, which are denoted by $p(t, x_0)$, $p(0, x_0) = x_0$. Together with the system (1) the positive-definite function $V_1(x)$ is considered with an assumption that it is of a class $C^2(Z)$; this assumption will be utilized only when considering the second derivative of V_1. For other considerations the assumption $V \in C^1(Z)$ is sufficient.

The performance measure of the Lyapunov function defined as

$$\lambda = \sup_{x \in Z/0} \dot{V}(x)/V(x)$$

allows the estimate

$$V(p(t,x)) \leq V(x)\exp(\lambda t)$$

and the value $(-1/\lambda)$ corresponds, as defined by Ogata [8], to the largest time constant of the system, relating to changes in the Lyapunov function $V(p(t, x))$.

3. Recursive Algorithm

In paper [9] the recursive algorithm for design of Lyapunov function was introduced by defining the sequence of functions

$$V_{i+1}(x) = \int_0^T V_i(p,t,x))dt \quad , \qquad i = 1,2,... \tag{2}$$

where $T > 0$ is some constant.

It was proved that the Lyapunov derivative of $V_i(x)$ was given by the formula

$$\dot{V}_{i+1}(x) = \int_0^T \dot{V}_i(p(t,x))dt = V_i(p(T,x)) - V_i(x) \quad , \qquad i = 1,2,... \tag{3}$$

4. Function $\Lambda_i(x)$ and Performance Measure

Assume that the function $V_1(x)$ is selected in such a manner that the function

$$\Lambda_1(x) = \dot{V}_1(x)/V_1(x)$$

exists and is bounded on the set $Z_1\backslash 0$. Define the functions $\Lambda_i(x)$, $i = 1,2,...$

$$\Lambda_i(x) = \dot{V}_i(x)/V_i(x) \tag{4}$$

The function $\Lambda_i(x)$ measures the Lyapunov function performance at point x. We define the performance measure on the $Z_1\backslash 0$ by introducing

$$\lambda_i = \sup_{x \in Z_1 \backslash 0} \Lambda_i(x)$$

In paper [9] the following inequality has been proved

$$\lambda_{i+1} \le \lambda_i$$

together with the fact that λ_i is lower bound. Thus, the sequence $\{\lambda_i\}$ has a limit λ_{lim}.

5. Function Λ_i as the Function of Time

Consider an arbitrary non-zero solution p(t, x), $x \in Z_1$, and the function $\Lambda_i(p(t,x))$ on the interval [0, T]. In paper [7] the following corollary was proved.

Corollary. Consider $x \in Z_2$ and the solution p(t, x) for $t \in [0,T]$. For each fixed x, the sequence of functions of time $\{\Lambda_i(p(t, x))\}$ has a limit $\Lambda_{lim}(p(t, x))$ such that

$$\Lambda_{lim}(p(t,x)) = C \text{ for } t \in [0,T]$$

Remark. The constant C depends on the selection of x; C = C(x).

6. Recursive Functions for Linear Systems

Consider a linear stationary system

$$\dot{x} = Ax \ , \ x \in R^n \tag{5}$$

with the solution $p(t, x_0)$, $p(0, x_0) = x_0$. Together with (5) consider the quadratic Lyapunov function

$$V = x^T P x$$

where P is a symmetric, positive-definite matrix. We get

$$\dot{V}(x) = x^T (A^T P + PA)x$$

Assume now that the system (5) is asymptotically stable. Then using the recursive Lyapunov function procedure we introduce the function V_{i+1} as the integral

$$V_{i+1}(x) = \int_0^T V_i(p(t,x))dt \quad , \qquad i = 1,2,... \tag{6}$$

The integral (6) is well defined and converges even when the upper integration limit T is infinite. Entering for the function V_i under the integral the corresponding quadratic form yields

$$V_{i+1}(x) = \int_0^T p^T (t,x)P_i p(t,x)dt$$

Integration of the above quadratic form, due to the fact that p(t, x,) is the linear form of x, results in $V_{i+1}(x)$ being also the quadratic form with some positive-definite matrix P_i, i.e.,

$$V_{i+1}(x) = x^T P_{i+1} x$$

By virtue of (3), utilizing the fact that (5) is asymptotically stable, we have for $T = \infty$

$$\dot{V}_{i+1} = V_i(p(\infty,x)) - V_i(x) = -x^T P_i x$$

Differentiating V_{i+1} along the solution of (5) we get

$$\dot{V}_{i+1} = x^T (A^T P_{i+1} + P_{i+1}A)x$$

for all x's and thus

$$A^T P_{i+1} + P_{i+1}A = -P_i \qquad i = 1,2,... \tag{7}$$

This equation is a specific form of well known Lyapunov equation. It represents the recursive algorithm to determine the sequence of matrices P_i. The matrix P_1, introduced to generate P_i, is selected as positive-definite, usually as unit matrix. This and the fact that the system is asymptotically stable is enough, [10], to ensure the positive-definiteness of all the matrices P_i, $i = 2,3,...$.

Due to proven properties of the recursive Lyapunov function, while solving Eq. (7) at each step we obtain the Lyapunov function with the better performance measure

$$\lambda_{i+1} \le \lambda_i \, .$$

7. Special Form of the System

We consider the case when the nxn matrix A is simple, i.e., it is similar to a diagonal matrix composed from ℓ complex-conjugate pairs $\mu_j \pm i\nu_j$ and k real μ_j eigenvalues. We assume further that the matrices A and $-A^T$ have no eigenvalues in common; thus we allow for positive real parts μ_j. As it is known [11], such an assumption is sufficient to have the unique solution P_{i+1} at each step of the procedure. Let the eigenvalues be ordered so that

$$0 < |\mu_1| < |\mu_2| \dots < |\mu_{\ell+k} \tag{8}$$

Denote by $a_j \pm ib_j$ the pair of eigenvectors corresponding to the pair $\mu_j \pm i\nu_j$ of eigenvalues and by a_k the eigenvector corresponding to real eigenvalue μ_k, where a_j, b_j are real vectors.

We now create the real, nonsingular nxn matrix E, columns of which are the vectors a_j, b_j ordered so that the vector b_j is inserted into the sequence $\{a_1,\dots,a_{\ell+k}\}$ directly after the vector a_j. As it is known [9], the matrix M obtained by

$$M = E^{-1}AE \tag{9}$$

is of the block form

$$M = \begin{bmatrix} M^{(1)} & & & & 0 \\ & \cdot & & & \\ & & \cdot & & \\ & & & M^{(\ell+k-1)} & \\ 0 & & & & M^{(\ell+k)} \end{bmatrix}$$

where if the number j corresponds to the complex-conjugate pair $\mu_j \pm \nu_j$ then

$$M^{(j)} = \begin{bmatrix} \mu_j & \nu_j \\ -\nu_j & \mu_j \end{bmatrix}$$

and if μ_j is real, then

$$M^{(j)} = [\mu_j]$$

8. Associated Recursive Lyapunov Equation and Its Solution

Introduce the recursive Lyapunov equation associated with Eq. (7)

$$A^T S^{(i+1)} + S^{(i+1)}A = 2\mu_1 S^{(i)}\ ,\ S^{(1)} = P_1\ , \qquad i = 1,2,... \tag{10}$$

The matrices $S^{(i)}$ and P_i are related by

$$P_i = -\frac{1}{(2\mu_1)^{(i-1)}} S^{(i)}$$

Denote

$$R^{(i)} = E^T S^{(i)} E \tag{11}$$

Then Eq. (10) transforms to

$$M^T R^{(i+1)} + R^{(i+1)}M = 2\mu_1 R^{(i)} \qquad i = 1,2,... \tag{12}$$

Two cases should be considered, namely the first one when we have $\mu_1 \pm i\nu_1$ as the complex-conjugate pair and the second one when μ_1 is a real eigenvalue. We discuss only the first case, stating that the discussion of the second one is similar, but easier. For the sake of simplicity we assume that, while fulfilling (8) the first 2ℓ eigenvalues are complex-conjugate pairs $\mu_j + i\nu_j$, $j = 1,...,\ell$, thus the real eigenvalues are $\mu_{\ell+j}$, $j = 1,...,k$. We rewrite matrix $R^{(i)}$ to the symmetric block form

$$R^{(i)} = \begin{bmatrix} D_{11}^{(i)}\ , & D_{12}^{(i)}\ , & \dots\ , & D_{1,\ell+k}^{(i)} \\ D_{12}^{(i)T}\ , & D_{22}^{(i)}\ , & \dots\ , & D_{2,\ell+k}^{(i)} \\ \cdot & \cdot & \dots\ , & \cdot \\ \cdot & \cdot & \dots\ , & \cdot \\ D_{1,\ell+k}^{(i)T}\ , & D_{2,\ell+k}^{(i)T}\ , & \dots\ , & D_{\ell+k,\ell+k}^{(i)} \end{bmatrix}$$

where blocks $D_{p,s}^{(i)}$ are 2x2 matrices for $p, s = 1,...,\ell$ $p \le s$; blocks $D_{p,s}^{(i)}$ are 2x1 matrices for $p \le \ell$ and $s > \ell$; and blocks $D_{p,s}^{(i)}$ are 1x1 matrices for $p \le s$, $p, s = \ell+1,...,\ell+k$. The blocks below diagonal are obtained by transposition of the corresponding above-diagonal blocks. Such a form of the matrix $R^{(i)}$ entered into Eq. (11) yields the decoupled equations on blocks $D_{ps}^{(i+1)}$

$$M^{(p)T} D_{ps}^{(i+1)} + D_{ps}^{(i+1)} M^{(p)} = 2\mu_1 D_{ps}^{(i)} \qquad \begin{matrix} p,s = 1,...,k+\ell \\ p \le s \\ i = 1,2,... \end{matrix} \tag{13}$$

Form of the block $D_{11}^{(i+1)}$. Denoting,

$$D_{11}^{(i+1)} = \begin{bmatrix} r_{11}^{(i+1)} & r_{12}^{(i+1)} \\ r_{12}^{(i+1)} & r_{22}^{(i+1)} \end{bmatrix}$$

we get the following expressions for the elements of the block

$$r_{11}^{(i+1)} = \frac{r_{11}^{(i)}(2\mu_1^2+\nu_1^2) - 2r_{12}^{(i)}\mu_1\nu_1 + r_{22}^{(i)}\nu_1^2}{2(\mu_1^2+\nu_1^2)} \tag{14}$$

$$r_{12}^{(i+1)} = \frac{r_{11}^{(i)}\mu_1\nu_1 + 2r_{12}^{(i)}\mu_1^2 - r_{22}^{(i)}\mu_1\nu_1}{2(\mu_1^2+\nu_1^2)} \tag{15}$$

$$r_{22}^{(i+1)} = \frac{r_{11}^{(i)}\nu_1^2 + 2r_{12}^{(i)}\mu_1\nu_1 + r_{22}^{(i)}(2\mu_1^2+\nu_1^2)}{2(\mu_1^2+\nu_1^2)} \tag{16}$$

To determine behavior of the elements when $i \to \infty$ we transform Eqs. (14)-(16). Firstly we find that the sum of diagonal elements is constant,

$$r_{11}^{(i+1)} + r_{22}^{(i+1)} = r_{11}^{(i)} + r_{22}^{(i)} = r_{11}^{(1)} + r_{22}^{(1)} \tag{17}$$

The difference of diagonal elements fulfills the difference equation

$$r_{11}^{(i+2)} - r_{22}^{(i+2)} - \frac{2\mu_1^2}{\mu_1^2+\nu_1^2}\left(r_{11}^{(i+1)} - r_{22}^{(i+1)}\right) + \frac{\mu_1^2}{\mu_1^2+\nu_1^2}\left(r_{11}^{(i+1)} - r_{22}^{(i+1)}\right) = 0$$

which has a solution

$$\lim_{i\to\infty}\left(r_{11}^{(i)} - r_{22}^{(i)}\right) = 0 \tag{18}$$

which is globally stable. Finally using (18) it is easy to see that the element $r_{12}^{(i)}$ tends to zero.

Form of the other diagonal 2x2 blocks $D_{jj}^{(i+1)}$, $j = 2,\ldots,\ell$. We have

$$r_{2j-1,2j-1}^{(i+1)} + r_{2j,2j}^{(i+1)} = \frac{\mu_1}{\mu_j}\left(r_{2j-1,2j-1}^{(i)} + r_{2j,2j}^{(i)}\right), \quad \frac{\mu_1}{\mu_j} < 1$$

Since, analogously as in Eq. (18), the difference between the diagonal elements tends to zero as does the out-diagonal element, we conclude that

$$\lim_{i\to\infty} D_{jj}^{(i)} = 0 \qquad j = 2,\dots,\ell$$

Similar reasoning, omitted due to the lack of space, applies to the remaining blocks of the matrix $R^{(i)}$. Inverting Eq. (11) we get

$$S^{(i)} = (E^{-1})^T R^{(i)} E^{-1}$$

The following Lemma concludes the discussion of associated equations.

Lemma. Let A be an nxn simple matrix having ℓ pairs, $\mu_j \pm v_j$, of complex-conjugate and k real, μ_j, eigenvalues, $2\ell + k = n$. Assume that $\mu_j \neq \mu_k$, $j \neq k$, and $\mu_j \neq 0$, $j = 1,\dots,\ell+k$. Let the eigenvalues be ordered so that

$$0 < |\mu_1| < |\mu_2| \dots < |\mu_{\ell+k}|$$

Then the solution of the associated recursive Lyapunov equation is given by

$$S^{(\infty)} = (E^{-1})^T R^{(\infty)} E^{-1} \tag{19}$$

where:

(a) If $v_1 \neq 0$ (i.e., $\mu_1 \pm iv_1$ is a pair of complex-conjugate eigenvalues) then there are only two non-zero elements of $R^{(\infty)}$,

$$\lim_{i\to\infty} R^{(i)} \triangleq R^{(\infty)} = \frac{r_{11}^{(1)} + r_{22}^{(1)}}{2} \begin{bmatrix} 1 & 0 & 0 \\ 0 & 1 & \\ & 0 & 0 \end{bmatrix} \tag{20}$$

(b) If $v_1 = 0$ (i.e., μ_1 is a real eigenvalue) then there is only one non-zero element of $R^{(\infty)}$,

$$\lim_{i\to\infty} R^{(i)} \triangleq R^{(\infty)} = \begin{bmatrix} r_{11}^{(1)} & 0 \\ 0 & 0 \end{bmatrix} \tag{21}$$

The following Conclusion 1 addresses an interesting property of the sequence of matrices-solutions to recursive equation (7).

Conclusion 1. Let P_{i+1} be an i-th step solution of (7). Then the relation

$$\lim_{i\to\infty} \left(P_{i+1} - \frac{1}{2\mu_1} P_i \right) = 0 \tag{22}$$

holds.

This relation allows for determination of μ_1 based on solving of Eq. (7).

9. Horizontal Shift of Poles of A

Let A fulfill the assumptions of the Lemma 1. Two cases should be considered, namely the first one when $\mu_1 \pm i\nu_1$ is the complex-conjugate pair and the second when μ_1 is a real eigenvalue. We discuss here only the first case, stating that the discussion of the second one is similar, but easier.

Introducing constant $\gamma > 0$ and considering the matrix $A-\gamma S^{(\infty)}$, we have

$$A-\gamma S^{(\infty)} = EME^{-1} - \gamma(E^T)^{-1}R^{(\infty)}E^{-1} = E[M-\gamma E^{-1}(E^T)^{-1}R^{(\infty)}]E^{-1} \tag{23}$$

Due to the relation (20) only the first two columns of the matrix $(E^{-1}(E^T)^{-1}R^{(\infty)}$ contain non-zero elements. Therefore, while the first two eigenvalues $\bar{\mu}_1 + i\bar{\nu}_1$, $\bar{\mu}_2 + i\bar{\nu}_2$ of the matrix $A-\gamma S^{(\infty)}$ differ from the corresponding eigenvalues of M, all the remaining are identical.

To calculate $\bar{\mu}_1$ and $\bar{\nu}_1$ we need to know only four elements of the matrix $E^{-1}(E^T)^{-1}R^{(\infty)}$, namely $[E^{-1}(E^T)^{-1}R^{(\infty)}]_{ij}$ for $i,j = 1,2$. Considering the matrix A adjoint eigenvalue problem, [12], denote the left eigenvectors of A by $c_1 + id_1, c_1 - id_1, \ldots, c_\ell + id_\ell, c_\ell - id_\ell, c_{\ell+1}, \ldots, c_{\ell+k}$. Assume that the right and left eigenvectors are normalized so that the transposed matrix of right eigenvectors is equal to the inverse of the left eigenvectors matrix. Utilizing the introduced vectors, after transformations we obtain the following formulas for the four aforementioned elements of the matrix $E^{-1}(E^T)^{-1}R^{(\infty)}$

$$[E^{-1}(E^T)^{-1}R^{(\infty)}]_{ij} = \begin{cases} 4c^2 r_{11}^{(\infty)} & \text{for } i=1,\ j=1 \\ -4cd\cos\alpha\ r_{11}^{(\infty)}, & i=1,\ j=2 \text{ and } i=2,\ j=1 \\ 4d^2 r_{11}^{(\infty)}, & i=2,\ j=2 \end{cases}$$

where c, d are magnitudes of the vectors c_1, d_1 and α is the angle between them.

The first two new eigenvalues $\bar{\mu}_1 + i\bar{\nu}_1$, $\bar{\mu}_2 + i\bar{\nu}_2$ of the matrix $A-\gamma S^{(\infty)}$ fulfill the equation

$$\det\begin{bmatrix} \mu_1 - 4\bar{\gamma}c^2 - \lambda & +4\bar{\gamma}cd\cos\alpha + \nu_1 \\ +4\bar{\gamma}\,cd\cos\alpha - \nu_1 & \mu_1 - 4\bar{\gamma}\,d^2 - \lambda \end{bmatrix} = 0$$

where $\bar{\gamma} = \gamma r_{11}^{(\infty)}$. We solve the equation assuming that γ is sufficiently small. The result is contained in the following conclusion, together with the result for the case, when μ_1 is a real eigenvalue.

Conclusion 2. Let A be a matrix satisfying assumptions of Lemma. Let $S^{(\infty)}$ be the solution of the associated recursive Lyapunov equation (10). Consider the matrix $A-\gamma S^{(\infty)}$.

(a) If $\mu_1 \pm i\nu_1$, is a pair of complex-conjugate eigenvalues then the matrix $A-\gamma S^{(\infty)}$ has all, besides the first two, eigenvalues identical with those of the matrix A. For sufficiently small γ the first two new eigenvalues remain complex-conjugate, $\bar{\mu}_1 = \bar{\mu}_2$, $\bar{\nu}_2 = -\bar{\nu}_1$ and

$$\bar{\mu}_1 = \mu_1 - 2\gamma(c^2+d^2)r_{11}^{(\infty)}$$

$$\bar{\nu}_1 = \nu_1 + 4\gamma^2(c^2+d^2)r_{11}^{(\infty)} \sim \nu_1$$

i.e. the corresponding pair of poles of the matrix $A-\gamma S^{(\infty)}$ is shifted horizontally with respect to poles of A.

(b) If μ_1 is a real eigenvalue then the matrix $A-\gamma S^{(\infty)}$ has all, except the first one, eigenvalues identical with those of matrix A. The eigenvalue $\bar{\mu}_1$ of $A-\gamma S^{(\infty)}$ remains real, fulfills

$$\bar{\mu}_1 = \mu - \gamma 4c^2 \, r_{11}^{(\infty)}$$

i.e. it is shifted horizontally with respect to this of A.

10. Null Space $N[S^{(\infty)}]$ of the Matrix $S^{(\infty)}$

Preserving previous assumptions on the matrix A let $x^{(i)}$, $i=1,\ldots,n$ denote its eigenvectors. We consider the case when the first two eigenvectors $x^{(1)}$, $x^{(2)}$ are complex-conjugate and the matrix $S^{(\infty)}$ has been accordingly determined. Consider the vector

$$y^{(i)} = S^{(\infty)} x^{(i)} \qquad i=3,\ldots,n$$

Using (19) we have

$$y^{(i)} = (E^{-1})^T R^{(\infty)} E^{-1} x^{(i)}$$

Let $e^{(i)}$ be the unit i-th vector of the base, having all, except the i-th component equal to zero. Then since E is a modal matrix of A we have

$$x^{(i)} = E \, e^{(i)}$$

and correspondingly

$$y^{(i)} = (E^{-1})^T R^{(\infty)} e^{(i)}$$

Recalling (20) we get

$$y^{(i)} = 0 \qquad i=3,...,n$$

which means that all, except first two, eigenvectors of A belong, [10], to the null space $\mathcal{N}[S^{(\infty)}]$ of $S^{(\infty)}$. The stated property leads to

Conclusion 3. Let assumptions of Conclusion 2 be satisfied. Consider the matrix $A-\gamma FS^{(\infty)}$, where F is an arbitrary matrix.

(a) If $\mu_1 \pm i\nu_1$, is a pair of complex-conjugate eigenvalues then the matrix $A-\gamma FS^{(\infty)}$ has all, besides first two, eigenvectors identical with these of the matrix A.

(b) If μ_1 is a real eigenvalue then the matrix $A-\gamma FS^{(\infty)}$ has all, except the first, eigenvectors identical with these of the matrix A.

11. Examples

Consider the state variable feedback system

$$\dot{x} = Ax + Bu \quad , \quad u = Kx$$

The gain matrix K is to be selected so that the assigned poles of the system are shifted horizontally to assigned positions. Consider 3-D system with

$$A = \begin{bmatrix} 0 & 1 & 0 \\ -1 & -3 & 1 \\ -2 & 0 & 0 \end{bmatrix}$$

The system eigenvalues are $\lambda_{1,2} = \mu_1 \pm i\nu_1 = -0.0533557 \pm 0.8298703i$, $\lambda_3 = -2.89335$.

Case 1. Horizontal shift to the left of a pair of complex conjugate poles -0.0533557 ± 0.8298703i. Matrix B is a 3x3 identity matrix.

$$K = -\gamma B^T S^{(\infty)}$$

where $S^{(\infty)}$ has been obtained by solving recursive Lyapunov equation in five steps. Table 1 shows that significant shift

$$(\mu_1)_{SHIFTED} = 4.43(\mu_1)_{INITIAL}$$

results in 1.5% error of imaginary part ν_1 and no change of λ_3 is observed.

Case 2. Same as Case 1 but B = $[1\ \ 0\ \ 0]^T$ (see Table 1). Again λ_3 does not change; for

$$(\mu_1)_{SHIFTED} = 3.72(\mu_1)_{INITIAL}$$

we have 1.6% error of imaginary part ν_1.

Case 3. Horizontal shift of the pole λ_3 with B being a 3x3 identity matrix (see Table 1). The auxiliary shift was first performed to create the matrix

$$A' = A + 2.5I$$

such that

$$|\lambda_3'| = |\lambda_3 + 2.5| = 0.39335 < |\mu_1 + 2.5| = 2.4466443$$

The matrix A' was entered into recursive Lyapunov equation and the matrix $S^{(\infty)}$ was found. For

$$(\mu_3)_{SHIFTED} = 1.31(\mu_3)_{INITIAL}$$

no change in location of $\lambda_{1,2}$ has been observed.

Case 4.

$$A = \begin{bmatrix} -0.4422 & -0.3053 & 0.1474 & -0.1276 \\ -0.8779 & -14.2805 & 0.4723 & 0.5166 \\ 3.4358 & 12.3956 & -1.8956 & -4.0314 \\ 0.000 & 0.000 & 1.000 & 0.000 \end{bmatrix}$$

and B is 4x4 identity matrix. Poles of A are $\lambda_1 = -0.612479$, $\lambda_{2,3} = \mu_2 \pm i\nu_2 = -0.640151 \pm 1.61852i$, $\lambda_4 = -14.7255$. To shift the poles $\lambda_{2,3}$ the auxiliary shift was first performed yielding

$$A' = A + 0.6400I$$

The eigenvalues λ_1, λ_4 do not change during the shift (see Table 1). The shift accuracy is satisfactory for

$$(\mu_2)_{SHIFTED} = 3.02\ (\mu_2)_{INITIAL}$$

namely at this range $(\nu_2)_{SHIFTED}$ is not less than 88% of $(\nu_2)_{INITIAL}$. For larger shift (last two positions in the table) the imaginary part of $\lambda_{2,3}$ attenuates quickly and finally the imaginary poles disappear.

Case 5. To improve accuracy of the Case 4 the shift has been accomplished in two steps. Firstly the matrix $S_4^{(\infty)}$ (same as at Case 4) was used with $\gamma_1 = 1\text{x}10^{-3}$ leading to the matrix $A^* = A - (1\text{x}10^{-3})BB^TS_4^{(\infty)}$.

Then the matrix A^* was inserted into recursive Lyapunov equation producing the new matrix $S_5^{(\infty)}$. The total shift was accomplished by utilizing the gain

$$K = -\gamma_1 B^T S_4^{(\infty)} - \gamma_2 S_5^{(\infty)}$$

Such procedure allowed for the shift

$$(\mu_2)_{SHIFTED} = 3.33\ (\mu_2)_{INITIAL}$$

with 8.5% error of ν_2. Larger shift (last position of the column) causes significant increase in error.

12. Final Remarks

While both regular and recursive Lyapunov equations require the same numerical procedure to solve (applied in loop for recursive equation), the solutions of these equations show different properties. It has been shown that at the solving process the value of the real part of in some sense dominant eigenvalue of the system matrix is disclosed. Further it has been found that the solution of the associated recursive Lyapunov equation may be used to establish a state-variable feedback resulting in a horizontal shift of selected either a real pole or a complex-conjugate pair of poles. Such a shift does not change the eigenvectors of the remaining poles, since the matrix $FS^{(\infty)}$ projects the vectors from R^n to its subspace spanned by eigenvectors corresponding to shifted eigenvalues. The theoretical results are confirmed by examples.

REFERENCES

1. Leitmann, G., "Deterministic Control of Uncertain Systems," Fourth International Conference on Mathematical Modelling, Zurich, August 1983.

2. Siljak, D.D., "Parameter Space Methods for Robust Control Design, a Guided Tour, *IEEE Trans. on Aut. Contr.*, Vol. 34, No. 7, 1989.

3. Barmish, B.R., Leitmann, G., "On Ultimate Boundedness Control of Uncertain Systems in the Absence of Matching Assumptions," *IEEE Trans. on Aut. Contr.*, Vol. 27, No. 1, 1982.

4. Juang, Y., Hong, Z., Wang, Y., "Lyapunov Approach to Robust Pole-Assignment Analysis, *Int. J. Control*, Vol. 49, No. 3, 1989.

5. Dib, H.M., "Linear Regulators With Eigenvalue Placement in Specified Regions and Applications to Robotics," Ph.D. Diss., Univ. of Houston, TX, 1987.

6. Jury, E.I. *Inners and Stability of Dynamic Systems*, Robert E. Krieger Publ. Co., Malabar, 1982.

7. Olas, A., "Recursive Lyapunov Functions: Properties, Linear Systems," Ch. 14 in *Advances in Control and Dynamic Systems*, Vol. XXXIV, Academic Press, 1990.

8. Ogata, K., *Modern Control Engineering*, Prentice-Hall, 1970.

9. Olas, A., "Recursive Lyapunov Functions," *ASME Journal of Dynamic Systems, Measurement, and Control*, Vol. III, No. 4, 1989, 641-645.

10. Skelton, R.E., *Dynamic Systems Control*, John Wiley and Sons, NY, 1988.

11. Lancaster, P., *Theory of Matrices*, Academic Press, 1969.

12. Meirovitch, L., *Elements of Vibration Analysis*, McGraw-Hill, 1986.

Table 1. Pole Placement Results.

Eigenvalues of the matrix $A\text{-}\gamma BB^T S^{(\infty)}$ when varying factor γ

Case 1 $\lambda_{1,2}$, λ_3 Shifting $\lambda_{1,2}$	Case 2 $\lambda_{1,2}$, λ_3 Shifting $\lambda_{1,2}$	Case 3 $\lambda_{1,2}$, λ_3 Shifting λ_3 (no change in $\lambda_{1,2}$ observed)	Case 4 λ_1, $\lambda_{2,3}$, λ_4	Case 5 λ_1, $\lambda_{2,3}$, λ_4
-0.0533557 ± 0.829703i -2.89329	-0.0533557 ± 0.829703i -2.89329	-0.0533557 ± 0.829703i -2.89329	-0.612479 -0.640151 ± 1.611852i -14.7255	—
-0.0716708 ± 0.829584i -2.89329	-0.0678791 ± 0.829577i -2.89329	-2.98282	-0.612480 -0.856075 ± 1.61361i -14.7255	—
-0.089986 ± 0.829223i -2.89329	-0.0824026 ± 0.829195i -2.89329	-3.16189	-0.612480 -1.28793 ± 1.57381i -14.7255	-0.612480 -1.28793 ± 1.57381i -14.7255
-0.108301 ± 0.828623i -2.89329	-0.0969257 ± 0.828559i -2.89329	-3.25142	0.612480 -1.93570 ± 1.43131i -14.7255	-0.612479 -1.5302 ± 1.56677i -14.7256
-0.126616 ± 0.827781i -2.89329	-0.111449 ± 0.827667i -2.89329	-3.34095	-0.612480 -2.79940 ± 1.01668i -14.7255	-0.612477 -2.13665 ± 1.48524i -14.7256
-0.144932 + 0.826698i -2.89329	-0.125973 + 0.82652i -2.89329	-3.43048	-0.612480 -4.85310 -2.90496 -14.7255	-0.61248 -2.74288 ± 1.2964i -14.7256
-0.163247 ± 0.825372i -2.89329	-0.140496 ± 0.825115i -2.89329	-3.52001		
-0.181562 ± 0.823803i -2.89329	-0.155019 ± 0.823452i -2.89329	-3.60955		
-0.198878 ± 0.821988i -2.89329	-0.169543 ± 0.821528i -2.89329	-3.69908		
-0.218193 ± 0.819926i -2.89329	-0.184066 ± 0.819343i -2.89329	-3.78861		
-0.236508 ± 0.817616i -2.89329	-0.198590 ± 0.816894i -2.89329			

STEADY STATE FEEDFORWARD COMPENSATION FOR DISCRETE–TIME MULTIVARIABLE SYSTEMS

H. A. Pak* and Rowmau Shieh†

* Assistant Professor

† Graduate Student

Department of Mechanical Engineering

University of Southern California

Los Angeles, CA 90089-1453.

Abstract

A class of discrete-time feedforward controllers which are designed for steady state disturbance rejection and command tracking are introduced for linear multivariable systems. The controllers use discretized future set points along time varying trajectories of the exogenous inputs described by linear models. This is in contrast to the use of the time derivatives of the instantaneous set points in the conventional asymptotic tracking controllers. As a result, apart from numerical noise reduction, the need for real-time generation of the derivative states of the exogenous input models are avoided for the purpose of realization of the feedforward controllers.

1 INTRODUCTION

In servo control, the use of feedforward precompensation has been a long established practical technique for the purpose of command tracking and disturbance rejection. In particular, for computer control systems several discrete-time feedforward design methods have been developed in recent years. For example, when the dynamical model of the exogenous inputs are unknown the *optimal finite length preview control* technique (Tomizuka 1975) may be used in order to improve the tracking response considerably. Also it has been shown that the *zero phase error tracking control* technique (Tomizuka 1987) leads to further improvements in the tracking response in applications where the inputs have high frequency contents relative the closed loop bandwidth.

In this paper we present a discrete-time design technique for the purpose of disturbance rejection and tracking control of multivariable systems subjected to time varying exogenous inputs with known models. In a related work a linear quadratic optimal control approach was used to design both the feedback and feedforward controllers simultaneously (Pak & Shieh 1989). The linear quadratic technique requires full state, or estimated state, feedback. In addition, due to the inclusion of the control effort in the cost function and its subsequent minimization, perfect asymptotic tracking or disturbance rejection is not achievable. In the present work, however, it is not required that the feedback controller should have any particular form. It is only assumed that the feedback controller is prespecified and it provides a stable closed loop response. Furthermore, the proposed feedforward control scheme results in perfect asymptotic tracking and disturbance rejection provided that the closed loop system exhibits linear time invariant behavior.

For continuous time control systems perfect asymptotic tracking controllers have been proposed using both the frequency domain (Seraji 1987) and the time domain (Davison 1973) design techniques. These controllers require the implementation of an additional smoothing prefilter to avoid multiple differentiations of the input signals which could otherwise cause unacceptable noise levels. Using a state space formulation this requirement is often met by the implementation of a tracking observer. Our proposed formulation avoids the need for the implementation of the tracking observer. Furthermore, the required knowledge of exogenous inputs is limited to an integer number of their future (preview) set points. In other words, the internal states of the desired trajectory model are not needed.

2 MODEL OF EXOGENOUS INPUTS

Consider a linear time-invariant system described by equations:

$$\mathbf{x}(k+1) = A\mathbf{x}(k) + B\mathbf{u}(k) + E\mathbf{w}(k) \tag{1}$$

$$\mathbf{y}(k) = C\mathbf{x}(k) + F\mathbf{w}(k) \tag{2}$$

where x(k) is an $n \times 1$ state vector, u(k) is an $m \times 1$ input vector, y(k) is a $q \times 1$ output vector, w(k) is a $r \times 1$ disturbance input vector, and A, B, C, E and F are matrices of appropriate dimensions for the system. The system is required to track a $q \times 1$ vector of reference trajectory,

$$\mathbf{y}_r(k) = \begin{bmatrix} y_{r_1}(k) & y_{r_2}(k) & \cdots & y_{r_q}(k) \end{bmatrix}^T,$$

and also to reject the known disturbances,

$$\mathbf{w}(k) = \begin{bmatrix} w_1(k) & w_2(k) & \cdots & w_r(k) \end{bmatrix}^T.$$

It is assumed that both y_{r_i}'s and w_i's belong to a class of signals which can be modeled as the output of linear time-invariant systems with different sets of initial conditions:

$$\begin{aligned} \bar{\mathbf{x}}_e(k+1) &= \bar{A}_e \bar{\mathbf{x}}_e(k) \\ \mathbf{y}_e(k) &= \bar{C}_e \bar{\mathbf{x}}_e(k), \qquad \bar{\mathbf{x}}_e(0) = \bar{\mathbf{x}}_{e0} \end{aligned} \tag{3}$$

Moreover each channel of exogenous signal should satisfy the same characteristic equation.

$$\lambda^p + a_p\lambda^{p-1} + \cdots + a_2\lambda + a_1 = 0$$

The above assumption for the exogenous signal is not as restrictive as it may appear. A wide range of commonly used signals such as polynomials, sinusoidals, exponentials, and their combinations fall within this category. Actually, each input channel may have a different type of exogenous signal, or the same type of signal but with a different set of parameters. However, in order to share the same characteristic equation, the selection of exogenous signal dynamic model should cover all of the eigenvalues of each individual input channel. Hence, when different channels are to follow different types of exogenous signals, the dimension of equation (3) will increase accordingly.

By definition the system described in equation (3) is observable, therefore a transformation can always be found to convert $(\bar{A}_e, \bar{C}_e)$ to the following observability canonical form:

$$\begin{aligned} \mathbf{x}_e(k+1) &= \begin{bmatrix} 0 & I & 0 & \cdots & 0 \\ 0 & 0 & I & \cdots & 0 \\ & & & \ddots & \vdots \\ & & & & I \\ -a_1 I & -a_2 I & -a_3 I & \ldots & -a_p I \end{bmatrix} \mathbf{x}_e(k) \\ \mathbf{y}_e(k) &= \begin{bmatrix} I & 0 & \cdots & 0 \end{bmatrix} \mathbf{x}_e(k), \qquad \mathbf{x}_e(0) = \mathbf{x}_{e0} \end{aligned} \tag{4}$$

which shows $\mathbf{x}_e(k) = \begin{bmatrix} \mathbf{y}_e(k) & \mathbf{y}_e(k+1) & \cdots & \mathbf{y}_e(k+p-1) \end{bmatrix}^T$. Thus, the states of exogenous signal model are expressed in terms of $p-1$ future (preview) set points of the signal itself.

The control input considered in this paper is

$$\mathbf{u}(k) = \mathbf{u}_{fb}(k) + G_r \mathbf{x}_r(k) + G_w \mathbf{x}_w(k). \tag{5}$$

We assume the feedback control action, $\mathbf{u}_{fb}(k)$, which is to stabilize the system and to achieve a specified transient response, has been already designed. The feedforward controller then consists of $G_r\mathbf{x}_r(k)$ and $G_w\mathbf{x}_w(k)$, where G_r and G_w are the constant gain matrices; $\mathbf{x}_r(k)$ and $\mathbf{x}_w(k)$ are the states of the reference trajectory model and disturbance model respectively. Furthermore, $\mathbf{x}_r(k)$ and $\mathbf{x}_w(k)$ as defined in equation

(4), are given by

$$\mathbf{x}_r(k) = \left[\begin{array}{cccc} \mathbf{y}_r(k) & \mathbf{y}_r(k+1) & \cdots & \mathbf{y}_r(k+p_r-1) \end{array} \right]^T,$$

and
$$\mathbf{x}_w(k) = \left[\begin{array}{cccc} \mathbf{w}(k) & \mathbf{w}(k+1) & \cdots & \mathbf{w}(k+p_w-1) \end{array} \right]^T,$$

where p_r and p_w are the number of the preview points for the reference trajectory and the disturbance respectively.

Since the system is linear, the feedforward controller design can be processed independently. In section 3, the feedforward gain matrix, G_w, for disturbance rejection with $\mathbf{y}_r = 0$ will be derived. This will be followed by the design of the tracking controller gain matrix, G_r, in section 4.

3 DISTURBANCE REJECTION CONTROLLER

Since the feedback controller is given, the system can be written as

$$\begin{aligned} \mathbf{x}(k+1) &= A_c\mathbf{x}(k) + B\mathbf{u}_{ff}(k) + E\mathbf{w}(k) \\ \mathbf{y}(k) &= C\mathbf{x}(k) + F\mathbf{w}(k), \qquad \mathbf{x}(0) = \mathbf{x}_0 \end{aligned} \tag{6}$$

where A_c represents the closed loop system matrix, and $\mathbf{u}_{ff}(k)$ is the feedforward control input. Then $\mathbf{u}_{ff}(k) = G_w\mathbf{x}_w(k)$; G_w has the dimension of $m \times p_w r$. Let

$$G_w = \left[\begin{array}{cccc} G_0^w & G_1^w & \cdots & G_{p_w-1}^w \end{array} \right]$$

with $G_j^w, j = 0, 1, \cdots, p_w - 1$ being $m \times r$ matrices. From the definition of $\mathbf{x}_w(k)$, we have

$$\mathbf{u}_{ff}(k) = \sum_{j=0}^{p_w-1} G_j^w \mathbf{w}(k+j) \tag{7}$$

As the dynamic model of the disturbance is known, $\mathbf{w}(k)$ can be shown in a general form in terms of its eigenvalues,

$$\mathbf{w}(k) = \sum_{i=1}^{\bar{n}} (\mathbf{c}_0^i + \mathbf{c}_1^i k + \cdots + \mathbf{c}_{m_i-1}^i k^{m_i-1})\lambda_i^k \tag{8}$$

where $\bar{n}$ is the total number of distinct eigenvalues, m_i is the multiplicity of the eigenvalue λ_i, and c's are $m \times 1$ coefficient vectors associated with the eigenvalue λ_i and its multiplicity. It also follows immediately that $\sum_{i=1}^{\bar{n}} m_i = p_w$.

Substituting equation (8) into equation (7) gives

$$\mathbf{u}_{ff}(k) = \sum_{i=1}^{\bar{n}} \lambda_i^k \left(\sum_{j=0}^{p_w-1} G_j^w \mathbf{c}_0^i \lambda_i^j + \sum_{j=0}^{p_w-1} G_j^w \mathbf{c}_1^i (k+j) \lambda_i^j + \cdots + \sum_{j=0}^{p_w-1} G_j^w \mathbf{c}_1^i (k+j)^{m_i-1} \lambda_i^j . \right)$$

After executing binomial expansion to each $(k+j)^l$, $l = 1, 2, \cdots, m_i - 1$, and sorting the coefficients according to k's power, we obtain

$$\begin{aligned}
\mathbf{u}_{ff}(k) &= \sum_{i=1}^{\bar{n}} \lambda_i^k \left\{ \sum_{j=0}^{p_w-1} G_j^w \left(\mathbf{c}_0^i + \mathbf{c}_1^i j + \cdots + \mathbf{c}_{m_i-1}^i j^{m_i-1} \right) \lambda_i^j + \right. \\
&\quad \left[\sum_{j=0}^{p_w-1} G_j^w \left(\mathbf{c}_1^i + \begin{pmatrix} 2 \\ 1 \end{pmatrix} \mathbf{c}_2^i j + \cdots + \begin{pmatrix} m_i - 1 \\ m_i - 2 \end{pmatrix} \mathbf{c}_{m_i-1}^i j^{m_i-2} \right) \lambda_i^j \right] k + \\
&\quad \left[\sum_{j=0}^{p_w-1} G_j^w \left(\mathbf{c}_2^i + \begin{pmatrix} 3 \\ 1 \end{pmatrix} \mathbf{c}_3^i j + \cdots + \begin{pmatrix} m_i - 1 \\ m_i - 3 \end{pmatrix} \mathbf{c}_{m_i-1}^i j^{m_i-3} \right) \lambda_i^j \right] k^2 + \\
&\quad \vdots \\
&\quad \left. \sum_{j=0}^{p_w-1} G_j^w \mathbf{c}_{m_i-1}^i \lambda_i^j k^{m_i-1} \right\} \\
&= \sum_{i=1}^{\bar{n}} (\boldsymbol{\beta}_0^i + \boldsymbol{\beta}_1^i k + \cdots + \boldsymbol{\beta}_{m_i-1}^i k^{m_i-1}) \lambda_i^k, \qquad (9)
\end{aligned}$$

where $\boldsymbol{\beta}_l^i$, $l = 1, 2, \cdots, m_i - 1$ correspond to the appropriate coefficients respectively. Let $\boldsymbol{\beta}_i = \begin{bmatrix} \boldsymbol{\beta}_0^i & \boldsymbol{\beta}_1^i & \cdots & \boldsymbol{\beta}_{m_i-1}^i \end{bmatrix}$, then we may write

$$\boldsymbol{\beta}_i = G_w \Lambda_i \Gamma_i \qquad (10)$$

where

$$\Gamma_i = \begin{bmatrix}
\mathbf{c}_0^i & \mathbf{c}_1^i & \mathbf{c}_2^i & \cdots & \cdots & \cdots & \mathbf{c}_{m_i-1}^i \\
\mathbf{c}_1^i & \begin{pmatrix} 2 \\ 1 \end{pmatrix} \mathbf{c}_2^i & \begin{pmatrix} 3 \\ 2 \end{pmatrix} \mathbf{c}_3^i & \cdots & \cdots & \begin{pmatrix} m_i - 1 \\ m_i - 2 \end{pmatrix} \mathbf{c}_{m_i-1}^i & 0 \\
\mathbf{c}_2^i & \begin{pmatrix} 3 \\ 1 \end{pmatrix} \mathbf{c}_3^i & \begin{pmatrix} 4 \\ 2 \end{pmatrix} \mathbf{c}_4^i & \cdots & \begin{pmatrix} m_i - 1 \\ m_i - 3 \end{pmatrix} \mathbf{c}_{m_i-1}^i & 0 & 0 \\
\vdots & & & & & & \vdots \\
\mathbf{c}_{m_i-1}^i & 0 & 0 & \cdots & \cdots & \cdots & 0
\end{bmatrix},$$

and

$$\Lambda_i = \begin{bmatrix} I & 0 & 0 & \cdots & 0 \\ \lambda_i I & \lambda_i I & \lambda_i I & \cdots & \lambda_i I \\ \lambda_i^2 I & 2\lambda_i^2 I & 2^2\lambda_i^2 I & \cdots & 2^{m_i-1}\lambda_i^2 I \\ \vdots & \vdots & \vdots & & \vdots \\ \lambda_i^{p_w-1} I & (p_w-1)\lambda_i^{p_w-1} I & (p_w-1)^2\lambda_i^{p_w-1} I & \cdots & (p_w-1)^{m_i-1}\lambda_i^{p_w-1} I \end{bmatrix}.$$

The output response for the closed loop system of equation (6) is given by

$$\mathbf{y}(k) = CA_c^k\mathbf{x}_0 + C\sum_{j=0}^{k-1} A_c^{k-j-1}B\mathbf{u}_{ff}(j) + C\sum_{j=0}^{k-1} A_c^{k-j-1}E\mathbf{w}(j) + F\mathbf{w}(k).$$

Since A_c is stable, $A_c^k \to 0$ as $k \to \infty$, then we have

$$\lim_{k\to\infty}\mathbf{y}(k) = \lim_{k\to\infty} C\sum_{j=0}^{k-1} A_c^{k-j-1}B\mathbf{u}_{ff}(j) + \lim_{k\to\infty} C\sum_{j=0}^{k-1} A_c^{k-j-1}E\mathbf{w}(j) + \lim_{k\to\infty} F\mathbf{w}(k).$$

Substituting $\mathbf{u}_{ff}(k)$ from equation (9) and $\mathbf{w}(k)$ from equation (8) gives

$$\begin{aligned} &\lim_{k\to\infty}\mathbf{y}(k) = \\ &\quad \lim_{k\to\infty}\sum_{i=0}^{\bar{n}} CA_c^{k-1}\sum_{j=0}^{k-1}(\lambda_i A_c^{-1})^j\left[B(\beta_0^i + \beta_1^i j + \cdots + \beta_{m_i-1}^i j^{m_i-1}) + \right. \\ &\quad \left. E(\mathbf{c}_0^i + \mathbf{c}_1^i j + \cdots + \mathbf{c}_{m_{i-1}}^i j^{m_{i-1}})\right] + \lim_{k\to\infty} F\sum_{i=1}^{\bar{n}}(\mathbf{c}_0^i + \mathbf{c}_1^i k + \cdots + \mathbf{c}_{m_{i-1}}^i k^{m_{i-1}})\lambda_i^k \end{aligned} \tag{11}$$

After a lengthy manipulation which is outlined in the Appendix, we obtain

$$\begin{aligned} \lim_{k\to\infty}\mathbf{y}(k) &= \sum_{i=1}^{\bar{n}}\left[(\delta_0^i + \delta_1^i k + \cdots + \delta_{m_i-1}^i k^{m_i-1}) + (\zeta_0^i + \zeta_1^i k + \cdots + \zeta_{m_i-1}^i k^{m_i-1})\right. \\ &\quad \left. F(\mathbf{c}_0^i + \mathbf{c}_1^i k + \cdots + \mathbf{c}_{m_{i-1}}^i k^{m_{i-1}})\right]\lambda_i^k \end{aligned} \tag{12}$$

with

$$\begin{aligned} \delta_0^i &= H_i\beta_0^i - D_1^i\beta_1^i + D_2^i\beta_2^i + \cdots + (-1)^{m_i-1}D_{m_i-1}^i\beta_{m_i-1}^i \\ \delta_1^i &= H_i\beta_1^i - \binom{2}{1}D_1^i\beta_2^i + \binom{3}{2}D_2^i\beta_3^i + \cdots + (-1)^{m_i}D_{m_i-2}^i\beta_{m_i-1}^i \end{aligned}$$

$$
\begin{aligned}
\delta_2^i &= H_i\beta_2^i - \begin{pmatrix} 3 \\ 1 \end{pmatrix} D_1^i\beta_3^i + \begin{pmatrix} 4 \\ 2 \end{pmatrix} D_2^i\beta_4^i + \cdots + (-1)^{m_i-1} D_{m_i-3}^i\beta_{m_i-1}^i \\
&\vdots \\
\delta_{m_i-1}^i &= H_i\beta_{m_i-1}^i \\
\zeta_0^i &= \mathcal{H}_i\mathbf{c}_0^i - W_1^i\mathbf{c}_1^i + W_2^i\mathbf{c}_2^i + \cdots + (-1)^{m_i-1}W_{m_i-1}^i\mathbf{c}_{m_i-1}^i \\
\zeta_1^i &= \mathcal{H}_i\mathbf{c}_1^i - \begin{pmatrix} 2 \\ 1 \end{pmatrix} W_1^i\mathbf{c}_2^i + \begin{pmatrix} 3 \\ 2 \end{pmatrix} W_2^i\mathbf{c}_3^i + \cdots + (-1)^{m_i-2}W_{m_i-2}^i\mathbf{c}_{m_i-1}^i \\
\zeta_2^i &= \mathcal{H}_i\mathbf{c}_2^i - \begin{pmatrix} 3 \\ 1 \end{pmatrix} W_1^i\mathbf{c}_3^i + \begin{pmatrix} 4 \\ 2 \end{pmatrix} W_2^i\mathbf{c}_4^i + \cdots + (-1)^{m_i-3}W_{m_i-3}^i\mathbf{c}_{m_i-1}^i \\
&\vdots \\
\zeta_{m_i-1}^i &= \mathcal{H}_i\mathbf{c}_{m_i-1}^i
\end{aligned}
$$

where

$$
\begin{aligned}
H_i &= C(\lambda_i I - A_c)^{-1}B, \\
D_l^i &= \lambda_i C V_l^i(\lambda_i I - A_c)^{-(l+1)}B, \qquad l = 1, 2, \cdots, m_{i-1} \\
\mathcal{H}_i &= C(\lambda_i I - A_c)^{-1}E, \\
W_l^i &= \lambda_i C V_l^i(\lambda_i I - A_c)^{-(l+1)}E, \qquad l = 1, 2, \cdots, m_{i-1}
\end{aligned}
$$

and $V_l^i = A_c^{l-1} + \gamma_1^l A_c^{l-2}\lambda_i + \cdots + \gamma_{l-2}^l A_c\lambda_i^{l-2} + \lambda_i^{l-1}I$,

in which γ's are obtained from the following recursive formula:

$$
\begin{aligned}
\gamma_1^l &= 1 + (-1)^l\left[\begin{pmatrix} l+2 \\ 3 \end{pmatrix} - 2^l\begin{pmatrix} l+2 \\ 4 \end{pmatrix} + \cdots + l^l(-1)^{l+1}\right] \\
\gamma_n^l &= n\gamma_{n-1}^l - \begin{pmatrix} n \\ 2 \end{pmatrix}\gamma_{n-2}^l + \cdots + (-1)^{n+1} + \\
&\quad (-1)^{l+n+1}\left[\begin{pmatrix} l+n+1 \\ n+2 \end{pmatrix} - 2^l\begin{pmatrix} l+n+1 \\ n+3 \end{pmatrix} + \cdots + l^l(-1)^{l+1}\right], n > 1
\end{aligned}
$$

Now for disturbance rejection, it is required that $\lim_{k\to\infty}\mathbf{y}(k) = 0$, i.e.,

$$
\delta_0^i + \zeta_0^i + F\mathbf{c}_0^i = 0
$$

$$\delta_1^i + \zeta_1^i + F\mathbf{c}_1^i = 0$$

$$\vdots$$

$$\delta_{m_i-1}^i + \zeta_{m_i-1}^i + F\mathbf{c}_{m_i-1}^i = 0$$

which gives

$$\begin{aligned}
\beta_{m_i-1}^i &= R_0^i \mathbf{c}_{m_i-1}^i \\
\beta_{m_i-2}^i &= R_0^i \mathbf{c}_{m_i-2}^i + \begin{pmatrix} m_i - 1 \\ 1 \end{pmatrix} R_1^i \mathbf{c}_{m_i-1}^i \\
\beta_{m_i-3}^i &= R_0^i \mathbf{c}_{m_i-3}^i + \begin{pmatrix} m_i - 2 \\ 1 \end{pmatrix} R_1^i \mathbf{c}_{m_i-2}^i + \begin{pmatrix} m_i - 1 \\ 2 \end{pmatrix} R_2^i \mathbf{c}_{m_i-1}^i \\
&\vdots \\
\beta_1^i &= R_0^i \mathbf{c}_1^i + \begin{pmatrix} 2 \\ 1 \end{pmatrix} R_1^i \mathbf{c}_2^i + \begin{pmatrix} 3 \\ 2 \end{pmatrix} R_2^i \mathbf{c}_3^i + \cdots + \begin{pmatrix} m_i - 1 \\ m_i - 2 \end{pmatrix} R_{m_i-2}^i \mathbf{c}_{m_i-1}^i \\
\beta_0^i &= R_0^i \mathbf{c}_0^i + R_1^i \mathbf{c}_1^i + \cdots + R_{m_i-1}^i \mathbf{c}_{m_i-1}^i
\end{aligned}$$

with the recursive formula

$$\begin{aligned}
R_0^i &= -H_i^\dagger (\mathcal{H}_i + F) \\
R_l^i &= H_i^\dagger \left[\begin{pmatrix} l \\ 1 \end{pmatrix} D_1^i R_{l-1}^i - \begin{pmatrix} l \\ 2 \end{pmatrix} D_2^i R_{l-2}^i + \cdots + (-1)^{l-1} D_l^i R_0^i + (-1)^{l-1} W_l^i \right], l > 1
\end{aligned}$$

where $H_i^\dagger = H_i^T (H_i H_i^T)^{-1}$, is the pseudoinverse of H_i.
Let $R_i = \begin{bmatrix} R_0^i & R_1^i & \cdots & R_{m_i-1}^i \end{bmatrix}$, then we have

$$\beta_i = R_i \Gamma_i \tag{13}$$

Comparing to equation (9), it follows

$$G_{ff} \Lambda_i = R_i. \tag{14}$$

To include all distinct eigenvalues, let $\Lambda = [\Lambda_1 \ \Lambda_2 \ \cdots \ \Lambda_{\bar{n}}]$ and $R = [R_1 \ R_2 \ \cdots \ R_{\bar{n}}]$, then

$$G_w \Lambda = R. \tag{15}$$

Since $\sum_{i=1}^{n} m_i = p_w$, Λ is a $p_w m \times p_w m$ matrix. Furthermore, by inspection of each submatrix Λ_i, we can conclude that Λ is nonsingular except for the trivial case of a trajectory with a pulse dynamics (zero eigenvalue). Therefore we can finally obtain the feedforward gain matrix for disturbance rejection; that is

$$G_w = R\Lambda^{-1}. \tag{16}$$

For realization of the above disturbance rejection control scheme the following points merit further discussion:

1. The number of future disturbance information required from the known generator model is always one less than the dimension of A_e.

2. In the above derivation of the feedforward controller, the feedforward gain matrix exists only if the rows of the $q \times m$ matrix H_i are linearly independent, which leads the result of $m \geq q$. Moreover, the solution for the feedforward gains is not unique, here the optimal least square solution (Strang 1980) was chosen; that is implied by the definition of $H_i^{\dagger} = H_i^T(H_i H_i^T)^{-1}$.

3. It is possible that the control system cannot achieve disturbance rejection if any of the eigenvalues in the disturbance dynamic model coincides with the transmission zeros of the closed loop system. However, this is unlikely in most applications for the following reasons. If the closed loop system has its transmission zeros within the unit disc (minimum phase behavior), then the possibility of pole-zero cancellation is practically avoided since the eigenvalues of the disturbance dynamic model are either on the unit disc or outside the unit disc in the complex z-plane. On the other hand, if the transmission zeros are outside the unit disc, they are likely to be on the negative real-axis due to the fast sampling phenomenon (Åström et al 1980). Now the eigenvalues of the disturbance dynamic model are unlikely to be on the negative real axis, as this would mean disturbance signals with frequency contents equal to one half the sampling frequency.

4. It is possible to use the open loop pair (A, B) instead of the closed loop pair (A_c, B) for the design of feedforward gain matrix G_w. However the use of the closed loop system dynamics ensures better performance in the presence of dynamic modeling error and unknown disturbances.

4 TRACKING CONTROLLER

Consider the system given in the equation (6) tracking the known trajectory $\mathbf{y}_r(k)$, but with $\mathbf{w}(k) = 0$. Then we have

$$\begin{aligned} \mathbf{x}(k+1) &= A_c\mathbf{x}(k) + B\mathbf{u}_{ff}(k) \\ \mathbf{y}(k) &= C\mathbf{x}(k), \qquad \mathbf{x}(0) = \mathbf{x}_0 \end{aligned} \tag{17}$$

where $\mathbf{u}_{ff}(k) = G_r\mathbf{x}_r(k)$; G_r has the dimension of $m \times p_r r$. Let

$$G_r = \begin{bmatrix} G_0^r & G_1^r & \cdots & G_{p_r-1}^r \end{bmatrix}$$

with $G_j^r, j = 0, 1, \cdots, p_r - 1$ being $m \times q$ matrices. From the definition of $\mathbf{x}_r(k)$, we obtain

$$\mathbf{u}_{ff}(k) = \sum_{j=0}^{p_r-1} G_j^r \mathbf{y}_r(k+j) \tag{18}$$

$\mathbf{y}_r(k)$ can also be expressed in terms of eigenvalues, which is

$$\mathbf{y}_r(k) = \sum_{i=1}^{\bar{n}} (\mathbf{c}_0^i + \mathbf{c}_1^i k + \cdots + \mathbf{c}_{m_i-1}^i k^{m_i-1})\lambda_i^k \tag{19}$$

Then following a similar procedure to the one outlined in section 3 for disturbance rejection, we can derive

$$\lim_{k\to\infty} \mathbf{y}(k) = \sum_{i=1}^{\bar{n}} (\delta_0^i + \delta_1^i k + \cdots + \delta_{m_i-1}^i k^{m_i-1})\lambda_i^k \tag{20}$$

in which δ's are derived from the same formula given in section 3. However, the eigenvalues will be different between the disturbance model and the reference trajectory model.

For asymptotic tracking, the coefficients of equations (19) and (20) should be equal, i.e.,

$$\mathbf{c}_0^i = \delta_0^i, \qquad \mathbf{c}_1^i = \delta_1^i, \qquad \cdots, \qquad \mathbf{c}_{m_i-1}^i = \delta_{m_i-1}^i.$$

As a result we can equate

$$\begin{aligned}
\beta^i_{m_i-1} &= Q^i_0 \mathbf{c}^i_{m_i-1} \\
\beta^i_{m_i-2} &= Q^i_0 \mathbf{c}^i_{m_i-2} + \begin{pmatrix} m_i - 1 \\ 1 \end{pmatrix} Q^i_1 \mathbf{c}^i_{m_i-1} \\
\beta^i_{m_i-3} &= Q^i_0 \mathbf{c}^i_{m_i-3} + \begin{pmatrix} m_i - 2 \\ 1 \end{pmatrix} Q^i_1 \mathbf{c}^i_{m_i-2} + \begin{pmatrix} m_i - 1 \\ 2 \end{pmatrix} Q^i_2 \mathbf{c}^i_{m_i-1} \\
&\vdots \\
\beta^i_1 &= Q^i_0 \mathbf{c}^i_1 + \begin{pmatrix} 2 \\ 1 \end{pmatrix} Q^i_1 \mathbf{c}^i_2 + \begin{pmatrix} 3 \\ 2 \end{pmatrix} Q^i_2 \mathbf{c}^i_3 + \cdots + \begin{pmatrix} m_i - 1 \\ m_i - 2 \end{pmatrix} Q^i_{m_i-2} \mathbf{c}^i_{m_i-1} \\
\beta^i_0 &= Q^i_0 \mathbf{c}^i_0 + Q^i_1 \mathbf{c}^i_1 + \cdots + Q^i_{m_i-1} \mathbf{c}^i_{m_i-1}
\end{aligned}$$

with the recursive formula

$$\begin{aligned}
Q^i_0 &= H_i^\dagger \\
Q^i_l &= Q^i_0 \left[\begin{pmatrix} l \\ 1 \end{pmatrix} D^i_1 Q^i_{l-1} - \begin{pmatrix} l \\ 2 \end{pmatrix} D^i_2 Q^i_{l-2} + \cdots + (-1)^{l-1} D^i_l Q^i_0 \right], l > 1
\end{aligned}$$

Let $Q_i = \begin{bmatrix} Q^i_0 & Q^i_1 & \cdots & Q^i_{m_i-1} \end{bmatrix}$, then we have

$$\beta_i = Q_i \Gamma_i \tag{21}$$

Comparing to equation (10), it follows

$$G_r \Lambda_i = Q_i. \tag{22}$$

To include all distinct eigenvalues, let $\Lambda = [\Lambda_1 \ \Lambda_2 \ \cdots \ \Lambda_{\bar{n}}]$ and $Q = [Q_1 \ Q_2 \ \cdots \ Q_{\bar{n}}]$, then

$$G_r \Lambda = Q. \tag{23}$$

Λ can only be singular in a trivial case when a trajectory has a pulse dynamics (zero eigenvalues). Therefore we can obtain the feedforward gain matrix for asymptotic tracking, which is

$$G_r = Q \Lambda^{-1}. \tag{24}$$

Note that similar remarks to those mentioned at the end of section 3 apply here for the practical realization of the above tracking controller.

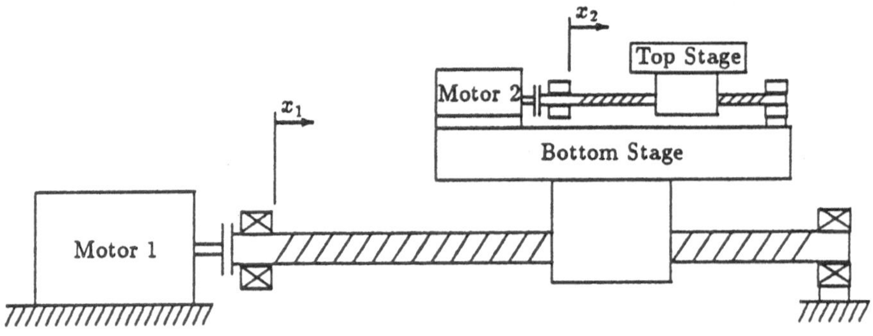

Figure 1: Schematic diagram of positioning control system

5 AN ILLUSTRATIVE EXAMPLE

To demostrate the response of a multivariable discrete-time system under the above control scheme we consider a 2×2 feeddrive system. The system consists of two actuators for a dual stage positioning of a carriage. A schematic diagram of the system is shown on figure 1, in which two direct current (dc) servo motors are connected to ballscrews for actuation purposes. The state space representation of the system is given by

$$\begin{bmatrix} \dot{x}_1 \\ \ddot{x}_1 \\ \dot{x}_2 \\ \ddot{x}_2 \end{bmatrix} = \begin{bmatrix} 0 & 1 & 0 & 0 \\ 0 & -\frac{k_{T_1} k_{b_1}}{M_1 R_1 g_1^2} & 0 & \frac{k_{T_2} k_{b_2}}{M_1 R_2 g_2^2} \\ 0 & 0 & 0 & 1 \\ 0 & \frac{k_{T_1} k_{b_1}}{M_1 R_1 g_1^2} & 0 & -\frac{k_{T_2} k_{b_2}}{R_2 g_2^2}\left(\frac{1}{M_1} + \frac{1}{M_2}\right) \end{bmatrix} \begin{bmatrix} x_1 \\ \dot{x}_1 \\ x_2 \\ \dot{x}_2 \end{bmatrix} + \begin{bmatrix} 0 & 0 \\ \frac{k_{T_1} k_{a_1}}{M_1 R_1 g_1} & -\frac{k_{T_2} k_{a_2}}{M_1 R_2 g_2} \\ 0 & 0 \\ -\frac{k_{T_1} k_{a_1}}{M_1 R_1 g_1} & \frac{k_{T_2} k_{a_2}}{R_2 g_2}\left(\frac{1}{M_1} + \frac{1}{M_2}\right) \end{bmatrix} \begin{bmatrix} u_1 \\ u_2 \end{bmatrix} \tag{25}$$

$$\mathbf{y} = \begin{bmatrix} 1 & 0 & 0 & 0 \\ 0 & 0 & 1 & 0 \end{bmatrix} \begin{bmatrix} x_1 \\ \dot{x}_1 \\ x_2 \\ \dot{x}_2 \end{bmatrix} \tag{26}$$

where x_1 represents the linear motion of the bottom stage with respect to the stationary frame, and x_2 the linear motion of the top stage with respect to the bottom stage. The system parameters are:

k_{T_i} = the ith motor torque constant

k_{b_i} = the back emf constant

R_i = the motor terminal resistance

k_{a_i} = the amplification gain

M_i = the reflected mass

g_i = the angular-to-linear displacement ratio

It is assumed that the motors' electrical transients are negligible.

In this example the closed loop contoller was designed under sampled-data control, for which the sampling period T_s was chosen as $5ms$. The system parameters were given as follows

$$\begin{array}{ll} M_1 = 750kg,\ M_2 = 75kg, & k_{T_1} = k_{T_2} = .13Nm/A \\ k_{b_1} = k_{b_2} = .13V/rad/s, & k_{a_1} = k_{a_2} = 7V/V \\ R_1 = R_2 = .65\Omega, & g_1 = g_2 = \frac{.005}{2\pi}m \end{array}$$

State feedback was used to provide acceptable transient response using

$$\mathbf{u}_{fb} = \begin{bmatrix} 2318.2 & 29.646 & 1263.1 & 10.742 \\ 1229.1 & 20.756 & 1180.2 & -9.223 \end{bmatrix} \begin{bmatrix} x_1 \\ \dot{x}_1 \\ x_2 \\ \dot{x}_2 \end{bmatrix},$$

This provided a closed loop system model described by

$$\begin{aligned} \mathbf{x}(k+1) &= A_c\mathbf{x}(k) + B\mathbf{u}_{ff}(k) + E\mathbf{w}(k) \\ \mathbf{y}(k) &= C\mathbf{x}(k) \end{aligned}$$

with

$$A_c = \begin{bmatrix} 9.555\times10^{-1} & 3.895\times10^{-3} - 1.810\times10^{-2} & -1.047\times10^{-4} \\ -1.863 & 5.516\times10^{-1} - 8.577 & -6.037\times10^{-2} \\ -1.514\times10^{-1} & -2.391\times10^{-3} - 8.430\times10^{-1} & 3.070\times10^{-3} \\ -4.70\times10 & -7.510\times10^{-1} - 4.810\times10 & 4.944\times10^{-1} \end{bmatrix}$$

$$B = \begin{bmatrix} 2.557\times10^{-5} & -1.204\times10^{-5} \\ 9.670\times10^{-3} & -3.082\times10^{-3} \\ -1.203\times10^{-5} & 1.459\times10^{-4} \\ -3.082\times10^{-3} & 4.405\times10^{-2} \end{bmatrix}$$

$$C = \begin{bmatrix} 1 & 0 & 0 & 0 \\ 0 & 0 & 1 & 0 \end{bmatrix}$$

$$E = \begin{bmatrix} 1 & 1 & 1 & 1 \\ 0 & -1 & 0 & -1 \end{bmatrix}^T$$

The system will be used to track two specified trajectory vectors under known disturbances.

Case 1: One of the eigenvalues of exogenous signals has multiplicity.

The reference trajectories are given as $\mathbf{y}_r = \begin{bmatrix} h_1 kT_s \\ h_2 \sin \omega kT_s \end{bmatrix}$. Then the eigenvalues of the trajectory dynamic model are 1 (for the ramp input), with the multiplicity $m = 2$, $\cos \omega T_s + j \sin \omega T_s$ and $\cos \omega T_s - j \sin \omega T_s$ (for the sinusoidal input). So

$$\Lambda = \begin{bmatrix} 1 & 0 & 1 & 1 \\ 1 & 1 & \cos \omega T_s + j \sin \omega T_s & \cos \omega T_s - j \sin \omega T_s \\ 1 & 2 & \cos 2\omega T_s + j \sin 2\omega T_s & \cos 2\omega T_s - j \sin 2\omega T_s \\ 1 & 3 & \cos 3\omega T_s + j \sin 3\omega T_s & \cos 3\omega T_s - j \sin 3\omega T_s \end{bmatrix}$$

and

$$Q = \Big[(C(I - A_c)^{-1}B)^{-1} \quad (C(I - A_c)^{-1}B)^{-1}C(I - A_c)^{-2}B(C(I - A_c)^{-1}B)^{-1}$$
$$(C[(\cos \omega T_s + j \sin \omega T_s)I - A_c]^{-1}B)^{-1} \quad (C[(\cos \omega T_s - j \sin \omega T_s)I - A_c]^{-1}B)^{-1} \Big].$$

For the case of $\omega = 4\pi$ the tracking controller gain matrix is

$$G_r = \begin{bmatrix} 15600 & -734.6 & -45380 & 2001 & 40290 & -135.0 & -8206 & 121.7 \\ -3946 & 4046 & 6632 & -10550 & -1890 & 9931 & 433.3 & -2246 \end{bmatrix}.$$

Assume the system is also subject to sinusoidal disturbances, which have the eigenvalues $\cos\omega_d T_s + j\sin\omega_d T_s$ and $\cos\omega_d T_s - j\sin\omega_d T_s$. Then

$$\Lambda = \begin{bmatrix} 1 & 1 \\ \cos\omega_d T_s + j\sin\omega_d T_s & \cos\omega_d T_s - j\sin\omega_d T_s \end{bmatrix}$$

and

$$R = \begin{bmatrix} -[C(\lambda_1 I - A_c)^{-1}B]^{-1}C(\lambda_1 I - A_c)E & -[C(\lambda_2 I - A_c)^{-1}B]^{-1}C(\lambda_2 I - A_c)E \end{bmatrix},$$

where $\lambda_1 = \cos\omega_d T_s + j\sin\omega_d T_s$ and $\lambda_2 = \cos\omega_d T_s - j\sin\omega_d T_s$. For the case of $\omega_d = 8\pi$ the feedforward gain matrix is

$$G_w = R\Lambda^{-1} = \begin{bmatrix} 3046.2 & 120.08 & -15976 & -40.59 \\ -4157.2 & 20.526 & -2606.6 & -8.467 \end{bmatrix}.$$

Then the total control input can be obtained as

$$\mathbf{u}(k) = \mathbf{u}_{fb}(k) + G_{ff}\begin{bmatrix} \mathbf{y}(k) \\ \mathbf{y}(k+1) \\ \mathbf{y}(k+2) \\ \mathbf{y}(k+3) \end{bmatrix} + G_w \begin{bmatrix} \mathbf{w}(k) \\ \mathbf{w}(k+1) \end{bmatrix}$$

The output response is given in figure 2, which shows perfect steady state disturbance rejection and asymptotic tracking behavior.

Case 2: None of the eigenvalues of exogenous signals has multiplicity.

The reference trajectories are given by a sinusoid and a cycloid as $\mathbf{y}_r = \begin{bmatrix} h_1 \sin\omega k T_s \\ h_2(\omega k T_s - \sin\omega k T_s) \end{bmatrix}$ The system is also subjected to sinusoidal disturbances whose eigenvalues are known. Then following the similar procedure as in Case 1, we can obtain

$$G_r = \begin{bmatrix} 15600 & -734.0 & -45380 & 2000 & 40300 & -134.4 & -8210 & 131.7 \\ -3945 & 4044 & 6631 & -10550 & -1890 & 9930 & 433.5 & -2247 \end{bmatrix}.$$

for $\omega = 5\pi$, and

$$G_w = R\Lambda^{-1} = \begin{bmatrix} 3010 & 120.0 & -16010 & -40.68 \\ -4168 & 20.49 & -2612 & -8.483 \end{bmatrix}.$$

for $\omega_d = 10\pi$.

The output response is given in figure 3, which also shows perfect disturbance rejection and tracking characteristics after the initial transients.

6 CONCLUSIONS

A class of feedforward controllers capable of steady state disturbance rejection and command tracking were formulated for linear multivariable discrete-time systems. While the control scheme's realizability is limited to disturbances and input trajectories with known linear models, it is of practical value is many applications where the exogenous input signals are predefined (eg. machine tools, robotic manipulators, cam and follower motion replacement devices, etc.) It was shown that, by formulating the input models in the observability canonical form, we can replace the need for the generation and the use of the derivative states of the inputs with their preview states (future inputs). This action should help reduce numerical noise and avoid extra numerical computation for the purpose of real-time implementation.

REFERENCES

Åström, K. J., Hagander, P. and Sternby, J., 1980, "Zeros of Sampled Systems," *Proc. 19th IEEE Conf. on Decision and Control*, Albuquerque, pp. 1077-1081

Davison, E. J., 1973, "The Feedforward Control of Linear Multivariable Time-Invariant Systems," *Automatica*, vol. 9, pp. 561-573

Pak, H. A. and Shieh, R., 1989, "Optimal Preview Controllers Based upon Explicit Trajectory Models," presented in *Second Workshop on Control Mechanics*, Univ. of Southern Calif., Los Angeles, Feb. 1989; also to appear on *Control and Dynamic Systems: Advances in Theory and Applications*, vol. 35, 1990

Seraji, H., 1987, "Design of Feedforward Controllers for Multivariable Plants," *Int. J. Control*, vol. 46, pp. 1633-1651

Strang, G., 1980, *Linear Algebra and its Applications*, 2nd ed., Academic Press, New York

Tomizuka, M., 1975, "Optimal Continuous Finite Preview Problem," *IEEE Trans. Auto. Control*, vol. 20, pp. 362-365

Tomizuka, M., 1987, "Zero Phase Error Tracking Algorithm for Digital Control," *ASME Journal of Dynamic Systems, Measurement and Control*, vol. 109, pp. 65-68

APPENDIX

Outline Derivation of Equation (12)

The steady-state response of the system (6) is as given in equation (11),

$$\lim_{k\to\infty} \mathbf{y}(k) = \lim_{k\to\infty} \mathbf{h}(k) + \lim_{k\to\infty} \mathbf{t}(k) + \lim_{k\to\infty} F\sum_{i=1}^{\bar{n}}(\mathbf{c}_0^i + \mathbf{c}_1^i k + \cdots + \mathbf{c}_{m_i-1}^i k^{m_i-1})\lambda_i^k$$

where

$$\begin{aligned} \mathbf{h}(k) &= \sum_{i=0}^{\bar{n}} CA_c^{k-1}\sum_{j=0}^{k-1}(\lambda_i A_c^{-1})^j B(\beta_0^i + \beta_1^i j + \cdots + \beta_{m_i-1}^i j^{m_i-1}) \\ \mathbf{t}(k) &= \sum_{i=0}^{\bar{n}} CA_c^{k-1}\sum_{j=0}^{k-1}(\lambda_i A_c^{-1})^j E(\mathbf{c}_0^i + \mathbf{c}_1^i j + \cdots + \mathbf{c}_{m_i-1}^i j^{m_i-1}). \end{aligned}$$

First considering $\lim_{k\to\infty} \mathbf{h}(k)$, we have

$$\begin{aligned} &\lim_{k\to\infty} \mathbf{s}(k) = \\ &\quad \sum_{i=1}^{\bar{n}} \Bigg[\lim_{k\to\infty} CA_c^{k-1}\sum_{j=1}^{k-1}(\lambda_i A_c^{-1})^j B\beta_0^i + \lim_{k\to\infty} CA_c^{k-1}\sum_{j=1}^{k-1} j(\lambda_i A_c^{-1})^j B\beta_1^i + \\ &\quad \cdots + \lim_{k\to\infty} CA_c^{k-1}\sum_{j=1}^{k-1} j^{m_i-1}(\lambda_i A_c^{-1})^j B\beta_{m_i-1}^i\Bigg] \\ &= \sum_{i=1}^{\bar{n}}\sum_{l=0}^{m_i-1}(-1)^l C(\alpha_0^l A_c^l + \alpha_1^l\lambda_i A_c^{l-1} + \cdots + \alpha_l^l\lambda_i^l I)(\lambda_i I - A_c)^{-(l+1)} B\beta_l^i\lambda_i^k \end{aligned}$$

where

$$\begin{aligned}
\alpha_0^l &= \begin{pmatrix} l+1 \\ 1 \end{pmatrix}(k-1)^l - \begin{pmatrix} l+1 \\ 2 \end{pmatrix}(k-2)^l + \cdots + (-1)^l(k-l-1)^l \\
\alpha_1^l &= -\begin{pmatrix} l+1 \\ 2 \end{pmatrix}(k-1)^l + \begin{pmatrix} l+1 \\ 3 \end{pmatrix}(k-2)^l + \cdots + (-1)^l(k-l)^l \\
&\vdots \\
\alpha_l^l &= (-1)^l(k-1)^l
\end{aligned}$$

The expression for α's are derived from the sum of finite series, $\sum_{j=1}^{k-1} j^l(\lambda_i A_c^{-1})^j$, $l = 1, 2, \cdots, m_i - 1$. Moreover, after going through a lengthy rearrangment, $\alpha_0^l A_c^l + \alpha_1^l \lambda_i A_c^{l-1} + \cdots + \alpha_l^l \lambda_i^l I$ can be transformed to

$$\eta_l^l k^l + \eta_{l-1}^l k^{l-1} + \cdots + \eta_0^l$$

with

$$\eta_{l-j}^l = (-1)^l \begin{pmatrix} l \\ j \end{pmatrix} \left[a_0^j A_c^l + a_1^j \lambda_i A_c^{l-1} + \cdots + a_{l-1}^j \lambda_i^{l-1} A_c^l + (-1)^l \lambda_i^l I\right], \; j = 0, 1, \cdots, l;$$

where

$$\begin{aligned}
a_0^j &= \begin{pmatrix} l+1 \\ 1 \end{pmatrix} - 2^j \begin{pmatrix} l+1 \\ 2 \end{pmatrix} + 3^j \begin{pmatrix} l+1 \\ 3 \end{pmatrix} + \cdots + (-1)^l(l+1)^j \\
a_1^j &= -\begin{pmatrix} l+1 \\ 2 \end{pmatrix} + 2^j \begin{pmatrix} l+1 \\ 3 \end{pmatrix} - 3^j \begin{pmatrix} l+1 \\ 4 \end{pmatrix} + \cdots + (-1)^l l^j \\
&\vdots \\
a_{l-1}^j &= (-1)^{l-1} \begin{pmatrix} l+1 \\ l \end{pmatrix} + (-1)^l 2^j
\end{aligned}$$

giving,

$$\begin{aligned}
\lim_{k\to\infty} s(k) &= \sum_{i=1}^{\bar{n}} \sum_{l=0}^{m_i-1} (-1)^l C(\eta_0^l + \eta_1^l k + \cdots + \eta_{l-1}^l k^{l-1} + \eta_l^l k^l)(\lambda_i I - A_c)^{-(l+1)} B\beta_l^i \lambda_i^k \\
&= \sum_{i=1}^{\bar{n}} \Big[C\eta_0^0(\lambda_i I - A_c)^{-1} B\beta_0^i - C(\eta_0^1 + \eta_1^1 k)(\lambda_i I - A_c)^{-2} B\beta_1^i + \cdots +
\end{aligned}$$

$$(-1)^{m_i-1}C(\eta_0^{m_i-1}+\eta_1^{m_i-1}k+\cdots+\eta_{m_i-1}^{m_i-1}k^{m_i-1})(\lambda_i I-A_c)^{-m_i}B\beta^i_{m_i-1}]$$

$$= \sum_{i=1}^{\bar n}\Bigg[\sum_{l=0}^{m_i-1}(-1)^l C\eta_0^l(\lambda_i I-A_c)^{-(l+1)}B\beta_l^i +$$

$$\left(\sum_{l=1}^{m_i-1}(-1)^l C\eta_1^l(\lambda_i I-A_c)^{-(l+1)}B\beta_l^i\right)k+\cdots+$$

$$\left(\sum_{l=m_i-2}^{m_i-1}(-1)^l C\eta_{m_i-2}^l(\lambda_i I-A_c)^{-(l+1)}B\beta_l^i\right)k^{m_i-2}+$$

$$(-1)^{m_i-1}C\eta_{m_i-1}^{m_i-1}(\lambda_i I-A_c)^{-m_i}B\beta^i_{m_i-1}k^{m_i-1}\Bigg].$$

Furthermore, η_{l-j}^l can be factorized as

$$\eta_{l-j}^l=\lambda_i\begin{pmatrix} l \\ j\end{pmatrix}V_j^i(A_c-\lambda_i I)^{l-j}$$

with $V_0^i=\lambda_i^{-1}I$, $V_1^i=I$, and

$$V_j^i = A_c^{j-1}+\gamma_1^j A_c^{j-2}\lambda_i+\cdots+\gamma_{j-2}^j A_c\lambda_i^{j-2}+\lambda_i^{j-1}I, \qquad j=2,3,\cdots,m_i-1$$

where γ's are from the following recursive formula:

$$\gamma_1^j = 1+(-1)^j\left[\begin{pmatrix} j+2 \\ 3\end{pmatrix}-2^j\begin{pmatrix} j+2 \\ 4\end{pmatrix}+\cdots+j^j(-1)^{j+1}\right]$$

$$\gamma_n^j = n\gamma_{n-1}^j-\begin{pmatrix} n \\ 2\end{pmatrix}\gamma_{n-2}^j+\cdots+(-1)^{n+1}+$$

$$(-1)^{j+n+1}\left[\begin{pmatrix} j+n+1 \\ n+2\end{pmatrix}-2^j\begin{pmatrix} j+n+1 \\ n+3\end{pmatrix}+\cdots+j^j(-1)^{j+1}\right], n>1$$

Hence,

$$\lim_{k\to\infty} s(k) = \sum_{i=1}^{\bar n}\Bigg[\sum_{l=0}^{m_i-1}(-1)^l\lambda_i C V_l^i(\lambda_i I-A_c)^{-(l+1)}B\beta_l^i+$$

$$\left(\sum_{l=1}^{m_i-1}(-1)^{l+1}\lambda_i\begin{pmatrix} l \\ 1\end{pmatrix}CV_{l-1}^i(\lambda_i I-A_c)^{-l}B\beta_l^i\right)k+\cdots+$$

$$\left(\sum_{l=m_i-2}^{m_i-1}(-1)^{l+m_i-2}\lambda_i\begin{pmatrix} l \\ m_i-2\end{pmatrix}CV_{l-m_i+2}^i(\lambda_i I-A_c)^{-(l-m_i+3)}B\beta_l^i\right)k^{m_i-2}+$$

$$C(\lambda_i I-A_c)^{-1}B\beta^i_{m_i-1}k^{m_i-1}\Bigg]\lambda_i^k$$

$$= \sum_{i=1}^{\bar n}(\delta_0^i+\delta_1^i k+\cdots+\delta_{m_i-1}^i k^{m_i-1})\lambda_i^k$$

with

$$\begin{aligned}
\delta_0^i &= C(\lambda_i I - A_c)^{-1} B\beta_0^i - \lambda_i C V_1^i (\lambda_i I - A_c)^{-2} B\beta_1^i + \cdots + \\
&\quad (-1)^{m_i - 1} \lambda_i C V_{m_i-1}^i (\lambda_i I - A_c)^{-m_i} B\beta_{m_i-1}^i \\
\delta_1^i &= C(\lambda_i I - A_c)^{-1} B\beta_1^i - \begin{pmatrix} 2 \\ 1 \end{pmatrix} \lambda_i C V_1^i (\lambda_i I - A_c)^{-2} B\beta_2^i + \cdots + \\
&\quad (-1)^{m_i} \begin{pmatrix} m_{i-1} \\ 1 \end{pmatrix} \lambda_i C V_{m_i-2}^i (\lambda_i I - A_c)^{-m_i+1} B\beta_{m_i-1}^i \\
&\vdots \\
\delta_{m_i-2}^i &= C(\lambda_i I - A_c)^{-1} B\beta_{m_i-2}^i - \begin{pmatrix} m_{i-1} \\ m_{i-2} \end{pmatrix} \lambda_i C V_1^i (\lambda_i I - A_c)^{-2} B\beta_{m_i-1}^i \\
\delta_{m_i-1}^i &= C(\lambda_i I - A_c)^{-1} B\beta_{m_i-1}^i
\end{aligned}$$

Similarly, we can derive

$$\lim_{k \to \infty} \mathrm{t}(k) = \sum_{i=1}^{\bar{n}} (\zeta_0^i + \zeta_1^i k + \cdots + \zeta_{m_i-1}^i k^{m_i-1}) \lambda_i^k$$

with

$$\begin{aligned}
\zeta_0^i &= C(\lambda_i I - A_c)^{-1} E c_0^i - \lambda_i C V_1^i (\lambda_i I - A_c)^{-2} E c_1^i + \cdots + \\
&\quad (-1)^{m_i - 1} \lambda_i C V_{m_i-1}^i (\lambda_i I - A_c)^{-m_i} E c_{m_i-1}^i \\
\zeta_1^i &= C(\lambda_i I - A_c)^{-1} E c_1^i - \begin{pmatrix} 2 \\ 1 \end{pmatrix} \lambda_i C V_1^i (\lambda_i I - A_c)^{-2} E c_2^i + \cdots + \\
&\quad (-1)^{m_i} \begin{pmatrix} m_{i-1} \\ 1 \end{pmatrix} \lambda_i C V_{m_i-2}^i (\lambda_i I - A_c)^{-m_i+1} E c_{m_i-1}^i \\
&\vdots \\
\zeta_{m_i-2}^i &= C(\lambda_i I - A_c)^{-1} E c_{m_i-2}^i - \begin{pmatrix} m_{i-1} \\ m_{i-2} \end{pmatrix} \lambda_i C V_1^i (\lambda_i I - A_c)^{-2} E c_{m_i-1}^i \\
\zeta_{m_i-1}^i &= C(\lambda_i I - A_c)^{-1} E c_{m_i-1}^i
\end{aligned}$$

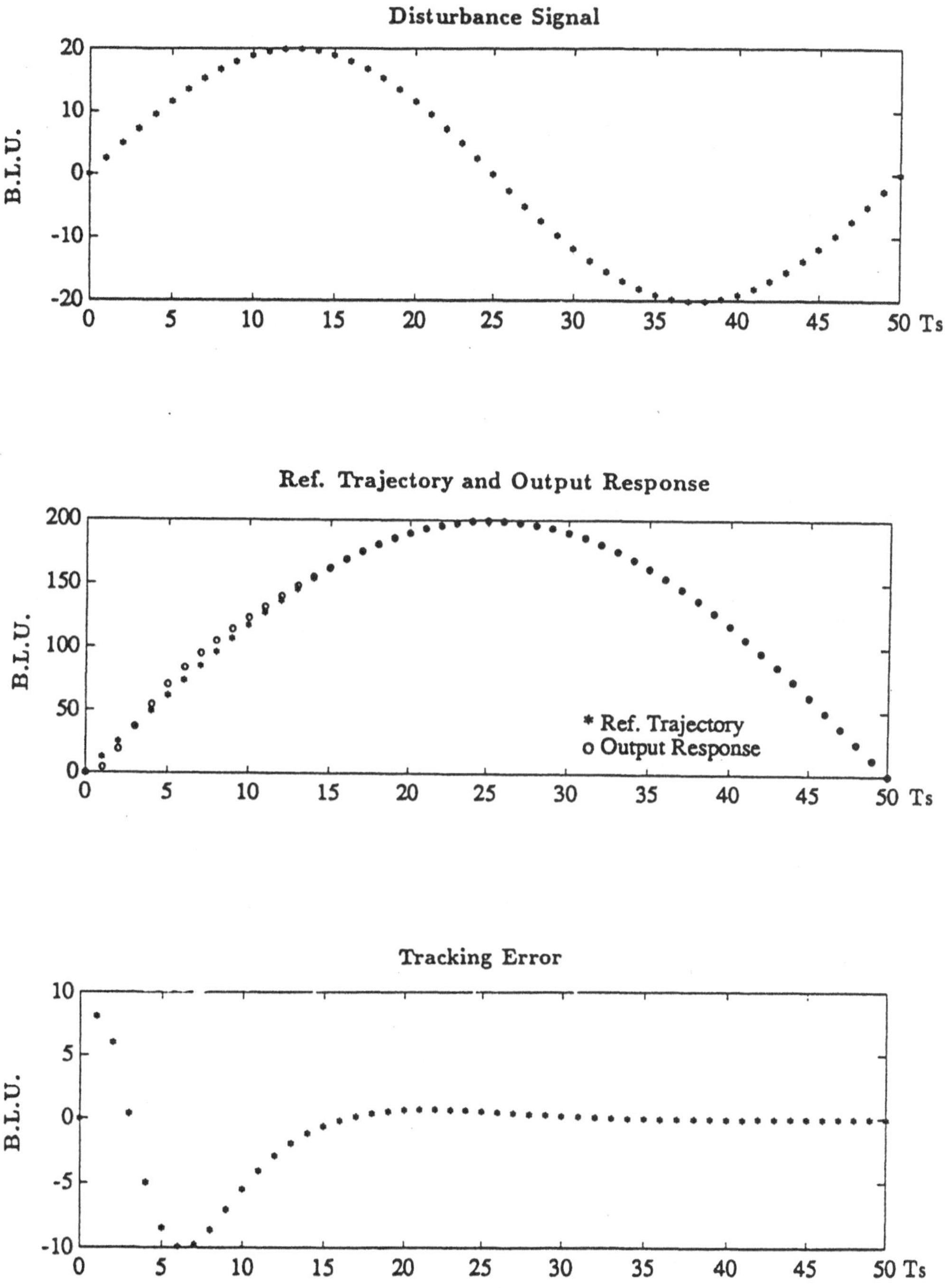

Figure 2a. Illustrative Example: Case 1 - Top Stage

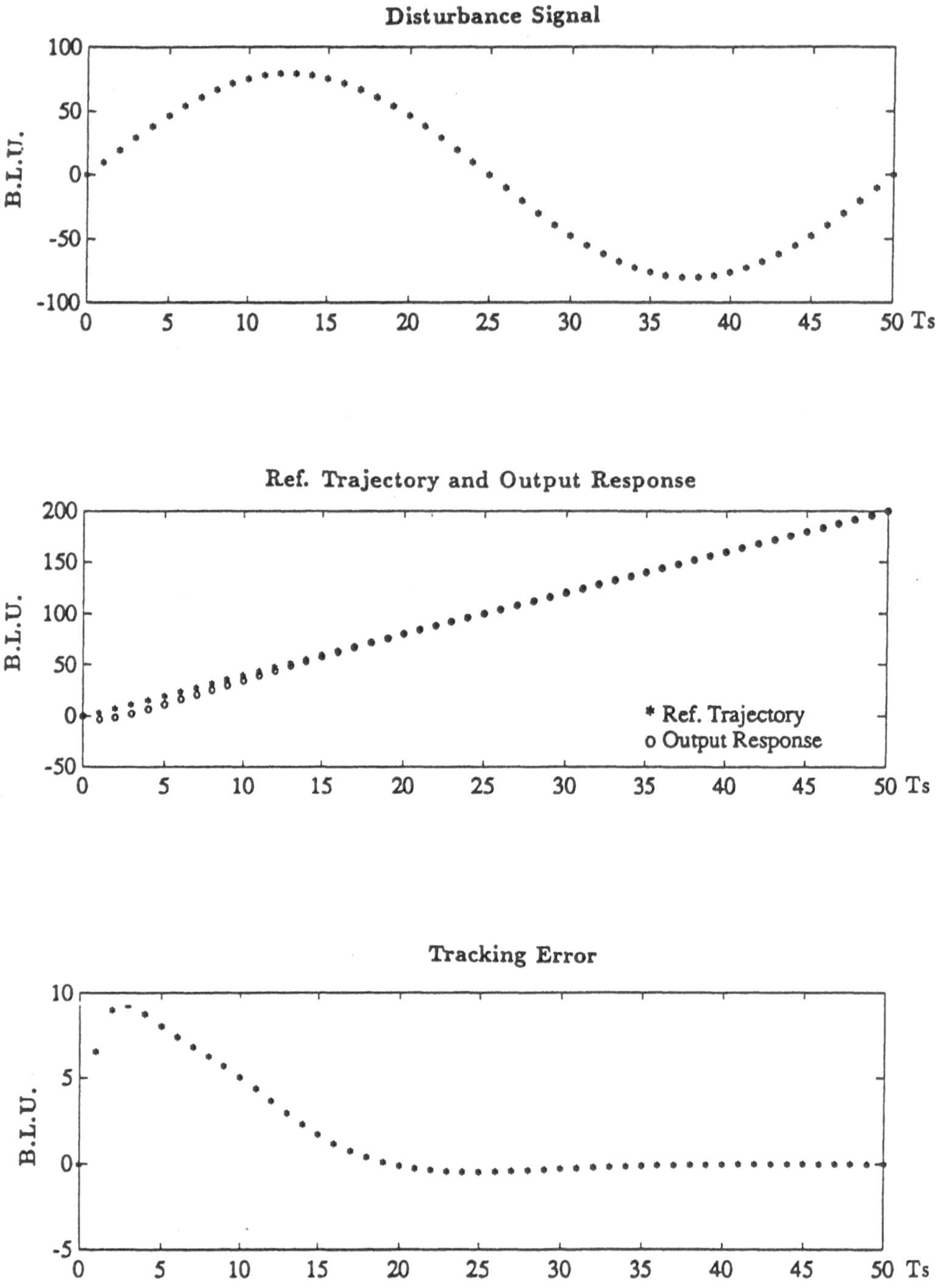

Figure 2b. Illustrative Example: Case 1 - Bottom Stage

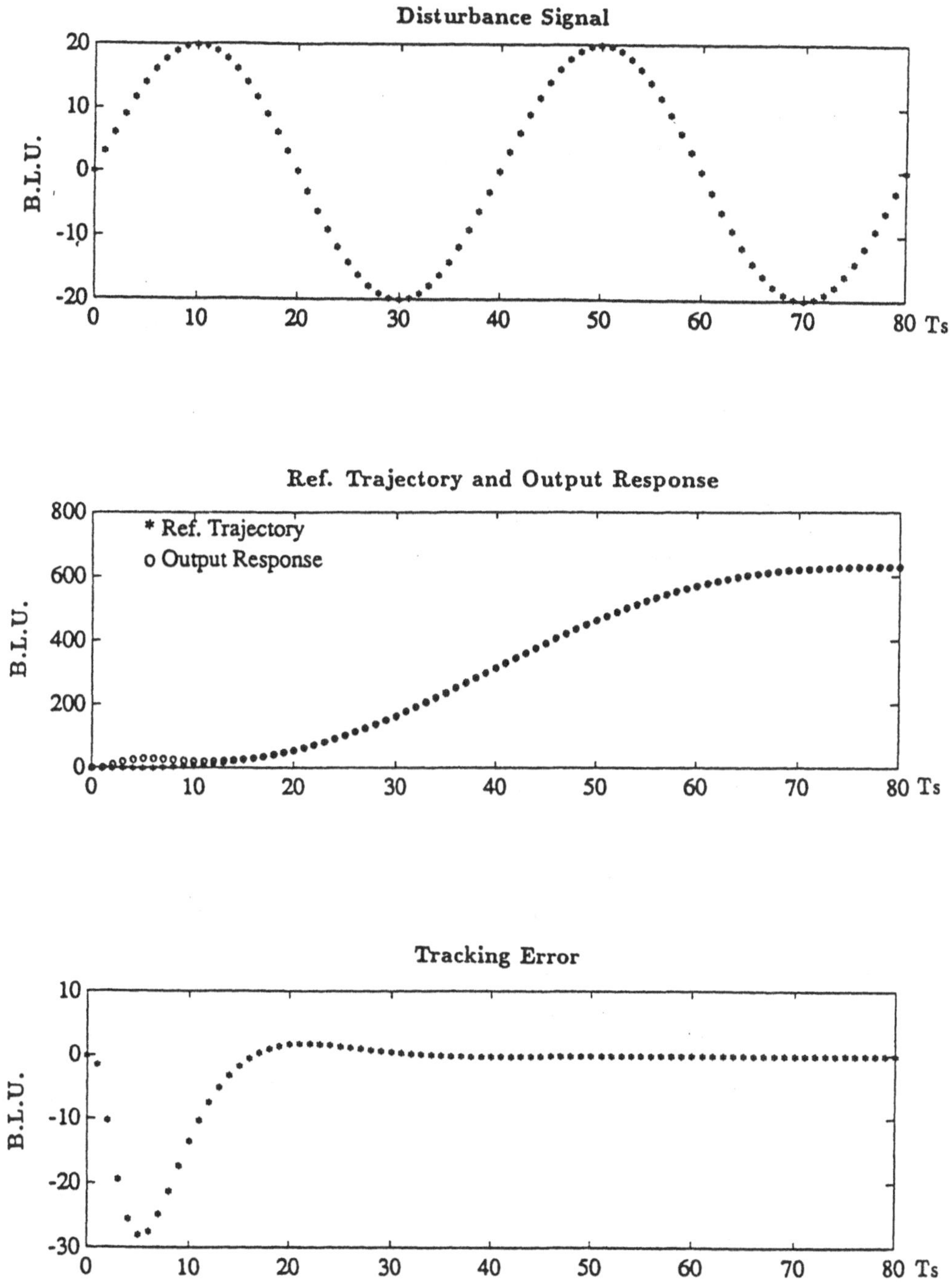

Figure 3a. Illustrative Example: Case 2 - Top Stage

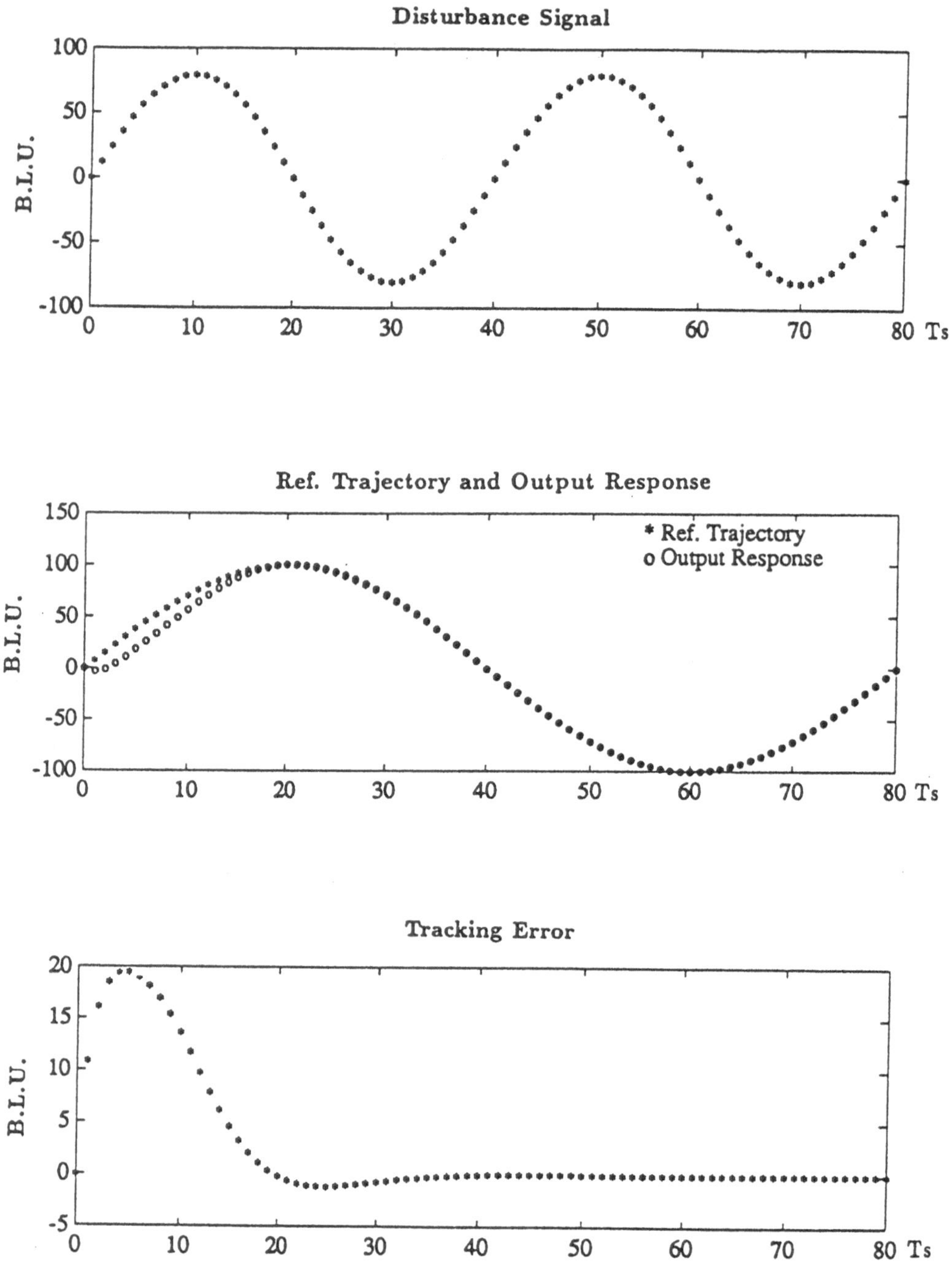

Figure 3b. Illustrative Example: Case 2 - Bottom Stage

DYNAMIC MODELING OF FLEXIBLE APPENDAGE AND SINGLE-LINK STRUCTURES WITH RIGID PAYLOAD

T. R. Parks*
Hughes Aircraft Company, Culver Cuty California
University of Southern California, Los Angeles California

H. A. Pak**
University of Southern California, Los Angeles California

ABSTRACT

The "exact" and pole/zero transfer functions are developed for a flexible beam-like single structural link with actuator on one end and payload on the other. This model represents a broader class of actuator hub and payload mass properties than is found in prior literature. The dependence of dynamics on the hub and payload are studied and graphs are provided to facilitate estimation of poles and zeroes for any similar plant. It is shown that a 10:1 reduction in fundamental frequency and substantial change in mode shape results from variations in payload through a practical range. Payload rotary inertia is shown to cause a qualitative change in mode shape resulting in loss of observability/controllability in some cases. Implications to plant and control design are discussed including sensor placement and accommodation of dynamic changes. Analytical results are compared to those measured on an experimental beam and show very good agreement in modal frequency and shape.

I INTRODUCTION

The topic of modeling and control of a flexible beam in bending is the focus of much current research. Early and ongoing work by Book[1], Cannon[2], and other researchers[3-6] has resulted in a great growth in understanding of the dynamic structure and the analytical and practical control issues relating to such systems. This effort is largely motivated by the need for lighter manipulator mechanisms which can provide precise end-effector tracking at increased speed and to accommodate heavier payloads. The impetus for this improved performance becomes even stronger in space-born appendages, and "arms" where time and weight are both very costly and where payloads are often relatively large[7]. It is easily seen that in attempting to provide lighter, faster mechanisms, structural frequencies become reduced while there is an increase in the high frequency content of the desired motion. This

---

* Section Head, / Doctorate Fellow
** AssistantProfessor of Mechanical Engineering

ultimately results in the excitation of structural oscillations and in the need for incorporating these dynamics into the plant model.

Control may be categorized as *colocated* (actuator and sensor at the same location) and *noncolocated* (sensor distant to the actuator). A noncolocated system can provide more precise tracking, improved disturbance rejection and increased bandwidth[4] at the sensor location (hence providing the above performance improvements), but is considerably more difficult to stabilize than a colocated system due to its nonminimum phase nature[4,9]. In implementing such control, an accurate plant model is essential so that confidence in analytical stability margins may be assured[6]. Among the various models found in the literature, none has fully accounted for the mass properties of a payload at the arm tip.

In this paper, the dynamics of a flexible link with a more general class of payload and actuator hub parameters is expressed in pole/zero transfer function form (Dynamic equations by modal expansion are derived for this system in Ref.(8)). In addition to providing insight into the relationship between payload and link dynamics, the model may be of use in understanding the behavior of multi-link systems where all other joints are locked (or are relatively slow moving) and where bending occurs primarily in the link under consideration. The effect of each parameter on the plant dynamics is studied and graphs are provided which allow the graphical determination of pole/zero frequencies of any beam-like plant sufficiently similar to this model. It is seen that the payload mass properties - in particular the rotary inertia - have a profound effect on the plant dynamics. This is manifested in not only an order of magnitude change in fundamental frequency, but also in great changes of mode shape. As a result, observability is lost for various sensor configurations in certain payload cases (In this paper, references to state space interpretations are made with respect to a modal coordinate realization. See Refs.(4) and (8)). Payload rotary inertia is seen to be solely responsible for the existence of imaginary zeroes in the hub torque to arm tip position transfer function and for a corresponding 180 degree phase shift. Existence of these zeroes has been unreported until recently[9]. In a related phenomenon, this inertia causes a disruption in the otherwise one-for-one increase in node count with mode number.

A discussion of the dynamic results provides guidelines for plant design. These include selection of rigid actuator hub radius and inertia and any fixed mass and rotary inertia at the arm tip. It is seen that the hub radius should generally be kept as large as possible, while inertia should be held to a minimum - this in order to mitigate certain observability problems as well as to maintain comparatively high plant pole frequencies. It is also shown that use of two sensors or less is generally inadequate for a high order noncolocated scheme with even modest payload changes. Further, it is recognized that such a scheme must allow for adjustability of the controller in order to provide acceptable performance and stability in operations involving substantial payload changes (e.g. "pick-and-place" maneuvers for a manipulator). Three control approaches are proposed. The paper concludes with a comparison between the analytically derived frequencies and mode shapes and those measured on an experimental beryllium copper beam. The results are seen to be in very close agreement for a variety of payload cases.

II DYNAMIC MODEL

The system used in this study is shown in Figure 1 where the hub and payload are assumed rigid. In developing the model, we employ Euler-Bernoulli beam theory for which the following assumptions are implicit [11]. Rotary motion, longitudinal motion, and shear strain of the beam fibers are negligible; beam material properties and cross section are symmetric with respect to the neutral bending axis; and structural damping is negligible. We further assume that the material properties and cross section do not depend on x. The system is described by the equation:

$$y^{(IV)}(x,t) + \frac{\rho A}{EI}\ddot{y}(x,t) = 0 \tag{1}$$

with boundary conditions:

$$J_h\ddot{y}(0,t) = R\,M(t) + REI[y''(0,t) - Ry'''(0,t)] \tag{2a}$$

$$y(0,t) = Ry'(0,t) \tag{2b}$$

$$J_p\ddot{y}'(L,t) + m_pL_{cg}\ddot{y}(L,t) = -EIy''(L,t)) \tag{2c}$$

$$m_p[\ddot{y}(L,t) + L_{cg}\ddot{y}'(L,t)] = EIy'''(L,t)) \tag{2d}$$

where : $\rho \triangleq$ mass density, $A \triangleq$ cross sectional area, $E \triangleq$ Young's modulus, $I \triangleq$ area moment of inertia ; are physical properties of the beam and J_p is the payload mass moment of inertia taken about the payload/beam interface.

The solution to the boundary value problem (Eq's(1,2)) will be expressed as an infinite product which is then truncated to provide a finite order approximation of the plant with exact transfer function poles and zeroes. (See also [3,6,10,12])

By applying separation of variables arguments and by taking the Laplace transform with respect to time, the solution to Eq.(1) has the form:

$$Y(x,\gamma) = C_1(\gamma)\sin\gamma x + C_2(\gamma)\cos\gamma x + C_3(\gamma)\sinh\gamma x + C_4(\gamma)\cosh\gamma x \tag{3}$$

where: $s = \pm i\sqrt{\frac{EI}{\rho A}}\gamma^2$ (4)

"s" is the transformed variable and $i = \sqrt{-1}$

We introduce the following normalized *boundary parameters* :

$$\mathcal{J}_h \triangleq \frac{J_h}{m_b L^2};\quad \mathcal{J}_p \triangleq \frac{J_p}{m_b L^2};\quad \mathcal{m}_p \triangleq \frac{m_p}{m_b};\quad \mathcal{R} \triangleq \frac{R}{L};\quad \mathcal{L}_{cg} \triangleq \frac{L_{cg}}{L} \tag{5}$$

where m_b is the flexible beam mass. These relate to *boundary factors* according to :

$$b_1 \triangleq \tfrac{1}{2}\frac{J_h}{\rho A}\gamma^3\ (= \tfrac{1}{2}\mathcal{J}_h\beta^3);\quad b_2 \triangleq \frac{J_p}{\rho A}\gamma^3\ (= \mathcal{J}_p\beta^3);\quad b_3 \triangleq \frac{m_p}{\rho A}\gamma\ (= \mathcal{m}_p\beta);$$

$$b_4 \triangleq R\gamma\ (= \mathcal{R}\,\beta);\quad b_5 \triangleq \frac{m_p L_{cg}}{\rho A}\gamma^2\ (= \mathcal{m}_p \mathcal{L}_{cg}\beta^2) \tag{6}$$

where $\beta = \gamma L$ (7)

We also define the *frequency normalization constant*, C_ω, and *normalized distance*, $\tilde{x}$, as :

$$C_\omega = \sqrt{\frac{EI}{\rho A L^4}} \tag{8}$$

$$\tilde{x} = \frac{x}{L} \tag{9}$$

Transforming the boundary conditions to the β domain results in :

$$2b_1\beta^3 Y(0,\beta) = \frac{-M(\beta)}{EI} + b_4[b_4 Y'''(0,\beta) - \beta Y''(0,\beta)] \tag{10a}$$

$$Y(0,\beta) = b_4 Y'(0,\beta) \tag{10b}$$

$$b_2\beta Y'(1,\beta) + b_5\beta^2 Y(1,\beta) = Y''(1,\beta) \tag{10c}$$

$$b_3\beta^3 Y(1,\beta) + b_5\beta^2 Y'(1,\beta) = -Y'''(1,\beta) \tag{10d}$$

$$s = \pm i C_\omega \beta^2 \tag{11}$$

Here " ' " denotes partial differentiation with respect to $\tilde{x}$ and $M(\beta)$ is the Laplace transform of $M(t)$ mapped into the fourth order β domain according to Eq.(11). Substituting Eq.(3) into each of Eqs.(10) and using Eq.(7) one can solve for the constants C_i . After considerable algebraic manipulation and simplification we obtain :

$$\frac{Y(\beta,\tilde{x})}{M(\beta)} = \frac{L^2}{2EI}\frac{N(\beta,\tilde{x})}{D(\beta)} \tag{12}$$

where :

$$N(\beta,\tilde{x}) = N_1(\beta)\sin(\beta\tilde{x}) + N_2(\beta)\cos(\beta\tilde{x}) + N_3(\beta)\sinh(\beta\tilde{x}) + N_4(\beta)\cosh(\beta\tilde{x}) \tag{13}$$

$$N_1(\beta) = (b_2b_3+2b_2b_4-b_5^2+2b_5-1)S\,Sh + (b_2b_3-2b_3b_4-b_5^2-2b_5-1)C\,Ch$$
$$+ (b_2b_3b_4-b_4b_5^2-2b_4b_5+2b_2-b_4)S\,Ch + (b_2b_3b_4-b_4b_5^2-2b_4b_5-2b_3-b_4)CSh - b_2b_3+b_5^2-1 \quad (14a)$$

$$N_2(\beta) = (-b_2b_3b_4+b_4b_5^2+2b_4b_5+2b_3+b_4)S\,Sh + (b_2b_3b_4-b_4b_5^2+2b_4b_5+2b_2-b_4)C\,Ch$$
$$+ (-b_2b_3+2b_3b_4+b_5^2+2b_5+1)S\,Ch + (b_2b_3+2b_2b_4-b_5^2+2b_5-1)CSh - b_2b_3b_4+b_4b_5^2-b_4 \quad (14b)$$

$$N_3(\beta) = (-b_2b_3-2b_2b_4+b_5^2+2b_5+1)S\,Sh + (b_2b_3+2b_3b_4-b_5^2+2b_5-1)C\,Ch$$
$$+ (-b_2b_3b_4+b_4b_5^2-2b_4b_5+2b_3+b_4)S\,Ch + (-b_2b_3b_4+b_4b_5^2+2b_4b_5+2b_2+b_4)CSh - b_2b_3+b_5^2-1 \quad (14c)$$

$$N_4(\beta) = (b_2b_3b_4-b_4b_5^2+2b_4b_5-2b_3-b_4)S\,Sh + (b_2b_3b_4-b_4b_5^2-2b_4b_5-2b_2-b_4)C\,Ch$$
$$+ (b_2b_3+2b_2b_4-b_5^2-2b_5-1)S\,Ch + (-b_2b_3-2b_3b_4+b_5^2-2b_5+1)CSh - b_2b_3b_4+b_4b_5^2-b_4 \quad (14d)$$

$$D(\beta) = 2(-b_2b_3b_4-b_2b_4^2+b_4b_5^2-2b_1b_5+b_3+b_4)S\,Sh + 2(-b_1b_2b_3+b_3b_4^2+b_1b_5^2+2b_4b_5+b_1+b_2)C\,Ch$$
$$+ (-b_2b_3b_4^2+b_4^2b_5^2-2b_1b_2+2b_3b_4+b_4^2+2b_5)(S\,Ch+S\,Ch) + (2b_5b_4^2+2b_1b_3+2b_2b_4+b_2b_3-b_5^2-1)$$
$$(CSh-S\,Ch) + 2b_1(b_2b_3-b_5^2+1) \quad (15)$$

and: $S \triangleq \sin\beta$; $C \triangleq \cos\beta$; $Sh \triangleq \sinh\beta$; $Ch \triangleq \cosh\beta$;

Note that $N(\beta,x)$ and $D(\beta)$ each possess only even powers in β. We may then transform Eq.(12) to the S-domain by taking the inverse of Eq.(4) along the primary value according to:

$$\beta = \frac{1+i}{\sqrt{2}}\sqrt{\frac{s}{C_\infty}} \quad (16)$$

The "exact" transfer function between torque at the hub and displacement at any specified location, $0 \leq \tilde{x} \leq 1$, is given by Eqs.(12) and (16). The frequency response, for example, is obtained by simply substituting $s=i\omega$ and evaluating through the desired frequency range. Further, the transfer function of an arbitrary order spatial derivative (e.g. beam slope or curvature) is readily found by performing the appropriate differentiation of Eq.(12).

In order to provide a finite order approximation of the given system useful for control analysis, we expand the transcendental transfer function into an infinite product whose convergence is assured by the Mittag-Leffler theorem [12,13].

The roots of the exact numerator and denominator expressions are usually (see Section IV for exceptions) either purely *imaginary-opposite* (imaginary pairs symmetric about the real axis) or purely *real-opposite* (real pairs symmetric about the imaginary axis) in the S-plane [6]. Each root type maps into the β plane (by Eq.(16)) along the four lines:

$$\beta = \alpha, -\alpha, i\alpha, -i\alpha; \qquad \alpha \in \Re \quad (17)$$

for imaginary-opposite, S-plane roots and

$$\beta = (1+i)\alpha,\ (1-i)\alpha,\ (-1+i)\alpha,\ (-1-i)\alpha\,; \qquad \alpha \in \Re \tag{18}$$

for real-opposite S-plane roots. Since each of these families has its roots symmetrically spaced about the origin, it suffices to solve for the roots of the appropriate expression along the primary branch only.

The product expansions of the numerator and denominator of Eq.(12), denoted here by F(β), are of the form:

$$F(\beta) = a\beta^{m_o}\prod_{i=1}^{\infty}\left(1-\left(\frac{\beta}{\beta_i}\right)^4\right) \tag{19}$$

where: $$a = \frac{1}{m_o!}\frac{\partial^{m_o}F(0)}{\partial \beta^{m_o}} \tag{20}$$

β_i is the ith root of the desired expression taken along the primary branch of Eq.(17) or (18).
'm_o' is the multiplicity of the root at the origin.

The roots of D(β) belong to the family of Eq.(17) and depend on all boundary parameters J_h, R, J_p, m_p, and L_{cg}

The numerator is a function of x as well as β. For control purposes the value x (or $\tilde{x}$) is viewed as the sensor location. For a position sensor at the hub, Eq.(13) is evaluated at $\tilde{x}=0$ resulting in :

$$N(\beta,0) = 2R\beta\,[2b_5 S\,Sh + (b_2b_3-b_5^2-1)C\,Ch + (b_2+b_3)S\,Ch + (b_2-b_3)C\,Sh - b_2b_3 + b_5^2 - 1] \tag{21}$$

This expression depends only on m_p, J_p, and L_{cg}. For the special case when R=0, the numerator is identically zero. It is, however, the roots *in* β as $\tilde{x}$ becomes arbitrarily close to zero that are of interest. Power series expansion of the numerator in $\beta\tilde{x}$ shows that for arbitrarily small $\tilde{x}$, these roots approach those of Eq.(21). (Equivalently, the roots of the beam slope expression at $\tilde{x}=0$ are those of Eq.(21) which is useful to observe since the colocated sensor would typically measure hub angle). This result is to be expected since the zeroes at $\tilde{x}=0$ correspond to frequencies at which the hub is motionless and therefore are independent of the hub radius.

For a position sensor at the beam tip, the numerator reduces to:

$$N(\beta,1) = 2[(b_2b_4-b_5-1)S-(b_4b_5+b_2+b_4)C+(b_2b_4+b_5-1)Sh\ (b_4b_5+b_2-b_4)Ch] \quad (22)$$

Here the zeroes depend only on R, J_p, and L_{cg}.

For a sensor at any arbitrary location, $(0 \leq \tilde{x} \leq 1)$, the roots of the numerator are those of $N(\beta,\tilde{x})$. Note that in all cases the numerator roots are independent of the hub inertia ($b_1=J_h\beta^3$). The set of numerator zeroes contains purely imaginary members at the hub and becomes populated with increasingly prevalent real-opposite S-plane roots with increasing $\tilde{x}$. At the tip ($\tilde{x}=1$), the set of zeroes contains a single pair of imaginary-opposite roots (these exist only for $J_p \neq 0$) with all other roots being real[1].

Applying Eqs.(19) and (20) to each of $D(\beta)$ and $N(\beta,\tilde{x})$ and transforming to the S-domain results in:

$$\frac{Y(s,\tilde{x})}{M(s)} = \frac{(R+\tilde{x})}{J_o s^2} \prod_{i=1}^{\infty} \frac{(1-(\frac{s}{z_i})^2)}{(1+(\frac{s}{p_i})^2)} \quad (23)$$

where the z_i's are the imaginary or real-opposite zeroes mapped from the roots of the appropriate numerator expression in β (e.g. Eq.(22)) via Eq.(11), the p_i's are the modulii of the imaginary-opposite poles mapped from the roots of Eq.(15), and J_o is the system polar moment of inertia about O

III SENSITIVITY TO BOUNDARY PARAMETER CHANGES

In this section, we characterize the behavior of the poles and zeroes of Eq.(23) as the boundary parameters J_h, J_p, m_p, & R vary. L_{cg} is taken equal to zero unless stated otherwise. For a more detailed account of the effect of all boundary parameters see Ref.(8). In the following discussion, the pole and zero values are in units of hertz and have been normalized by $C_\omega=1$. The frequencies given below may then be used to predict those of any similar system by using Eq.(8).

a) *Poles*

Figure 2 shows the relative effect of each parameter on the poles while all other parameters are held at zero value. It is clearly seen that changes in the payload and hub inertias have a much stronger effect than do changes in hub radius and payload mass both in terms of total change in the poles and in the rate with which they change. This rate of

---

[1]The complex zeroes discussed in Section IV appear to occur at positions near, but not at the beam ends.

change increases dramatically with root number. (e.g. in Fig 2b the second root changes much faster than the first for J_p<.02).

The pole frequency changes under the coupled effects of the boundary parameters is given in Figure 3. The pole values have been computed as a function of payload inertia since this parameter is shown to be generally of the greatest concern in beam tip motion control. As can be seen, frequencies again tend to monotonically decrease with the increase of any parameter value for a fixed set of the other parameters. Exceptions to this are that for sufficiently large J_h an increase in R causes an increase in frequency and for further increases in J_h (approaching a cantilevered base) the poles become insensitive to R. There are also precise sets of J_h, J_p and R (corresponding to a node at the beam tip - see part d, this section) which render an associated pole insensitive to changes in m_p. The effect of J_h and J_p is again much stronger than that of the other parameters and the general ranking of the severity of effect remains as in the uncoupled case of Fig.2.

b) *Zeroes*

Since the transfer function zeroes vary with position along the beam's length, a thorough characterization of the zeroes would require that the boundary parameter/frequency relationship be determined at appropriately closely spaced locations $\tilde{x}$. A technique for characterizing the zeroes continuously along the beam's length was developed in Refs.(6&9). Here we consider only the *colocated* (i.e. those at $\tilde{x}$=0) and *beam tip* ($\tilde{x}$=1) *zeroes.*

The colocated zeroes are independent of R and J_h and are purely imaginary. Figure 4 shows the parameter/zero relationship at this location. For the first (lowest frequency) zero, the root tends asymptotically to zero value as tip mass becomes infinite. This root is plotted for m_p=10 to indicate the rather slow approach to the limiting value. Comparison of the figure with the pole frequency plots shows that for a given m_p, the zero frequencies all lie in between those of the poles (see also Refs.(3,9)). The only exception to this alternation is that for an infinite hub inertia, the poles and zeros are identical. This results in a node at the hub and corresponds to a cantilevered configuration. A striking feature of Figure 4 is that for each root, there is a tip inertia value (this occurs at $J_p=\infty$ for the first root) for which the zero's frequency is independent of tip mass. This phenomenon has important implications to control system analysis as discussed later in this section.

We consider next the zeroes at the beam tip which are independent of m_p and J_h. The modulii of the real and imaginary zeroes is shown in Figure 5. The real zeroes show a moderate increase in frequency with tip inertia and a decrease with hub radius. The consequence of reduced modulii in the real zeroes is that additional gain will be obtained at frequencies beyond these modulii with phase being unaffected (since these are real-opposite pairs). As is shown in Refs.(4) and (6), however there is a time delay and beam tip motion reversal associated with these roots. These nonminimum phase effects are, of course, increased by any reduction in real root modulii. Also, as was discussed in Ref.4, the maximum achievable closed loop bandwidth increases with the positive real values of the nonminimum phase zeroes. Thus it may be possible to obtain an increase in the attainable bandwidth with increased payload inertia Note that the region in which the real zeroes undergo their most rapid transition (frequency vs. J_p) coincides with the traverse of the beam tip imaginary zero locus. Thus the increase in attainable bandwidth may be interpreted in terms of the 180° phase lead associated with this imaginary-opposite pair. The poles also experience their most rapid change in this region.

The imaginary zero pair is seen to have infinite frequency for zero tip inertia but rapidly transitions into the lower frequency range as inertia is increased. This zero pair shows high sensitivity to payload inertia at all plotted inertia values. For J_p values less than .05, this zero is nearly independent of hub radius. This zero pair shows high sensitivity to payload inertia at all plotted inertia values. Note that the reduction in frequency with inertia may make it impossible to realize the bandwidth increase suggested above.

c) *Simultaneous Tip / Hub Zeroes*

We now return to the observation that for certain tip inertia values, a given colocated zero is independent of tip mass. This suggests that at each such value and corresponding frequency there is no translation (there may be angular displacement) at the tip - i.e. there is an imaginary zero. In order for this physical explanation to hold, the loci of imaginary tip zeroes must all intersect these points of colocated zero m_p insensitivity (This in turn implies that the pair of imaginary tip zeroes is itself independent of hub radius at these values of frequency and inertia). As seen in Figure 6 this is indeed the case. The colocated loci approach and depart from the intersections in an "hourglass" fashion while the tip loci cross at these points

The insensitivities discussed above imply that the first partial derivative of the colocated numerator (Eq.(21)) with respect to m_p and that of the tip numerator (Eq.(22)) with respect to R are each equal to zero at the intersections. The simultaneous solution of these two expressions results in intersections at:

$$\beta^* = n\pi \ ; \qquad n = 0,1,2,\ldots \tag{24}$$

and

$$J_p^* = \frac{1}{(\beta^*)^3}\left(\frac{\sin\beta^* + \sinh\beta^*}{\cosh\beta^* - \cos\beta^*}\right) \tag{25}$$

Figure 7 shows the beam's "response shape" - that is the spatial boundary value solution - for the first four finite solutions. As can be seen, sensors located at the tip and hub would both have no output under these payload inertia values at the corresponding frequencies.

d) *Nodes*

A node exists when a zero at a given location $\tilde{x}$ occurs at a pole of the system. This is where the corresponding mode shape crosses the line of zero displacement. Such locations and pole/zero values deserve particularly close attention since they render the output at that location unobservable.

Of particular interest is the case of a node at the beam tip since design philosophy often directs placing a sensor at the location of objective control. These nodes also have some special properties. Recall that the beam tip zeroes are independent of m_p. Furthermore, the poles can be shown to be insensitive to changes in m_p according to:

$$\frac{\partial D}{\partial m_p} = \beta(2-2b_2b_4)S\,Sh + (2b_4^2-2b_1b_2)C\,Ch + (2b_4-b_2b_4)(S\,Ch+C\,Sh)$$
$$+ (2b_1+b_2)(C\,Sh-S\,Ch) = 0 \tag{26}$$

We also have that the beam tip zero expression, Eq.(22), must be satisfied. Simultaneous solution of these two equations yields a relationship between J_h, J_p, R and β for which the transfer function denominator is insensitive to changes in m_p at a tip zero frequency. Thus, <u>for a given R and J_h, specific values of J_p exist which will produce nodes at the beam tip for all values of m_p</u>. For the special case of R=0 the two equations become :

$$\beta^3 J_h(S+Sh-S^2Ch-C^2Sh) - (S\,Ch+C\,Sh)(Sh-S) = 0 \tag{27}$$

$$J_p = \frac{S + Sh}{\beta^3(Ch-C)} \tag{28}$$

In this case, specification of J_h alone, and solution of Eqs.(27,28) provide the necessary tip inertia(s) and resulting frequency(s) of the tip node condition.

By superimposing the imaginary zero loci of Figure 5 on the pole loci of Figure 3, a beam tip node may be predicted. Figure 8 shows such a case for parameter values of J_h=m_p=R =.10 and full range of m_p (The resemblance that this Figure has to that of Figure 6 is largely superficial.). As can be seen, for J_p values of .173, .0202, .00384, and .00118 - which correspond to normalized frequencies of 0.539, 2.06, 6.49 and 14.3 (Hz) - nodes exist in the 1st, 2nd, 3rd,and 4th modes respectively.

Figure 9 shows the behavior of the first four flexible mode shapes as J_p increases from .010 to .250 (m_p=.10). In Figure 9a, J_p is varied through the range for which the second pole undergoes its most rapid change. As seen in subfigure (a,i), the modes have in succeeding order, 1,2,2 and 3 nodes for $J_p = .010$. As tip inertia is increased to .020 the second mode's outer node moves to the beam tip corresponding to an end-point zero. Further increases in inertia cause this node to vanish leaving 1,1,2 and 3 nodes in the respective modes. Notice that the first, third, and fourth nodes change little during this transition. We also note that the higher frequency mode shapes become nearly cantilevered due to the presence of hub inertia.

Figure 9b shows the case where further increase in J_p causes the first mode to transition through an end-point zero resulting in a change in node count from 1,1,2,3 to 0,1,2,3. In this case only the first mode undergoes substantial change of shape. We now see that in both cases, only the mode which is near a tip-node condition (i.e whose modal frequency is nearest the beam tip imaginary zero) is highly sensitive to tip inertia change. By comparison of the pole/zero transfer function (23) with the more standard modal summation form (see Refs.(4,8,13)), it is seen that the zeroes along the beam length are also most sensitive in this frequency region (Note that mode shape relates closely to transfer function zeroes as seen by N(β,x) having the modal shape when evaluated at the modal β value). This has implications to sensor placement at locations other than at the beam hub and tip, i.e. the locations of pole-zero cancellation (nodes) also change greatly with J_p at these frequencies.

In general, at any beam location where an imaginary zero crosses a pole, a node will exist at the common frequency. These may be predicted by superimposing that location's zero-parameter loci on the plant pole loci.

e) *Transfer Function Sensitivity*

Employing the conventional definition of logarithmic sensitivity to the exact transfer function Eq.(12) we have for the case of beam tip inertia (explicit expressions are given in Ref.(8)) :

$$S_{J_p}^{TF} \triangleq \frac{\partial(TF)}{\partial J_p}\frac{J_p}{TF}$$

$$= J_p \frac{\frac{\partial N}{\partial J_p} D - \frac{\partial D}{\partial J_p} N}{N D} \tag{29}$$

where: TF ≜ Right Hand Side of Eq.(12)

By evaluating the resulting expression at real values of β we obtain, via the transformation Eq.(16), the frequency distribution of the sensitivity for a given value of J_p.

As is shown in Figure 10, the magnitude of this distribution has an *averaged* maximum in the frequency region of the beam tip imaginary zero. (These distributions were shown for $J_h = m_p = R = .10$, $J_p = .01, .10$ & 1.0 at beam locations of $\tilde{x}=1.0, .75$ & 0 in Ref.(8) with the result that the *averaged* maximum is in the beam tip imaginary zero frequency region for all cases)

The sensitivity of the transfer function to changes in tip mass is expressed similarly to Eq.(29). Here the parameter effect is less profound and less frequency selective than for the J_p case.

IV DISCUSSION of DYNAMIC RESULTS and CONTROL IMPLICATIONS

One general result of this study is that there appears to be no "special subspace" of boundary parameter values which makes the plant particularly insensitive to further parameter changes.

a) *Hub Parameter Design Considerations*

From the pole-vs-parameter results we conclude that J_h should generally be kept as low as practicable. This is necessary not only for the more obvious reason of maintaining pole frequencies as high as possible, but also and perhaps more importantly, for assurance of observability (discussed below). There will always be some amount of hub inertia associated with the motor rotor and other drive components. This may ,in some cases, prove beneficial in reducing sensitivity of the higher modal frequencies to further changes in J_h

For a design scenario where the distance from the center of hub rotation to the beam tip is specified, an increase in hub radius, R, would result in a one-for-one reduction in flexible beam length. By Eqs.(8,11) we have that (aside from the effect of changes in R as a plant parameter and the effects listed below) a given system frequency is then increased according to :

$$\omega = \omega_0\left(\frac{1}{1-R}\right)^2 \tag{30}$$

where ω_0 corresponds to a given frequency value when R=0. Also, since the beam properties, L, and m_b are reduced proportionally, the remaining boundary parameters are increased according to Eqs.(5). An additional increase in J_h, generally proportional to R^2, would also normally result as a physical consequence. It is shown in Ref.(8) that for a wide variety of parameter sets, these effects tend to offset one another so that neither the poles nor zeroes are highly sensitive to changes in R under this scenario.

b) *Payload Parameter Design Considerations*

Since the imaginary zero varies widely with L_{cg} through all frequencies, it is impossible to place a fixed inertia at the beam tip (e.g. end effector or other rigid member) sufficient to make the system insensitive to payload changes at low frequencies. It may in some cases be advantageous, however, to provide sufficient fixed inertia, say J_p=.04, so that sensitivity is limited to the lowest frequencies and a beam tip node is only possible for the first mode (this mode would generally be "well observed" by an angle sensor at the hub). Similarly, a small amount (unavoidable) of initial beam tip mass would provide a modest reduction in sensitivity.

c) *Special Role of Payload Inertia*

From previous observations, we have that with fixed hub parameters: 1)the beam tip node, 2) the maximum rate of mode shape change, 3) the fastest rate of pole frequency change, 4) the fastest rate of beam tip real zero change , 5) the fastest rate of colocated zero change, and 6) the greatest overall transfer function sensitivity ; take place at or around the beam tip imaginary zero frequency. We therefore conclude that for a given hub:

Sensitivity of plant dynamics to payload inertia changes is most extreme in the frequency region of the beam tip imaginary zero. This zero is itself dependent only on the payload inertia, J_p

(Recall that sensitivity to J_p is much greater than to m_p) This observation relates to the rather intuitive result that for relatively high payload mass property values, all poles and zeroes will have undergone substantial frequency change with the lower frequency poles and zeroes being most sensitive to further changes in payload. For relatively small payload mass properties, the lower frequency poles and zeroes will be essentially unchanged (from their unloaded-beam values) with significant change and sensitivity to further change occurring in the higher frequency poles and zeroes.

d) *Sensor Placement Issues*

The tip-node condition results in a "loss of observability" with respect to output at that location. This does not necessarily mean a loss of *system* observability but rather that the beam tip sensor, which is relied upon for precise positioning and disturbance rejection, is unable to identify all states. In the case of simultaneous tip/hub zeroes we recognize that at these frequencies there will be no sensor output at the tip and hub locations. This will be the case regardless of inner loop control structure since the numerator of the final closed loop transfer function will always contain these zeroes. The result for any estimator design which uses only these system outputs is that its performance would be no better than an open loop observer at the zero frequency. We further see that relatively small values of hub inertia lower the higher modal frequencies so that they approach those of a cantilevered plant. (e.g. in Fig. 3a, for $J_h > .01$ the third and higher roots are very nearly at the cantilevered frequency). Thus the simultaneous tip-hub zeroes (independent of hub inertia) nearly become simultaneous nodes at these locations. This results in an effective loss of system observability when only beam tip position and hub angle (or position) sensors are employed and strongly indicates the use of at least two noncolocated sensors. Placement of additional sensors should be determined in light of the results of section III d) regarding the change of node locations with payload. To this end, the observability figures of merit given in Ref.(16) may be evaluated along the beam length for a design range of payload properties.

e) *Implications of Payload Changes to Controller Design*

Aside from what may be implied in d) above, it would be premature at this point to rely on a state estimator to perform satisfactorily in the presence of even small payload changes since a fixed-payload plant model would generally deviate substantially from the in-service one. This in turn violates the basic assumption of the separation principal for state space regulator design i.e. that the estimator plant model is accurate.

From Figure 3 we see that the fundamental plant pole pair undergoes an order of magnitude change as payload mass properties vary through their practical range. The imaginary zeroes also undergo substantial change with payload. The most troublesome for a fixed controller design may be the beam-tip imaginary zero pair which changes from $\omega = \infty$ @ $J_p = 0$ to $\omega = 0.22$ @ $J_p = 1$ (R=0) and which bears an associated 180° phase shift (Consider, for example, a design whose stability relies on the 180° lead which may be suddenly lost as a result of payload changes). The difficulty associated with this zero is compounded by its high sensitivity to J_p for all payloads. Clearly, even modest payload variations may result in a substantial change of plant dynamics and it may be difficult for a noncolocated control scheme based on a fixed system model could provide adequate performance and stability under such conditions (It may be possible to find a fixed low gain controller which is stable for a variety of payload properties, but this would tend to perform "sluggishly" under the more massive of these payloads[17]).

It is unlikely that any direct form of model reference adaptive control (MRAC) may be applied to this system due to the positive realness requirements which would be difficult to satisfy due to the strongly nonminimum phase plant[18,19]. It appears then that any adaptive scheme must involve indirect techniques. Three suggested control approaches are outlined below. The first addresses operations where payload change occurs so rapidly (e.g. maniupulator "pick and place") that the plant changes are seen as rapidly time varying and are therefore unsuitable for on-line adaptation. This scheme may also be practical where repeated tasks are performed with similar payloads (re-identification is unnecessary) or where a single maneuver requires precise positioning/tracking. The second and third approaches are applicable to operations where payload is constant for at least moderate duration relative to the adaptation time. The second involves on-line identification of the plant parameters and control gain adjustment i.e. indirect adaptive control. This represents a considerable computational load for all but those lower (model) order plants which are slow moving. The third approach lessens the computational load by reducing the number of identified parameters as motivated by observations of the plant dynamic behavior.

1) Using established identification techniques (e.g.recursive least squares), plant parameters may be identified "near-line", i.e.immediately prior to path following operation for the payload-attached beam condition (presumably the unloaded beam control is established). Some autonomous gain selection (e.g.pole placement or LQG may then be employed. Thus the operational sequence involves object attachment, persistent excitation and identification, control gain selection, and operation by switching control gains as payload is removed and attached.
2) Assuming persistence of excitation exists, the plant is identified on-line with the parameter changes resulting from the sudden payload variation seen as errors in the identifier's initial conditions. The Bezout Identity (pole placement)[20] or Algebraic Riccati Equation (LQ)[21] are then solved on-line to adjust control gains.
3) Provided that moderately accurate a priori knowledge of payload mass properties exists, maneuvers may be initiated based on a gain scheduled controller. On-line identification proceeds but is limited to the pole(s) and zero(s) in the frequency region of payload variation sensitivity (see section IVc) above). Thus indirect adaptive control may be employed based on identification of a small subset of the most uncertain plant parameters

f) *Other Uses of Dynamic Modeling Results*

Finally, the observation - for a given beam hub, the beam tip imaginary zero, the simultaneous tip/hub zero, and the tip-node conditions,depend solely on J_p - may be of some general use in applied dynamics. Examples might involve inertia measurement, or any oscillating beam with ends supported by bearings whose loads are to be minimized at a particular frequency.

V EXPERIMENTAL VERIFICATION of RESULTS

A flexible beam has been constructed for the experimental research of this topic. The design is functionally similar to that used by Cannon and Schmitz[2,4]. The arm consists of two parallel beryllium copper side sheets separated by light weight yet relatively stiff aluminum cross members. These members provide high torsional stiffness and structural stability with respect to vertical bending thereby effectively restraining motion to the horizontal plane. The design - with cross section shown in Figure 11 - provides near zero friction and stiffness coupling at the side sheet / stiffener interface. This allows the arm to be modeled as a simple Euler-Bernoulli beam with stiffness provided by the side plates and stiffener mass distributed along the length. A payload fixture allows a wide variation of mass properties and provisions exist for changing parameters at the hub as well. Figure 12 is a photograph of the beam.

Preliminary experiments have been performed to characterize the system dynamics. The frequency normalization constant, C_ω, was first found by measuring the fundamental cantilevered frequency of the unloaded beam and by using Eq.(11) and solving $D(\beta)=0$.(Arbitrarily large J_p, all other parameters =0 (For this cantilevered case, the well known result is: $\beta_{C_1}=1.8751$). Theoretical predictions of modal frequency are then made by solving $D(\beta)=0$ for the specific case, and using the above C_ω (and Eq.(11)). The pinned modal frequencies were found as resonance values under a sinusoidal torque applied at the hub actuator. Cantilevered modal frequencies were found as the anti-resonances (The colocated pinned zeroes are the cantilevered poles). This approach is valid where adequate frequency separation from adjacent poles and zeroes exists. Resonances and anti-resonances were taken as frequencies corresponding to maximum and minimum amplitudes measured by a high resolution encoder under a constant amplitude sinusoidal current driven through the DC actuator. The agreement of mode shapes with those predicted by the model was evaluated in terms of node locations since these are accurately measurable. Node locations were found by driving the actuator at the previously measured modal frequency under a colocated PD control (thereby eliminating the rigid body drift). A noncontacting linear proximity sensor was then used to establish the locations of minimum deflection. In cases where a mode has no node, agreement was assessed from the ratio of amplitude at the beam tip to that at the midpoint. Also, in some modes a "quasi-node" (i.e. a point with no translation during modal oscillation) occurs in the payload. These were also measured and compared with those analytically predicted. (Note that mode shape relates closely to transfer function zeroes as seen by $N(\beta,x)$ having the modal shape when evaluated at the modal β value.)

Figures 13 and 14 compare the measured modal frequencies and node locations with those predicted for a representative payload case. As can be seen, both the frequencies and mode shapes agree well with the model. Several other intermediate payload cases have also been tested (not shown) with similar agreement (see Ref.[8]). The average frequency disagreement is 1.4% while worst case is 5.2%. The average node location disparity (normalized by beam length) is 0.4% with a worse case of 1.9%. Experiments were also done to determine the agreement between measured and theoretical theoretical beam tip zero frequencies. Measurement involved determining the PD controlled actuator frequency at which the beam tip had no perceptible translation. Four payload parameter sets corresponding to theoretical frequencies of 0.57, 1.75, 5.90, and 10.10 Hz were tested with a resulting average disagreement of 2.6% and maximum of 3.6%. This agreement is meaningful since several of the results above pertain explicitly to this imaginary zero.

These favorable results are an indication that the beam design objectives of conformity with the Euler-Bernoulli assumptions and of providing near ideal boundary conditions, were largely satisfied. They should not be interpreted as an indication that the model would adequately represent a less ideal system (Refs.[6,16]). The model, as verified experimentally, is in the least useful however in providing qualitative insight into dynamic behavior and control issues.

VI CONCLUSION

The dynamic equations given in this paper characterize a broader class of hub actuator and payload configurations than is found in prior literature. A sensitivity study showed that very large changes in plant dynamics may occur as a result of payload variations and that large changes in the higher order poles and zeroes occur upon the addition of relatively minute payload. The payload mass moment of inertia was seen to cause qualitative changes in the hub torque to beam tip position transfer function, including the occurrence of a purely imaginary zero pair whose frequency varies widely with this inertia. This qualitative change was also manifested in the mode shapes. Thus it was shown from both pole-zero and modal interpretations that system observability (for a modal coordinate realization) may be lost in the presence of payload variations if only colocated and beam tip sensors are used. The location of additional sensor(s) must also be judiciously selected in order to assure acceptable observation. It is concluded that in order to provide adequate performance and stability margins in the presence of even moderate payload changes, some provision for controller adjustment may be necessary. Three proposed control methods were outlined based on indirect adaptive control. The dynamic model developed herein was verified by test on an experimental single-link flexible beam where measured mode shapes and frequencies agreed closely with those predicted by the model.

REFERENCES

[1] Book, W. J.,Modeling,"Design and Control of Flexible Manipulator Arms", Ph.D. Thesis, MIT, Department of Mechanical Engineering, April 1974.

[2] Cannon, R. H., Jr., and Schmitz, E., "Initial experiments on the End-Point Control of a Flexible One-Link Robot" *The Journal of Robotics Research,* 3(3): pp.62-75, Fall 1984.

[3] Wie, B., "On The Modeling And Control Of Flexible Space Structures",Ph.D. Thesis, Stanford University, SUDAAR 525, June, 1981.

[4] Schmitz, E., "Experiments On The End-point Position Control Of A Very Flexible One-link Manipulator", Ph.D. Thesis, Stanford University, SUDAAR No. 548, June, 1985.

[5] Kane, T. R., Ryan, R. R., and Banerjee, A. K., "Dynamics of a Cantilevered Beam Attached to a Moving Base", *Journal of Guidance, Control, and Dynamics* ,Vol. 10,No.2, March-April 1987, pp.139-151

[6] Spector, V. A., and Flashner, H., Sensitivity of Structural Models for Noncollocated Control Systems, *Journal of Dynamic Systems, Measurement, and Control,* Vol. 111, Dec. 1989, pp 646-655

[7] Nguyen, P. K., Ravindran, R., Carr, R., and Gossain, D. M., Structural Flexibility Of The Shuttle Remote Manipulator System Mechanical Arm, *Proceedings Of The Guidance And Control Conference,* AIAA Paper No. 82-1536, August 198

[8] Parks, T. R. , Ph D Thesis, To Be Accepted, University of Southern California, Department of Mechanical Engineering, Fall, 1989

[9] Spector, V. A., Ph.D. Thesis, "Modeling of Flexible Systems for Control System Design", University of Southern California, Department of Mechanical Engineering, December 1988

[10] Graff, K. F., "Wave Motion In Elastic Solids", Ohio State University Press,1975.

[11] Goodson, R. E., "Distributed System Simulation Using Infinite Product Expansions", *Simulation,* December, 1970, pp. 255-263.

[12] Jeffreys, H. and Jeffreys, B.S., "Methods Of Mathematical Physics", 2nd Edition,Cambridge University Press, 1950.

[13] Mierovitch, L. M. "Analytical Methods In Vibrations", The Macmillan Co.,1967.

[14] Hughes, P. C., Space Structure Vibration Modes: "How Many Exist? Which Ones Are Important?", *Proceedings Of The Workshop On Applications Of Distributed System Theory To The Control Of Large Space Structures,* JPL Publication 83-46, July 1, 1983.

[15] Spector, V. A. and Flashner, H., "Flexible Manipulator Modeling For Control System Development", *AIAA Paper 87-2264,* August, 1987.

[16] Hughes, P.C., and Skelton, R.E., "Controllibility and Observability for Flexible Spacecraft", *Proceedings of the Second VPI&SU / AIAA Symposium on Dynamics and Control of Large Flexible Spacecraft,* Blacksburg, Va., June 1979.

[17] Menq, C., and Chen, J. "Payload Adaptation of a Flexible One-Link Manipulator," *Proceedings of the American Control Conference,* Atlanta, June, 1988, pp. 69-73

[18] Astrom, K.J., and Wittenmark, B., "Adaptive Control", Addison - Wesley Publishing Company, 1989

[19] Meldrum, D.R., and Balas, M.J., "Application of Model Reference Adaptive Control to a Flexible Remote Manipulator Arm", *Proceedings of the American Control Conference*, summer 1986, pp. 825-832

[20] Ioannou, P. and Sun, J. "The Theory and Design of Robust Direct and Indirect Adaptive Control Schemes", *International Journal of Control*, Vol. 47, 1988.

[21] Sun, J. and Ioannou, P., Robust Adaptive LQ Control Schemes" Accepted for publication, *Proceedings of the American Control Conference*, Pittsburgh, 1989

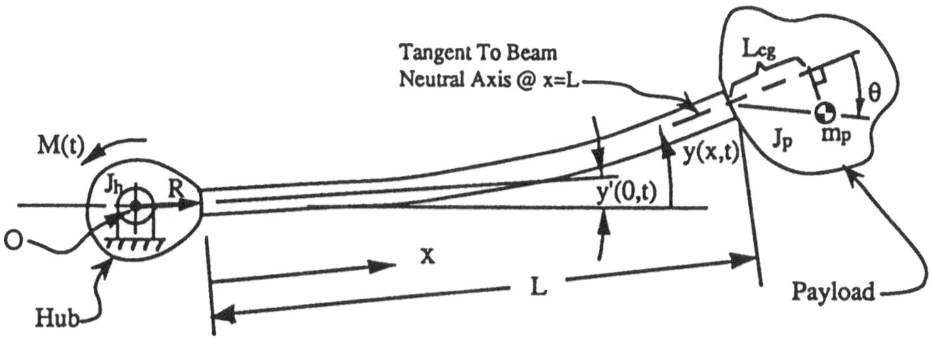

FIGURE 1. ONE LINK FLEXIBLE SYSTEM

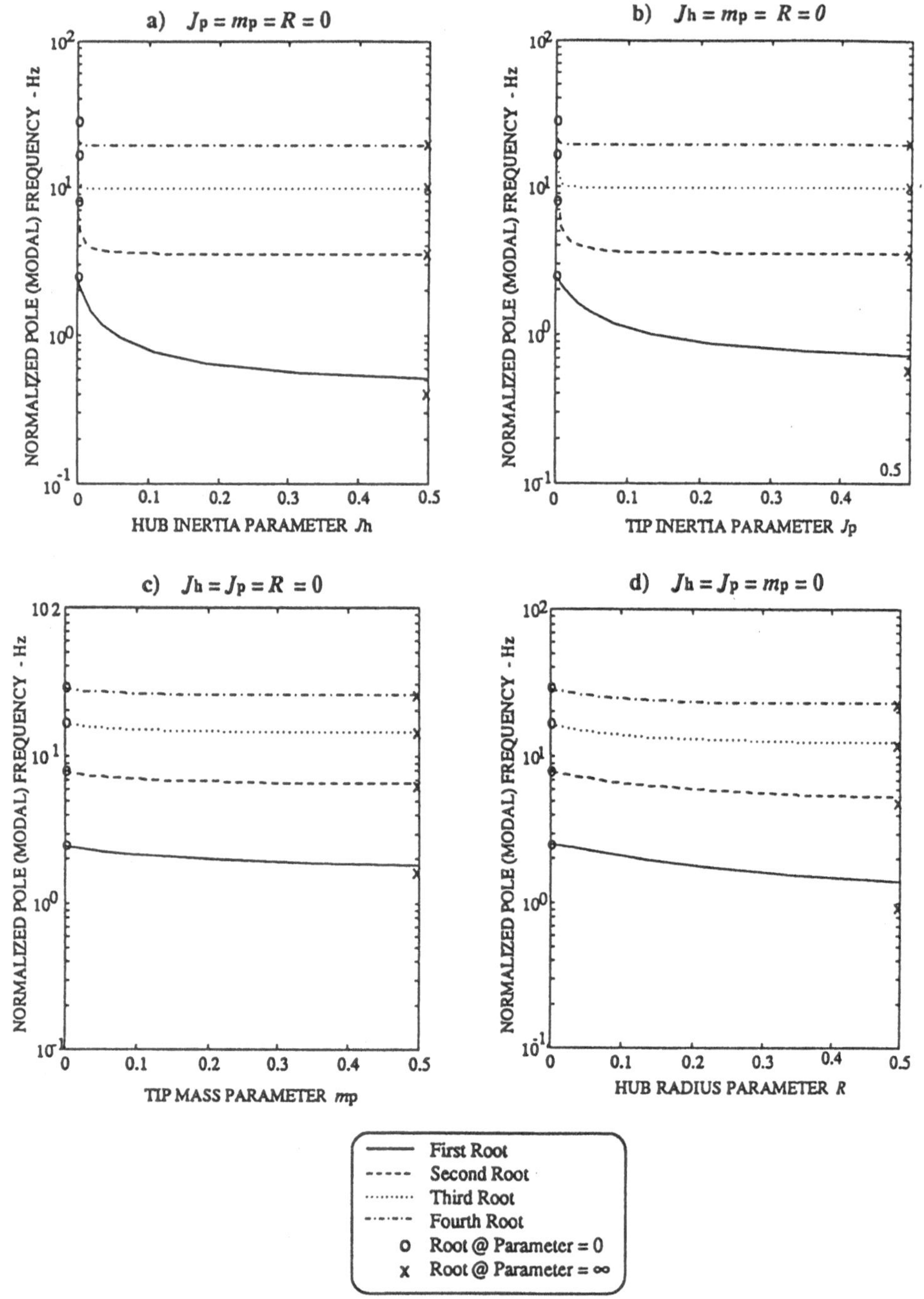

FIGURE 2. DEPENDENCE OF POLES ON BOUNDARY PARAMETERS - UNCOUPLED

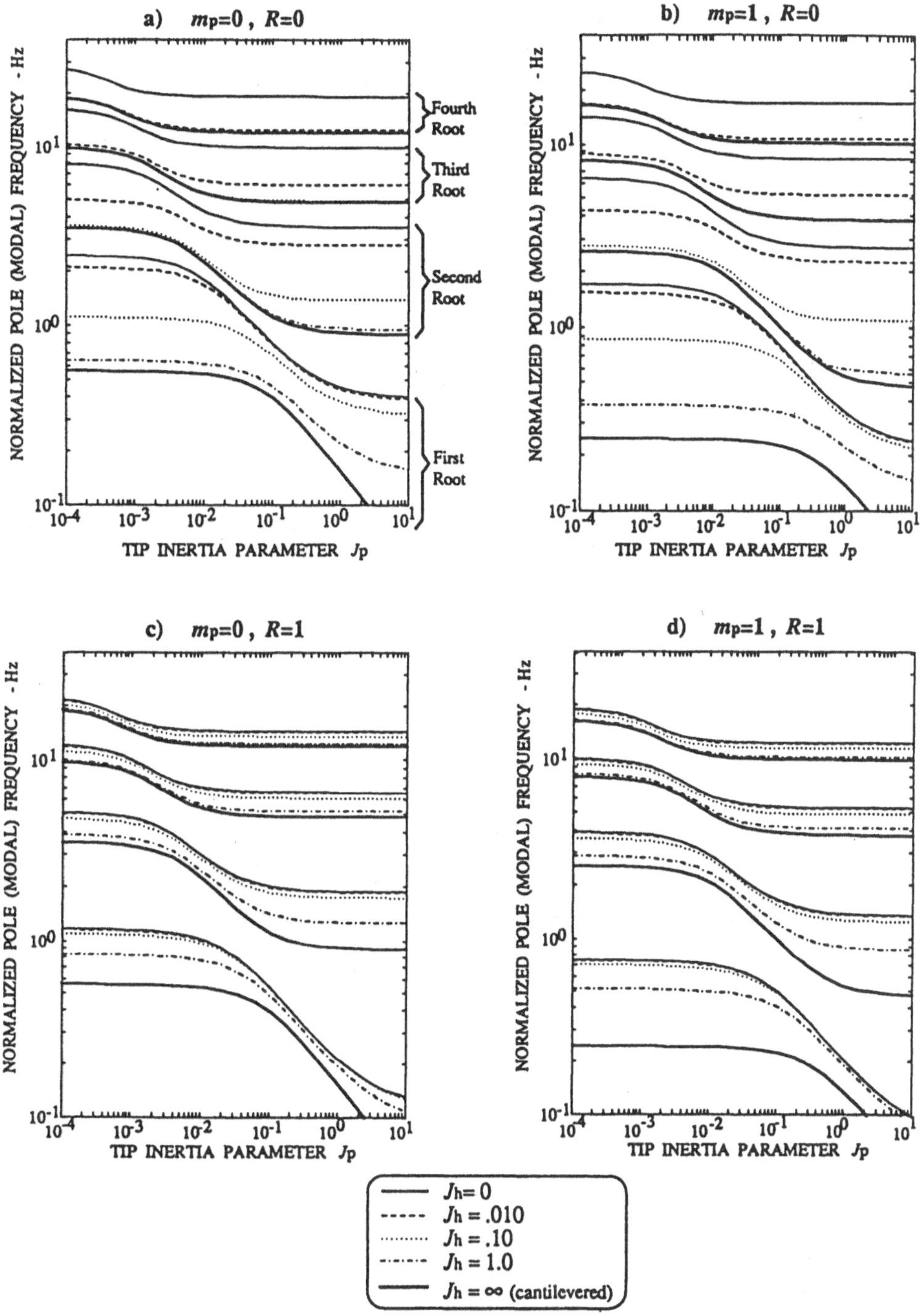

FIGURE 3. DEPENDENCE OF POLES ON BOUNDARY PARAMETERS

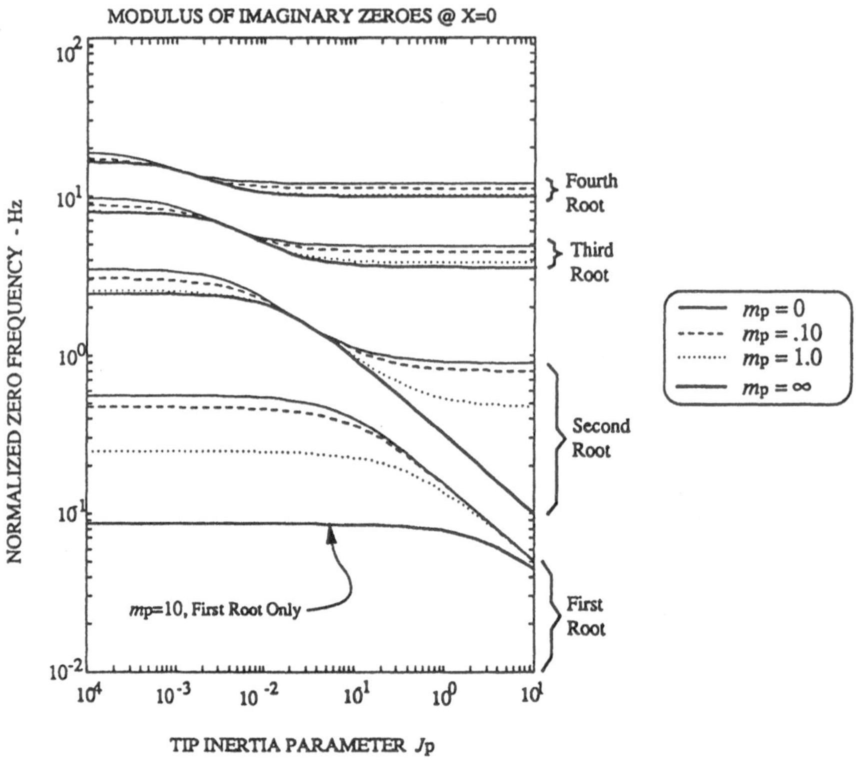

FIGURE 4. DEPENDENCE OF COLOCATED (x=0) ZEROES ON BOUNDARY PARAMETERS

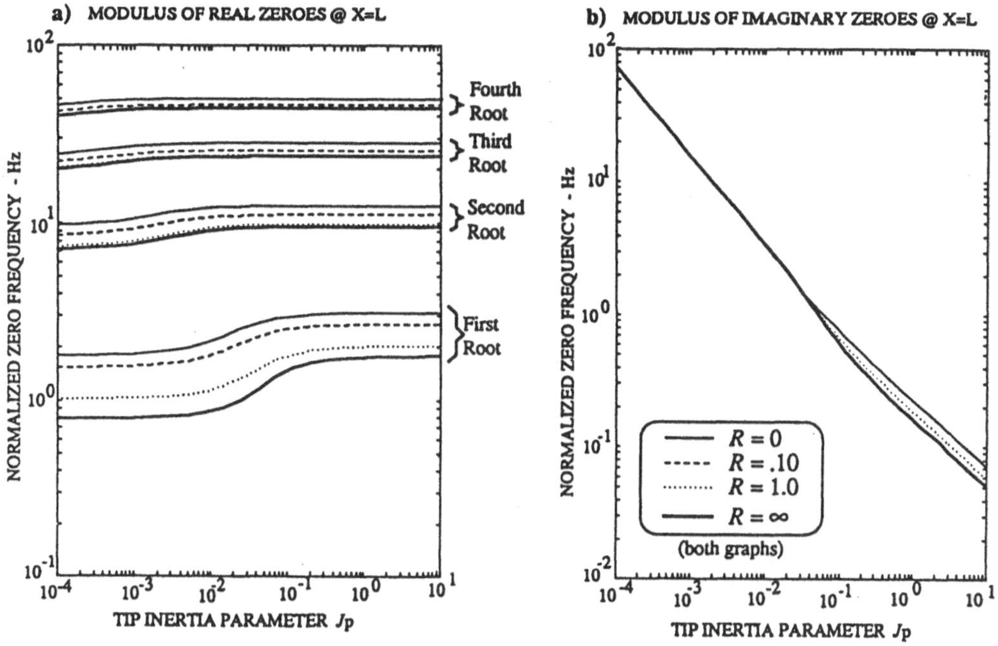

FIGURE 5. DEPENDENCE OF BEAM TIP ZEROES ON BOUNDARY PARAMETERS

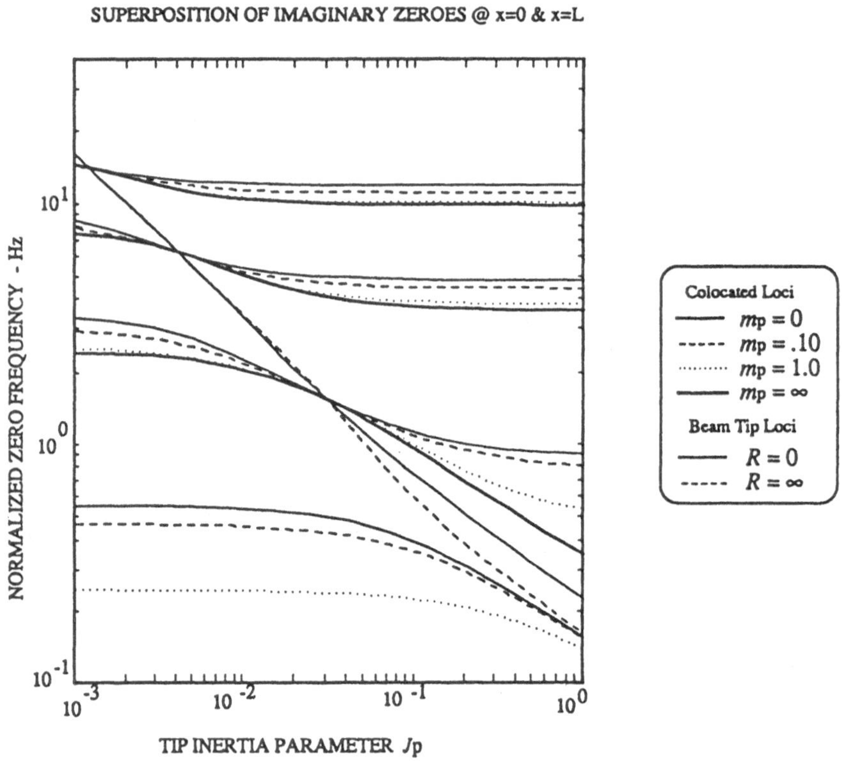

FIGURE 6 SIMULTANEOUS TIP/HUB ZEROES

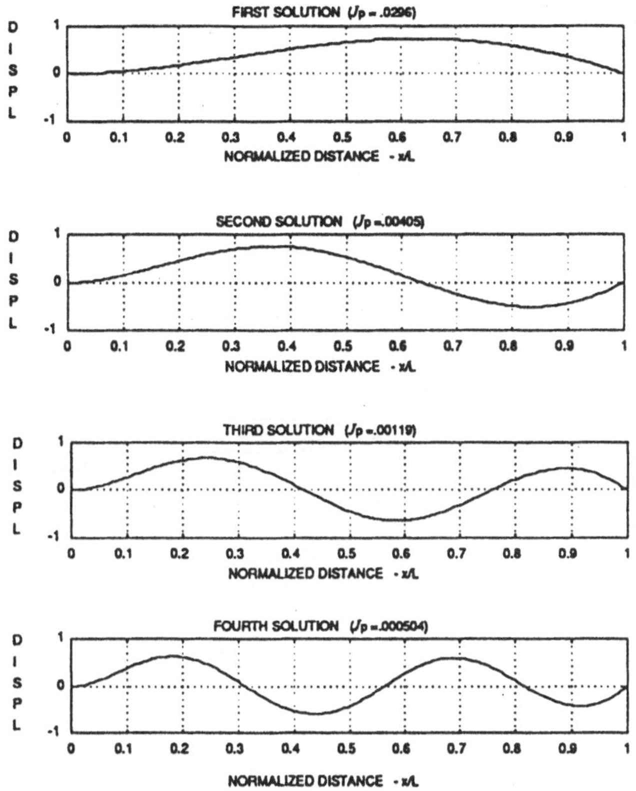

FIGURE 7. SIMULTANEOUS TIP/HUB ZERO RESPONSE SHAPE

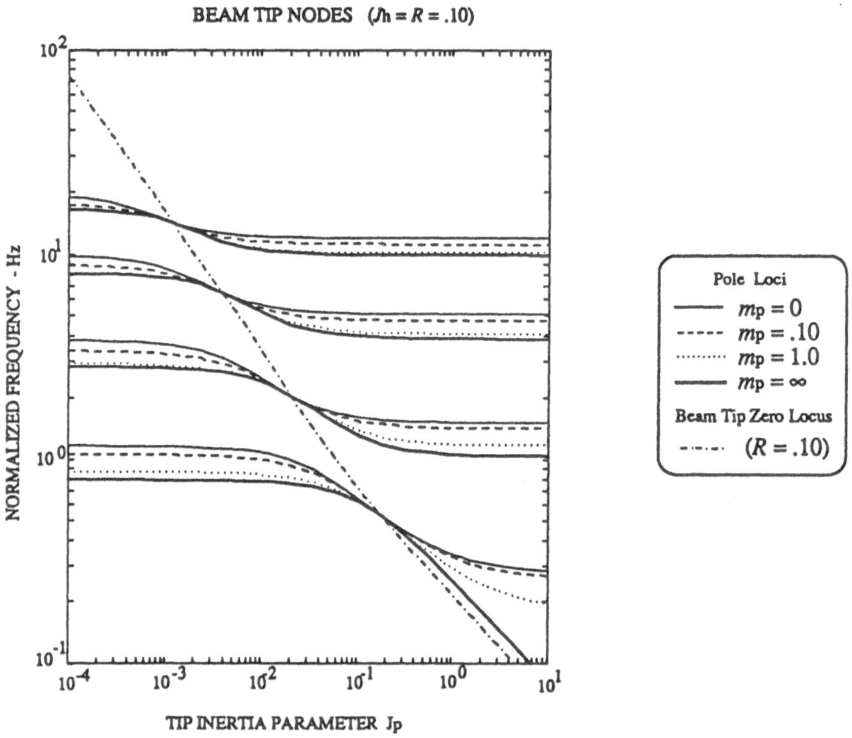

FIGURE 8. INTERSECTION OF POLES AND BEAM TIP IMAGINARY ZERO

$J_p = .010$

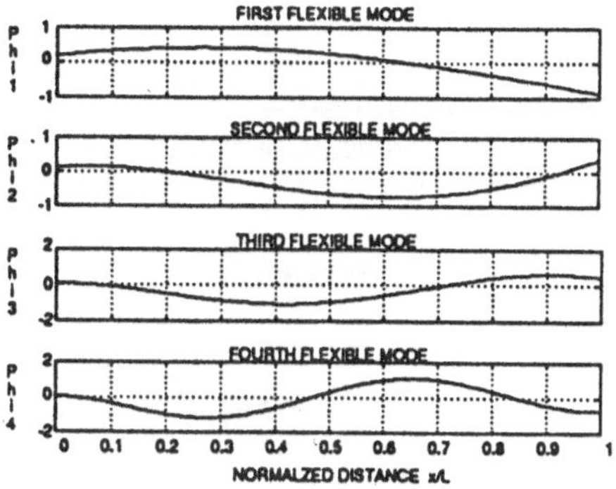

$J_p = .020$

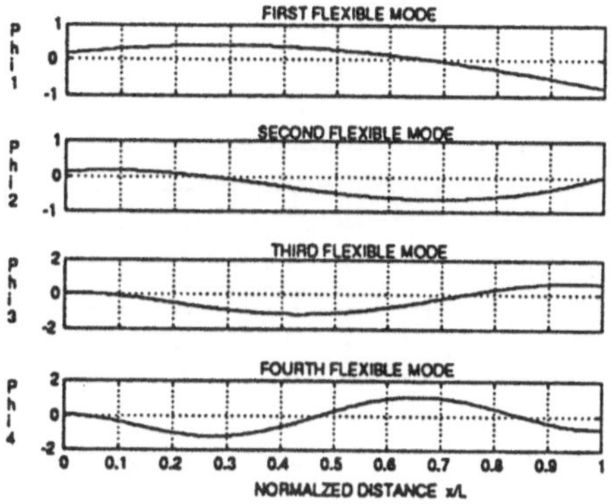

$J_p = .030$

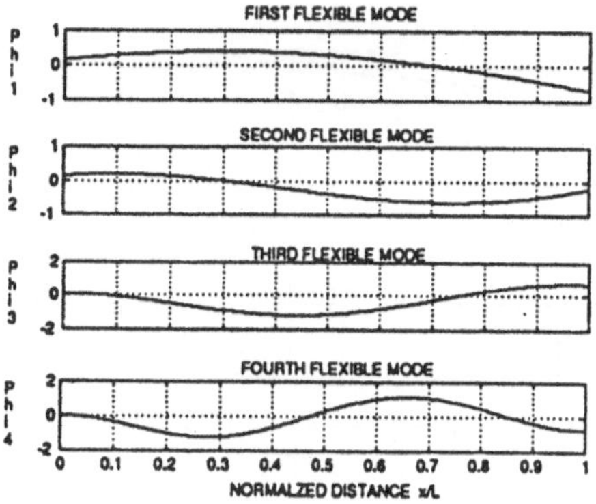

a) Transition Through Tip-Node In Second Mode

FIGURE 9. CHANGE IN MODESHAPES AS TIP INERTIA IS VARIED ($J_h = R = m_p = .10$)

$Jp = .10$

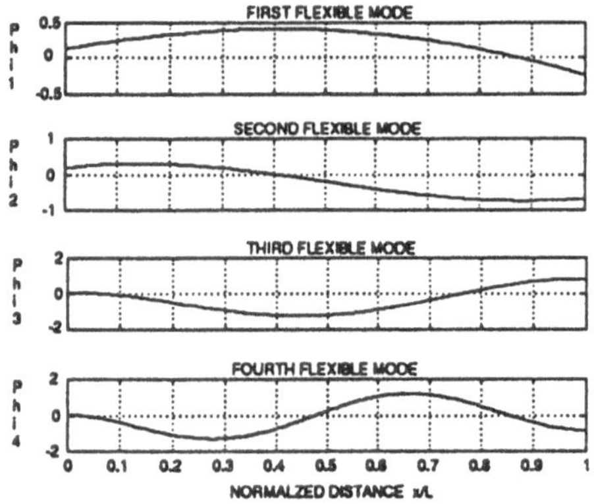

$Jp = .17$

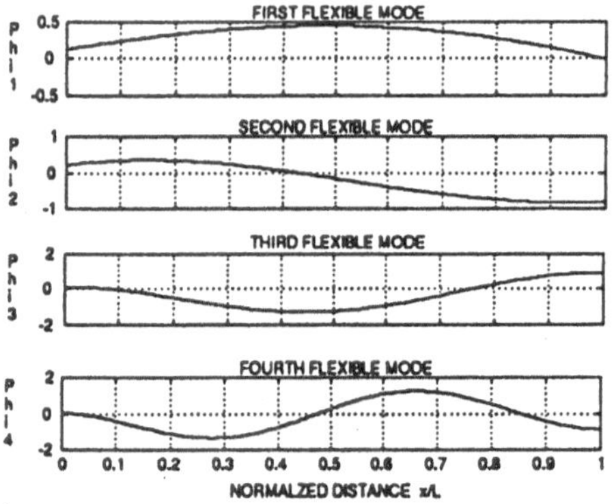

$Jp = .25$

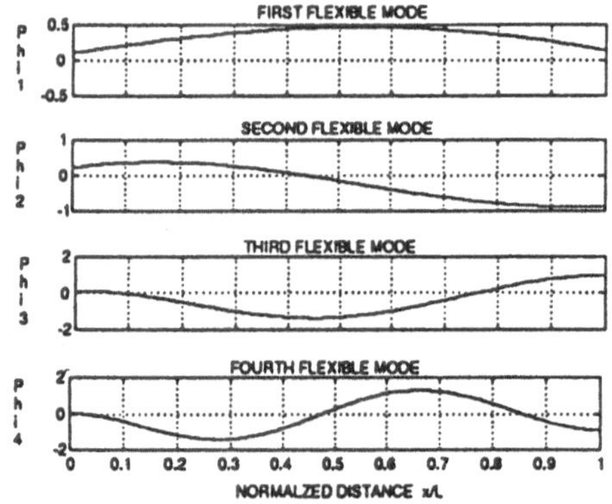

b) Transition Through Tip-Node In First Mode

FIGURE 9. CHANGE IN MODESHAPE AS TIP INERTIA IS VARIED ($Jh = R = mp = .10$)

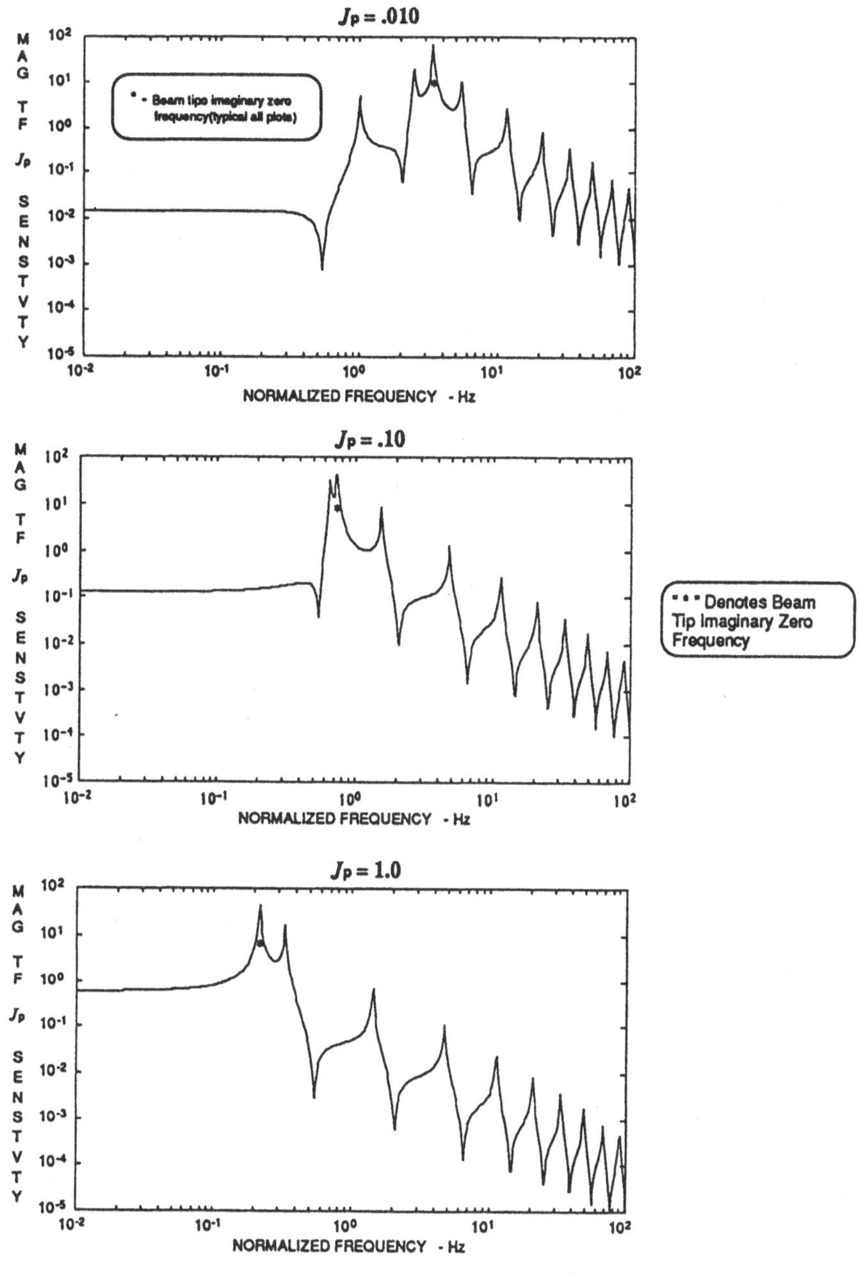

FIGURE 10. TRANSFER FUNCTION (X=L) SENSITIVITY TO PAYLOAD INERTIA ($J_h = m_p = R = 0.10$)

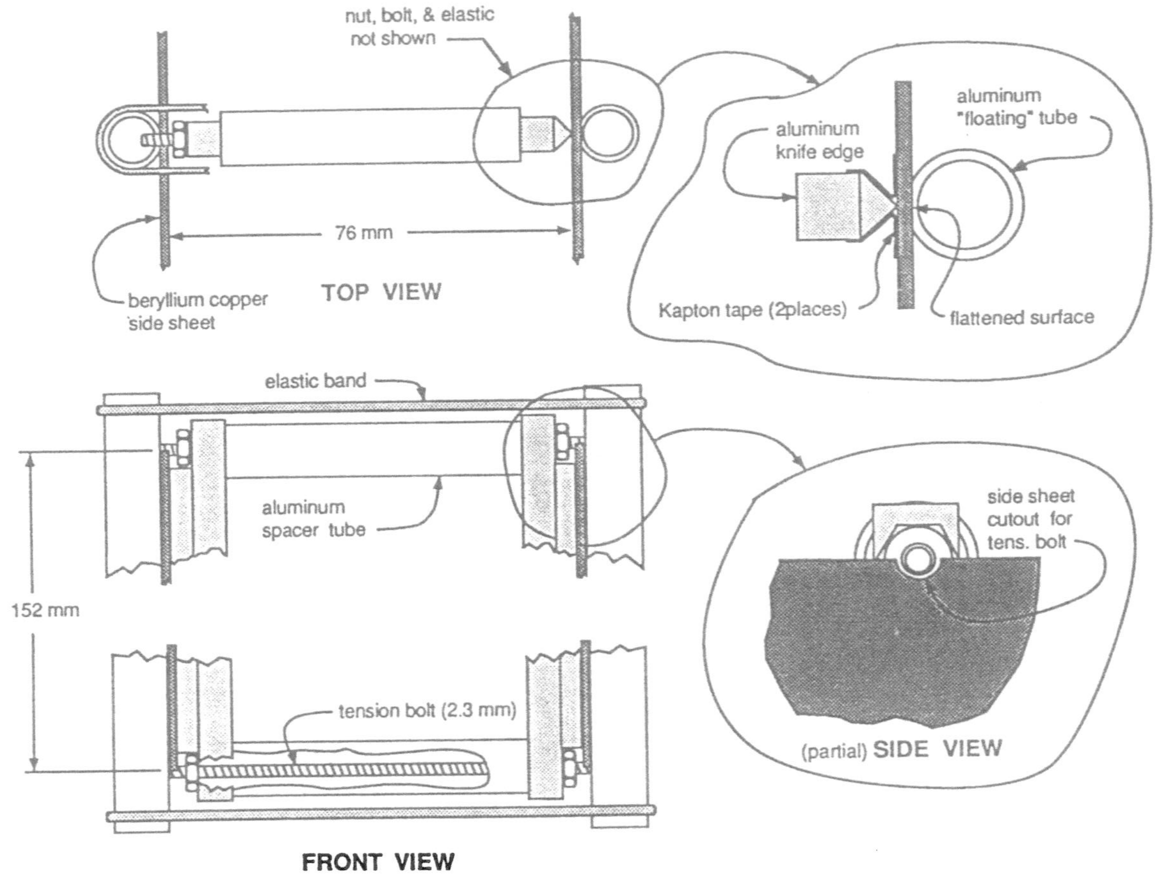

FIGURE 11. EXPERIMENTAL ARM - CROSS MEMBER DETAIL

FIGURE 12. EXPERIMENTAL ARM WITH PAYLOAD

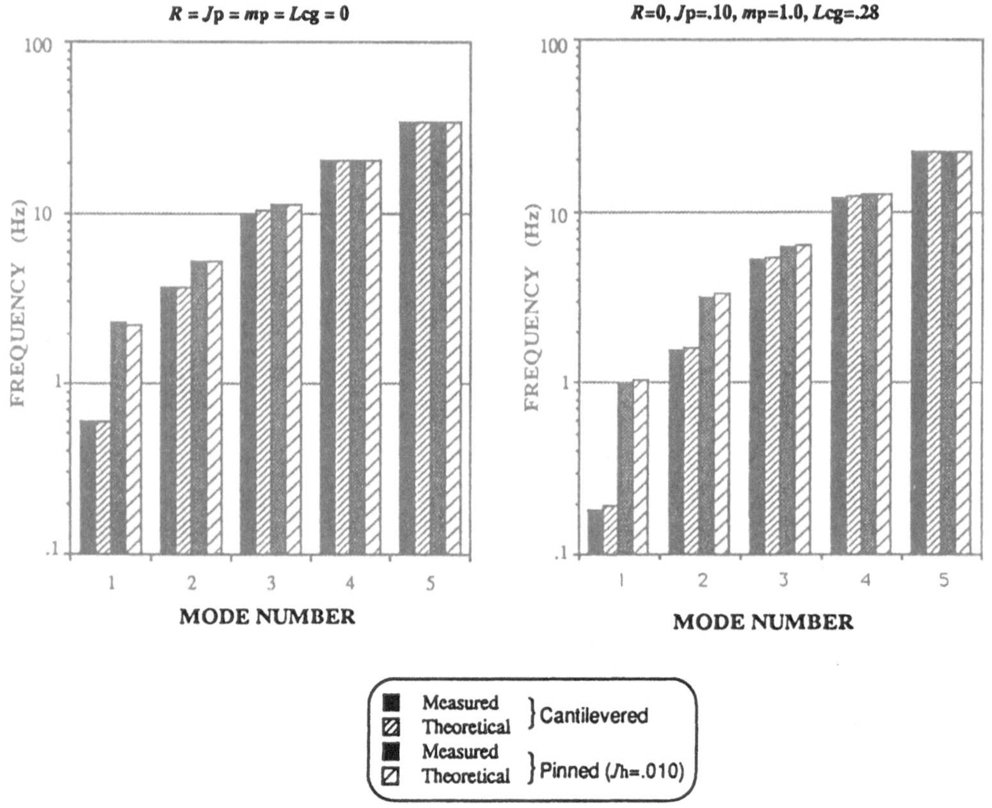

FIGURE 13. MEASURED AND THEORETICAL MODAL FREQUENCIES

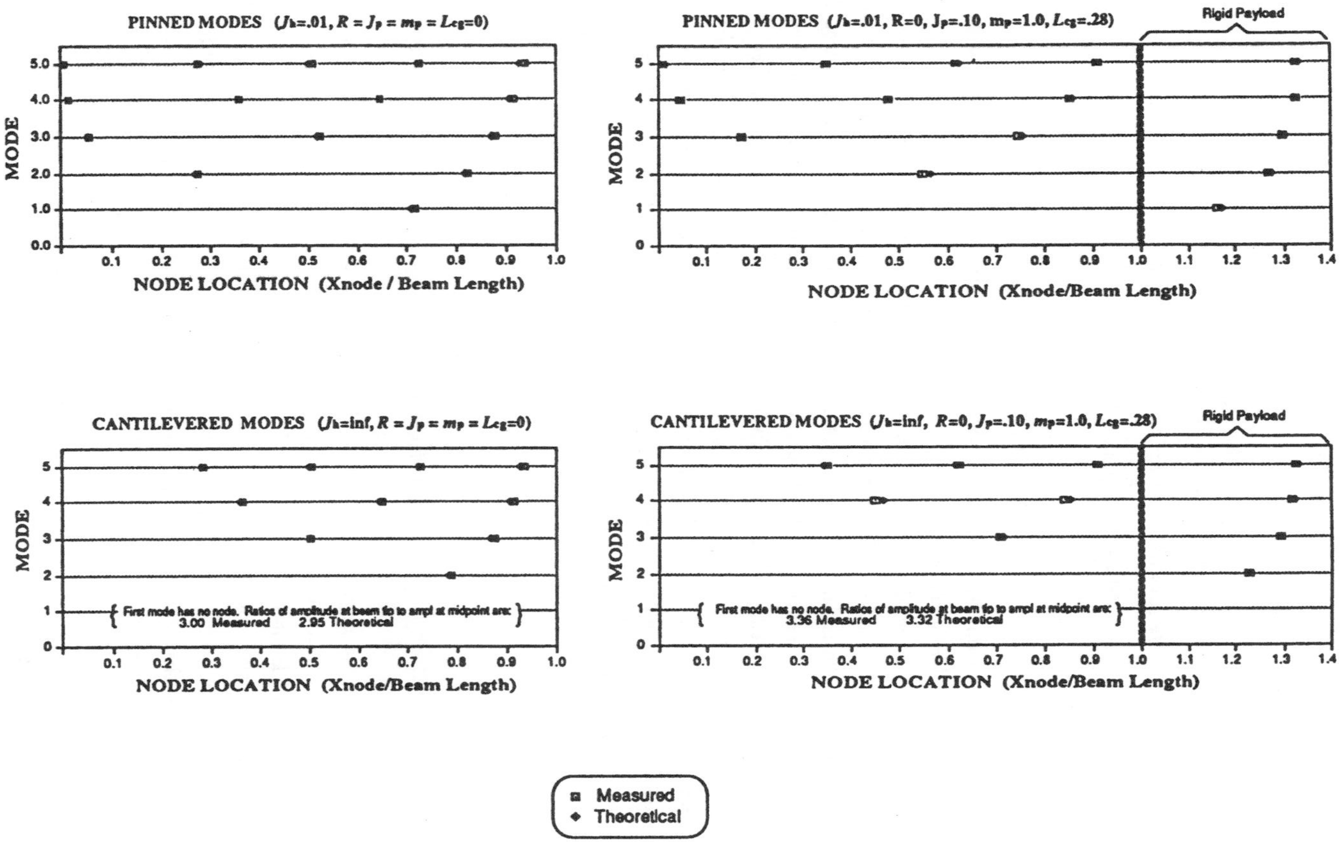

FIGURE 14. COMPARISON OF MEASURED AND THEORETICAL MODE SHAPES

MULTIVARIABLE CONTROL

OF AN

INTEGRATED PROPULSION - AIRFRAME SYSTEM

Ronald A. Perez
Osita D. I. Nwokah

School of Mechanical Engineering
Purdue University
West Lafayette, IN 47907

ABSTRACT

The development of an Integrated Control scheme to enhance the performance of a generic interconnected multivariable dynamical system, consisting of a turbo fan engine and an airframe, in the presence of predominantly destructive dynamical interactions over the flight envelope is considered in this paper. The control scheme consists of two components : a simple static forward loop or feedback loop precompensator to improve the interactions followed by a forward or feedback loop controller to improve the performance. System performance specifications dictate zero steady state errors in the engine and airframe controlled output variables as well as minimal overshoot with rapid and smooth acceleration and deceleration profiles. Furthermore the system must be tolerant to soft and hard output sensor failures by means of analytic redundancy only. A control methodology to satisfy the above specifications is presented here. Necessary and sufficient conditions are presented in order to achieve stable closed loop performance of the overall system by tuning every loop separately (i.e. decentralized stability). This leads to very simple control structures, but even for these rather simple control schemes, a significant improvement over previous integration schemes is obtained.

1. INTRODUCTION

The movement towards improved maneuverability, and the need for reduction of pilot work load, has led to the gradual integration of propulsion and airframe control systems. The advent of STOL and VTOL aircraft will further increase system complexity and accelerate the push towards more extensive control integration. Concepts such as variable geometry inlets, along with thrust vectoring and reversing provide a degree of interaction between propulsive and aerodynamic forces that requires a more complete integration of the airframe and the propulsion systems. This trend towards integration of subsystem controls has motivated the development of models and methods to analyze the behavior of dynamically coupled subsystems and to design control laws for the improvement of subsystem cooperation and the enhancement of the overall system performance.

---

† This work is supported partially by Allison Gas Turbine, Division of General Motors and Purdue Research Foundation, David Ross Grant No. 690-1423.

The interactions are mainly a consequence of airframe forces and moments induced by the propulsion system, and variations of inlet/engine operation with flight condition (Sain et al.[1]). Subsystem interaction has a direct effect upon the aircraft stability, control and performance.

The interaction between loops, or between inputs and outputs could have a constructive effect, if the result of the interaction augments the control effort. It could also have a destructive effect, if the result of the interaction either destabilizes the system, or degrades it's performance.

One of the objectives of integrating the Propulsion and the Airframe systems is to maximize any constructive interactions, while minimizing the destructive interactions, since otherwise, those negative effects would have to be countered by an additional control effort.

Although the engine model used in this project is the Pratt & Whitney F100, and the airframe model was supplied by NASA to Fennell & Black[2], the present work is aimed at developing methods to analyze and minimize the overall destructive interaction, while enhancing any constructive interaction that may exist between generic subsystems, and to design suitable robust controllers to improve the Engine/Airframe operation.

As a first step, in this paper it is assumed that either all the interactions are destructive or that the negative effect of the destructive interactions outweights the positive effect of the constructive interactions. This is indeed the case in the particular case study analyzed here. Figures 6. through 9. show typical responses of the aircraft to perturbations of 1% of the nominal values of the airframe states. We see the destructive effect of the overall negative interactions, when the airframe is coupled with the engine.

Therefore the problem posed here is the following : given a dynamical system represented by the coupled engine - airframe model at a given operating point, design a simple forward loop or feedback loop controller, such that the effect of the interaction is minimized over a pre-specified frequency interval. Furthermore determine if diagonal controllers are feasibly for performance enhancement.

Advanced propulsion systems operate at, or near, design limits with tight control of speed, pressure, temperature and airflow to achieve maximum performance while maintaining engine durability and stability. An accurate and reliable control system is required to ensure needed engine performance and operational stability throughout the flight envelope. The control system must sense pilots commands, airframe requirements, as well as critical engine parameters. It must then compute the necessary schedules and actuate system variables for total engine control over the full range of operation. Mission, airframe and engine requirements are combined to generate the following prioritized control criteria list :

- Engine Protection

  Temperature limits

  Speed limits

  Pressure limits

Structural stability

- Engine Stability

    Fan and compressor stall margins

    Augmentor spikes

    Engine fluctuations

- Steady state performance and accuracy

    Control sensitivity

        Deterioration

        Component aging

        Unmodelled dynamics

    Thrust modulation

    Thrust and fuel consumption requirements

    Repeatability

- Transient requirements

    Thrust

    Acceleration and deceleration times

    Combustion stability

- Start / Transition capability

These criteria may differ in some details for different propulsion system applications, but the general character will be the same. Engine protection is at the top of the list because of concern for aircraft / pilot safety. In cases where flight safety depends on engine dynamics (for example, VTOL and STOL), transient response would move to the top of the list along with engine protection.

2.SYSTEM MODELS

In this section we present the models used in this work. We emphasize that the methods presented in this work are not dependent upon the plant. We selected an engine model of high enough order to make the analysis and design realistic. The airframe that was used has been analyzed by Fennell & Black[2], and was supplied to them by NASA.

2.1. Engine model

In order to generate an accurate dynamic model of a gas turbine engine we need detailed steady state component descriptions (usually referred to as component maps) of all major gas path hardware, as well as dynamic elements to represent gas volume, thermal capacitance and rotor inertial effects. Prior to engine hardware availability, these maps are estimated as a technology extension from similar existing hardware, and then updated when the engine test data becomes available. Figure 1. show the gas path of the Pratt & Whitney F100 engine.

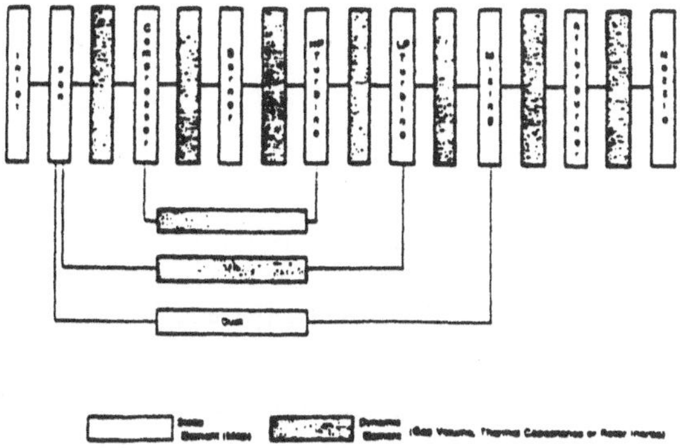

Figure 1. : Pratt & Whitney F100 Gas Path.

In order to generate a dynamical model (linear or nonlinear) an operating (or nominal) point(s) must be selected within the flight envelope (Figure 2. shows a typical flight envelope). The guidelines that are most commonly used to select the operating point(s) at which a model is to be obtained are the following :

- Important steady - state engine ratings
- Flight envelope extremes where mechanical or aerothermodynamic limits are observed
- Aircraft system limitations
- Engine model considerations which reveal engine behavioral differences that depend on engine power setting and inlet air density

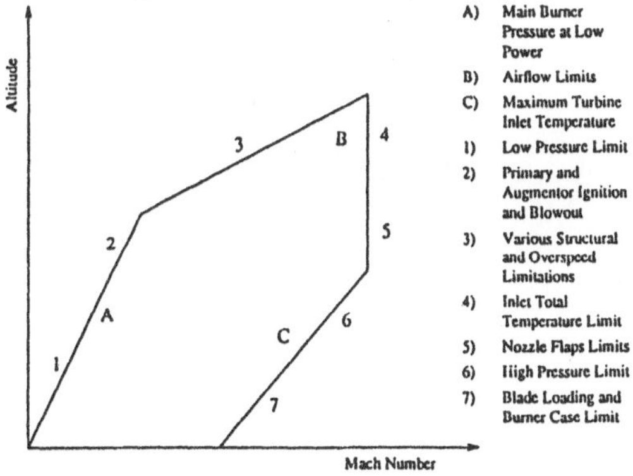

Figure 2. Flight Envelope with Constraint Limits for the F100 Turbofan Engine

Using conservation laws of thermodynamics and fluid dynamics (see Figure 1. for a flow diagram) we can obtain a set of nonlinear equations (at a given operating point) such that :

$$\dot{x} = f(x, u, \delta) \tag{2.1}$$

$$y = g(x, u, \delta) \tag{2.2}$$

where x, y, u, δ are the states, outputs, inputs and ambient parameters (for example, ambient temperature, pressure, air density, etc.) respectively. $x \in \mathbb{R}^m$, $y \in \mathbb{R}^n$, $u \in \mathbb{R}^p$ and $\delta \in \mathbb{R}^r$

The objective is to obtain a set of linear equations (at a given operating point) that can be represented by :

$$\left.\begin{aligned} \dot{x}(t) &= Ax(t) + Bu(t) \\ y(t) &= Cx(t) + Du(t) \end{aligned}\right\} \tag{2.3}$$

$x \in \mathbb{R}^m$, $u \in \mathbb{R}^n$, $y \in \mathbb{R}^n$, A, B, C and D are partitioned constant matrices of appropriate dimensions.

The method used to linearize the equations of the nonlinear model is the "derivative offset with forced steady state match". To apply this technique, we perturb equations (2.1) and (2.2), obtaining :

$$\Delta \dot{x} = \frac{\partial f(x,u,\delta)}{\partial x}\bigg|_{x_o,y_o,\delta_o} \Delta x + \frac{\partial f(x,u,\delta)}{\partial u}\bigg|_{x_o,y_o,\delta_o} \Delta u \tag{2.4}$$

$$\Delta y = \frac{\partial g(x,u,\delta)}{\partial x}\bigg|_{x_o,y_o,\delta_o} \Delta x + \frac{\partial g(x,u,\delta)}{\partial u}\bigg|_{x_o,y_o,\delta_o} \Delta u \tag{2.5}$$

where the subscript o denotes a steady state value. Holding the input constant in equations (2.4) and (2.5) :

$$\Delta \dot{x} = \frac{\partial f(x,u,\delta)}{\partial x}\bigg|_{x_o,y_o,\delta_o} \Delta x \tag{2.6}$$

$$\Delta y = \frac{\partial g(x,u,\delta)}{\partial x}\bigg|_{x_o,y_o,\delta_o} \Delta x \tag{2.7}$$

Therefore :

$$A \triangleq \frac{\partial f(x,u,\delta)}{\partial x}\bigg|_{x_o,y_o,\delta_o} \tag{2.8}$$

$$C \triangleq \frac{\partial g(x,u,\delta)}{\partial x}\bigg|_{x_o,y_o,\delta_o} \tag{2.9}$$

At steady state, x is a constant (x_o), therefore $\dot{x} = 0$. Equations (2.4) and (2.5) then become :

$$0 = A\ \Delta x + \left.\frac{\partial f(x,u,\delta)}{\partial u}\right|_{x_o,y_o,\delta_o} \Delta u \tag{2.10}$$

$$\Delta y = C\ \Delta x + \left.\frac{\partial g(x,u,\delta)}{\partial u}\right|_{x_o,y_o,\delta_o} \Delta u \tag{2.11}$$

Therefore :

$$B \triangleq \left.\frac{\partial f(x,u,\delta)}{\partial u}\right|_{x_o,y_o,\delta_o} = (-A\ \Delta x)(\Delta u)^{-1} \tag{2.12}$$

$$D \triangleq \left.\frac{\partial g(x,u,\delta)}{\partial u}\right|_{x_o,y_o,\delta_o} = (\Delta y - C\ \Delta x)(\Delta u)^{-1} \tag{2.13}$$

From the above the constant matrices A, B, C and D of equation (2.3) are obtained. The model used in this work was derived from a low order F100 engine model, which was obtained through a model reduction process (DeHoff et al.[13] , Miller and Hackney[14] and Lallman[15]). A full power engine operating point is assumed. This condition corresponds to a Fan Pressure Ratio of 2.9, a Compressor Pressure Ratio of 7.93, and a Burner Temperature of 1559 K, at a flight condition with a Mach Number of 0.9 and Altitude of 13.72 km. . The linearized equations of motion are of the form :

$$\dot{x}_e = A_e x_e + B_e u_e + C_{ea} y_a + C_{ei} y_i \tag{2.14}$$

$$y_e = F_e x_e + G_e u_e + H_{ea} y_a + H_{ei} y_i \tag{2.15}$$

where : $x_e \in \mathbb{R}^5$, $u_e \in \mathbb{R}^5$, $y_a \in \mathbb{R}^2$, $y_e \in \mathbb{R}^6$, $y_i \in \mathbb{R}^1$. A_e, B_e, C_{ea} , C_{ei} , F_e , G_e , H_{ea}, H_{ei}, are constant matrices of appropriate dimensions. The subscripts e, a, ea, and ei denote : engine, airframe, engine due to airframe, and engine due to inlet, respectively.

The state variables and their nominal values are :

- x_{e_1} , fan speed (9785 rpm)
- x_{e_2} , compressor speed (12,401 rpm)
- x_{e_3} , augmentor pressure (74.89 kpa)
- x_{e_4} , main burner fuel flow (2.31 N/sec)
- x_{e_5} , compressor discharge pressure (673.8 kpa)

The control variables and their nominal values are :

u_{e_1} , command fuel flow (2.31 N/sec)

u_{e_2} , nozzle area (0.259 m^2)

u_{e_3} , inlet guide vane (0.997 deg)

u_{e_4} , rear compressor variable vane (2.99 deg)

u_{e_5} , compressor bleed (0.997 %)

y_i , inlet pressure recovery ratio (1.0)

The output variables and their nominal values are :

y_{e_1} , net thrust per engine (12,833 N)

y_{e_2} , fan airflow (27.58 kg/sec)

y_{e_3} , turbine inlet temperature (1,590 K)

y_{e_4} , fan stall margin (0.14164)

y_{e_5} , compressor stall margin (0.1445)

y_{e_6} , relative fan exit pressure change (0.9973)

The system matrices from equations (2.14) and (2.15) are listed in Appendix 1.

2.2. Airframe model

The model used in this work was supplied by NASA to Fennell & Black[2] and corresponds to a twin engine, advanced fighter aircraft model, in which it is assumed that the thrust acts parallel to the aircraft centerline and that the engines are located a short distance below the center of gravity. The linearized equations of motion are of the form :

$$\dot{x}_a = A_a x_a + B_a u_a + C_{ae} y_e \tag{2.16}$$

$$y_a = F_a x_a \tag{2.17}$$

where $x_a \in \mathbb{R}^5$, $u_a \in \mathbb{R}^1$, $y_e \in \mathbb{R}^6$, $y_a \in \mathbb{R}^2$. A_a, B_a, C_{ae}, and F_a, are constant matrices of appropriate dimensions. The subscripts e, a, ae denote : engine, airframe, and airframe due to engine respectively.

The state variables and their nominal values are :

x_{a_1} , velocity (265.6 m/sec)

x_{a_2} , angle of attack (0.0761 rad)

x_{a_3} , pitch rate (0.0 rad)

x_{a_4} , pitch attitude (0.0761 rad)

x_{a_5} , altitude (13.72 km)

The control variables and their nominal values are :

u_{a_1} , horizontal stabilator (-0.0346)

y_{e_1} , net thrust per engine (12,833 N)

The output variables and their nominal values are :

y_{a_1} , mach number (0.9)

y_{a_2} , altitude (13.72 km)

The system matrices from equations (2.16) and (2.17) are listed in Appendix 2.

Figure 3., shows the relationship between the parameters of the Engine and the Airframe.

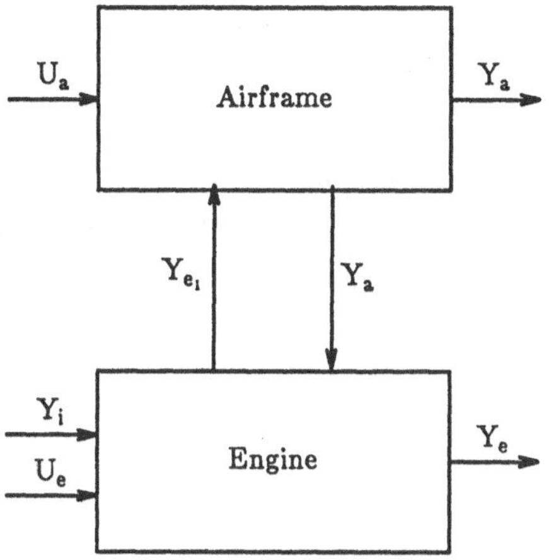

Figure 3. : Engine - Airframe interconnection.

2.3. Engine / Airframe model

The integrated model is obtained by combining the previous uncoupled models.

Upon substitution of equation (2.15) into equation (2.16), we obtain :

$$\dot{x}_a = (A_a + C_{ae}H_{ea}F_a)x_a + (C_{ae}F_e)x_e + B_a u_a + (C_{ae}G_e)u_e + (C_{ae}H_{ei})y_i \tag{2.18}$$

Similarly upon substitution of equation (2.17) into equation (2.14), we have :

$$\dot{x}_e = (C_{ea}F_a)x_a + A_e x_e + B_e u_e + C_{ei} y_i \tag{2.19}$$

Substituting equation (2.17) into equation (2.15) also gives :

$$y_e = (H_{ea}F_a)x_a + F_e x_e + G_e u_e + H_{ei} y_i \tag{2.20}$$

Let : $x \triangleq [x_a\ x_e]^T,\ \ x \in \mathbb{R}^{10}$, $u \triangleq [u_a\ u_e\ y_i]^T,\ \ u \in \mathbb{R}^7$, $y \triangleq [y_a\ y_e]^T,\ \ y \in \mathbb{R}^8$.

Then equations (2.18), (2.19), (2.17) and (2.20) become :

$$\dot{x}(t) = Ax(t) + Bu(t) \tag{2.21}$$

$$y(t) = Cx(t) + Du(t) \tag{2.22}$$

where :

$$A = \begin{bmatrix} A_a + C_{ae}H_{ea}F_a & C_{ae}F_e \\ \hline C_{ea}F_a & A_e \end{bmatrix}, \quad B = \begin{bmatrix} B_a & C_{ae}G_e & C_{ae}H_{ei} \\ \hline 0 & B_e & C_{ei} \end{bmatrix} \tag{2.22a}$$

$$C = \begin{bmatrix} F_a & 0 \\ \hline H_{ea}F_a & F_e \end{bmatrix}, \quad D = \begin{bmatrix} 0 & 0 & 0 \\ \hline 0 & G_e & H_{ei} \end{bmatrix} \tag{2.22b}$$

From equations (2.21) and (2.22) we can obtain the open loop transfer function matrix :

$$G(s) = C(sI - A)^{-1}B + D \tag{2.23}$$

It is important to note that the values of $\{A, B, C, D\}$ will vary continuously over the whole flight envelope, but the basic structure of the matrices remains unchanged. This implies that the set of all $\{G(s)\}$ which can be generated over some domain ψ is connected. Each quadruple $(A, B, C, D) \in \{A, B, C, D\}$ will correspond to the model derived from a different operating point on the flight envelope as shown in Figure 2.

3. INTERACTION ANALYSIS

The interaction optimization aimed for in this study strongly depends on a suitable definition of an interaction index (Nwokah[3]). This in turn depends on some fundamental properties of non-negative matrices.

Let Z (s) be a complex matrix belonging to $\mathbb{C}^{n\times n}$. Define the comparison matrix M (Z) of Z(s) as :

$$m_{ij} = -\mid z_{ij} \mid, \qquad i \neq j, \ i,j = 1,2, \cdots\cdots, n$$

$$m_{ii} = \mid z_{ii} \mid, \qquad i = 1,2, \cdots\cdots, n$$

Let D_z be given by :

$$D_z = \mathrm{diag}(m_{ii}), \qquad i = 1,2, \cdots\cdots, n$$

and

$$C_z = D_z - M(Z)$$

Let λ_z be the maximum eigenvalue of the non-negative matrix $D_z^{-1}C_z$.

By the Frobenius - Perron Theorem ana the Gershgorin's circle Theorem (Berman and Plemmons[4], Lancaster[5], Fiedler and Ptak[6]) we can obtain :

$$| z_{kk} - \hat{z}_{kk}^{-1} | \leq \lambda_z | z_{kk} |$$

$\hat{z}_{kk}^{-1} = \frac{1}{\hat{z}_{kk}}$ is the inverse of the kth diagonal element of $\hat{Z}$, and λ_z is called the Perron root of the interaction matrix of Z. The Perron root λ_z can be viewed as a loop interaction index, since the off-diagonal elements $|z_{kl}|$ go to zero as λ_z monotonically approaches zero

One of the drawbacks of using the Perron root as a measure of system interaction is the fact that it relies on square matrices. Almost every transfer matrix obtained for engine-airframe systems is nonsquare. It is thus necessary to square down the system by a suitable combination of the outputs before applying this interaction measure. The only requirement of this "square down" matrix is that any open loop zeros† that may be added by this process must lie in some suitable region of the left half of the complex plane. Current studies include the design of an algorithm that will square down a system (with feedthrough element) and place any added zeros at pre-specified locations.

Consider the linear time invariant system described by equation (2.3). Assuming that the square down operation has been accomplished, the open loop transfer matrix, relating u(s) and y(s) is :

$$G(s) = C(sI - A)^{-1}B + D \tag{3.1}$$

where $G(s) \in \mathbb{C}^{nxn}$. Our objective is to minimize the level of interaction of the resultant system. An obvious first step is to try to re-configure the input-output pairing, by doing column or row permutations (there are n ! possible input-output combinations) to determine if there exists a permutation matrix L_j, such that $\lambda^j_{G(s)L_j}(s) < 1 \ \forall \ s \in [\omega_o \ \omega_1]$; where :

$$\lambda^j_{G(s)L_j}(s) = \min \left\{ \max_{1 \leq k \leq n!} \left\{ \lambda_k(\overline{G(s)L_k}) \right\} \quad \forall \ s \in [\omega_o \ \omega_1] \right\} . \tag{3.2}$$

Here $\overline{G(s)L_j}$ is the interaction matrix of $G(s)L_j$. While this procedure does not guarantee that the system will be less interacting, if it reduces the level of interaction, we will use the new system (namely : $G(s)L_j$), as the nominal dynamical system, provided that such permutations are physically admissible. If the resultant structure produces a reconfigured plant whose interaction index is still not sufficiently small as to allow for independent tuning of the control loops, then simple forward or feedback loop multivariable controllers whose main purpose is to reduce the interaction to acceptable levels will need to be designed. Acceptable interaction levels usually imply $\lambda_z << 1$ over the given frequency interval. Three different methods of designing simple compensators to minimize the interaction index over a given frequency interval will be presented. Since the simplest multivariable controllers are invariably constant gain (zero dynamics) compensators, the methods given below produce minimal interaction

† See Appendix 4 for a brief discussion on multivariable zeros.

constant compensators. The first two methods will yield a forward loop compensator, while the third method will yield a feedback loop compensator. The three methods that were considered are :

1. The Hawkins[7] reduction method at a given frequency ω_p.
2. The Hawkins reduction method extended over a certain frequency interval.
3. A new method that minimizes the off-diagonal terms, while maximizing the diagonal elements.

Our proposed control structure is a two degree of freedom feedback structure; where the diagonal pre-filter is external to the feedback loop. By regarding interaction as diagonal loop uncertainty, the controller K is selected to reduce this uncertainty to a level where the pre-filter F can be used to affect input-output matching. This is in contradiction to Rosenbrock's Nyquist array design technique, where open loop diagonal dominance is sought, and further diagonal feedback is needed to satisfy performance and stability requirements.

This feedback dominance arrangement further insures robustness of the dominance condition to parameter variations, and falls directly into the QFT design methodology Horowitz[23] , which is needed for control over the flight envelope. The major reason for taking this route, is that open loop diagonal dominance in no way implies closed loop diagonal dominance. Closed loop diagonal dominance is required to deduce closed loop performance from the performance of the individual loops.

Consider the following configuration (used for the first two methods) :

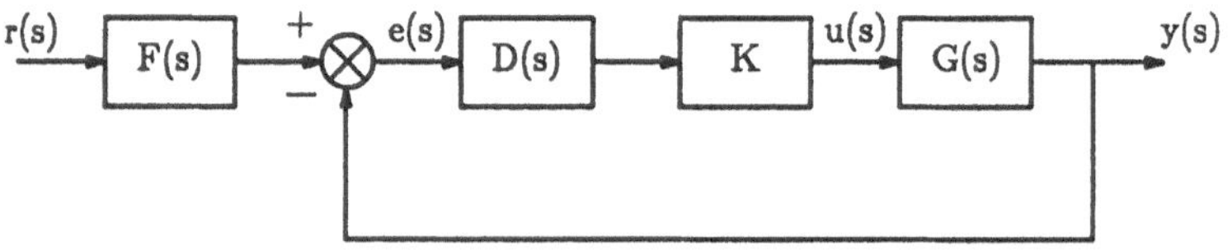

Figure 4. : Closed loop structure, compensator in the forward loop.

where $G(s) \in \mathbb{C}^{n \times n}$ is assumed to have no unstable hidden modes, so that diagonal stabilization is feasible, $D(s) \in \mathbb{C}^{n \times n}$ is a diagonal controller (preferable PID), $K \in \mathbb{R}^{n \times n}$, $\Omega \triangleq [\omega_o \; \omega_1]$. Let the inner closed loop transfer matrix from Figure 4. (with $D(s) = I$) be denoted by :

$$Q(s) = (I + G(s)K)^{-1} G(s)K \tag{3.3}$$

Our objective is to design K, such that Q(s) is stable, and :

$$\lambda(\,|\overline{Q}(s)|\,) < 1 \quad \forall\; s \in \Omega \quad ,(\text{if possible}), \tag{3.4}$$

where $\overline{Q}(s)$ is the interaction matrix of Q(s) and $|\overline{Q}(s)|$ is the matrix whose elements are the magnitude functions of the elements of $\overline{Q}(s)$. Let us consider the inverse of Q(s) given by :

$$Q^{-1}(s) = \hat{Q}(s) = \hat{K}\hat{G}(s) + I \;. \tag{3.5}$$

Where $(\hat{\cdot})$ denotes the inverse of $(\cdot)$. Since I is already a diagonal matrix, is sufficient to guarantee generalized diagonal dominance for $K\hat{Q}(s)$ in order to have diagonal dominance for $\hat{Q}(s)$. In addition (Araki et al.[8]) have shown empirically that if $\lambda(\,|\overline{Q}(s)|) < 0.35$, then $\lambda(\,|\overline{Q}(s)|) < 1$.

Let

$$\hat{q}_{lj}(s) \triangleq \alpha_{lj}(s) + i\beta_{lj}(s) \tag{3.6}$$

where $\alpha_{lj}(s) = \text{real}_{lj}(\hat{K}\hat{G}(s))$, and $\beta_{lj}(s) = \text{imag}_{lj}(\hat{K}\hat{G}(s))$. Therefore we need to design K such that $\lambda(\,|\overline{Q}(s)|) < 0.35 \quad \forall\ s \in \Omega$, which in turn would guarantee that $\lambda(\,|\overline{Q}(s)|) < 1 \quad \forall\ s \in \Omega$.

3.1. Method 1 : The Hawkins method (Hawkins[7]) minimizes the sum of the squares of the off-diagonal terms. This concept was first suggested by Rosenbrock[9] in terms of columns, and applied here in terms of rows. Following Hawkins[7], we have :

$$A_j = (a_{il}^{(j)}) = \sum_{\substack{k=1 \\ k \neq j}}^{m} [\alpha_{ik}(\omega_p)\alpha_{lk}(\omega_p) + \beta_{ik}(\omega_p)\beta_{lk}(\omega_p)] \quad l = 1,\ 2,\ \cdots\cdots,\ m \tag{3.7}$$

and

$$A_j\hat{K}_j^{\,T} - \lambda\hat{K}_j^{\,T} = 0 \tag{3.8}$$

which is the standard eigenvector problem. Therefore the controller K is designed by rows, where each row corresponds to the eigenvector of the smallest eigenvalue of A_j.

3.2. Method 2 : The Hawkins method (Hawkins[7]) is extended to cover an arbitrary frequency interval $\Omega = [\omega_o\ \omega_1]$.

By re-working Hawkins' formulation, but instead minimizing :

$$J_j = \int_{\omega_o}^{\omega_1} [\sum_{\substack{k=1 \\ k \neq j}}^{m} |\sum_{i=1}^{m} \hat{k}_{ji}(\alpha_{ik}(\omega) + i\beta_{ik}(\omega))|^2 + \lambda(1 - \sum_{i=1}^{m} \hat{k}_{ji}^{\,2})]d\omega \tag{3.9}$$

we obtain :

$$A_j = (a_{il}^{(j)}) = \frac{\int_{\omega_o}^{\omega_1} [\sum_{\substack{k=1 \\ k \neq j}}^{m} (\alpha_{ik}(\omega)\alpha_{lk}(\omega) + \beta_{ik}(\omega)\beta_{lk}(\omega))]d\omega}{\omega_1 - \omega_o} \quad ,l = 1,2,\ \cdots\cdots,\ m \tag{3.10}$$

and

$$A_j\hat{K}_j^{\,T} - \lambda\hat{K}_j^{\,T} = 0 \tag{3.11}$$

3.3 Method 3 : Consider the following configuration :

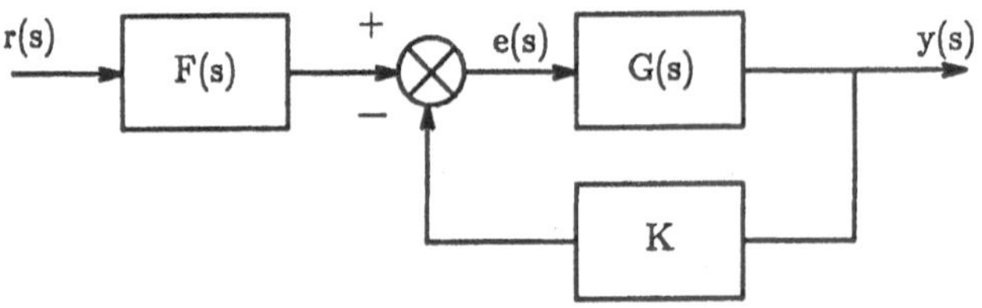

Figure 5. : Closed loop structure, compensator in the feedback loop.

where $G(s) \in \mathbb{C}^{nxn}$, $K \in \mathbb{R}^{nxn}$, $\Omega = [\omega_o \ \omega_1]$. Let the inner closed loop transfer function from Figure 5. (assumed stable) be denoted by :

$$Q(s) = (I + G(s)K)^{-1}G(s) \tag{3.12}$$

Our objective is to design K, such that :

$$\lambda(\,|\overline{Q}(s)|) < 1 \quad \forall\ s \in \Omega \tag{3.13}$$

where $\overline{Q}(s)$ is the interaction matrix of Q(s). Let us consider the inverse of Q(s) given by :

$$Q^{-1}(s) = \hat{Q}(s) = \hat{G}(s) + K \tag{3.14}$$

Let

$$\hat{g}_{ij}(s) \triangleq \alpha_{ij}(s) + i\beta_{ij}(s) \tag{3.15}$$

where $\alpha_{ij}(s) = \text{real}(\hat{g}_{ij}(s))$ and $\beta_{ij}(s) = \text{imag}(\hat{g}_{ij}(s))$. This method minimizes the off - diagonal elements of $\hat{Q}(s)$, by minimizing the following expression :

$$J = \int_{\omega_o}^{\omega_1} |\hat{g}_{ij}(\omega) + k_{ij}|^2 d\omega \qquad i \neq j \tag{3.16}$$

and maximizes the diagonal elements of $\hat{Q}(s)$, by maximizing the following expression :

$$R = \int_{\omega_o}^{\omega_1} |\hat{g}_{ii}(\omega) + k_{ii}|^2 d\omega \tag{3.17}$$

The resulting controller is :

$$k_{ij} = \frac{\int_{\omega_o}^{\omega_1} \alpha_{ij}(\omega)d\omega}{\omega_o - \omega_1} \qquad i \neq j \tag{3.18}$$

and

$$k_{ii} = \int_{\omega_o}^{\omega_1} \alpha_{ii}(\omega)d\omega \tag{3.19}$$

In order to implement the above controller on a computer the following equivalent forms were used :

$$k_{ij} = \frac{\displaystyle\lim_{n \to +\infty} \left[\sum_{k=1}^{n} \alpha_{ij}(w_k)\Delta\omega \right]}{\omega_o - \omega_1} \qquad i \neq j \tag{3.20}$$

$$k_{ii} = \lim_{n \to +\infty} \left[\sum_{k=1}^{n} \alpha_{ii}(w_k)\Delta\omega \right] \tag{3.21}$$

where $w_k = \omega_o + k(\Delta\omega)$, and $w_\infty = \omega_1$.

We proceed to apply the three methods just outlined to minimize the overall interaction of the Airframe (example 1) and the integrated Engine - Airframe (example 2). Plots of the overall closed loop interaction index will be presented at different stages of the design.

Example 1 : Interaction between Airframe input-output pairs

Equation (2.16) is re-written as :

$$\dot{x}_a = A_a x_a + B_n u \tag{3.22}$$

where $u \triangleq [u_a \; y_{e_1}]^T \in \mathbb{R}^2$. Following Fennell and Black[2], equation (2.17) is redefined as :

$$Y_a = F_n x_a + G_n u \tag{3.23}$$

the matrices B_n, F_n and G_n are listed in Appendix 3.

The output variables are :

y_{a_1} , Velocity

y_{a_2} , Glide Path Angle

y_{a_3} , Pitch Rate

The transfer matrix G(s) of equations (3.22) and (3.23) is of dimension 3x2. A "square down" matrix L that adds stable zeros was selected as :

$$L = \begin{bmatrix} 0 & 0 & 1 \\ 1 & 0 & 0 \end{bmatrix} \tag{3.24}$$

Let $T_1(s) \triangleq LG(s)$.

Zeros of G(s)	Zeros of $T_1(s)$
-5.81132367E-17	-5.81132367E-17
	-5.57558450E-01
	-2.81520641E-03

Thus two stable zeros have added. From Theorem A4.1, it is shown that the zero of G(s) remains invariant under several static transformations, amongst them a similarity transformation. The interaction index (Perron root) of $T_1(s)$ is plotted in Figure 10., and we see that $T_1(s)$ is already diagonally dominant. Column permutation was performed but did not yield a lower interaction index. This implies that the flight dynamics control loops have sufficiently weak interaction as to make independent flight control tuning feasible. The fact that the interaction index was very small at low frequencies and even smaller at high frequencies, indicates that no interaction reduction controller is needed.

Example 2 : Interaction between Airframe and Engine

From equations (2.21) and (2.22), the transfer function matrix G(s) is of dimension 8x7. A "square down" matrix L that adds stable zeros was selected as :

$$L = \left[\begin{array}{cc|cccccc} 1 & 0 & 0 & 0 & 0 & 0 & 0 & 0 \\ 0 & 1 & 0 & 0 & 0 & 0 & 0 & 0 \\ \hline 0 & 0 & 1 & 0 & 0 & 0 & 0 & 0 \\ 0 & 0 & 0 & 1 & 0 & 0 & 0 & 0 \\ 0 & 0 & 0 & 0 & 1 & 0 & 0 & 0 \\ 0 & 0 & 0 & 0 & 0 & 7 & 1 & 0 \\ 0 & 0 & 0 & 0 & 0 & 0 & 0 & 1 \end{array}\right] \tag{3.25}$$

Let $T_3(s) \triangleq LG(s)$. $U(s) = [u_a\ u_{e_1}\ u_{e_2}\ u_{e_3}\ u_{e_4}\ u_{e_5}\ y_i]^T$ are the inputs of the integrated engine-airframe model G(s), and $Y(s) = [y_{a_1}\ y_{a_2}\ y_{e_1}\ y_{e_2}\ y_{e_3}\ y_{e_4}\ y_{e_5}\ y_{e_6}]^T$ are the outputs. Let $Y_3(s)$ be the outputs of $T_3(s)$, then $Y_3(s) = [y_{a_1}\ y_{a_2}\ y_{e_1}\ y_{e_2}\ y_{e_3}\ (7y_{e_4} + y_{e_5})\ y_{e_6}]^T$.

The interaction index of $T_3(s)$ is plotted in Figure 11. (o). It is clear that the system is highly interactive. Column permutations are performed, and a permutation matrix controller P which gives the minimum interaction over the specified frequency interval ([0.01, 1000] rad/sec) is selected as :

$$P = \left[\begin{array}{cc|ccccc} 1 & 0 & 0 & 0 & 0 & 0 & 0 \\ 0 & 1 & 0 & 0 & 0 & 0 & 0 \\ \hline 0 & 0 & 0 & 0 & 0 & 0 & 1 \\ 0 & 0 & 0 & 1 & 0 & 0 & 0 \\ 0 & 0 & 0 & 0 & 0 & 1 & 0 \\ 0 & 0 & 1 & 0 & 0 & 0 & 0 \\ 0 & 0 & 0 & 0 & 1 & 0 & 0 \end{array}\right] \tag{3.26}$$

Let $T_4(s) \triangleq T_3(s)P = LG(s)P$. Let $Y_4(s)$ be the outputs of $T_4(s)$. Then $Y_4(s) = Y_3(s)$. If $U_4(s)$ is the input to $T_4(s)$, then $U_4(s) = [u_a\ y_i\ u_{e_1}\ u_{e_3}\ u_{e_5}\ u_{e_2}\ u_{e_4}]^T$.

The interaction index of $T_4(s)$ is plotted in Figure 11. (Δ). The improvement achieved through column permutation is now self evident.

To further reduce interaction, we proceed to design 3 static controllers. Two were designed in the forward loop path (methods 1 and 2), and one was designed in the feedback loop path (method 3).

Applying method 1, a static forward loop controller K_1 was designed as :

$$K_1 = \left[\begin{matrix} 0.5485089d{-}01 & -0.1209326d{+}00 & -0.5795665d{+}02 & -0.3320227d{+}02 \\ 0.1312813d{-}02 & -0.2858392d{-}02 & -0.2797350d{+}02 & 0.1334577d{+}03 \\ 0.3740014d{+}01 & -0.3123984d{+}01 & 0.3442183d{+}03 & 0.9700278d{+}03 \\ -0.4045065d{+}02 & -0.2089491d{+}03 & 0.1563837d{+}06 & 0.2912969d{+}05 \\ -0.1011868d{+}01 & 0.2057017d{+}01 & -0.9485329d{+}02 & -0.1113487d{+}03 \\ -0.7444981d{+}00 & -0.6707829d{-}01 & 0.7032627d{+}03 & 0.4038604d{+}03 \\ 0.5350879d{+}04 & 0.2174727d{+}04 & -0.7641792d{+}07 & -0.4942301d{+}07 \end{matrix}\right.$$

$$\left.\begin{matrix} -0.2099165d{+}03 & 0.3009426d{+}03 & 0.8591297d{+}00 \\ -0.1667616d{+}03 & 0.5820937d{+}02 & 0.1774553d{+}02 \\ 0.9425349d{+}03 & -0.2273436d{+}04 & 0.9848557d{+}02 \\ 0.5900658d{+}06 & -0.7743105d{+}06 & -0.7169094d{+}04 \\ -0.3213225d{+}03 & 0.5285466d{+}03 & -0.7453121d{+}01 \\ 0.2547548d{+}04 & -0.3654028d{+}04 & 0.2176579d{+}01 \\ -0.2744698d{+}08 & 0.4004688d{+}08 & -0.9361850d{+}05 \end{matrix}\right] \tag{3.27}$$

The inner closed loop transfer function (from Figure 4., with $D(s) = I$) is :

$$Q_1(s) = (I + (LG(s)P)K_1)^{-1}(LG(s)P)K_1 \tag{3.28}$$

Applying method 2, a static forward loop controller K_2 was designed as :

$$K_2 = \left[\begin{matrix} 0.9952791d{-}02 & -0.1160403d{-}03 & -0.4465834d{+}02 & -0.6396568d{+}02 \\ -0.1005372d{+}04 & -0.1938614d{+}01 & 0.3449683d{+}06 & 0.7716022d{+}06 \\ -0.5077116d{+}01 & -0.3517859d{-}03 & -0.9627707d{+}06 & 0.4163255d{+}05 \\ 0.2667976d{+}08 & 0.4325375d{+}04 & -0.2934303d{+}09 & -0.6937518d{+}09 \\ 0.1595190d{+}05 & -0.1093173d{+}01 & 0.6470843d{+}07 & 0.9571792d{+}07 \\ -0.1062968d{+}06 & -0.1223774d{+}02 & -0.2391462d{+}07 & -0.2973849d{+}07 \\ 0.1400242d{+}10 & 0.1641234d{+}06 & 0.3073402d{+}11 & 0.3640040d{+}11 \end{matrix}\right.$$

$$\left.\begin{matrix} -0.5215237d{+}03 & 0.6311377d{+}03 & 0.9103599d{-}07 \\ 0.4846055d{+}07 & -0.5961620d{+}07 & 0.2376196d{+}01 \\ -0.5869426d{+}07 & 0.6790569d{+}07 & 0.7753885d{+}01 \\ -0.4588081d{+}10 & 0.5548585d{+}10 & -0.1626695d{+}04 \\ 0.7711499d{+}08 & -0.9317358d{+}08 & 0.6762350d{+}00 \\ -0.2607098d{+}08 & 0.3154259d{+}08 & 0.1450176d{+}01 \\ 0.3284987d{+}12 & -0.3970333d{+}12 & -0.1565817d{+}05 \end{matrix}\right] \tag{3.29}$$

The inner closed loop transfer function (from Figure 4., with $D(s) = I$) is :

$$Q_2(s) = (I + (LG(s)P)K_2)^{-1}(LG(s)P)K_2 \tag{3.30}$$

Applying method 3, a static feedback loop controller K_3 was designed as :

$$K_3 = \left[\begin{array}{llll} 0.2599733d{+}05 & 0.2406834d{+}07 & 0.5533280d{-}04 & 0.2672510d{-}06 \\ 0.2599733d{+}02 & 0.9607989d{+}15 & 0.5533279d{-}07 & 0.2672510d{-}09 \\ 0.2599733d{+}02 & 0.2406834d{+}04 & 0.7340059d{+}05 & 0.2672510d{-}09 \\ 0.2599733d{+}02 & 0.2406834d{+}04 & 0.5533279d{-}07 & -0.1627873d{+}05 \\ 0.2599733d{+}02 & 0.2406834d{+}04 & 0.5533279d{-}07 & 0.2672510d{-}09 \\ 0.2599733d{+}02 & 0.2406834d{+}04 & 0.5533279d{-}07 & 0.2672510d{-}09 \\ 0.2599733d{+}02 & 0.2406834d{+}04 & 0.5533279d{-}07 & 0.2672510d{-}09 \end{array}\right.$$

$$\left.\begin{array}{lll} -0.9163896d{-}08 & 0.4905722d{-}05 & 0.5954730d{-}05 \\ -0.9163896d{-}11 & 0.4905722d{-}08 & 0.5954730d{-}08 \\ -0.9163896d{-}11 & 0.4905722d{-}08 & 0.5954730d{-}08 \\ -0.9163896d{-}11 & 0.4905722d{-}08 & 0.5954730d{-}08 \\ -0.1411562d{+}00 & 0.4905722d{-}08 & 0.5954730d{-}08 \\ -0.9163896d{-}11 & 0.5241113d{+}03 & 0.5954730d{-}08 \\ -0.9163896d{-}11 & 0.4905722d{-}08 & 0.2438532d{+}06 \end{array}\right] \tag{3.31}$$

The inner closed loop transfer function (from Figure 5.) is :

$$Q_3(s) = (I + (LG(s)P)K_3)^{-1}(LG(s)P) \tag{3.32}$$

The interaction indices of $Q_1(s)$, $Q_2(s)$ and $Q_3(s)$ are plotted in Figure 12.; the interaction index of $Q_3(s)$ is also plotted in Figure 13. The significant reduction of interaction over the useful frequency range, which is obtained by using the proposed interaction optimization method is now clear.

4.DECENTRALIZED STABILITY

If the interaction can be reduced to a level where it does not create a major design difficulty; then single loop controllers ($D(s) = \mathrm{diag}(d_i(s))$ can be used to improve loop performance, provided that the addition of such controllers do not create new stability problems.

In this section we study under what conditions the sum of the encirclements of the origin by the Nyquist diagrams of the diagonal elements of a square rational function matrix is equal to the number of encirclements of the origin by the Nyquist diagram of the determinant of the matrix (i.e. diagonally decentralized stability). Let Z(s) be an nxn rational function matrix. Let D_N be a closed elementary contour (often the Nyquist Contour) which satisfies the following properties. D_N is constructed such that no pole or zero of det Z(s) or of $z_{k\ell}(s)$ $k,\ell = 1,2,\cdots,n$ lies on D_N. Furthermore every finite pole or zero of det Z(s) and $z_{kk}(s)$, $k = 1,2,\cdots,n$ lying in the closed night half plane lies inside D_N. The construction of such D_N is possible because det Z(s) and $z_{k\ell}(s)$ are rational functions with isolated poles and zeros. Let $\mathbb{C}_{+e}$ be the extended right half plane.
Write

$$Z(s) = D(s) + C(s) = D(s)\,[I + D^{-1}(s)\,C(s)] \tag{4.1}$$

where:

$$D(s) = \operatorname{diag}\,(z_{11}(s)\,,\ z_{22}(s)\,,\ \cdots\,,\ z_{nn}(s)) \tag{4.2}$$

and C(s) satisfies:

$$\begin{cases} c_{kk}(s) = 0\ ,\ k = 1,2,\ \cdots,n \\ \\ c_{k\ell}(s) = z_{k\ell}(s)\ ,\ k \neq \ell\ ,\ k,\ell = 1,2,\ \cdots,n\ . \end{cases} \tag{4.3}$$

Define the interaction matrix of Z(s) as:

$$M(s) = D^{-1}(s)\ C(s)\ . \tag{4.4}$$

Then :

$$\det Z(s) = \det D(s) \cdot \det(I + M(s))\ , \tag{4.5}$$

or

$$\frac{\det\ Z(s)}{\prod\limits_{k=1}^{n} z_{kk}(s)} = \det\,(I + M(s)) \tag{4.6}$$

Let det Z(s) map D_N into Γ and let Γ encircle the origin N times. Also let $z_{kk}(s)$ map D_N into Γ_k and encircle the origin N_k times, all in the clockwise direction for $k = 1,2,\cdots,n$.

Theorem 4.1. (Nwokah and Perez[17])
Let M(s) be the interaction matrix of the nxn rational function matrix Z(s). Let $z_{kk}(s)$ map D_N into Γ_k and encircle the origin N_k times clockwise. Also let det Z(s) map D_N into Γ and encircle the origin N times clockwise. Then,

$$N = \sum_{k=1}^{n} N_k \tag{4.7}$$

if and only if det(I + M(s)) is bounded and non vanishing on $\mathbb{C}_{+e}$; or equivalently $\inf\limits_{s \in \mathbb{C}_{+e}} |\det(I + M(s))| > 0$.

□

Theorem 4.1 has the following important implications:

a) The number of unstable poles and zeros in det Z(s) must correspond to the number of unstable poles and zeros in $\prod\limits_{k=1}^{n} z_{kk}(s)$ in order for M(s) to be bounded on D_N.

b) The fixed modes of Z(s), which correspond to those poles of the system that are not affected by the modifications of the poles of $z_{kk}(s)$, $k = 1,2,\cdots,n$ in a feedback control or otherwise, must be stable. (Nwokah and Thompson[18]).

c) The results presented here can be carried over to block matrices i.e. to the situation where $z_{kk}(s)$ is not an element but a square matrix of dimension $n_k \times n_k$ and where $\sum_{k=1}^{n} n_k = n$, where n is the total number of blocks, by replacing $z_{kk}(s)$ by det $z_{kk}(s)$, (see Nwokah[19]).

d) It can be shown (Nwokah[3]) that the diagonally scale invariant norm of the interaction matrix : $\inf_{D \in \mathcal{D}/0} \|D^{-1}M(s)D\| = \lambda_0(s)$, is a measure of system interaction. Therefore $\lambda_0(s)$ is called the interaction index of Z(s). Consequently diagonal control is feasible only if Z(s) has a finite interaction index on D_N. Notice however that it is only when $\lambda_0(s) < 1$ on D_N that the system behavior is dominated by the diagonal elements.

Corollary 4.1. (Rosenbrock[9])
Let Z(s) be row or column diagonally dominant on D_N. Let $z_{kk}(s)$ map D_N into Γ_k and let Γ_k encircle the origin N_k times. Let det Z(s) map D_N into Γ and let Γ encircle the origin N times. Under these conditions,

$$N = \sum_{k=1}^{n} N_k \ . \tag{4.8}$$

Proof:
We merely recognize that a diagonally dominant matrix is already an H-matrix (Araki and Nwokah[20]). For an H-matrix, the interaction matrix satisfies $\rho(M(s)) < 1, \ \forall \ s \in D_N$, (Nwokah[21]), where $\rho(\cdot)$ is the spectral radius of $(\cdot)$.

But

$$|\lambda_k(s)| \leq \rho(M(s)) < 1 \ , \ k = 1,2, \cdots, n. \tag{4.9}$$

Hence:

$$\inf_{s \in \mathbb{C}_{+e}} |1 + \lambda_k(s)| > 0 \tag{4.10}$$

This in turn implies that:

$$\inf_{s \in \mathbb{C}_{+e}} |\det(I + M(s)| = \inf_{s \in \mathbb{C}_{+e}} \prod_{k=1}^{n} |1 + \lambda_k(s)| > 0 \ . \tag{4.11}$$

Invoking Theorem 4.1 then completes the proof.

□

Corollary 4.2 (Koussiouris[22])
Let Z(s) satisfy:

$$\left| \prod_{k=1}^{n} z_{kk}(s) \right| > \left| \det Z(s) - \prod_{k=1}^{n} z_{kk}(s) \right| \ , \forall \ s \in \mathbb{C}_{+e} \tag{4.12}$$

Let $z_{kk}(s)$ map D_N into Γ_k and let Γ_k encircle the origin N_k times. Let det Z(s) map D_N into Γ and let it encircle the origin N times. Under these conditions :

$$N = \sum_{k=1}^{n} N_k \tag{4.13}$$

Proof:

$$| \det Z(s) - \prod_{k=1}^{n} z_{kk}(ss) | = | \prod_{k=1}^{n} z_{kk}(s) - \det Z(s) |$$

$$\geq | \prod_{k=1}^{n} z_{kk}(s) | - | \det Z(s) | \tag{4.14}$$

The given condition (4.12) then implies that :

$$| \prod_{k=1}^{n} z_{kk}(s) | > | \prod_{k=1}^{n} z_{kk}(s) | - | \det Z(s) | \tag{4.15}$$

Dividing both sides of the inequality (4.15) by the LHS, and simplifying, produces the inequality:

$$\frac{| \det Z(s) |}{\prod_{k=1}^{n} | z_{kk}(s) |} > 0 \tag{4.16}$$

But by the construction of D_N, both det Z(s) and $z_{kk}(s)$ are finite on D_N. Hence

$$0 < \frac{| \det Z(s) |}{\prod_{k=1}^{n} | z_{kk}(s) |} < \infty \tag{4.17}$$

This shows that det(I + M(s)) is bounded an non vanishing on D_N and by Theorem 4.1, the result follows.

□

We proceed to apply the diagonally decentralized stability conditions just presented, to examples 1 and 2; by identifying the respective return difference matrices (subject to diagonal dynamic controllers) with Z(s).

Example 3 : Interaction between Airframe input-output pairs

From example 1, the transfer function $T_1(s)$ is given by :

$$T_1(s) = L\left[F_n(sI - A_a)^{-1}B_n + G_n\right] \tag{4.18}$$

From Theorem 4.1 :

$$\det(I + M(s)) = \frac{\det T_1(s)}{\prod_{k=1}^{n} T_{1_{kk}}(s)} \tag{4.19}$$

Figure 14. shows the plot of the previous equation $\forall\ \omega \in [0\ 100]$. We can see, as expected, that the locus does not pass through or enclose the origin (0,0), thus, from Theorem 4.1 this guarantees decentralized stability. Recall that $T_1(s)$ was made diagonally dominant in example 4.1.

Example 4 : Interaction between Airframe and Engine

From example 2, the transfer functions $Q_1(s)$, $Q_2(s)$, and $Q_3(s)$ are given by :

$$Q_1(s) = (I + (LG(s)P)K_1)^{-1}(LG(s)P)K_1 \tag{4.20}$$

$$Q_2(s) = (I + (LG(s)P)K_2)^{-1}(LG(s)P)K_2 \tag{4.21}$$

$$Q_3(s) = (I + (LG(s)P)K_3)^{-1}(LG(s)P) \tag{4.22}$$

where

$$G(s) = C(sI - A)^{-1}B + D \tag{4.23}$$

A, B, C and D are partitioned matrices given by equations (2.22a) and (2.22b). From Theorem 4.1 we obtain :

$$\det(I + M(s)) = \frac{\det Q_i(s)}{\prod_{k=1}^{n} Q_{i_{kk}}(s)} \qquad i = 1,\ 2,\ 3 \tag{4.24}$$

Figures 15., 17. and 18. show the plot of the previous equation for $i = 1$, 2 and 3 and $\forall\ \omega \in [0\ 200]$, $\forall\ \omega \in [0\ 200]$ and $\forall\ \omega \in [0\ 500]$ respectively. Figure 16. shows the plot of the previous equation for $\forall\ \omega \in [20\ 200]$, we can see, as expected, that the origin (0,0) is encircled by the locus of $Q_1(s)$ and $Q_2(s)$ but not by the locus of $Q_3(s)$. Recall that $Q_1(s)$ and $Q_2(s)$ are not diagonally dominant, as opposed to $Q_3(s)$. Thus from Theorem 4.1 we conclude at once that diagonal control is feasible for $Q_3(s)$ and this configuration is tolerant to sensor and actuator failures, we also conclude that for $Q_1(s)$ and $Q_2(s)$, stability of the diagonal elements does not imply the stability of the plant.

5.DISCUSSION CONCLUSIONS

The work reported here can be applied to any linear multivariable system. By performing a careful interaction analysis a priori, we can minimize the negative effect of the destructive interactions. We have shown how to select permutation controllers to reduce the level of interaction, as well as two methods to design static controllers that will minimize the overall negative effect of the destructive interactions. We have also presented necessary and sufficient conditions to guarantee decentralized stabilizability; these conditions are quick, and easy to compute. Further developments of the present work is in the direction of designing dynamic diagonal controllers to insure sensor failure tolerance.

6. ACKNOWLEDGMENTS

The authors would like to thank Allison Gas Turbine, Division of General Motors and Purdue Research Foundation for their support of this study. The authors are also indebted to Dr. Ben Wong and Mr. Tom Scott from Allison Gas Turbine for their willingness to discuss several key aspects of gas turbines.

This work is part of on ongoing joint effort between Allison Gas Turbine and Purdue University in Propulsion - Airframe Dynamics and Control.

7. REFERENCES

[1]Sain, M. K., Peczkowski, J. L. and Melsa, J. L., *Alternatives for Linear Multivariable Control*, National Engineering Consortium Inc., Chicago, Illinois, 1978.

[2]Fennell, R. E. and Black, S. B., *Integrated Airframe Propulsion Control*, NASA-CR-3806.

[3]Nwokah, O. D. I., *The Convergence and Local Minimality of Bounds for Transfer Functions*, Int. J. Control, 1979, Vol. 30, No. 2.

[4]Berman, A. and Plemmons, R. J., *Non-Negative Matrices in The Mathematical Sciences*, Academic Press Inc., New York, 1979.

[5]Lancaster, P., *Theory of Matrices*, Academic Press Inc., New York, 1969.

[6]Fiedler, M. and Ptak, V., *On Matrices with Non-Positive Off-Diagonal Elements and Positive Principal Minors*, Czech. Math. J. 87, 382-400, 1962.

[7]Hawkins, D. J., *Pseudodiagonalisation and the Inverse Nyquist Array Method*, Proc. I.E.E. Vol. 119, 3, March 1972.

[8]Araki, M., Kondo, B. and Yamamoto, K., *GG-Pseudo Band Method for The Design of Multivariable Control Systems*, reprints 8th Triennial World Congress, IFAC, Kyoto, Japan, Vol. 3, Paper 15.4, 1981.

[9]Rosenbrock, H. H., *Computer Aided Control System Design*, Academic Press Inc. (London) Ltd., 1974.

[10]Kouvaritakis, B. and Macfarlane, A. *Geometric Approach to the analysis and Synthesis of System Zeros, Part 2. Non-Square Systems*, Int. J. Control, 1976, Vol. 23, No. 2.

[11]Nwokah, O. D. I., *Neo-Classical Multivariable Feedback Control*, Lecture Notes for Course ME 597n, August 1986.

[12]Gantmacher, F. R., *The Theory of Matrices*, Vol. I and II, Chelsea Publishing Co., New York 1959.

[13]DeHoff, R. L., Hall, W., Adams, R. J., and Gupta, N. K., *F100 Multivariable Control Synthesis Program*, System Control Inc., AFAPL-TR-77-35 Vol. I and II, June 1977.

[14]Miller, R, J. and Hackney, R. D., *F100 Multivariable Control System Engine Models Design Criteria*, Pratt & Whitney Aircraft, AFAPL-TR-76-74, August 1976.

[15]Lallman, F. J., *Simplified Off-Design Performance Model of a Dry Turbofan Engine Cycle*, NASA TM-83204, September 1981.

[16]Perez, Ronald A., *Integrated Propulsion-Airframe Dynamics & Control, a Ph.D. Thesis Proposal*, Department of Mechanical Engineering, Purdue University, April 1989.

[17]Nwokah, O. D. I., and Perez, R. A., *On Multivariable Stability in the Gain Space*, submitted to A.S.M.E. Journal of System Dynamics, Measurement and Control, 1990.

[18]Nwokah O.D.I., Thompson D.F., *Algebraic and topological aspects of quantitative feedback theory*, Int. J. Control (In Press).

[19]Nwokah O.D.I., *The robust decentralized stabilization of complex feedback systems*, Proc. I.E.E., Ser. D., 1359, 43-47, 1987.

[20]Araki M., Nwokah O.D.I., *Bounds for closed loop transfer functions of multivariable systems*, I.E.E.E. Trans. Autom. Control, AC-17, 666-670, 1975.

[21]Nwokah O.D.I., *The robust decentralized stabilization of complex feedback systems*, Proc. I.E.E., Ser. D., 1359, 43-47, 1987.

[22]Koussiouris T.G., *A new stability theorem for multivariable systems*, Int. J. Control, 32, 435-441, 1980.

[23]Horowitz, I., *Quantitative Feedback Theory*, IEE Proc., Pt.D, 129, 215-226, 1982.

8. APPENDICES

Appendix 1 : Engine Matrices

The matrices that were used in the Engine model (equations (2.14) and (2.15)) are the following :

$$A_e = \begin{bmatrix} -9.6060d-01 & -5.9660d-01 & -1.6111d+02 & 1.0820d+03 & 1.8810d+01 \\ 7.9550d-01 & -1.6440d+00 & -3.8410d+01 & -1.5460d+02 & 9.8230d+00 \\ 1.7860d-02 & -3.5720d-02 & -8.8860d+00 & 4.1260d+01 & 5.7560d-01 \\ 0.0000d+00 & 0.0000d+00 & 0.0000d+00 & -1.0000d+01 & 0.0000d+00 \\ -5.1280d+00 & 1.2520d+01 & 6.5720d+02 & 9.6260d+03 & -1.6970d+02 \end{bmatrix}$$

$$B_e = \begin{bmatrix} -1.9360d+02 & -3.8860d+03 & -2.7230d+01 & -8.0010d+00 & 2.8000d+03 \\ 3.3290d+01 & 9.1136d+02 & 1.7470d+01 & -1.4170d+01 & 7.1940d+03 \\ -3.6540d+00 & -1.5410d+03 & 6.2880d-01 & -5.1300d-01 & -2.5590d+01 \\ 1.0000d+01 & 0.0000d+00 & 0.0000d+00 & 0.0000d+00 & 0.0000d+00 \\ -1.6400d+03 & -1.4370d+04 & -8.5430d+01 & 9.7290d+01 & -1.3270d+05 \end{bmatrix}$$

$$C_{ea} = \begin{bmatrix} 0.0000d+00 & 0.0000d+00 \\ 0.0000d+00 & 0.0000d+00 \\ 1.0660d+02 & -9.5940d+01 \\ 0.0000d+00 & 0.0000d+00 \\ 4.2370d+04 & -3.5050d+07 \end{bmatrix} \qquad C_{ei} = \begin{bmatrix} 0.0000d+00 \\ 0.0000d+00 \\ 1.0660d+02 \\ 0.0000d+00 \\ 4.2370d+04 \end{bmatrix}$$

$$F_e = \begin{bmatrix} 1.6490d{+}00 & -2.0280d{+}00 & -2.7850d{-}02 & -2.7050d{+}02 & 2.7790d{+}01 \\ 1.9620d{-}03 & 8.6410d{-}07 & -8.6710d{-}04 & 8.1530d{-}02 & 2.8780d{-}07 \\ 8.3060d{-}02 & -5.0020d{-}02 & -2.9120d{+}00 & 1.8870d{+}02 & -8.5330d{-}02 \\ 2.0680d{-}04 & 1.8620d{-}06 & -8.7330d{-}03 & 9.4280d{-}03 & 4.2630d{-}05 \\ -6.2040d{-}05 & 1.5240d{-}04 & 6.9600d{-}03 & 6.7220d{-}02 & -1.6060d{-}03 \\ 5.7050d{-}05 & -3.7290d{-}05 & -6.3700d{-}03 & -3.5380d{-}02 & 1.1570d{-}04 \end{bmatrix}$$

$$G_e = \begin{bmatrix} 1.8820d{+}01 & 1.1410d{+}04 & 4.0250d{+}01 & -1.3780d{+}01 & 1.5030d{+}04 \\ 2.3330d{-}04 & 9.9600d{-}01 & 8.1600d{-}02 & 1.6980d{-}04 & 7.0130d{-}01 \\ 6.2000d{+}00 & 3.0670d{+}02 & 1.3570d{+}00 & -2.4410d{-}01 & 1.0720d{+}03 \\ 7.7750d{-}04 & 2.7700d{-}01 & 7.3550d{-}04 & 1.4060d{-}04 & 5.9090d{-}01 \\ -2.0630d{-}02 & -2.1430d{-}01 & -1.0700d{-}03 & -1.6210d{-}03 & -4.0190d{-}01 \\ 1.1480d{-}02 & 3.0740d{-}01 & 1.9010d{-}03 & -2.3970d{-}04 & 2.6330d{-}01 \end{bmatrix}$$

$$H_{ea} = \begin{bmatrix} -5.5570d{+}03 & -4.6070d{+}02 \\ 2.0210d{+}01 & -5.8780d{+}00 \\ 0.0000d{+}00 & 0.0000d{+}00 \\ 0.0000d{+}00 & 0.0000d{+}00 \\ 0.0000d{+}00 & 0.0000d{+}00 \\ 0.0000d{+}00 & 0.0000d{+}00 \end{bmatrix} \qquad H_{ei} = \begin{bmatrix} 7.2930d{+}03 \\ 3.7260d{+}01 \\ 0.0000d{+}00 \\ 0.0000d{+}00 \\ 0.0000d{+}00 \end{bmatrix}$$

Appendix 2 : Airframe Matrices

The matrices that were used in the Airframe model (equations (2.16) and (2.17)) are the following :

$$A_a = \begin{bmatrix} -1.5729d{-}02 & -1.1923d{+}01 & 0.0000d{+}00 & -9.8065d{+}00 & 2.0029d{-}01 \\ -3.6756d{-}04 & -5.9740d{-}01 & 1.0000d{+}00 & 0.0000d{+}00 & 5.8608d{-}03 \\ -3.1733d{-}03 & -4.8424d{+}00 & -7.4522d{-}01 & 0.0000d{+}00 & 5.0486d{-}03 \\ 0.0000d{+}00 & 0.0000d{+}00 & 1.0000d{+}00 & 0.0000d{+}00 & 0.0000d{+}00 \\ 0.0000d{+}00 & -2.6556d{-}01 & 0.0000d{+}00 & 2.6556d{-}01 & 0.0000d{+}00 \end{bmatrix}$$

$$B_a = \begin{bmatrix} -2.3459d{+}00 \\ -6.4267d{-}02 \\ -7.4235d{+}00 \\ 0.0000d{+}00 \\ 0.0000d{+}00 \end{bmatrix}$$

$$C_{ae} = \begin{bmatrix} 1.2246d{-}04 & 0.0000d{+}00 & 0.0000d{+}00 & 0.0000d{+}00 & 0.0000d{+}00 & 0.0000d{+}00 \\ -3.5175d{-}08 & 0.0000d{+}00 & 0.0000d{+}00 & 0.0000d{+}00 & 0.0000d{+}00 & 0.0000d{+}00 \\ 2.5561d{-}06 & 0.0000d{+}00 & 0.0000d{+}00 & 0.0000d{+}00 & 0.0000d{+}00 & 0.0000d{+}00 \\ 0.0000d{+}00 & 0.0000d{+}00 & 0.0000d{+}00 & 0.0000d{+}00 & 0.0000d{+}00 & 0.0000d{+}00 \\ 0.0000d{+}00 & 0.0000d{+}00 & 0.0000d{+}00 & 0.0000d{+}00 & 0.0000d{+}00 & 0.0000d{+}00 \end{bmatrix}$$

$$F_a = \begin{bmatrix} 3.3887d{-}03 & 0.0000d{+}00 & 0.0000d{+}00 & 0.0000d{+}00 & 0.0000d{+}00 \\ 0.0000d{+}00 & 0.0000d{+}00 & 0.0000d{+}00 & 0.0000d{+}00 & 1.0000d{+}00 \end{bmatrix}$$

Appendix 3

The matrices that were used in equations (3.22) and (3.23) are the following :

$$B_n = \begin{bmatrix} -2.3459d{+}00 & 1.2246d{-}04 \\ -6.4267d{-}02 & -3.5175d{-}08 \\ -7.4235d{+}00 & 2.5561d{-}06 \\ 0.0000d{+}00 & 0.0000d{+}00 \\ 0.0000d{+}00 & 0.0000d{+}00 \end{bmatrix}$$

$$F_n = \begin{bmatrix} 1.00 & 0.00 & 0.00 & 0.00 & 0.00 \\ 0.00 & -1.00 & 0.00 & 1.00 & 0.00 \\ 0.00 & 0.00 & 1.00 & 0.00 & 0.00 \end{bmatrix} \qquad G_n = \begin{bmatrix} 0.00 & 0.00 \\ 0.00 & 0.00 \\ 0.00 & 0.00 \end{bmatrix}$$

Appendix 4 : Multivariable Zeros

Consider the following time-invariant linear system :

$$\left.\begin{aligned} \dot{x}(t) &= Ax(t) + Bu(t) \\ y(t) &= Cx(t) + Du(t) \end{aligned}\right\} \qquad \text{(A4.1)}$$

$x \in \mathbb{R}^n$, $u \in \mathbb{R}^\ell$, $y \in \mathbb{R}^m$, $m > \ell$.

A way of looking at the zeros is in terms of state-space parameters, to that effect we construct the system matrix P(s) (Rosenbrock[9]) :

$$P(s) \triangleq \begin{bmatrix} (sI - A) & B \\ -C & D \end{bmatrix} \qquad \text{(A4.2)}$$

The zeros of the system are the values s_o for which the $(n + m) \times (n + \ell)$ system matrix (for a proper system) loses rank. i.e. :

$$| P(s_o) | = \det \begin{bmatrix} (s_oI - A) & B \\ -C & D \end{bmatrix} = 0 \qquad \text{(A4.3)}$$

when $m \neq \ell$ such values (s_o) can be found by determining, for a P(s) of normal rank $[n + \min(\ell,m)]$ what values of "s" make all minors of order $[n + \min(\ell,m)]$ vanish. Considerations of the forms of such minors shows that they will contain among them one whose value is a polynomial of degree $[n - \max(\ell,m)]$. We thus note that the maximum degree of Z(s) can not be more than $[n - \max(\ell,m)]$ (Kouvaritakis and Macfarlane[10]).

Theorem A4.1. : The zeros of a linear time invariant multivariable system, are invariant under the following static transformations : similarity, output feedback and state feedback.

Proof :

We will only sketch the proof. A detailed proof is given in Perez[16].

Consider a system represented by equation (A4.1). Construct its "system matrix" as defined previously :

$$P(s) \triangleq \begin{bmatrix} (sI - A) & B \\ -C & D \end{bmatrix} \tag{A4.4}$$

The zeros of the system are the values s_o such that $| P(s_o) |$ will loose rank. i.e. :

$$| P(s_o) | = \det \begin{bmatrix} (s_oI - A) & B \\ -C & D \end{bmatrix} = 0 \tag{A4.5}$$

Similarity Transformation : Consider the following transformation (assume K is a nonsingular matrix) :

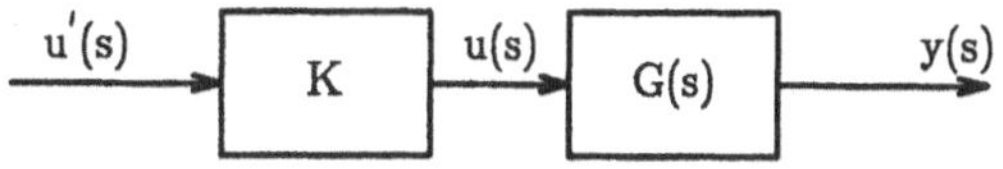

Figure A4.1 : Similarity Transformation.

therefore equation (A4.5) will become :

$$| P(s_o) | = \det \begin{bmatrix} (s_oI - A) & BK \\ -C & DK \end{bmatrix} = 0 \tag{A4.6}$$

Output Feedback : $u(t) = u'(t) - Hy(t)$

therefore equation (A4.5) will become :

$| P(s_o) | =$

$$\det \begin{bmatrix} (s_oI - A + BH(I + DH)^{-1}C) & (B - BH(I + DH)^{-1}D) \\ -(I + DH)^{-1}C & (I + DH)^{-1}D \end{bmatrix} = 0 \tag{A4.7}$$

State Feedback : $u(t) = u'(t) - Lx(t)$

therefore equation (A4.5) will become :

$$| P(s_o) | = \det \begin{bmatrix} (s_oI - A + BL) & B \\ -(C - DL) & D \end{bmatrix} = 0 \qquad \text{(A4.8)}$$

Therefore we need to show that equations (A4.5), (A4.6), (A4.7) and (A4.8) are all equivalent.

Applying Schur's formula for partitioned matrices (Gantmacher[12]), equations (A4.5), (A4.6), (A4.7) and (A4.8) can be shown to be equivalent, thus completing the proof.

□

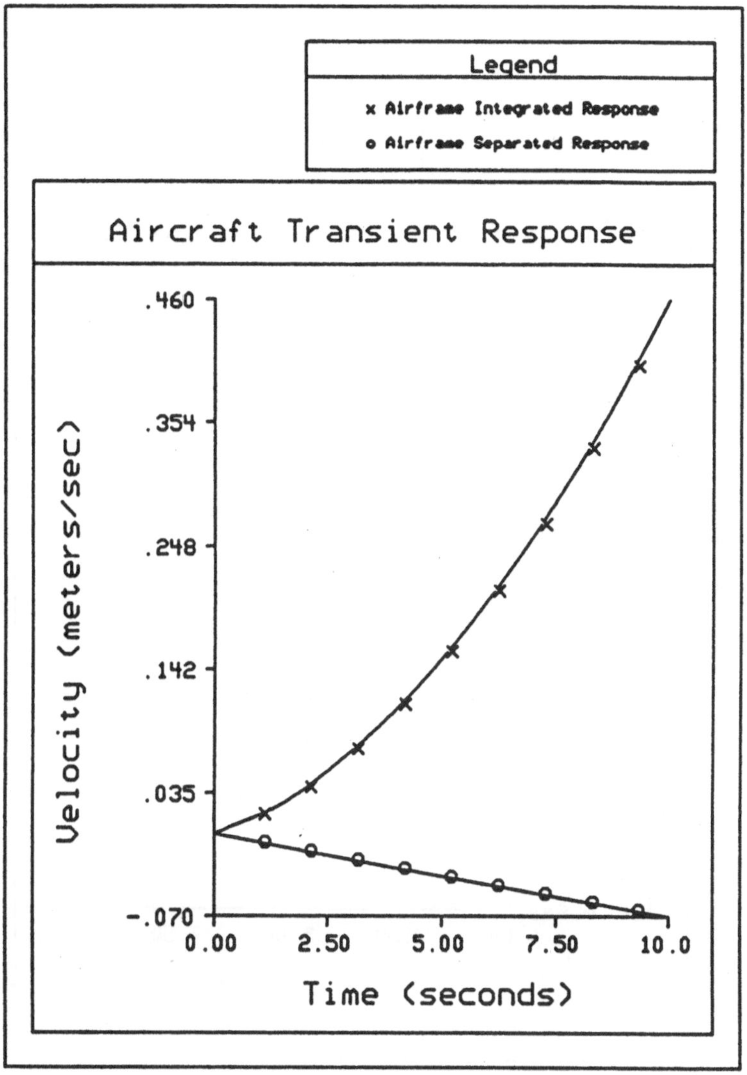

Figure 6. : Response due to a 1% offset about the nominal value of x_{a_4} .

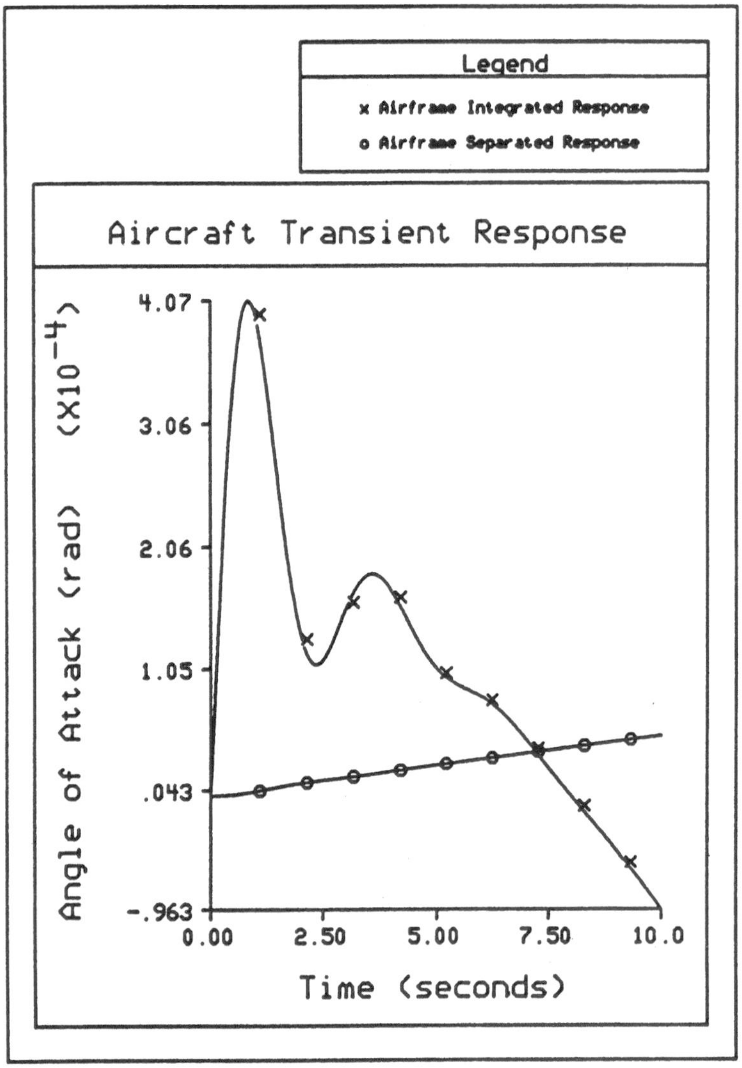

Figure 7. : Response due to a 1% offset about the nominal value of x_{a_4} .

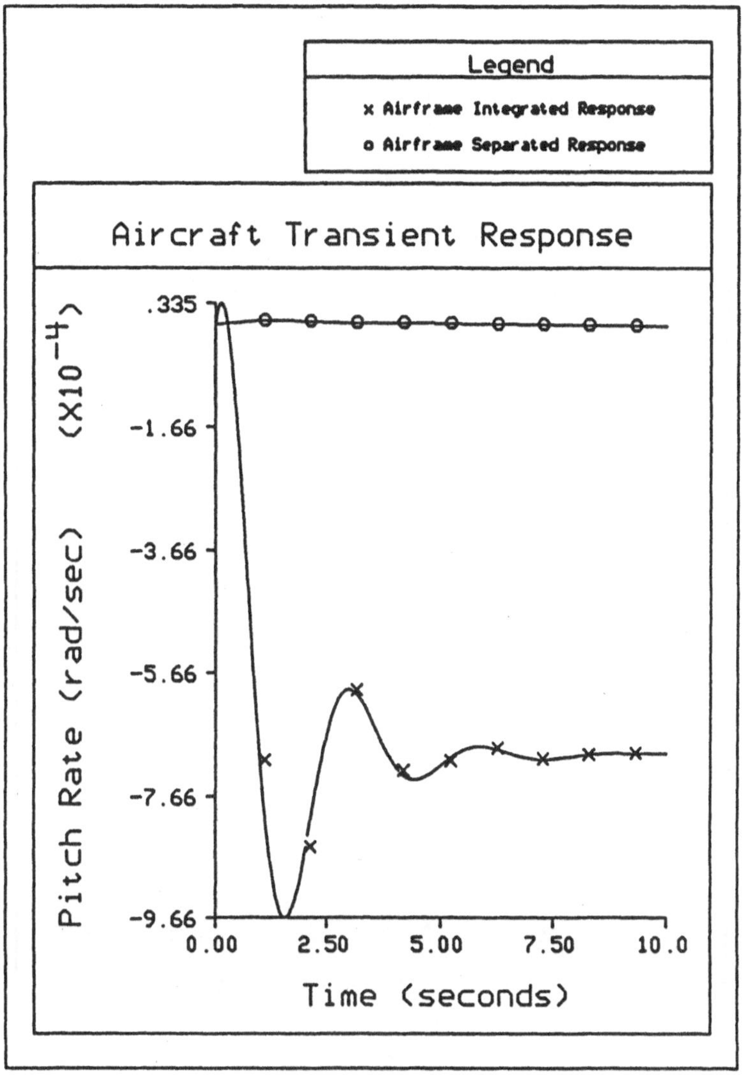

Figure 8. : Response due to a 1% offset about the nominal value of x_{a_4} .

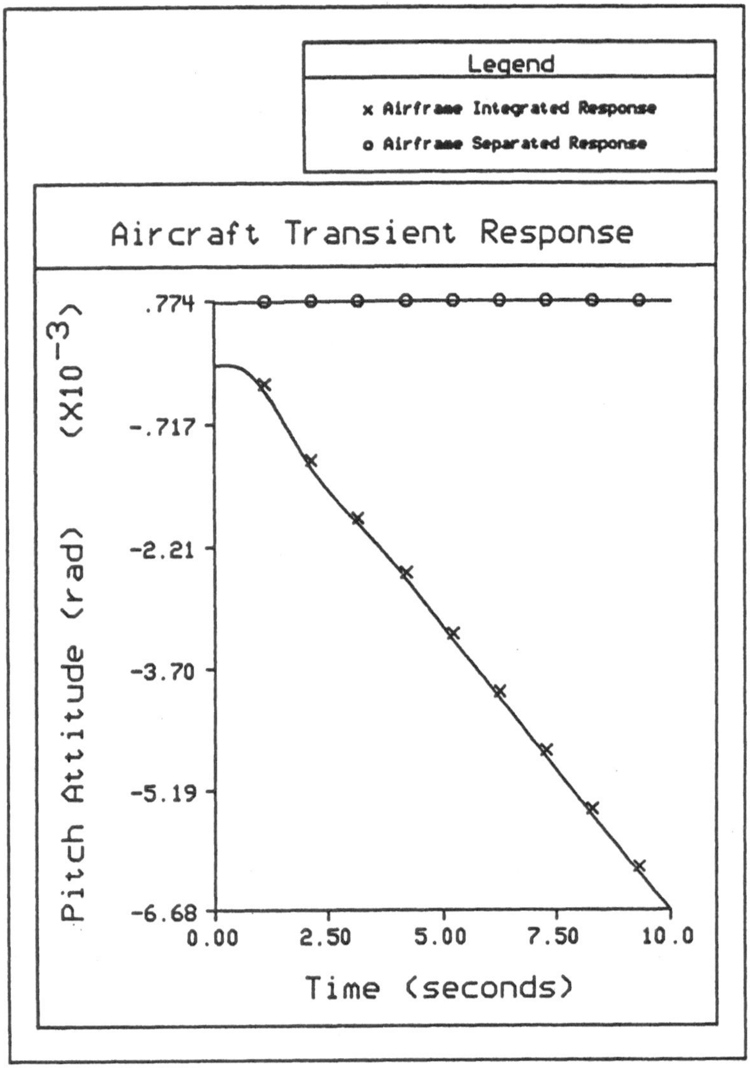

Figure 9. : Response due to a 1% offset about the nominal value of x_{a_4} .

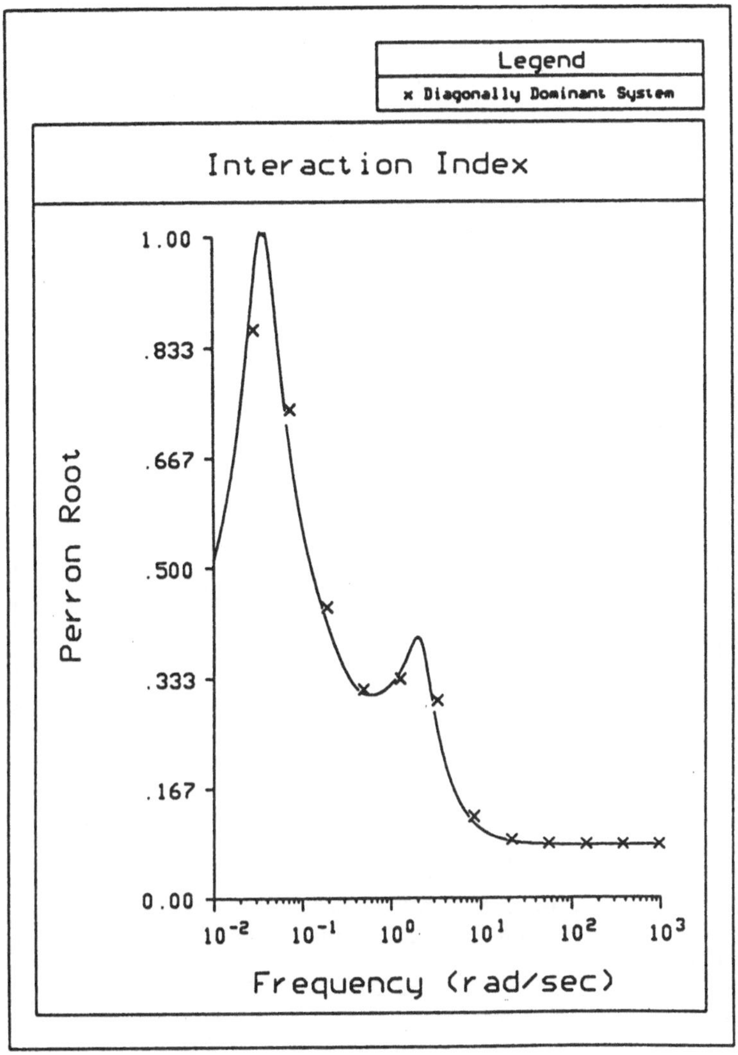

Figure 10. Interaction Index of Diagonally Dominant System

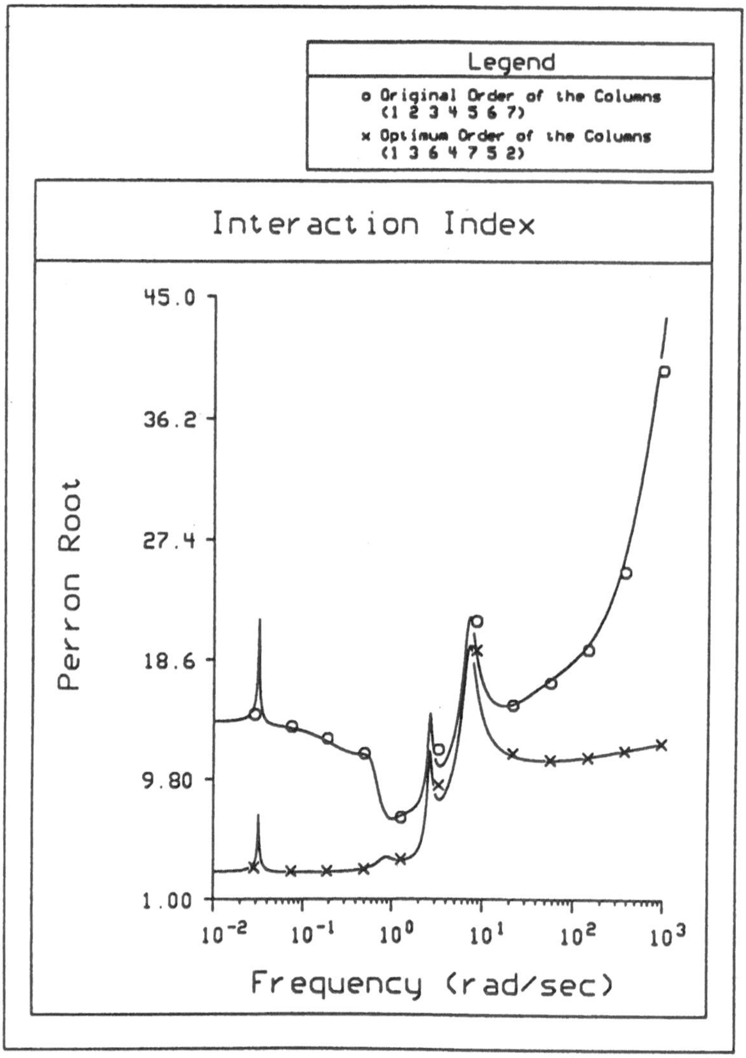

Figure 11. Interaction Index of the Original and Optimum Order of the Columns of the Integrated Engine-Airframe System

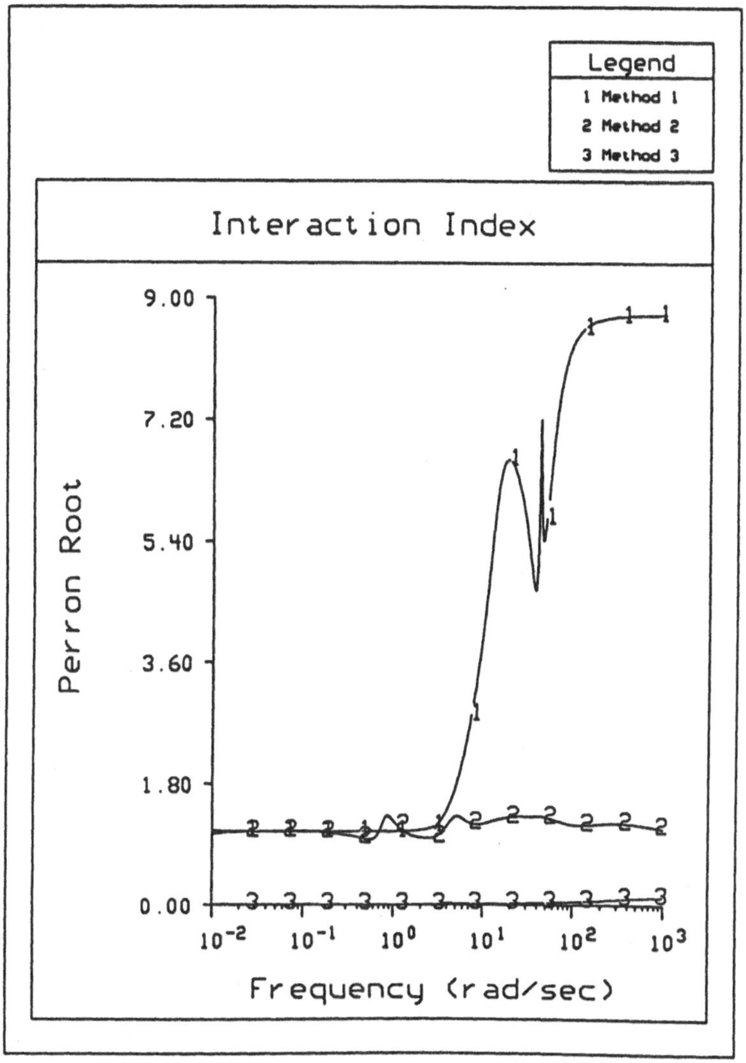

Figure 12. Interaction Indices of the Integrated Engine-Airframe System with 3 Interaction Reduction Controllers

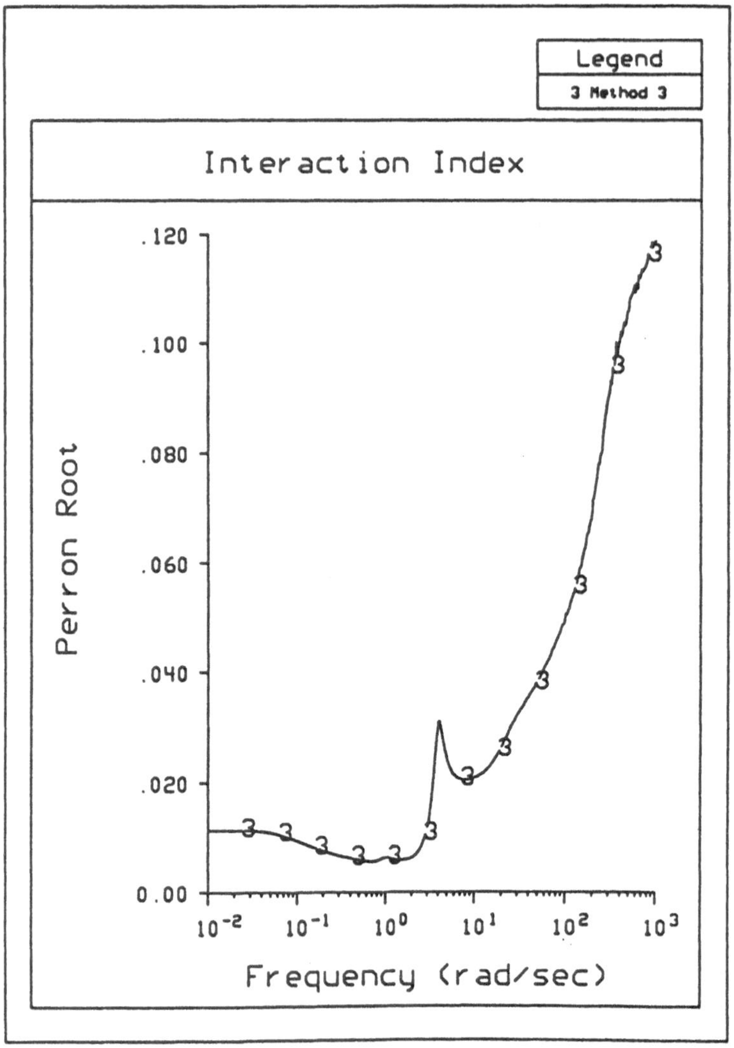

Figure 13. **Interaction Index of the Integrated Engine-Airframe System with the Optimum Interaction Reduction Controller**

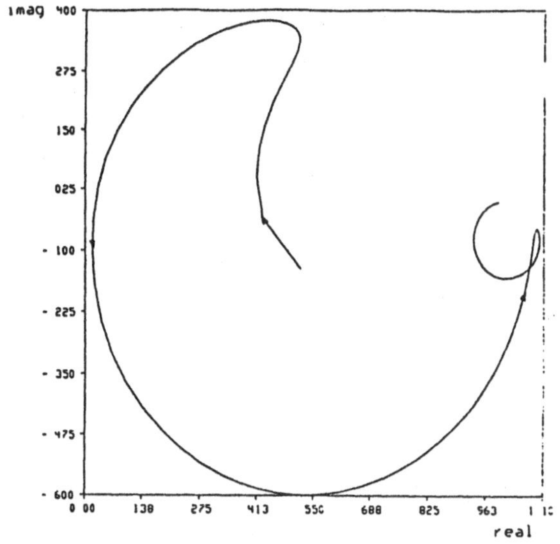

Figure 14. : Airframe : $\omega \in [0\ 200]$rad/sec.

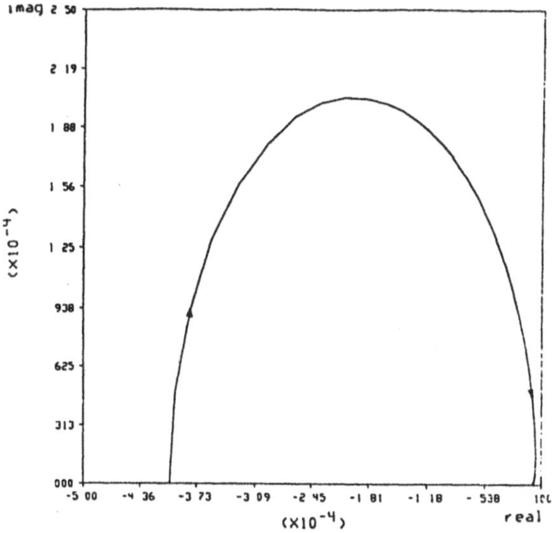

Figure 15. : Engine - Airframe and K_1 : $\omega \in [0\ 200]$rad/sec.

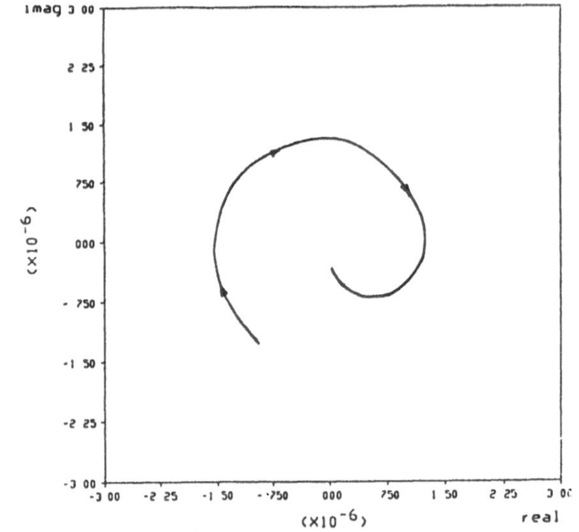

Figure 16. : Engine - Airframe and K_1 : $\omega \in [20\ 200]$rad/sec.

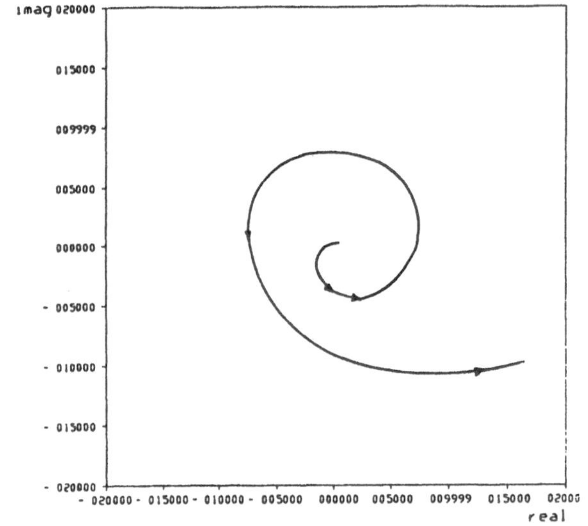

Figure 17. : Engine - Airframe and K_2 : $\omega \in [20\ 200]$rad/sec.

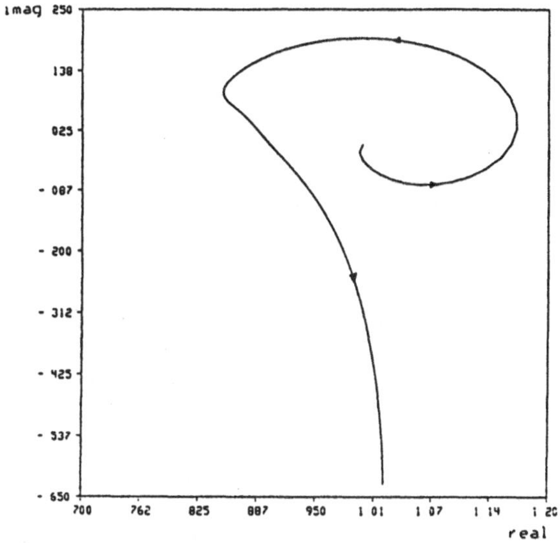

Figure 18. : Engine - Airframe and K_3 : $\omega \in [20\ 500]$rad/sec.

FINITE-TIME STABILIZATION OF UNCERTAIN NONLINEAR PLANAR SYSTEMS

E.P. RYAN
School of Mathematical Sciences
University of Bath
Claverton Down
Bath BA2 7AY
United Kingdom

Dedicated to George Leitmann
on the occasion of his 65th birthday

Abstract

A stabilizer is described which renders $0 \in \mathbb{R}^2$ a global finite-time attractor for a class of nonlinear uncertain two-dimensional dynamical systems. The stabilizer is of discontinuous feedback form: the analytic framework is that of differential inclusions. An adaptive version of the stabilizer is also developed which, for a larger class of systems, guarantees global asymptotic attractivity of the origin.

Keywords: Adaptive control; differential inclusions; discontinuous feedback; stability; universal stabilization.

1 INTRODUCTION

As motivation, consider the controlled Duffing-type system with extraneous periodic forcing and unknown real parameters α ($\neq 0$), a_i, β, ω:

$$\alpha\ddot{x}(t) + a_1\dot{x}(t) + a_2x(t) + a_3x^3(t) + a_4\sin\omega t + \beta u(t) = 0. \tag{1}$$

In the absence of control ($u(\cdot) = 0$), this system exhibits highly irregular dynamic behaviour, with attracting sets of fractal structure (in time-phase space), for ranges of parameter values [1,2]. If the unknown vector of parameters $a = (a_1, a_2, a_3, a_4) \in \mathbb{R}^4$ lies in a known compact set A, then solutions of the non-autonomous uncertain system (1) are, *a fortiori*, solutions of the autonomous differential inclusion

$$\alpha\ddot{x}(t) + \beta u(t) \in \mathcal{G}(x(t), \dot{x}(t)), \tag{2}$$

where the set-valued map $\mathcal{G}$ is given by

$$\mathcal{G} : (x, y) \mapsto (1 + |y| + |x| + |x|^3)\bar{a}\bar{B},$$

with $\bar{a} := \max\{|a| : a \in A\}$. Here $|a| := \max\{|a_1|, |a_2|, |a_3|, |a_4|\}$, and $\bar{B}$ denotes the closure of the open unit ball B centred on the origin in $\mathbb{R}^2$.

Clearly, (1) is but one particular example of a non-autonomous uncertain system which can be subsumed by a differential inclusion of form (2). More generally, any scalar two-dimensional non-autonomous uncertain system of the form

$$\alpha\ddot{x}(t) + g(t, x(t), \dot{x}(t)) + \beta u(t) = 0$$

can be embedded in an autonomous inclusion of form (2) with known right-hand side provided that the unknown function (assumed measurable in its first argument) is bounded by a known function γ in the sense that, for almost all $t \in \mathbb{R}$,

$$|g(t, x, y)| \leq \gamma(x, y) \quad \forall\ (x, y) \in \mathbb{R}^2.$$

In such cases, we may simply define $\mathcal{G}$ as the map $(x,y) \mapsto \gamma(x,y)\bar{B}$.

Our contribution is to the study of stabilizability properties for systems of form (2), with the underlying assumption that the set-valued map $\mathcal{G}$ is known, is continuous on $\mathbb{R}^2$ and takes non-empty, convex and compact values (compact intervals of $\mathbb{R}$).

We first consider, in Section 3, the case in which $\alpha \neq 0$ is known (we may assume $\alpha = 1$ without loss of generality) and β is bounded from below by a known constant $b > 0$. Under these conditions, we demonstrate that the inclusion system is *finite-time stabilizable*: in particular, a discontinuous state feedback strategy is constructed which guarantees that, for every initial-data pair $(x(0), \dot{x}(0)) = (x^0, y^0) \in \mathbb{R}^2$, every state solution $(x(\cdot), \dot{x}(\cdot))$ of (2) has maximal interval of existence $[0, \infty)$ and there exists a (calculable) scalar $T(x^0, y^0) \geq 0$ such that $(x(t), \dot{x}(t)) = (0,0)$ for all $t \geq T(x^0, y^0)$.

In Section 4, we weaken the hypotheses on α and β. There, we assume only that both parameters are non-zero. In this case, an adaptive modification to the strategy of Section 3 is proposed which preserves the zero state as a global attractor. However, we are only able to establish that the nature of the attractor is asymptotic: whether or not it is finite-time or, indeed, even exponentially attractive is an open question.

2 FORMULATION

The class of planar systems to be considered is the following:

$$\dot{x}(t) = y(t), \quad \alpha\dot{y}(t) + \beta u(t) \in \mathcal{G}(x(t), y(t)), \quad (x(0), y(0)) = (x^0, y^0) \in \mathbb{R}^2. \tag{3}$$

We suppose that the full state $(x(t), y(t))$ is available for feedback.

Assumption 1: The map $(x,y) \mapsto \mathcal{G}(x,y) \subset \mathbb{R}$ is continuous on $\mathbb{R}^2$ and takes non-empty, convex and compact values.

Remark: $\mathcal{G}$ is continuous on $\mathbb{R}^2$ if it is both upper and lower semicontinuous at every $(\bar{x}, \bar{y}) \in \mathbb{R}^2$: $\mathcal{G}$ is upper semicontinuous at $(\bar{x}, \bar{y})$ if, for each $\epsilon > 0$, there exists $\delta > 0$ such that $\mathcal{G}(x,y) \subset \mathcal{G}(\bar{x}, \bar{y}) + \epsilon B$ for all $(x,y) \in (\bar{x}, \bar{y}) + \delta B$; $\mathcal{G}$ is lower semicontinuous at $(\bar{x}, \bar{y})$ if, for every sequence $\{(x_n, y_n)\}$ converging to $(\bar{x}, \bar{y})$ and for every $\phi \in \mathcal{G}(\bar{x}, \bar{y})$, there exists a sequence $\{\phi_n \in \mathcal{G}(x_n, y_n)\}$ converging to ϕ (see [3] for further details).

The primary objective is to determine a feedback strategy which renders the zero state of (3) globally attractive.

3 NON-ADAPTIVE FEEDBACK STRATEGY: FINITE-TIME STABILIZATION

In this section we impose the following on the parameters $\alpha, \beta \in \mathbb{R}$.

Assumption 2: (i) α is non-zero and known; (ii) $\beta \geq b > 0$, b known.

Without loss of generality, we may assume that $\alpha = 1$, in which case system (3) becomes

$$\dot{x}(t) = y(t), \quad \dot{y}(t) + \beta u(t) \in \mathcal{G}(x(t), y(t)), \quad (x(0), y(0)) = (x^0, y^0). \tag{4}$$

Our approach to synthesis of a finite-time stabilizer for (4) is based on the solution to a particular optimal control problem which we first describe.

3.1 An optimal control problem

Suppose, for the moment, that $\mathcal{G}(\cdot,\cdot) = \{0\}$ and $\beta = b$. Then (4) reduces to the double integrator system

$$\dot{x}(t) = y(t), \quad \dot{y}(t) = -bu(t), \quad (x(0), y(0)) = (x^0, y^0). \tag{5}$$

For this system, consider the free-end-time optimal control problem

$$\mathcal{P}_1: \qquad \text{minimize } I(u) = \int_0^T (x^2(t) + \tfrac{1}{12}y^4(t))dt, \quad T \text{ free},$$

subject to $|u(t)| \leq b^{-1}$ almost everywhere and $(x(T), y(T)) = (0,0)$.
As shown in [4], the solution to problem $\mathcal{P}_1$ is identical to that of the familiar minimum-time control problem

$$\mathcal{P}_2: \qquad \text{minimize } J(u) = \int_0^T dt$$

subject to the same constraints as above. Thus, the optimal control for $\mathcal{P}_1$ is bang-bang and has the following feedback characterization (well-known in the context of $\mathcal{P}_2$)

$$u(t) = b^{-1}\sigma(x(t), y(t)), \tag{6}$$

$$\sigma: (x,y) \mapsto \begin{cases} \operatorname{sgn}(x + \frac{1}{2}y|y|), & x + \frac{1}{2}y|y| \neq 0 \\ \operatorname{sgn}(y), & x + \frac{1}{2}y|y| = 0, \quad y \neq 0 \,. \\ 0, & (x,y) = (0,0) \end{cases}$$

The value function for problem $\mathcal{P}_2$ is continuous but non-smooth: in fact, not even locally Lipschitz. By contrast, the value function $V : \mathbb{R}^2 \to [0,\infty)$ for problem $\mathcal{P}_1$ is continuously differentiable and is given by

$$V: (x,y) \mapsto x^2 y\sigma(x,y) + \tfrac{2}{3}xy^3 + \tfrac{3}{20}y^5\sigma(x,y) + \tfrac{4}{5}(x\sigma(x,y) + \tfrac{1}{2}y^2)^{5/2}.$$

Level surfaces of V are depicted in Figure 1.

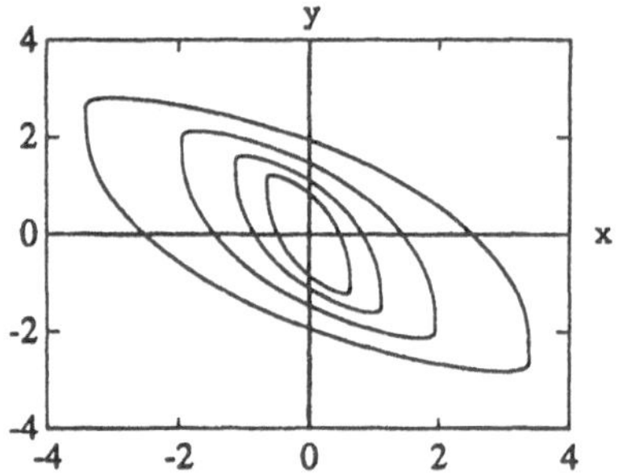

Figure 1 : Level surfaces of V

We note, for later use, that V satisfies the Hamilton-Jacobi-Bellman equation on $\mathbb{R}^2$

$$yV_x(x,y) - |V_y(x,y)| + x^2 + \tfrac{1}{12}y^4 = 0. \tag{7}$$

In essence, our approach to the stabilization problem for (4) is to modify the "minimum-time" feedback control (6) by incorporating a state-dependent gain designed to counteract the worst-case effects of $\mathcal{G}$. Stability of the resulting feedback-controlled inclusion is analysed using the C^1 function V as a Lyapunov function candidate. The following observation will be invoked in that analysis.

Proposition 1. *Define* $c := \max\{V(x,y) : x^2 + \frac{1}{12}y^4 = 1\}$. *Then*

$$V(x,y) \leq c(x^2 + \tfrac{1}{12}y^4)^{5/4} \quad \forall\, (x,y) \in \mathbb{R}^2.$$

Proof. Note initially that the continuous function V has the following homogeneity property: for all $(x,y) \in \mathbb{R}^2$,

$$V(k^2x, ky) = k^5 V(x,y) \quad \forall\, k > 0.$$

A direct consequence of this is

$$\sup_{(x,y)\neq(0,0)} \frac{V(x,y)}{(x^2 + \frac{1}{12}y^4)^{5/4}} = \max\,\{V(x,y) : x^2 + \tfrac{1}{12}y^4 = 1\},$$

whence the result. Q.E.D.

3.2 Finite-time stabilizer

Define $\gamma : \mathbb{R}^2 \to [0, \infty)$ by

$$\gamma(x, y) = \max \{|v| : v \in \mathcal{G}(x, y)\}.$$

By continuity of $\mathcal{G}$, it is easily seen that γ is also continuous. Our claim is that the discontinuous state feedback strategy

$$u(t) = b^{-1}(1 + \gamma(x(t), y(t)))\sigma(x(t), y(t)),$$

renders the zero state of (4) globally finite-time attractive.

We first interpret this strategy in the following generalized sense

$$u(t) \in \Psi(x(t), y(t)), \quad \Psi(x, y) := b^{-1}(1 + \gamma(x, y))\psi(x, y) \tag{8}$$

$$\text{where} \qquad \psi(x, y) := \begin{cases} \{+1\}, & x + \frac{1}{2}y|y| > 0 \\ [-1, 1], & x + \frac{1}{2}y|y| = 0 \\ \{-1\}, & x + \frac{1}{2}y|y| < 0 \end{cases}.$$

(Notation: for $c, d \in \mathbb{R}$ and $S \subset \mathbb{R}$, $c + dS$ denotes the set $\{c + ds : s \in S\}$.)

Writing $z(t) = (x(t), y(t))$, the feedback-controlled initial-value problem (4,8) is now embedded in the following

$$\dot{z}(t) \in \mathcal{F}(z(t)), \quad z(0) = z^0 = (x^0, y^0), \tag{9}$$

$$\mathcal{F}(z) := \{y\} \times \{v - \beta u : v \in \mathcal{G}(x, y),\ u \in \Psi(x, y)\} \ \forall\, z = (x, y) \in \mathbb{R}^2.$$

It is clear that $\mathcal{F}$ takes convex and compact values. Continuity of γ and upper semicontinuity of ψ ensure that Ψ is upper semicontinuous which, together with continuity of $\mathcal{G}$, implies upper semicontinuity of $\mathcal{F}$. Therefore, for each $z^0 \in \mathbb{R}^2$, (9) admits a solution and every such solution has a maximal extension[1] (see, for example, [3,5,6]).

Theorem 1. *Let $z : [0, \omega) \to \mathbb{R}^2$ be a maximal solution of (9). Then (i) $\omega = \infty$; (ii) $z(t) = 0$ for all $t \geq T(z^0) := 5c^{\frac{4}{5}}V^{\frac{1}{5}}(z^0)$, where c is as in Proposition 1.*

Proof. By direct calculation, we find that for all $(u, v) \in \Psi(x, y) \times \mathcal{G}(x, y)$

$$V_y(x, y)[v - \beta u] \leq |V_y(x, y)|[(1 - \beta b^{-1})\gamma(x, y) - \beta b^{-1}] \leq -|V_y(x, y)|.$$

Therefore, in view of (7), for almost all $t \in [0, \omega)$ we have

$$\frac{d}{dt}V(x(t), y(t)) \leq y(t)V_x(x(t), y(t)) - |V_y(x(t), y(t))| = -x^2(t) - \tfrac{1}{12}y^4(t).$$

It follows that the solution $z(\cdot)$ is bounded, whence assertion (i).

By Proposition 1, we may conclude that

$$\frac{d}{dt}V(x(t), y(t)) \leq -(c^{-1}V(x(t), y(t)))^{\frac{4}{5}} \quad a.e.,$$

which, on integration, confirms assertion (ii). Q.E.D.

---

[1]By a solution of (9) we mean a function $t \mapsto z(t)$, defined on some interval $[0, \omega)$, which is absolutely continuous on compact sub-intervals, and which satisfies the initial condition and the differential inclusion almost everywhere: a solution z is *maximal* if it does not have a proper extension which is also a solution: if z is a *bounded* maximal solution, then $\omega = \infty$.

3.3 Example

Consider again the Duffing-type system (1). We suppose that Assumption 2 holds with $\alpha = 1 = b$, and that the vector of unknown parameters $a = (a_1, a_2, a_3, a_4)$ lies in the unit cube in $\mathbb{R}^4$. Then (1) may be embedded in (4) with

$$\mathcal{G}(x,y) := \gamma(x,y)\bar{B}, \quad \gamma(x,y) := 1 + |y| + |x| + |x|^3.$$

For purposes of illustration, let $a_1 = -0.4$, $a_2 = 1$, $a_3 = -1$, $a_4 = 0.4$ and $\beta = 1 = \omega$, in which case Figure 2 depicts the uncontrolled and controlled phase plane behaviour. Figure 3 depicts the controlled state evolution, wherein finite-time attainment of the equilibrium is clearly evident.

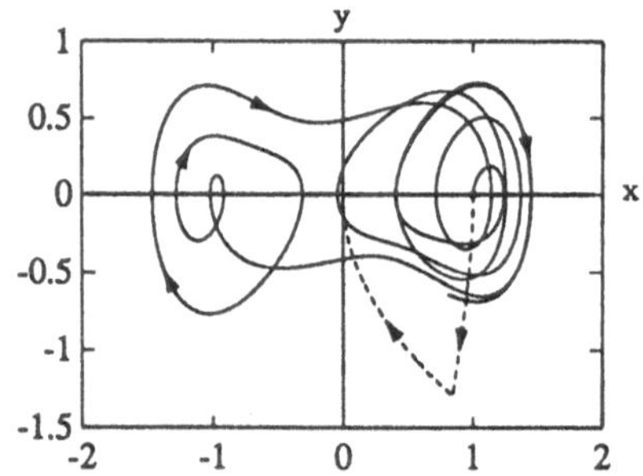

Figure 2 : Uncontrolled (solid) and controlled (dashed) behaviour

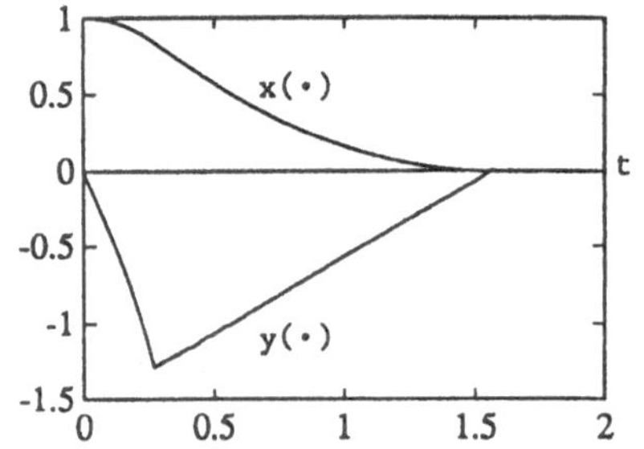

Figure 3 : Controlled state evolution

4 ADAPTIVE FEEDBACK STRATEGY: ASYMPTOTIC STABILIZATION

In this section we weaken the assumption on the parameters α and β of Section 3 to the following.

Assumption 3: $\alpha\beta \neq 0$.

Let ν be any continuous function of Nussbaum type [7-10]: precisely, any continuous function $\mathbb{R} \to \mathbb{R}$ with the properties

$$\limsup_{s\to\infty} \frac{1}{s}\int_0^s \nu = \infty, \qquad \liminf_{s\to\infty} \frac{1}{s}\int_0^s \nu = -\infty. \tag{10}$$

For example, the function $\nu : \theta \mapsto \theta^2 \cos\theta$ suffices. Writing

$$z(t) = (x(t), y(t), \kappa(t)),$$

our claim is that the following adaptive version of the generalized feedback (8) renders the zero state of (3) asymptotically attractive:

$$u(t) \in \Psi(z(t)), \quad \Psi : \ z = (x,y,\kappa) \mapsto \nu(\kappa)(1+\gamma(x,y))\psi(x,y), \tag{11}$$

where $\kappa(\cdot)$ is generated by the adaption law

$$\dot{\kappa}(t) = K(x(t), y(t)), \qquad \kappa(0) = \kappa^0 \tag{12}$$

$$K(x,y) := (1+\gamma(x,y))|V_y(x,y)|.$$

Observe that K is continuous; furthermore, $K(x,y) \geq 0$ for all (x,y), with equality holding on the set $\Sigma = \{(x,y):\ x + \frac{1}{2}y|y| = 0\}$, and so $\kappa(\cdot)$ is monotone increasing.

We embed the overall adaptively-controlled initial-value problem (3,11-12) in the following

$$\dot{z}(t) \in \mathcal{F}(z(t)), \quad z(0) = z^0 = (x^0, y^0, \kappa^0), \tag{13}$$

where now $(x,y,\kappa) = z \mapsto \mathcal{F}(z)$ is the set-valued map defined on $\mathbb{R}^3$ by

$$\mathcal{F}(z) := \{y\} \times \{\alpha^{-1}(v - \beta u):\ v \in \mathcal{G}(x,y),\ u \in \Psi(x,y,\kappa)\} \times \{K(x,y)\}$$

and is upper semicontinuous with convex and compact values.

In establishing stability properties of system (13) in Theorem 3, we will invoke the following extension [6] of LaSalle's principle ([11,12]) relating to asymptotic behaviour of solutions of initial-value problems for differential inclusions.

Theorem 2. *Consider the initial-value problem on* $\mathbb{R}^N$

$$\dot{z}(t) \in F(z(t)), \quad z(0) = z^0,$$

where $z \mapsto F(z) \subset \mathbb{R}^N$ *is upper semicontinuous on* $\mathbb{R}^N$ *with non-empty convex and compact values. Let* $V: \mathbb{R}^N \to \mathbb{R}$ *be continuously differentiable. Define*

$$q: \mathbb{R}^N \to \mathbb{R}, \quad z \mapsto q(z) := \max\{\langle \nabla V(z), \phi\rangle:\ \phi \in F(z)\}.$$

Suppose that

$$q(z) \leq 0 \quad \forall\, z.$$

If $z(\cdot)$ *is a bounded solution of the initial-value problem, then, for some* $c \in \mathbb{R}$, z *approaches the maximal quasi-invariant*[2] *set in* $\Sigma \cap V^{-1}(c)$, *where* $\Sigma = \{z \in \mathbb{R}^N:\ q(z) = 0\}$.

We now prove the main result of this section.

Theorem 3. *Let* $z(\cdot) = (x(\cdot), y(\cdot), \kappa(\cdot)) : [0,\omega) \to \mathbb{R}^3$ *be a maximal solution of (13). Then (i)* $\omega = \infty$; *(ii)* $\lim_{t\to\infty} \kappa(t)$ *exists and is finite; (iii)* $\lim_{t\to\infty} \|(x(t), y(t))\| = 0$.

Proof. Let V be as before and let $W:\ z = (x,y,\kappa) \mapsto V(x,y)$, a positive semi-definite form on $\mathbb{R}^3$. Write $\kappa^* = \alpha\beta^{-1}\max\{1, |\alpha|^{-1}\}$. Then a straightforward calculation yields the inequality

$$\begin{aligned}
\langle \nabla W(z), \phi\rangle &\leq yV_x(x,y) + |\alpha|^{-1}\gamma(x,y)|V_y(x,y)| \\
&\quad -\alpha^{-1}\beta\nu(\kappa)(1+\gamma(x,y))|V_y(x,y)| \\
&\leq yV_x(x,y) - |V_y(x,y)| - \alpha^{-1}\beta(\nu(\kappa) - \kappa^*)K(x,y) \\
&= -x^2 - \tfrac{1}{12}y^4 - \alpha^{-1}\beta(\nu(\kappa) - \kappa^*)K(x,y) \qquad \forall\, \phi \in \mathcal{F}(z)
\end{aligned}$$

which is valid for all $z = (x,y,\kappa) \in \mathbb{R}^3$. Therefore, for the maximally extended solution $z(\cdot)$, we conclude that

$$\frac{d}{dt} W(z(t)) \leq -\alpha^{-1}\beta(\nu(\kappa(t)) - \kappa^*)K(x(t), y(t)) \quad \text{for almost all } t,$$

[2]A set S is quasi-invariant if, for each $z^0 \in S$, there exists at least one maximal solution with trajectory in S.

which, on integration, gives for all $t, \tau \in [0,\omega)$ with $t \geq \tau$

$$0 \leq W(z(t)) \leq W(z(\tau)) + \alpha^{-1}\beta\kappa^*(\kappa(t)-\kappa(\tau)) - \alpha^{-1}\beta\int_{\kappa(\tau)}^{\kappa(t)} \nu \ . \tag{14}$$

We can now show that $\kappa(\cdot)$ is bounded. Seeking a contradiction, suppose that the monotone function $\kappa(\cdot)$ is unbounded. Fix $\epsilon > 0$ and let $\tau \in [0,\omega)$ be such that $\kappa(\tau) \geq \epsilon$. Then

$$0 \leq \liminf_{t\uparrow\omega} \frac{W(z(t))}{\kappa(t)} \leq \epsilon^{-1}W(z(\tau)) + \alpha^{-1}\beta\kappa^* + \liminf_{s\to\infty} \frac{-\alpha^{-1}\beta}{s}\int_{\kappa(\tau)}^{s} \nu$$

which, depending on the sign of $\alpha^{-1}\beta$ ($\neq 0$), contradicts one or the other of properties (10). Therefore, $\kappa(\cdot)$ is bounded and so $\lim_{t\uparrow\omega}\kappa(t)$ exists and is finite. From (14) it follows that $W(z(\cdot))$ is bounded, whence boundedness of $x(\cdot)$ and $y(\cdot)$. We have now shown that $z(\cdot) = (x(\cdot), y(\cdot), \kappa(\cdot))$ is bounded and so $\omega = \infty$. It remains only to prove assertion (iii). Let $U : \mathbb{R}^3 \to \mathbb{R}$ be given by

$$U: \ z = (x,y,\kappa) \mapsto W(z) - \alpha^{-1}\beta(\kappa^*\kappa - \int_0^{\kappa} \nu\ ).$$

Then, in view of (13) and (14) we see that, for all $z = (x,y,\kappa) \in \mathbb{R}^3$,

$$\langle \nabla U(z), \phi\rangle \leq -x^2 - \tfrac{1}{12}y^4 \quad \forall\, \phi \in \mathcal{F}(z).$$

By Theorem 2, it follows that the solution $z(\cdot) = (x(\cdot), y(\cdot), \kappa(\cdot))$ approaches the set

$$\{z = (x,y,\kappa) \in \mathbb{R}^3 : \ x = 0 = y\}.$$

Q.E.D.

4.1 Example

Consider again the Duffing-type system (1). We suppose that Assumption 3 holds. Defining

$$\mathcal{G}: \ (x,y) \ \mapsto \ 1 + |y| + |x| + |x|^3,$$

then the system can be embedded in (3). Adopting the same illustrative parameter values as in Section 3.3, Figure 4 depicts the adaptively-controlled state evolution; Figure 5 depicts the evolution of the adapting parameter and associated Nussbaum gain.

While Theorem 3 establishes the zero state as a global *asymptotic* attractor, Figures 4 and 5 would seem to suggest that the zero state may, in fact, have the stronger property of finite-time attractivity (as in the non-adaptive case of Section 3). However, we have not been able to prove this conjecture.

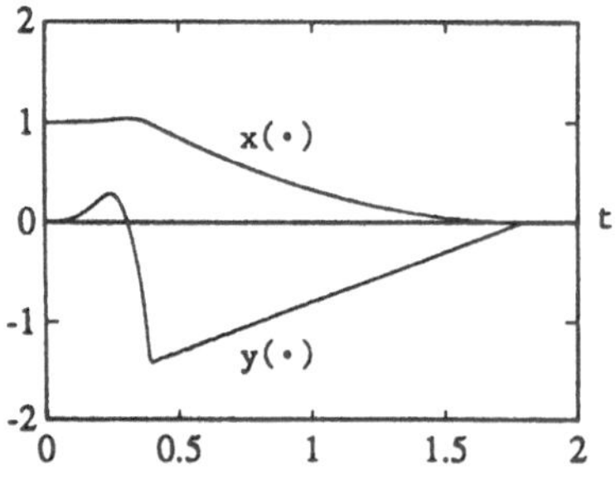

Figure 4 : Adaptively controlled state evolution

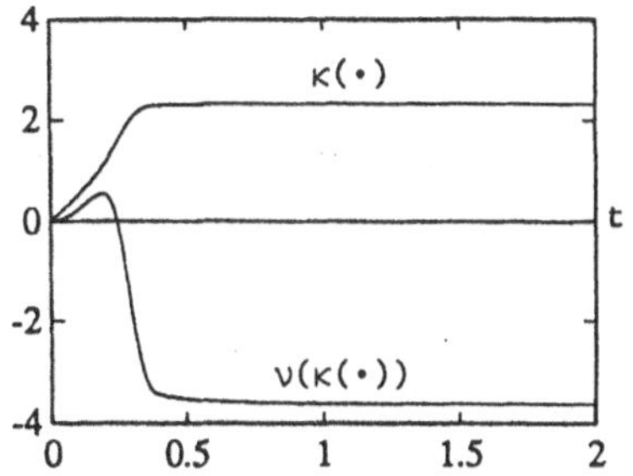

Figure 5 : Evolution of Nussbaum gain

5 DISCUSSION

In Section 3 of the paper, under relatively weak *a priori* assumptions we have constructed a stabilizing *discontinuous* feedback strategy for a class of nonlinear planar systems. The basic philosophy coincides with that underlying the Lyapunov design methodology of [13,14] (which, has since developed in a variety of directions - see [15] for an overview and an extensive bibliography). A novel feature of our approach is the adoption, as candidate Lyapunov function, of the non-quadratic value function V for a particular optimal control problem. With this choice of C^1 function V, the associated control design guarantees that the feedback system exhibits a strong stability property, namely, *finite-time* attractivity of the zero state.

Under weaker *a priori* assumptions on the system, an adaptive version of the discontinuous feedback strategy has been developed in Section 4. This strategy incorporates gains of Nussbaum type and is essentially a nonlinear analogue of the *universal stabilizers* for single-input linear systems developed in, for example, [7-10] (see, also [16-18]): however, the highly nonlinear nature of the system considered in the present paper necessitates a different analytical framework in which an extension [6], to differential inclusions, of LaSalle's principle plays a central role. The adaptive strategy renders the zero state a global *asymptotic* attractor: whether or not the attraction is, in fact, of a finite-time nature remains an open question (simulation suggests that it might be so).

REFERENCES

[1] J. Guckenheimer & P. Holmes, *Nonlinear Oscillations, Dynamical Systems, and Bifurcations of Vector Fields*, Springer-Verlag, New York, 1983.

[2] J.M.T. Thompson & H.B. Stewart, *Nonlinear Dynamics and Chaos*, Wiley, New York, 1986.

[3] J.P. Aubin & A. Cellina, *Differential Inclusions*, Springer-Verlag, New York, 1984.

[4] E.P. Ryan, *Optimal Relay and Saturating Control System Synthesis*, Peregrinus, London, 1982

[5] A.F. Filippov, *Differential Equations with Discontinuous Righthand Sides*, Kluwer, Dordrecht, 1988.

[6] E.P. Ryan, "Discontinuous feedback and universal adaptive stabilization", in *Control of Uncertain Systems* (D. Hinrichsen & B. Mårtensson, eds), Birkhäuser, to appear.

[7] R.D. Nussbaum, "Some remarks on a conjecture in parameter adaptive control", Systems & Control Letters, **3** (1983), 243-246.

[8] C.I. Byrnes & J.C.Willems, "Adaptive stabilization of multivariable linear systems", Proc 23rd IEEE Conf. Decision & Control (1984), 1574-1577.

[9] J.C. Willems & C.I. Byrnes, "Global adaptive stabilization in the absence of information on the sign of the high frequency gain", Lecture Notes in Control and Information Sciences (Springer-Verlag), **62** (1984), 49-57.

[10] A.S. Morse, "A three-dimensional universal controller for the adaptive stabilization of any strictly proper minimum-phase system with relative degree not exceeding two", IEEE Trans Autom Control, AC-**30** (1985), 1188-1191.

[11] J.P. LaSalle, "Stability theory for ordinary differential equations", J Diff Eqns., **4** (1968), 57-65.

[12] J.P. LaSalle, *The Stability of Dynamical Systems*, SIAM, Philadelphia, 1976.

[13] G. Leitmann, "Stabilization of dynamical systems under bounded input disturbance and parameter uncertainty", in *Differential Games and Control Theory, III*, (P.-T. Liu & E. Roxin, eds), Marcel Dekker, New York, 1978.

[14] M. Corless & G. Leitmann, "Continuous state feedback guaranteeing uniform ultimate boundedness for uncertain dynamic systems", IEEE Trans Autom Control, **AC-26** (1981), 1139-1144.

[15] G. Leitmann, "Deterministic control of uncertain systems via a constructive use of Lyapunov stability theory", *Proc. SIAM Conf. on Control in the 90's*, 1989.

[16] B. Mårtensson, "The order of any stabilizing regulator is sufficient a priori information for adaptive stabilization", Systems & Control Letters, **6** (1985), 87-91.

[17] B. Mårtensson, "Adaptive stabilization of multivariable linear systems", Contemporary Mathematics, **68** (1987), 191-225.

[18] A. Ilchmann, D.H. Owens & D. Prätzel-Wolters, "High-gain robust adaptive controllers for multivariable systems", Systems & Control Letters, **8** (1987), 397-404.

STABILIZING UNCERTAIN SYSTEMS WITH BOUNDED CONTROL†

A. G. Soldatos
Department of Mechanical Engineering
University of California - Berkeley
Berkeley, CA 94720

M. Corless
School of Aeronautics and Astronautics
Purdue University
West Lafayette, IN 47907

G. Leitmann
School of Engineering
University of California - Berkeley
Berkeley, CA 94720

ABSTRACT

We consider the problem of stabilizing an uncertain system when the norm of the control input is bounded by a prespecified constant. We treat continuous-time dynamical systems whose nominal part is linear and whose uncertain part is norm-bounded by a known constant.

---

† Supported by the U.S. National Science Foundation and the U.S. Air Force of Scientific Research under grants ECS-8703586 and MSM-8706927.

1. INTRODUCTION

In recent years, much effort has been devoted to the problem of obtaining stabilizing controllers for uncertain systems. In much of this research, the uncertainties are modelled deterministically, rather than stochastically, and they are characterized by certain structural conditions and known bounds; see, e. g., [3-6,12-21,23-26] and the references therein. Such uncertainties could be due to uncertain disturbance inputs, uncertain parameters, or nonlinear elements which are difficult to characterize.

In most of the literature, no consideration is given to control constraints. Here, as in [6,19-21,26], we consider the problem of stabilizing an uncertain system when the norm of the control input is bounded by a prespecified constant.

We treat continuous-time dynamical systems whose nominal part is linear and whose uncertain part is norm-bounded by a known constant. Given a ball of initial states, we consider the problem of obtaining controllers which yield "practical stability" with a region of attraction which includes the given ball.

The results are illustrated by two examples. The first example is a simple scalar example. In the second example, we consider an inverted pendulum subject to a bounded control torque. Treating the nonlinear term as an "uncertain" term, we demonstrate that given *any* ball of initial states, one can construct bounded stabilizing controllers which yield "practical stability" with the given ball as a region of attraction.

2. PROBLEM STATEMENT

Consider an uncertain dynamical system described by

$$\dot{x}(t) = Ax(t) + Bu(t) + Dv(t, x(t)) \tag{1}$$
$$x(t_0) = x_0$$

where $t \in \mathbb{R}$ is the "time", $x(t) \in \mathbb{R}^n$ is the state, $u(t) \in \mathbb{R}^m$ is the control input, and $v : \mathbb{R}\times\mathbb{R}^n \to \mathbb{R}^l$ is unknown and bounded. The constant matrices $A \in \mathbb{R}^{n\times n}$, $B \in \mathbb{R}^{n\times m}$, and $D \in \mathbb{R}^{n\times l}$ are known.

We make the following assumptions.

A1. The pair (A, B) is stabilizable [2].

A2. There is a matrix F such that

$$D = BF \quad . \tag{2}$$

A3. The function $v : \mathbb{R} \times \mathbb{R}^n \rightarrow V$ is continuous where V is a known, non-empty, compact subset of $\mathbb{R}^l$.

In what follows, we shall consider the control input to be generated by a *memoryless, state-feedback controller,* i.e.,

$$u(t) = p(x(t)) \tag{3}$$

where $p : \mathbb{R}^n \rightarrow \mathbb{R}^m$. Substituting (3) into (1) yields the *closed-loop system*

$$\begin{aligned} \dot{x}(t) &= Ax(t) + Bp(x(t)) + Dv(t, x(t)) \\ x(t_0) &= x_0 \end{aligned} \tag{4}$$

Previous activity, see e.g. [3-5, 12-18, 23-25], considered the problem of choosing p so that, for all uncertainties satisfying A3, system (4) has the property of *global uniform practical stability*[(1)]; in these results, no constraint is imposed on u(t). Here we impose the norm constraint

$$\|u(t)\| \leq \bar{\rho} \qquad \forall \; t \geq t_0 \tag{5}$$

where $\bar{\rho} > 0$ is specified. If the nominal uncontrolled system

$$\dot{x} = Ax \tag{6}$$

is unstable, one usually cannot obtain a bounded controller which assures global practical stability. Hence, we consider the following problem. Given a ball

$$R(r) \triangleq \{x_0 \in \mathbb{R}^n : \|x_0\| \leq r\} \tag{7}$$

of initial states, obtain a controller p (if one exists) which assures that the closed-loop system (4) is *uniformly practically stable with a region of attraction*[(1)] *which contains* R(r).

3. A SPECIFIC CLASS OF CONTROLLERS

In this section, we ignore the control constraint and presents controllers which assure global uniform practical stability of (4). In the next section we impose the norm constraint.

---

(1) The Appendix contains a definition.

Let

$$\rho \triangleq \max \{||Fv|| : v \in V\} \tag{8}$$

and choose any matrix $K \in \mathbb{R}^{m \times n}$ such that

$$\bar{A} \triangleq A + BK \tag{9}$$

is asymptotically stable. Choose now any symmetric, positive definite matrix $Q \in \mathbb{R}^{n \times n}$ and let $P \in \mathbb{R}^{n \times n}$ be the unique solution to the Lyapunov equation

$$P\bar{A} + \bar{A}^T P + Q = 0 \ ; \tag{10}$$

P is symmetric and positive definite; see [9,22]. For any $\varepsilon > 0$, the proposed controllers are given by

$$p(x) = p^{\varepsilon}(x) \triangleq Kx - \rho s(\varepsilon^{-1} B^T P x) \tag{11}$$

where the *saturation function* $s(\cdot)$ satisfies

$$s(y) = \begin{cases} y & \text{if} \quad ||y|| \le 1 \\ ||y||^{-1} y & \text{if} \quad ||y|| > 1 \end{cases} \tag{12}$$

To demonstrate that the above controllers achieve the desired behavior, consider the function W, given by

$$W(x) = x^T P x \ , \tag{13}$$

as a Lyapunov function candidate for (4). One can readily show that, along any solution $x(\cdot)$ of (4), controller (11) results in

$$\frac{dW(x(t))}{dt} \le L(x(t)) \tag{14a}$$

where

$$L(x) \triangleq -x^T Q x + 2\varepsilon\rho \ . \tag{14b}$$

Using the inequality

$$x^T Q x \ge \lambda_{min}(P^{-1}Q) x^T P x \ , \tag{15}$$

and letting

$$k^{\varepsilon} \triangleq 2\varepsilon\rho / \lambda_{min}(P^{-1}Q) \ , \tag{16}$$

it follows from (14) that

$$\frac{dW(x(t))}{dt} \le -\lambda_{min}(P^{-1}Q)[W(x(t)) - k^{\varepsilon}] \; ; \qquad (17)$$

hence, all solutions of (4) asymptotically approach the Lyapunov ellipsoid $X(k^{\varepsilon})$ where, for any $k \ge 0$,

$$X(k) \triangleq \{x \in \mathbb{R}^n : x^T P x \le k\} \; . \qquad (18)$$

Note that the "size" of $X(k^{\varepsilon})$ can be made arbitrarily small by choosing ε sufficiently small.

In this paper, we shall choose K in the following fashion. Let $Q_R \in \mathbb{R}^{n \times n}$ be any symmetric, positive definite matrix. Then, the Riccati equation

$$PA + A^T P - PBB^T P + Q_R = 0 \qquad (19)$$

has a unique positive definite solution $P \in \mathbb{R}^{n \times n}$; see [8]. Letting

$$K \triangleq -B^T P \qquad (20)$$

the matrix $\bar{A}$ defined by (9) is asymptotically stable. Note that the solution P of (19) satisfies (10) with the positive definite matrix

$$Q \triangleq Q_R + K^T K \; . \qquad (21)$$

Thus, the controllers under consideration are given by

$$p(x) = p^{\varepsilon}(x) \triangleq -B^T P x - \rho s(\varepsilon^{-1} B^T P x) \qquad (22)$$

where $P > 0$ solves (19).

4. CONSTRAINED CONTROLLERS

We return now to constraint (5) and make the following assumption

A4. $$\bar{\rho} > \rho \; . \qquad (23)$$

The controllers proposed in this section are simply saturating versions of those proposed in the previous section. Consider any controller p^{ε} given by (22), (19). A proposed controller is given by

$$p(x) = \bar{p}^{\varepsilon}(x) \triangleq s(\bar{\rho}^{-1} p^{\varepsilon}(x)) \qquad (24)$$

where $s(\cdot)$ is defined by (12). If we define a *constrained control region* $\bar{C}(p^{\varepsilon})$ as that region of $\mathbb{R}^n$ in which $||p^{\varepsilon}(x)|| \le \bar{\rho}$, then

$$\bar{p}^{\varepsilon}(x) = p^{\varepsilon}(x) \qquad\qquad \forall\, x \in \bar{C}(p^{\varepsilon}) \ . \tag{25}$$

Recalling (22), (12), it follows that

$$\|p^{\varepsilon}(x)\| = \begin{cases} (1+\varepsilon^{-1}\rho)\|B^T P x\| & \text{if} \quad \|B^T P x\| \le \varepsilon \\ \|B^T P x\| + \rho & \text{if} \quad \|B^T P x\| > \varepsilon \end{cases} . \tag{26}$$

If we choose ε to satisfy

$$\varepsilon \le \bar{\rho} - \rho \ , \tag{27}$$

then

$$1 + \varepsilon^{-1}\rho \le \varepsilon^{-1}\bar{\rho}$$

and

$$\|B^T P x\| \le \varepsilon \ \Rightarrow \ \|p^{\varepsilon}(x)\| \le \varepsilon^{-1}\bar{\rho}\|B^T P x\| \le \bar{\rho} \ ;$$

hence

$$\bar{C}(p^{\varepsilon}) = C(c) \triangleq \{x \in \mathbb{R}^n : \|B^T P x\| \le c\} \tag{28}$$

where

$$c \triangleq \bar{\rho} - \rho \ . \tag{29}$$

We have now the following result.

Lemma 1. Suppose there exists $\bar{k} > 0$ such that

$$R(r) \subset X(\bar{k}) \subset C(c) \ , \tag{30}$$

then, provided

$$k^{\varepsilon} < \bar{k} \ , \tag{31}$$

the closed-loop system (4),(24),(14) is uniformly practically stable with $R(r)$ as a region of attraction.

Proof. This result follows from the results of the previous section. With W defined by (13), it follows that along any solution of (4),(24),(14)

$$\frac{dW(x(t))}{dt} \le -\lambda_{min}(P^{-1}Q)\,[W(x(t)) - k^{\varepsilon}]$$

whenever $x(t) \in C(c)$. Since $X(\bar{k}) \subset C(c)$ and (31) holds, (4),(24),(14) is practically stable with region of attraction $X(\bar{k})$. Since $R(r) \subset X(\bar{k})$, $R(r)$ is also a region of attraction. □

Note that (31) can readily be satisfied by choosing ε sufficiently small; i.e.

$$\varepsilon < \lambda_{min}(P^{-1}Q)\bar{k}/2\rho \quad . \tag{32}$$

The following lemma provides a condition which is equivalent to (30) but which is easier to verify.

Lemma 2. There exists $\bar{k} > 0$ such that

$$R(r) \subset X(\bar{k}) \subset C(c) \tag{33}$$

iff

$$\lambda_{max}(P)\lambda_{max}(BB^TP) \leq (c/r)^2 \quad . \tag{34}$$

Proof. Noting that

$$\max\{x^TPx : \|x\| \leq r\} = \lambda_{max}(P)r^2 \ ,$$

it follows that $R(r) \subset X(\bar{k})$ iff

$$\lambda_{max}(P)r^2 \leq \bar{k} \quad .$$

Similarly,

$$\begin{aligned} \max\{\|B^TPx\|^2 : x^TPx \leq \bar{k}\} &= \lambda_{max}(P^{-1}PBB^TP)\bar{k} \\ &= \lambda_{max}(BB^TP)\bar{k} \, ; \end{aligned}$$

thus $X(\bar{k}) \subset C(c)$ iff

$$\lambda_{max}(BB^TP)\bar{k} \leq c^2 \quad .$$

Hence, there exists $\bar{k} > 0$ such that (33) holds iff

$$\lambda_{max}(P)\lambda_{max}(BB^TP)r^2 \leq c^2 \ ,$$

i.e.,

$$\lambda_{max}(P)\lambda_{max}(BB^TP) \leq (c/r)^2$$ □

Remark 4.1. Utilizing Lemma 2, the problem under consideration reduces to that of obtaining a positive definite matrix Q_R so that the positive definite solution P of (19) satisfies (34).

5. EXAMPLES

5.1. A Scalar Example.

Consider a scalar system described by

$$\dot{x}(t) = ax(t) + bu(t) + bv(t, x(t)) \tag{35}$$
$$x(t_0) = x_0$$

where all quantities are scalars;

$$a \geq 0 ,$$
$$| v(t, x) | \leq \rho \ ;$$

and ρ is known. For this example, Riccati equation (19) reduces to

$$2aP - b^2P^2 + Q_R = 0 \ . \tag{36}$$

The positive solution to (36) is given by

$$P = a/b^2 + [(a/b^2)^2 + Q_R]^{1/2} \ . \tag{37}$$

Thus, by appropriate choice of $Q_R > 0$, we can obtain any P which satisfies

$$P > 2a/b^2 \ .$$

Condition (34) becomes

$$b^2P^2 \leq (c/r)^2 \ ;$$

hence, an appropriate P exists iff

$$r < |b| c/2a \ .$$

5.2. Stabilization of an Inverted Pendulum with a Bounded Controller.

Consider a pendulum *P* (or one-link manipulator) which is subject to a bounded control torque T; see Figure 1. Its motion is described by

$$I\ddot{\theta} - mgl \sin\theta = T \tag{38}$$

where I is the moment of inertia of *P* about its axis of rotation, m is the mass of *P*, *l* is the distance between the center of mass of *P* and O, and g is the gravitational acceleration constant of the planet (or moon) in which *P* resides. We consider the situation in which

$$|T(t)| \leq \bar{T} \tag{39}$$

where the control torque bound $\bar{T}$ is specified and satisfies

$$\bar{T} > mgl \ . \tag{40}$$

If we let

$$x_1 \triangleq \theta, \ \ x_2 \triangleq \dot{\theta}, \ \ u \triangleq T/I \tag{41}$$

then

$$\begin{aligned} \dot{x}_1 &= x_2 \\ \dot{x}_2 &= \rho \sin x_1 + u \end{aligned} \tag{42}$$

where

$$\rho \triangleq mgl/I \tag{43}$$

and

$$|u(t)| \leq \bar{u} \triangleq \bar{T}/I \ . \tag{44}$$

Considering

$$v(t,x) = \rho \sin x_1 \tag{45}$$

description (42) is in the form of (1) with

$$A = \begin{bmatrix} 0 & 1 \\ 0 & 0 \end{bmatrix}, \quad B = D = \begin{bmatrix} 0 \\ 1 \end{bmatrix};$$

also

$$|v(t,x)| \leq \rho \qquad \forall \, t \in \mathbb{R}, \, x \in \mathbb{R}^n \, .$$

Utilizing the results of the previous sections, we demonstrate that, given *any* ball $R(r)$ of initial states, there is a controller which satisfies constraint (44) on $u(t)$ and which assures that the closed loop-system is practically stable with region of attraction $R(r)$.

It should be clear that Assumptions A1-A4 are satisfied. With

$$Q_R = \begin{bmatrix} \alpha^2 & 0 \\ 0 & \alpha^2 \end{bmatrix}, \quad \alpha > 0$$

the positive definite solution to Riccati equation (19) is given by

$$P = \begin{bmatrix} p_{11} & p_{12} \\ p_{21} & p_{22} \end{bmatrix}$$

$$p_{11} = \alpha(\alpha^2 + 2\alpha)^{1/2}$$

$$p_{12} = p_{21} = \alpha$$

$$p_{22} = (\alpha^2 + 2\alpha)^{1/2} \quad .$$

Thus,

$$\lambda_{max}(BB^T P) = (\alpha^2 + 2\alpha)^{1/2} \quad ,$$

$$\lambda_{max}(P) = [(\alpha + 1)(\alpha^2 + 2\alpha)^{1/2} + (\alpha^4 + \alpha^2 + 2\alpha)^{1/2}]/2 \quad .$$

Clearly,

$$\lim_{\alpha \to 0} \lambda_{max}(P)\lambda_{max}(BB^T P) = 0 \, .$$

Hence, given any ball R(r) of initial conditions, condition (34) can be satisfied by choosing $\alpha > 0$ sufficiently small. Choosing any $\varepsilon > 0$ which satisfies (32), a controller which assures practical stability with region of attraction R(r) is given by (24),(22).

6. APPENDIX

Consider a system described by

$$\dot{x}(t) = f(t, x(t)) \tag{46}$$

where $t \in \mathbb{R}$, $x(t) \in \mathbb{R}^n$, $f: \mathbb{R} \times \mathbb{R}^n \to \mathbb{R}^n$ and suppose that $A \subset \mathbb{R}^n$ contains a neighborhood of 0.

Definition 6.1. System (46) is *uniformly practically stable with region of attraction* A iff there exists a compact set $\mathcal{N} \subset \mathbb{R}^n$ containing 0 such that the following hold.

(i) *Existence of solutions.* For each $t_0 \in \mathbb{R}$, $x_0 \in A$, there exists a solution $x(\cdot)$: $[t_0, t_1) \to \mathbb{R}^n$ of (46) with $x(t_0) = x_0$.

(ii) *Indefinite extension of solutions.* Each solution $x(\cdot)$: $[t_0, t_1) \to \mathbb{R}^n$ of (46), with $x(t_0) \in A$, has an extension $\bar{x}(\cdot)$: $[t_0, \infty) \to \mathbb{R}^n$, i.e., $\bar{x}(t) = x(t)$ for all $t \in [t_0, t_1)$.

(iii) *Uniform boundedness of solutions.* For each $x_0 \in A$, there exists a bound $\beta(x_0)$ such that, if $x(\cdot)$: $[t_0,t_1) \to \mathbb{R}^n$ is any solution of (46) with $x(t_0) = x_0$, then $||x(t)|| \le \beta(x_0)$ for all $t \ge t_0$.

(iv) *Uniform stability to within N.* For each neighborhood $N^\eta \supset N$, there exists $\delta > 0$ such that if $x(\cdot)$: $[t_0,t_1) \to \mathbb{R}^n$ is any solution of (46) with $||x(t_0)|| < \delta$, then $x(t) \in N^\eta$ for all $t \ge t_0$.

(v) *Uniform attractivity of N.* For each neighborhood $N^\eta \supset N$ and each $x_0 \in A$, there exists $T(N^\eta,x_0)$ such that if $x(\cdot)$: $[t_0,t_1) \to \mathbb{R}^n$ is any solution of (46) with $x(t_0) = x_0$, then $x(t) \in N^\eta$ for all $t \ge t_0 + T(N^\eta,x_0)$.

Remark 6.1. If (46) satisfies the requirements of the above definition with $N = \{0\}$, then it is *uniformly asymptotically stable* with region of attraction A. The "size" of N can be regarded as a measure of the distance of the system behavior from that of asymptotic stability. If N is "small", then for practical purposes, the system might be considered asymptotically stable.

Definition 6.2. System (46) is *globally uniformly practically stable* iff it is practically stable with $\mathbb{R}^n$ as a region of attraction.

REFERENCES

[1] CESARI, L., "Asymptotic Behavior and Stability Problems in Ordinary Differential Equations", Springer-Verlag, New York, 1971.

[2] CHEN, C. T., "Linear System Theory and Design", Holt, Rinehart and Winston, 1970.

[3] CORLESS, M., and LEITMANN, G., "Continuous State Feedback Guaranteeing Uniform Ultimate Boundedness for Uncertain Dynamic Systems", IEEE Transactions on Automatic Control, Vol. AC-26, pp. 1139-1144, 1981.

[4] CORLESS, M. and LEITMANN, G., "Controller Design for Uncertain Systems via Lyapunov Functions", Proceedings of the American Control Conference, Atlanta, Georgia, 1988.

[5] CORLESS, M., LEITMANN, G., and RYAN, E. P., "Tracking in the Presence of Bounded Uncertainties", Proceedings of the 4th International Conference on Control Theory, 1984.

[6] HACHED, M., MADANI - ESFAHANI, S. M., and ZAK, S. H., "Stabilization of Uncertain Systems Subject to Hard Bound on Control with Application to a Robot Manipulator", IEEE Journal of Robotics and Automation, Vol. 4, pp. 310-323, 1988.

[7] HAHN, W., "Stability of Motion", Springer-Verlag, Berlin, 1967.

[8] KAILATH, T., "Linear Systems", Prentice-Hall, Englewood Cliffs, N. J., 1980.

[9] KREISSELMEIER, G., "A Solution of the Bilinear Matrix Equation AY+YB = -Q", SIAM Journal of Applied Mathematics, Vol. 23, 1972.

[10] LaSALLE, J. P., "The Stability of Dynamical Systems", SIAM, 1976.

[11] LaSALLE, J. P., and LEFCHETZ, S., "Stability by Liapunov's Direct Method with Applications", Academic Press, New York, 1961.

[12] LEITMANN, G., "Deterministic Control of Uncertain Systems", Acta Astronautica, Vol. 7, p. 1457, 1980.

[13] LEITMANN, G., "Feedback and Adaptive Control for Uncertain Dynamical Systems", in "New Mathematical Advances in Economic Dynamics", edited by D. F. Batten and P. F. Lessa, Croom Helm, London, England 1985.

[14] LEITMANN, G., "Guaranteed Ultimate Boundedness for a Class of Uncertain Linear Dynamical Systems", IEEE Transactions on Automatic Control, Vol. AC-23, p. 1109, 1978.

[15] LEITMANN, G., "On the Efficacy of Nonlinear Control in Uncertain Linear Systems", Journal of Dynamic Systems, Measurement, and Control, Vol. 102, pp. 95-102, 1981.

[16] LEITMANN, G., KELLY, J. M., and SOLDATOS, A. G., "Robust Control of Base Isolated Structures Under Earthquake Excitation", Journal of Optimization Theory and Applications, Vol. 53, pp. 159-180, 1987.

[17] LEITMANN, G., KELLY, J. M., and SOLDATOS, A. G., "Seismic Protection of Structures Using Base Isolation and Active Control," Proceedings of the American Control Conference, Minneapolis, Minnesota, 1987.

[18] LEITMANN, G., RYAN, E. P., and STEINBERG, A., "Feedback Control of Uncertain Systems: Robustness with Respect to Neglected Actuator and Sensor Dynamics", International Journal of Control, Vol. 43, p. 1243, 1986.

[19] MADANI - ESFAHANI, S. M., "Analysis and Synthesis of Uncertain Variable Structure Systems with Bounded Controllers", Ph.D. thesis, Purdue University, 1990.

[20] MADANI - ESFAHANI, S. M., HUI, S., and ZAK, S. H., "On the Estimation of Sliding Domains and Stability Regions of Variable Structure Control Systems with Bounded Controllers", Proceedings of the 26th Allerton Conference on Communication, Control, and Computing, Monticello, IL, 1988.

[21] MADANI - ESFAHANI, S. M., and ŻAK, S. H., "Variable Structure Control of Dynamical Systems with Bounded Controllers", Proceedings of the American Control Conference, 1987.

[22] ROTHSCHILD, D., JAMESON, A., "Comparison of Four Numerical Algorithms for Solving the Lyapunov Matrix Equation", International Journal of Control, Vol. 2, pp. 181-198, 1970.

[23] SLOTINE, J. J., and SASTRY, S. S., "Tracking Control of Non-Linear Systems Using Sliding Surfaces", International Journal of Control, Vol. 38, 1983.

[24] SOLDATOS, A. G., "Interim Report on Actively Controlled, Base-Isolated Structures under Earthquake Excitation", University of California, Berkeley, California, 1986.

[25] UTKIN, V. I., "Variable Structure Systems with Sliding Mode: A Survey", IEEE Transactions on Automatic Control, Vol. AC-22, pp. 212-222, 1977.

[26] VINCENT, T. L., "Control Design for Magnetic Suspension", Optimal Control Applications and Methods, Vol. 1, pp. 41-53, 1980.

[27] YASUDA, K., and HIRAI, K., "Upper and Lower Bounds on the Solution of the Algebraic Riccati Equation", IEEE Transactions on Automatic Control, Vol. AC-24, pp. 483-487, 1979.

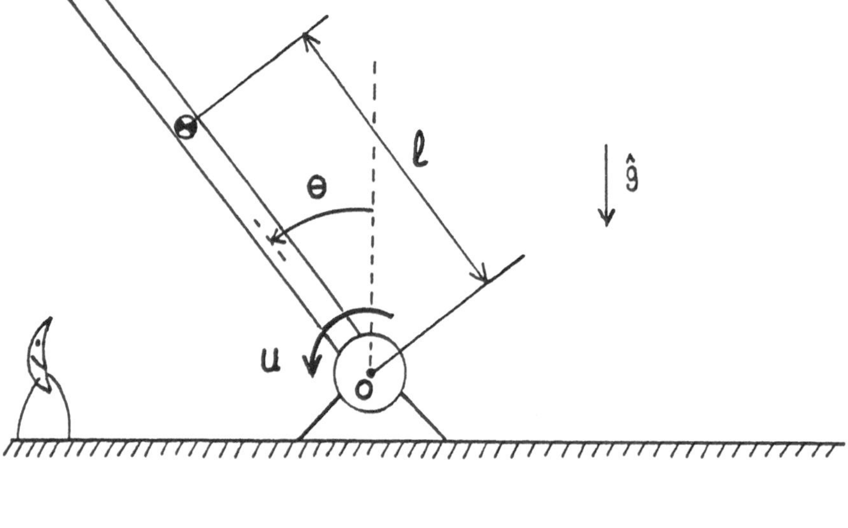

FIGURE 1

CONTROLLABILITY IMPLICATIONS OF NEWTON'S THIRD LAW

W. Stadler

San Francisco State University
San Francisco,CA 94132

Summary

For the mixed problem in particle dynamics some aspects of both the motion and the system forces are specified. A solution consists of the complete specification of internal and external forces as functions of time and of the motions of the particles. Such problems may be viewed as control problems with the as of yet undetermined forces as controls. Controllability implications may then be used to establish whether these forces are sufficient to sustain the required motions. This viewpoint here is used to illustrate, by means of examples, that central internal forces are insufficient to sustain certain motions whereas these same motions may be sustained by noncentral internal forces. The implications are that the postulation of the strong form of Newton's third law is too restrictive and should be dropped in favor of the weak form which is more in keeping with Newton's original intent.

1. Introduction

The mixed problem in classical particle dynamics deals with a partial specification of motions and forces, a solution consisting of the determination of the remaining forces and motions as function of time. The interaction of motion and forces is taken to be governed by the usual Newtonian axioms. Presumably, these provide enough conditions to allow one to uniquely solve for the motions and forces, the main difficulties being ascribed to questions of integrability.

To motivate the discussion, suppose the motion of a particle P with mass m is specified as

$$\underline{r}(t) = 2t^2\hat{i} + t^3\hat{j}$$

with the usual meaning for $\hat{i}$ and $\hat{j}$. One may then ask whether a force of the form $\underline{F}(t) = F(t)\hat{i}$ is capable of producing and maintaining such a motion. The applicability of Newton's second law is taken for granted, yielding

$$\underline{F}(t) = m(4\hat{i} + 6t\hat{j}).$$

This leads to

$$F(t) = 4m$$

and the contradiction

$$0 = 6tm.$$

Generally, the Newtonian axioms are taken to be established truth, the contradiction resting with the assumed type of force. That is, the proposed force cannot sustain the prescribed motion.

More generally, suppose the motion is specified as

$$\underline{r}(t) = x(t)\hat{i} + y(t)\hat{j}$$

and one now poses the question: Can this motion be sustained by a force of the form $\underline{F}(t) = u(t)\hat{i}$? The application of Newton's second law now yields

$$\ddot{x} = \frac{1}{m}\, u(t) \qquad \text{and } \ddot{y} = 0.$$

With $x_1 = x$, $x_2 = \dot{x}$, $x_3 = y$ and $x_4 = \dot{y}$, these equations may be written in standard first order form as

$$\begin{vmatrix} \dot{x}_1 \\ \dot{x}_2 \\ \dot{x}_3 \\ \dot{x}_4 \end{vmatrix} = \begin{bmatrix} 0 & 1 & 0 & 0 \\ 0 & 0 & 0 & 0 \\ 0 & 0 & 0 & 1 \\ 0 & 0 & 0 & 0 \end{bmatrix} \begin{vmatrix} x_1 \\ x_2 \\ x_3 \\ x_4 \end{vmatrix} + \begin{vmatrix} 0 \\ 1 \\ 0 \\ 0 \end{vmatrix} u$$

In this form the problem may be viewed as a linear problem in control theory: The adequacy of the chosen force distribution becomes a question in controllability.

The Kalman matrix for this system is

$$\underset{\sim}{C} = \begin{bmatrix} 0 & 1 \\ 1 & 0 \\ 0 & 0 \\ 0 & 0 \end{bmatrix},$$

which clearly has rank 2. Thus, the system is not completely controllable, as one might have expected. That is, there is at least one state which cannot be reached by a control (force) of this type. Indeed, no terminal state of the form

$$\underline{x}^* = \begin{vmatrix} 0 \\ 0 \\ x_3^* \\ x_4^* \end{vmatrix}$$

is reachable, since $\underline{x}^{*T}\underset{\sim}{C} = \underline{0}$.

In dynamics problems one is routinely left with a set of unknown forces with the implication that these, when solved for, will be capable of sustaining an accompanying motion. The redundancy or sufficiency of a prescribed force system for sustaining a motion is rarely discussed prior to advanced courses dealing with Lagrangian dynamics. In the following, it is shown that the solution of problems in particle dynamics may or may not exist depending on whether one admits noncentral internal forces or insists on their centrality as part of the statement of Newton's third law.

2. Controllability and the Newtonian Axioms

Consider a system N of n particles P_i with masses m_i and with motions $\underline{r}_i(t)$. Denote the resultant external force on the particle P_i by $\underline{F}_i(t)$ and let $\underline{F}_{ij}(t)$ be the internal force on the particle P_i due to the particle P_j. Then, the usual formulation of Newtonian particle dynamics consists of the following postulates:

(i) Newton's second law in the form

$$\underline{F}_i(t) + \sum_{j=1}^{n} \underline{F}_{ij}(t) = m_i \ddot{\underline{r}}_i(t) \qquad (2.1)$$

for each particle P_i in the system N.

(ii) Newton's third law

$$\text{(a)} \quad \underline{F}_{ij}(t) + \underline{F}_{ji}(t) = \underline{0} \quad , \quad \underline{F}_{ii}(t) \equiv \underline{0} \; , \qquad (2.2)$$

for $i = 1, \ldots, n; j = 1, \ldots, n$.

(b) Centrality of the internal forces:

$$[\underline{r}_i(t) - \underline{r}_j(t)] \times \underline{F}_{ij}(t) = \underline{0} \tag{2.3}$$

for $i = 1, .., n; j = 1, \ldots, n$.

It is usual to term (a) by itself the <u>weak</u> form of Newton's third law and to refer to (a) and (b) together as the <u>strong</u> form of the third law.

<u>Remark 2.1</u> The statement (i) clearly includes Newton's first law. From a historical point of view, Galileo already included routine statements of Newton's third law, the weak form is more in keeping with Newton's own intent, and the inclusion of the centrality of the internal forces appears to be due to Daniel Bernoulli in his attempt of deriving the statement of the law of moment of momentum (angular momentum) from Newton's second law. A law of moment of momentum, as an independent postulate, was first proposed by Jakob Bernoulli in 1686 and it was first proven correctly as a theorem hypothesizing Newton's laws, by Poisson in 1833.

For any given particle P_i, the problem to be solved generally belongs to one of the following categories:

(i) The <u>motion</u> $\underline{r}_i(t)$ is completely specified.
(ii) The forces acting on the particle $\underline{P}_i$ are specified.
(iii) Some of the forces are prescribed together with classes of permissible velocities or accelerations.

A problem is considered solved, when the motions $\underline{r}_i(t)$, and any unknown forces (internal or external) have been obtained as functions of time.

In the following, it is useful to view the internal forces as controls used to implement motion. The main question to be addressed is whether a particular internal force distribution has any chance of sustaining (controlling) a partially prescribed set of motions. When phrased in this fashion, one has a question of the controllability of a system.

The standard definition for the controllability of time - invariant linear systems suffices, since only the linearized (local) dynamical problem will be considered. The standard linear control system has the form

$$\dot{\underline{x}}(t) = \underset{\sim}{A}\, x(t) + \underset{\sim}{B}\, \underline{u}(t) \text{ on } [t_o,\infty) , \tag{2.4}$$

where $\underline{x} \in R^n$ is the state, $\underline{u} \in R^r$ is the control, $\underset{\sim}{A}$ is the $n \times n$ state matrix, and $\underset{\sim}{B}$ is the $n \times r$ input matrix.

Definition 2.1 Complete Controllability. The system (2.4) is completely state controllable at $t = t_o$ if and only if, for every x_o, $x_1 \in R^n$ there exists a finite time $t_1 \geq t_o$ and a control $\underline{u}(.)$ such that the solution $\underline{x}\ (t_i;t_o,\underline{x}_o,\underline{u}(.)) = \underline{x}_1$.

One immediately has the following Kalman condition for controllability.

Theorem 2.1 (Kalman Condition). The system (2.4) is

completely state controllable if and only if the n x rn matrix

$$\underset{\sim}{C} = [\underset{\sim}{B} \mid \underset{\sim}{A}\underset{\sim}{B} \mid \underset{\sim}{A}^2 \underset{\sim}{B} \mid \ldots \mid \underset{\sim}{A}^{n-1}\underset{\sim}{B}]$$

has rank n.

<u>Corollary 2.1</u> A state $\underline{x}_1$ is not controllable at $t = t_o$ if and only if $\underline{x}_1^T \underset{\sim}{C} = \underline{0}$.

<u>Remark 2.2</u> Generally, when using a linearization of a non-linear problem in the above context, it is required that there exist asymptotically stable control bringing the system to the locale implied by the linearization. Here, it suffices to simply consider all the results within a sufficiently small neighborhood of R^2. When a system is constrained to move in a subspace, then controllability within that subspace must be considered.

The following two problems serve to illustrate the connection between controllability and Newton's third law.

3. The Examples

The examples are problems in particle dynamics. In both, there are three particles with nonzero mass which maintain a rigid configuration during their motion. This rigidity is assumed to be maintained by internal forces acting between the particles. The manner in which such internal forces might be supplied physically is totally irrelevant within the present context. The governing equations of motion for each system will be based primarily on Newton's second law applied to each particle and on the use of either the strong or the weak form of Newton's third law.

Example 3.1 The first problem consists of three particles which are to maintain the fixed triangular configuration shown in Figure 3.1. The particle P_o of mass m_o is to remain fixed, while the particles P_1 and P_2 with masses m_1 and m_2, respectively, are to move on concentric circles in a gravitational field g in such a way that the whole system behaves like a compound pendulum. The use of Newton's second law and the strong form of Newton's third law (that is, including the centrality assumption) yields the equations of motion in the form:

P_0: $\underline{F} + \underline{W}_o + \underline{F}_{02} + \underline{F}_{01} = \underline{0}$

$$\hat{e}_r: \quad F^r + m_o g \cos\theta_1 + F_{20}\cos\Psi + F_{10} = 0 \tag{3.1}$$

$$\hat{e}_\theta: \quad F^\theta - m_o g \sin\theta_1 + F_{20}\sin\Psi = 0$$

P_1: $\underline{W}_1 + \underline{F}_{12} + \underline{F}_{10} = m_1\ddot{\underline{r}}_1$,

$$\hat{e}_r: m_1 g\cos\theta_1 - F_{12}\cos\phi - F_{10} = -m_1 a_1\dot{\theta}_1^2 \tag{3.2}$$

$$\hat{e}_\theta: -m_1 g\sin\theta_1 + F_{12}\sin\phi = m_1 a_1\ddot{\theta}_1$$

P_2: $\underline{W}_2 + \underline{F}_{21} + \underline{F}_{20} = m_2\ddot{\underline{r}}_2$

$$\hat{e}_r: m_2 g\cos\theta_1 - F_{12}\cos\phi - F_{20}\cos\Psi = -m_2 a_2\dot{\theta}_2^2\cos\Psi - m_2 a_2\ddot{\theta}_2\sin\Psi \tag{3.3}$$

$$\hat{e}_\theta: m_2 g\sin\theta_1 + F_{12}\sin\phi + F_{20}\sin\Psi = m_2 a_2\dot{\theta}_2^2\sin\Psi - m_2 a_2\ddot{\theta}_2\cos\Psi$$

The intended meaning of the symbols is clear from Figure 3.1 and the vector equations.

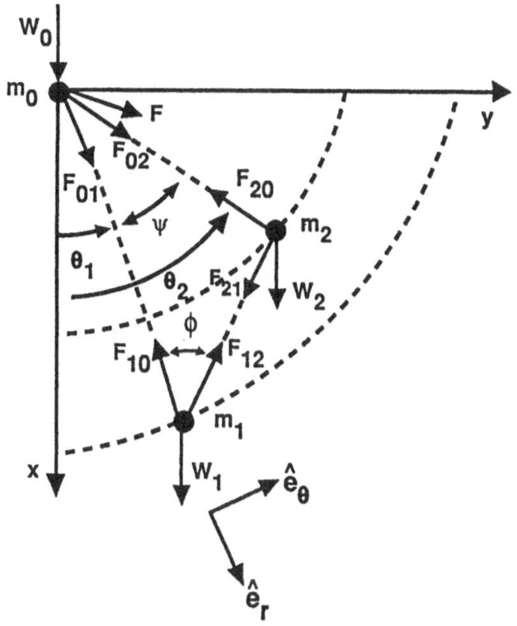

Figure 3.1 Moving Triangle with Central Forces

Note that there are six equations in six unknowns. Proceeding the traditional way, one would now attempt to solve for the unknown internal forces, the support reaction $\underline{F}$, and for the motions. Rather than attempting to solve the problem, however, suppose one now asks whether the postulated internal force system can sustain the motion at all. Since the system is a one-degree-of-freedom system, it suffices to consider controllability within a two-dimensional subspace; that is, it suffices to consider the equation

$$m_1 a_1 \ddot{\theta}_1 = -m_1 g \sin \theta_1 + F_{12} \sin \phi$$

or, upon linearization with, $x_1 = \theta_1$ and $x_2 = \dot{\theta}_1$

$$\begin{pmatrix} \dot{x}_1 \\ \dot{x}_2 \end{pmatrix} = \begin{bmatrix} 0 & 1 \\ \frac{-g}{a_1} & 0 \end{bmatrix} \begin{pmatrix} x_1 \\ x_2 \end{pmatrix} + \begin{pmatrix} 0 \\ \frac{\sin\phi}{m_1 a_1} \end{pmatrix} F_{12} \tag{3.4}$$

Conditions for controllability will now be used in a somewhat unorthodox manner; namely, to show <u>existence</u> or <u>nonexistence</u> of a solution with respect to a postulated internal force system (the internal forces are viewed as controls). Suppose the system turns out to be completely controllable. Since a control then exists to transfer the system from any initial state to any final state, this must include the final state implied by the actual solution of the system equations. That is, it may be asserted that a solution based on this force configuration <u>exists.</u> Conversely, suppose the system is completely uncontrollable, then the state implied by the actual motion cannot be reached either, and one is forced to conclude that <u>no solution exists</u> for the specified internal force system. In summary, one may deduce existence implications from the complete controllability or complete uncontrollability of the system.

The Kalman matrix for the system (3.4) is

$$\underset{\sim}{C} = \begin{bmatrix} 0 & \frac{\sin\phi}{m_1 a_1} \\ \frac{\sin\phi}{m_1 a_1} & 0 \end{bmatrix}$$

Since $\underset{\sim}{C}$ has rank two, the system is completely controllable and <u>a central internal force system suffices to sustain this motion.</u>

The next step in this example consists of a look at the limiting configuration as ϕ tends to zero, so that the rigid configuration to be maintained by the internal forces consists of three particles in a straight line (see Figure 3.2). The result is $\underset{\sim}{C} = 0$, and it follows that the system is completely uncontrollable; that is, $\underline{x}^T\underset{\sim}{C} = \underline{0}$ for every state $\underline{x}$. Since no state is reachable, the natural motion of the system is not possible either. That is, a <u>central internal force system cannot sustain this motion</u>. Equivalently, one <u>cannot</u> postulate the strong form of Newton's third law in posing this problem.

<u>Example 3.2</u> Suppose the problem now is reformulated for the rigid straight line configuration.

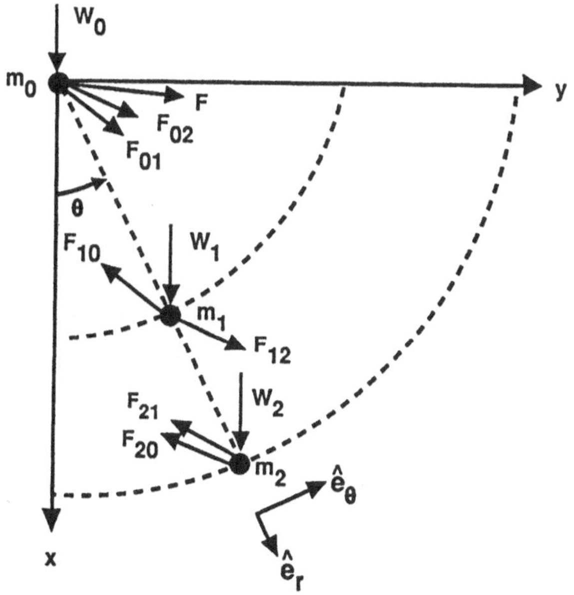

Figure 3.2 Moving Line with Noncentral Internal Forces

Based on Newton's second law together with the weak form of Newton's third law the equations of motion are:

$$\underline{W}_0 + \underline{F} - \underline{F}_{10} - \underline{F}_{20} = \underline{0},$$
$$\underline{W}_1 + \underline{F}_{10} + \underline{F}_{12} = m_1 \ddot{\underline{r}}_1, \tag{3.5}$$
$$\underline{W}_2 - \underline{F}_{12} + \underline{F}_{20} = m_2 \ddot{\underline{r}}_2,$$

a set of six equations in nine unknowns when the kinematic constraint relating the accelerations is included. Since the directions of the internal forces are no longer specified within the third law, the internal forces can no longer be solved for independently without further assumptions about their character. Without such assumptions, only the resultant internal force on a particle may be obtained. Denoting these internal resultants by

$$\underline{R}_1 = \underline{F}_{10} + \underline{F}_{12} \text{ and } \underline{R}_2 = -\underline{F}_{12} + \underline{F}_{20},$$

the equations (3.5) may be rewritten in the component form

$$F^r = -(m_1 a_1 + m_2 a_2)\,\dot{\theta}^2 + (m_o + m_1 + m_2)\,g \cos\theta,$$
$$F^\theta = (m_1 a_1 + m_2 a_2)\,\ddot{\theta} + (m_o + m_1 + m_2)\,g \sin\theta \tag{3.6}$$
$$R_1^r = -m_1 (a_1\dot{\theta}^2 + g \cos\theta) \text{ and } R_2^r = m_2 (a_2\dot{\theta}^2 + g \cos\theta),$$
$$R_1^\theta = m_1 (a_1\ddot{\theta} + g \sin\theta) \text{ and } R_2^\theta = -m_2 (a_2\ddot{\theta} + g \cos\theta)$$

From the controllability point of view, it again suffices to consider

$$R_1^\theta = m_1 (a_1\ddot{\theta} + g \sin\theta), \tag{3.7}$$

whose linearized state-space form is

$$\begin{pmatrix} \dot{x}_1 \\ \dot{x}_2 \end{pmatrix} = \begin{bmatrix} 0 & 1 \\ \frac{-g}{a_1} & 0 \end{bmatrix} \begin{pmatrix} x_1 \\ x_2 \end{pmatrix} + \begin{pmatrix} 0 \\ \frac{1}{m_1 a_1} \end{pmatrix} R_1^{\theta} \qquad (3.8)$$

The Kalman matrix for this system is

$$\underset{\sim}{C} = \begin{bmatrix} 0 & \frac{1}{a_1 m_1} \\ \frac{1}{a_1 m_1} & 0 \end{bmatrix} \qquad (3.9)$$

which clearly has rank 2. Thus, the systems is completely controllable. It follows that a noncentral internal force system <u>is</u> capable of sustaining the prescribed pendular motion. Equivalently, the <u>weak</u> form of Newton's third law is the appropriate postulate to be used in formulating this problem.

There are some additional difficulties in actually attempting to solve this simple problem. Equations (3.6) constitute a system of six equation in seven unknowns. Thus, a condition replacing the centrality assumption is needed in order to be able to solve the system of equations. This additional condition is usually taken to be an independently postulated <u>law</u> of moment of momentum. Such a postulate in particle dynamics, however, entails the added complication that this law turns out to be nonlocal in the sense that its statement has a different form for each subsystem of the original system of n particles. These difficulties with Newton's third law and an independently postulated law of moment of momentum were first noted in two articles by Stadler (Ref. 2) and by Casey and Stadler (Ref.3).

4. Conclusions

Controllability arguments have been used to show that the centrality assumption usually included in statements of Newton's third law becomes untenable for certain problems in particle dynamics. That is, its a priori postulation omits certain possible motions from consideration. As expected, it was found that a truss-like particle configuration was completely controllable with central internal forces. The limiting case when all of the particles are to move as a collinear pendulum was completely uncontrollable with central internal forces and completely controllable when noncentral internal forces were used.

It is suggested that a more appropriate formulation of basic postulates in particle dynamics is given by Newton's second law together with the weak form of Newton's third law and an independently postulated law of moment of momentum. This approach would be in keeping with current usage, since problems of this type are generally solved by an incorrect application of a theorem of moment of momentum, the basic hypothesis of centrality of the internal forces not being satisfied for the ensuing motion. The statement of theorem and axiom are the same for the n particle system (although not for its subsystems). We have thus been fortunate to obtain the correct answers for the wrong reasons, a good fortune from which all of us have surely benefitted at one time or another.

Acknowledgement

The Author wishes to acknowledge many useful discussions with Professor J. Dauer concerning the use of controllability in this context.

References

1. Lee, E.G. and L. Markus, "Foundations of Optimal Control Theory," John Wiley & Sons, Inc., New York, New York, 1967.

2. Stadler, W., On the Inadequacy of the Usual Newtonian Formulation for Certain Problems in Particle Mechanics, American Journal of Physics Vol. 50, No. 7, 1982, pp. 595-598.

3. Casey, J. and W. Stadler, A Remark on the Principle of Angular Momentum for Systems of Particles, Zeitschrift fur Angewandte Mathematik und Mechanik (ZAMM), Vol. 66, No. 3, 1986, pp. 190-192.

CONTROL OF CHAOS IN NONLINEAR DYNAMICAL SYSTEMS

Aynur Ünal
School of Engineering, Santa Clara University
Santa Clara, CA 95053

1. Abstract

CHAOTIC regimes of nonlinear dynamical systems are becoming more and more important as we find new applications. When we are interested in designing the operation of a dynamical system so as to avoid its chaotic regimes, we must know in advance the ranges of parameters for which the system's response will be chaotic.

A dynamical system will respond according to the values of the physical parameters in play. We can always specify a space of physical parameters over which the system operates and imagine traversing paths within such a space. On any segment of a path, the system's dynamical response will be characterized by the long-term behavior of its trajectories in a phase space, and its characterization may change at critical points on the path. For example, over a certain segment of parameter space, the system's response may become steady at long time. Corresponding trajectories in a phase space will approach a fixed point. We say that trajectories are *attracted* to such a point; accordingly, we call it a fixed-point attractor. On the other hand, over another segment of parameter space, the system's long-term behavior may be periodic in time; corresponding trajectories in a phase space will approach a closed curve or limit cycle. Over yet another segment of parameter space, the system's long-term response in time may contain two incommensurate periods. Corresponding trajectories in phase space will be attracted to a two-dimensional torus. Finally, one or more segments of parameter space may exist where the system's long-term response appears to be random or *chaotic*. Corresponding trajectories in phase space will be attracted to a limiting set, the dimension of which is noninteger. We call this attracting set a *strange attractor*. To control the onset of chaos, we must know the ranges

*Professor, School of Engineering, Santa Clara, CA 95053

of parameters over which the system's response is characterizable by a strange attractor.

In this paper, we shall seek necessary and sufficient conditions for the existence of a point in parameter space defining the onset of a chaotic regime in a given dynamical system and hence the control to such regimes.

2. Key Words

Nonlinear dynamical systems, chaos, control.

3. Introduction

The question of determining the range of parameters for a given nonlinear dynamical system for which the system enters into the chaotic regimes is of practical importance in aerospace related industries and health related applications as in the cases of aeroplane wing flutter and heart pacers. The author has been very much interested in developing the ability to predict the onset of chaos as one varies the parameters. Along this line, we have two possible approaches. If we can tell the sign of the largest Lyapunov coefficient for a given set of parameters then we do not need to compute the Lyapunov Spectra or the time series but still sense the onset of chaos. Unfortunately, at this time we do not know how to tell the sign of the largest Lyapunov Exponent. Lyapunov exponents are only available through numerical analysis and one has to repeat the analysis for each set of data which is inexhaustible. The second approach is through the use of Laplace-Borel transforms and the observation that the chaotic trajectories must have zero autocorrelation. Laplace-Borel transforms are discussed in [1-3] and provide a significant tool in the nonlinear analysis similar to integral transformations for linear systems.

Laplace-Borel transforms are closely connected with repeated integrals discussed in [4-6]. There does exist a formal correspondence between the Laplace-Borel transforms and a noncommutative algebra as shown in [7] and [8]. We use this correspondence to recast the dynamical system from its repeated-integral form to an algebraic form in the Laplace-Borel transform domain. The resulting algebraic equations can be solved, in principle, for the physical parameters responsible for the chaotic regime. Parameter control depends crucially on mapping out the parameter space in terms of patches on the surface for which the system is chaotic or not so that if one would like to avoid the chaos (aeroplane wing flutter problem) or if one would like to have the chaos (brain functions are normally chaotic, periodic functioning corresponds to pathological cases) one can choose the parameters.

4. Mathematical Analysis

The most crucial observation is that only for chaotic regimes, the auto correlation function vanishes and for other regimes like constant steady or periodic or quasiperiodic regimes the autocorrelation function is different than zero. Intuitively, the autocorrelation function is a measure of the internal similarity of the trajectories where internal similarities can be interpreted as accumulated information represented by the trajectories. For chaotic regimes, as the system evolves the internal similarity is lost whereas for other regimes a certain degree of similarity is kept.

4.1 Autocorrelation Function in Time Domain

The autocorrelation function is a measure of the similarity of a trajectory at a given time t with its value at a later time $t+\tau$. We define $AC(\tau)$ as the arithmetic mean of a large number of products such as $x(t) \cdot x(t+\tau)$ i.e.

$$AC(\tau) = \frac{1}{t_2 - t_1} \int_{t_1}^{t_2} x(t) \cdot x(t+\tau) dt$$

or in a more compact way, we can write:

$$AC(\tau) = \langle x(t) \cdot x(t+\tau) \rangle$$

where $AC(\tau)$ is a *temporal autocorrelation function.* We construct $AC(\tau)$ by varying the interval. It defines the degree of similarity of the trajectory $x(t)$ with itself as time evolves.

The Wiener-Kintchine theorem states that $AC(\tau)$ is the Fourier transform of the power spectrum. As a result of this we see that for the regimes represented by an attractor which is: (a) a fixed point; (b) a limit cycle; (c) a n-torus we have

$$\lim_{\tau \to \infty} AC(\tau) \neq 0,$$

since in these cases the power spectrum is formed of distinct rays. In other words, periodic or quasiperiodic trajectories keep their internal similarity with the evolution of time. This means that the behavior of the system is predictable. In constrast, for a chaotic regime, where the power spectrum has a broad band,

$$\lim_{\tau \to \infty} AC(\tau) = 0,$$

(c.f. Fig. 1).

MAIN THEOREM. The following algebraic criterion has to be satisfied for the onset of chaos:

$$\lim_{\substack{\tau\to\infty \\ x_0\to\infty}} \left(\sum_{k\geq 0} \frac{\tau^k}{k!x_0^k}\right) \mathcal{G} \amalg \mathcal{G} = 0$$

where $\mathcal{G}$ is the generating power series for the trajectories of the nonlinear dynamical system. In other words, $\mathcal{G}$ is the Laplace-Borel transformation of the trajectories of the given nonlinear dynamical system.

Proof. We adopt as definition of the autocorrelation function

$$AC(\tau) = \lim_{T\to\infty} \frac{1}{2T} \int_{-T}^{T} x(t) \cdot s(t+\tau)dt.$$

As we discussed previously, we must have

$$\lim_{\tau\to\infty} AC(\tau) = 0$$

for the onset of chaos. Let us take the Laplace-Borel transformation of both sides;

$$\mathcal{LB}[AC(\tau)] = \lim_{x_0\to\infty} x_0 \mathcal{LB}[x(t)] \amalg \mathcal{LB}[x(t+\tau)]. \tag{1}$$

Notice that we have utilized a result in [1] to take the Laplace-Borel transformation of a product of two functions. The Laplace-Borel transformation of the product is equal to the shuffle product of the individual transformations. Let us denote $\mathcal{LB}[x(t)]$ by $\mathcal{G}$. Then by the shifting theorem [1] we have:

$$\mathcal{LB}[x(t+\tau)] = \left(\sum_{k\geq 0} \frac{\tau^k}{k!x_0^k}\right) \mathcal{G}.$$

Hence the criterion for the onset of chaos becomes

$$\lim_{\substack{\tau\to\infty \\ x_0\to\infty}} \left(\sum_{k\geq 0} \frac{\tau^k}{k!x_0^k}\right) \mathcal{G} \amalg \mathcal{G} = 0$$

which completes the proof of the main theorem.

5. Concluding Remarks

The criterion obtained as a main theorem is a computer-algebraic one. It is noteworthy that the algebra required in obtaining the two generating power series and their shuffle product, together with the two limits, all can be done on a computer with a symbolic language. At the moment we have the option of using the following symbolic languages: PLI, REDUCE, MACSYMA, LISP and MATHEMATICA.

Although the criterion is simple in concept, its application by no means will be a trivial affair. A considerable effort doubtless will be necessary to handle the difficulty of bifurcation points. These are points in parameter space where analyticity will be lost, and hence where the generalized series expansions will not be valid. Such points will reveal themselves beforehand by slowing down the rate of convergence of the iterations.

Our criterion, which is computer-algebraic, can be utilized to characterize the ranges of the physical parameters for which chaotic regimes will take place. Hence, it can be used in designing the operation of a nonlinear dynamical system to avoid such regimes. In other words, we can *control* chaos in our nonlinear dynamical system by staying outside of the ranges defined by our criterion.

6. References

[1] Ünal, A., An algebraic criterion for the onset of chaos in nonlinear dynamical system, *J. of Nonlinear Analysis, Theory, Methods & Applications,* Vol. 13, No. 7, pp. 753-765, 1989.

[2] Ünal A. & Tobak M., *Use of Nonlinear Functional Expansions In Nonlinear Aerodynamics,* Abstr. Am. Math. Soc. 6, 421 (1985).

[3] Ünal A. & Tobak M., *Chaotic Solutions of A Nonlinear Second Order Ordinary Differential Equation,* Abstr. Am. Math. Soc. 7 (1986).

[4] Fliess M., *Fonctionnelles causales nonlinéaires et indétérminées noncommutatives,* Bull, Soc. Math, France 109, 3-40 (1981).

[5] Chen K.T., *Integration of paths, geometric invariants and a generalized Baker-Hausdorff formula,* Ann. Math. 65, 163-178 (1957).

[6] Chen, K.T., *Algebraic paths,* J. Algebra 10, 8-36 (1986).

[7] Chen, K.T., *Iterated path integrals,* Bull. Am. math. Soc. 83, 831-879 (1977).

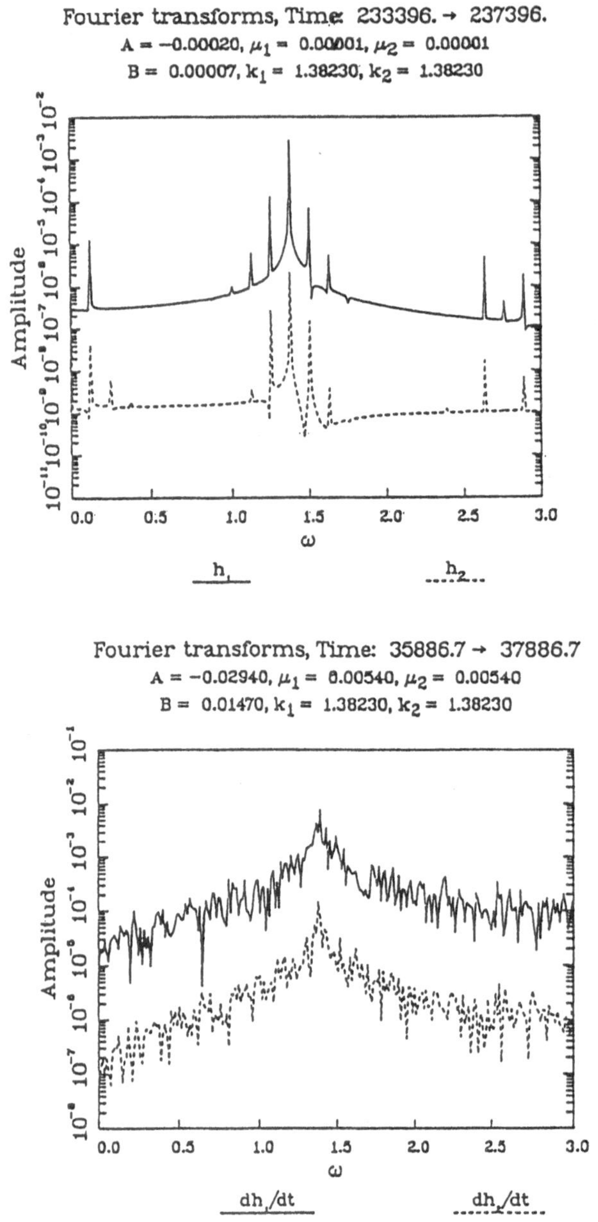

Figure 1. Power Spectrum of a Quasiperiodic and Chaotic Regime, respectively.

CONTROL OF A CHAOTIC SYSTEM

Thomas L. Vincent and Jianzu Yu

Department of Aerospace and Mechanical Engineering
The University of Arizona
Tucson, AZ 85721

Abstract. *The Lorenz equations are well known for their ability to produce chaotic motion. We investigate here the Lorenz system subject to a control input. Two different controllers are designed for this system, one based on linear methods and one based on a nonlinear analysis. The objective of the controller is to drive the system to one of the unstable equilibrium points associated with uncontrolled chaotic motion. Each controller is able to produce stable motion. However, the character of this motion may differ considerably, depending on adjustment of "gains" used in the controller. In particular, the motion may contain chaotic transients. It is possible to create a system with intermediate-term-sensitive dependence on initial conditions, but with no such long-term dependence.*

INTRODUCTION

Over the past 20 years, deterministic ***chaos*** has become a very fascinating subject of scientific study. Such systems are extremely interesting because of a paradox: They are deterministic systems whose motion leads to completely chaotic trajectories. A large number of books and papers have discussed a variety of the strange, seemingly unpredictable, phenomena associated with chaotic systems.

Today, modern science owes its success to its ability to predict natural phenomena. This, in turns, allows us a degree of control over our surroundings. The steady increase in man's preditive power has enabled the building of today's "high-tech" society. In recent years, researchers have discovered that order exists within chaos. Many chaotic systems exhibit such a high degree of order that they become useful paradigms for modeling unpredictable phenomena. There are, however, few books and papers on the control of chaotic systems. Can man also control chaotic systems in the usual sense? For example, can we develop a control program to maintain a chaotic system at a given equilibrium state? What methods do we use to develop the control algorithm?

This paper is devoted to an investigation of the design of a controller for a particular chaotic system described by the Lorenz equations. This is a well-understood system that has recently been used to model systems subjected to control.

THE LORENZ EQUATIONS

In 1963, E. N. Lorenz wrote a remarkable paper. In it, he described a three-parameter family of three nonlinear first-order ordinary differential equations that, when integrated numerically on a computer, appeared to have extremely complicated solutions. These equations are now known as the Lorenz equations. He was searching for a three-dimensional set of ordinary differential equations that would model some of the unpredictable behavior that we normally associate with the weather (Lorenz, 1979). The equations that he eventually used were derived from a model for fluid convection. They are:

$$\frac{dx_1}{dt} = -\sigma x_1 + \sigma x_2 \tag{1}$$

$$\frac{dx_2}{dt} = r x_1 - x_2 - x_1 x_3 \tag{2}$$

$$\frac{dx_3}{dt} = x_1 x_2 - b x_3 \, , \tag{3}$$

where σ, r, and b are three real positive parameters.

Briefly, the original derivation (Lorenz, 1963) can be described as follows. A two-dimensional fluid cell is warmed from below and cooled from above such that the resulting convective motion can be modeled by a partial differential equation. The variables in this partial differential equation are expanded into an infinite number of modes, all but three of which are then set identically to zero. The three remaining modes are given by equations (1)-(3). The variable x_1 is proportional to the intensity of the convective motion, x_2 is proportional to the temperature difference between the ascending and descending currents, and x_3 is proportional to the distortion of the vertical temperature profile from being linear. The three parameters σ, r, and b are respectively proportional to the Prandtl number, the Rayleigh number, and some physical proportions of the region under consideration. Consequently, all three are taken to be positive.

For a wide range of values for the parameters, solutions to Lorenz equations calculated on a computer look extremely complicated. Figure 1 shows the three-dimensional state-space trajectory for (1)-(3) when $\sigma = 10$, $b = 8/3$, and $r = 28$. Figure 2 shows the behavior of x_2 for the first 30 seconds. The trajectories shown in these figures are not periodic, but rather, as Lorenz's work indicated, they depict complicated "turbulent" behavior. Since apparent turbulent behavior can be generated by just these few nonlinear determinisitc differential equations, the Lorenz equations have been studied by many authors in the years since 1963. Some authors have sought to discover other real-world problems for which the Lorenz equations are an accurate model. Haken (1975) derives the Lorenz equations from a problem of irregular spiking in lasers, Malkus (1972) and Yorke and Yorke

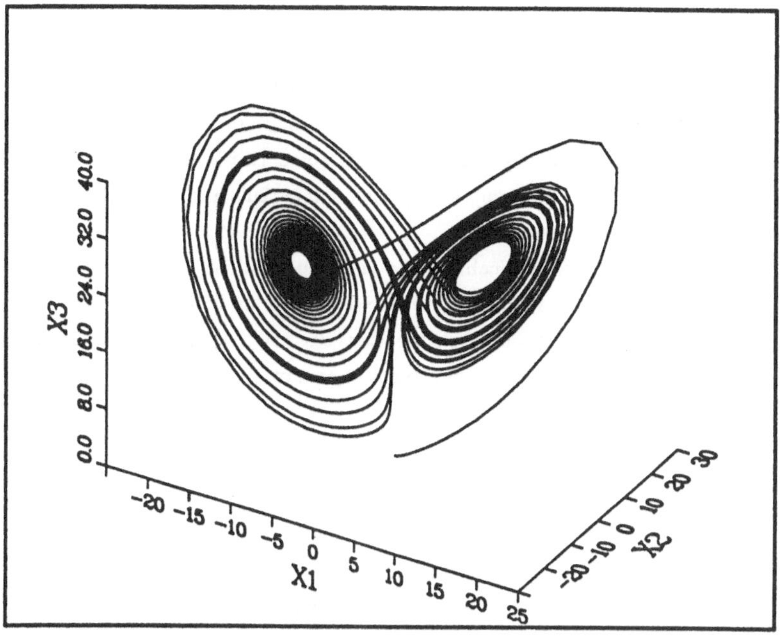

Figure 1. A state-space trajectory produced by the Lorenz system starting from $x = (0, 1, 0)$.

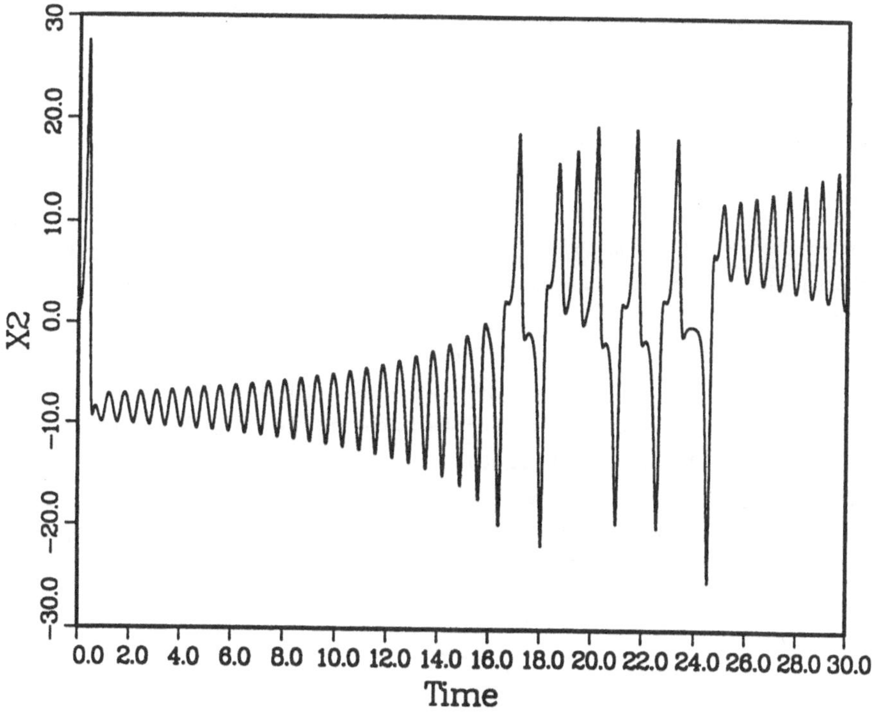

Figure 2. A graph of x_2 for the situation depicted in Figure 1.

(1979) both studied a problem of convection in a toroidal region, and Knobloch (1981) discusses a derivation from a disc dynamo. Malkus has constructed a laboratory water wheel that behaves in a fashion similar to (1)-(3) (Lorenz, 1979). Pedlosky (1972) and Pedlosky and Frenzen (1980) have derived the Lorenz equations from a study of the dynamics of a weakly unstable, finite amplitude, baroclinic wave (two-layer model). Brindley and Motoz (1980) obtain the equations in a similar problem (continuously stratified model), and Gibbon and McGuinness (1980) discuss both the two-layer baroclinic model and a laser problem.

Because of their practical applicability and since applications for the Lorenz equations may also include control inputs (Ehrhard and Müller, 1990), we choose the Lorenz equations as our starting point for the study of the control of chaotic systems. Consider now (1)-(3) with a control input added to the left side of (2) to give us a Lorenz system subject to control:

$$\frac{dx_1}{dt} = -\sigma x_1 + \sigma x_2 \tag{4}$$

$$\frac{dx_2}{dt} = r x_1 - x_2 - x_1 x_3 + u \tag{5}$$

$$\frac{dx_3}{dt} = x_1 x_2 - b x_3 \,. \tag{6}$$

For the natural convection in the closed-loop experiment of Ehrhard and Müller (1990), u corresponds to the tilt angle of the loop from the vertical.

STABILITY OF THE LINEARIZED LORENZ SYSTEM

In this section we use Lyapunov's first (or indirect method) for stability analysis of the uncontrolled Lorenz system (1)-(3). This method provides local stability properties by examining the linearized equations of motion corresponding to small perturbations in initial conditions.

From equations (1)-(3), we immediately obtain the equilibrium point $\bar{x}_1 = \bar{x}_2 = \bar{x}_3 = 0$. When $r > 1$, equations (1)-(3) possess two additional steady-state solutions: $\bar{x}_1 = \bar{x}_2 = \pm\,[b(r-1)]^{1/2}$, $\bar{x}_3 = (r-1)$. We will designate these equilibrium points by C1 and C2, respectively. Following Lorenz (1963), we shall let $\sigma = 10$, $b = 8/3$, and $r = 28$. The states of steady convection are then represented by the points C1 = (8.4852, 8.4852, 27) and C2 = (-8.4852, -8.4852, 27), while the state of no convection corresponds to the origin (0, 0, 0). Note that the Lorenz equations (1)-(3) have natural symmetry; replacing (x_1, x_2, x_3) with $(-x_1, -x_2, x_3)$, yields the same equations.

Assume that a state of steady convection is a point of interest (i.e., C1). We will consider just C1 for control analysis since corresponding results for C2 are easily deduced from the symmetry. To study the system behavior near the equilibrium point C1, we first linearize (1)-(3) in terms of the state perturbation equations about C1:

$$\frac{dz_1}{dt} = -\sigma z_1 + \sigma z_2 \tag{7}$$

$$\frac{dz_2}{dt} = z_1 - z_2 - \sqrt{b(r-1)}\, z_3 \tag{8}$$

$$\frac{dz_3}{dt} = \sqrt{b(r-1)}\, z_1 + \sqrt{b(r-1)}\, z_2 - bz_3 \ , \tag{9}$$

where

$$z_1 = x_1 - \sqrt{b(r-1)} \tag{10}$$

$$z_2 = x_2 - \sqrt{b(r-1)} \tag{11}$$

$$z_3 = x_3 - (r-1) \tag{12}$$

represent small changes from the equilibrium point C1. The characteristic equation for the system (7)-(9) is given by

$$\lambda^3 + (\sigma + b + 1)\lambda^2 + (r + \sigma)b\lambda + 2\sigma b(r-1) = 0 \ . \tag{13}$$

Details on conditions the parameters must satisfy for stability are given by Lorenz (1963) and Sparrow (1982). For the case considered here ($\sigma = 10$, $b = 8/2$, $r = 28$), the roots are given by

$$\lambda_1 = - 13.85457 \tag{14}$$

$$\lambda_2 = 0.09395 + 10.19451j \tag{15}$$

$$\lambda_3 = 0.09395 - 10.1495j \tag{16}$$

so that the equilibrium point C1 is unstable.

THE LINEARIZED LORENZ SYSTEM SUBJECT TO CONTROL

Linearizing the controlled system (4)-(6) using perturbation variables defined by (10)-(12) yields

$$\frac{dz_1}{dt} = - \sigma z_1 + \sigma z_2 \tag{17}$$

$$\frac{dz_2}{dt} = z_1 - z_2 - \sqrt{b(r-1)}\, z_3 - u \tag{18}$$

$$\frac{dz_3}{dt} = \sqrt{b(r-1)}\, z_1 + \sqrt{b(r-1)}\, z_2 - bz_3 \ . \tag{19}$$

In this case the perturbation control u is the same as in (5) since the nominal control is taken to be zero. To maintain this system at the desired equilibrium point C1, only partial state feedback control u is required. Let

$$u = - kz_1 \ , \tag{20}$$

where k is an appropriate feedback gain. From (14)-(16) we have

$$\frac{dz_1}{dt} = - \sigma z_1 + \sigma z_2 \tag{21}$$

$$\frac{dz_2}{dt} = (1-k)z_1 - z_2 - \sqrt{b(r-1)}\, z_3 \tag{22}$$

$$\frac{dz_3}{dt} = \sqrt{b(r-1)}\, z_1 + \sqrt{b(r-1)}\, z_2 - bz_3 \,. \tag{23}$$

The characteristic equation for this system with $\sigma = 10$, $b = 8/3$, and $r = 28$ is given by

$$\lambda^3 + 13.6667\, \lambda^2 + (101.3333 + 10k)\lambda + 26.6667k + 1440 = 0 \,. \tag{24}$$

We use the Routh-Hurwits criterion to determine the critical value of k for stability. The Routh tabulation of (24) is

λ^3	1	$101.333 + 10k$
λ^2	13.6667	$26.6667k + 1440$
λ^1	$\dfrac{13.6667(101.3333 + 10k) - (26.6667k + 1440)}{13.6667}$	0
λ^0	$26.6667k + 1440$	0

For all the roots of (24) to possess negative real parts, all coefficients in the first column of the Routh tabulation must have the same sign. The third term will be positive when $k > 0.5$, and the fourth term will be positive when $k > -54$. It is apparent that k must satisfy

$$k > 0.5 \,. \tag{25}$$

Figures 3 and 4 illustrate two typical trajectories in z-space with a feedback control gain of $k = 1$ and $k = 10$, respectively. Clearly, the system is very weakly stable for the smaller values of k. This will prove to be the more interesting case in what follows.

LINEAR FEEDBACK CONTROL OF THE LORENZ SYSTEM

Now, consider applying the linear feedback controller as given by (20) to the nonlinear Lorenz system (4)-(6). In terms of the original state variables, the control law is given by

$$u = -k(x_1 - \sqrt{b(r-1)} \,. \tag{26}$$

Under this control law, the system has the controlled equilibrium point C1. In the neighborhood of the equilibrium point, the behavior of a nonlinear system will be similar to that of a linearized system. Thus, there must be some region about this point for which the system will be stable with an appropriate feedback gain.

Figure 5 illustrates the behavior of (4)-(6) under the control law (26) with $\sigma = 10$, $b = 8/3$, $r = 28$, and $k = 10$. The system is seen to quickly converge to the C1 equilibrium point (8.4852, 8.4852, 27). As with the linearized system, $k = 10$ corresponds to strong stability and the Lorenz system is stable from any point in state space. In this sense, the Lorenz system behaves like a deterministic one.

Now, consider decreasing the feedback gain to $k = 1$, which corresponded to a weakly stable case for the linearized system. Figure 6 illustrates this case for various starting conditions. Here, we

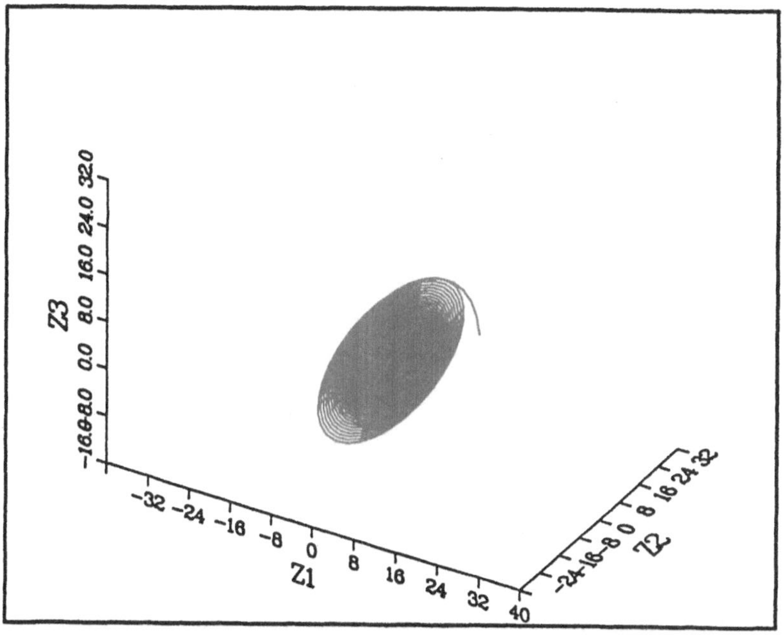

Figure 3. A trajectory in z-space produced by the controlled ($k = 1$) linearized Lorenz system starting from $x = (20, 20, 30) \Rightarrow z = (-11.5147, -11.5147, 3)$.

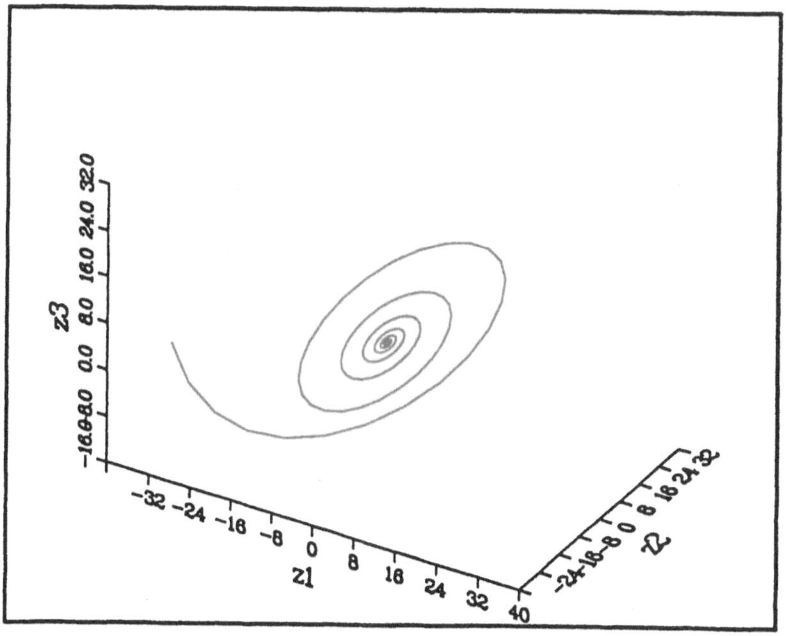

Figure 4. A trajectory in z-space produced by the controlled ($k = 10$) linearized Lorenz system starting from $x = (-20, 20, 30) \Rightarrow z = (-28.4853, -28.4853, 3)$.

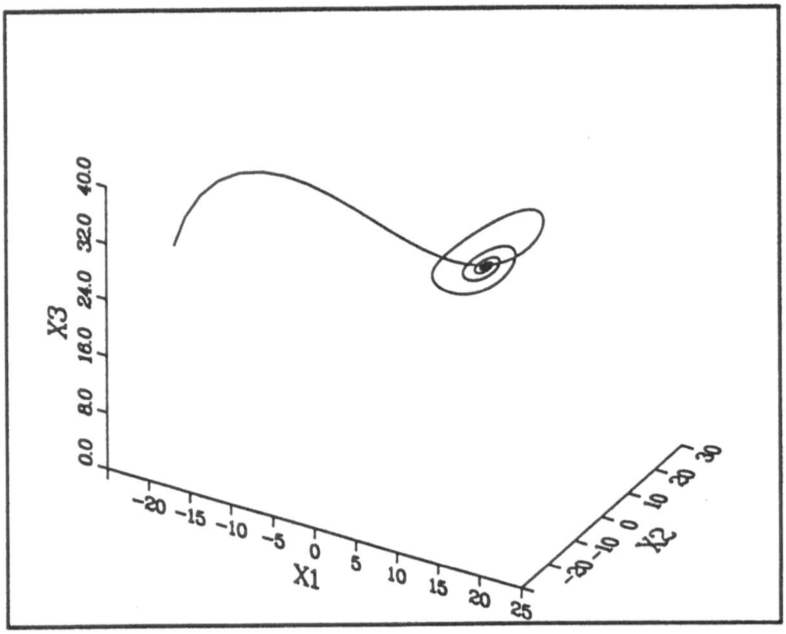

Figure 5. The Lorenz system under linear feedback control (k = 10) starting from x = (-20, -20, 30).

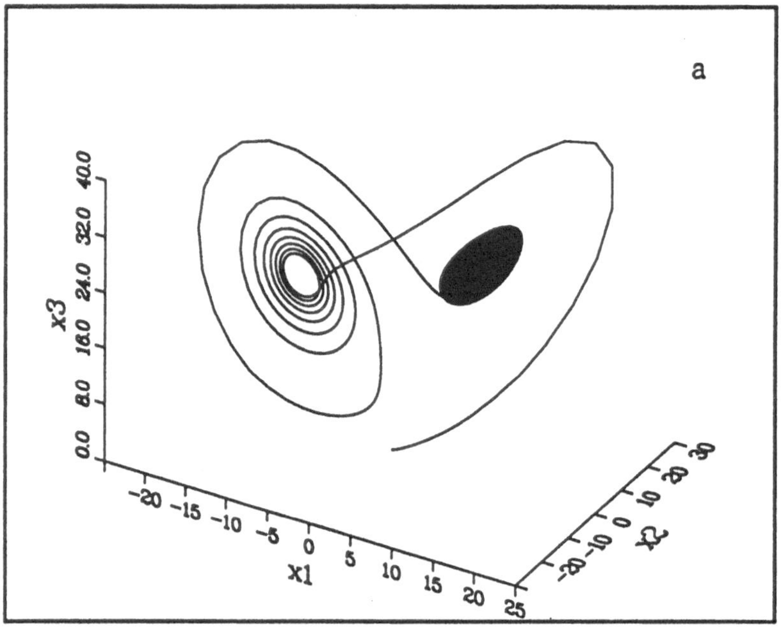

Figure 6. The Lorenz system under linear feedback control (k = 1) starting from (a) x = (0, 1, 0), (b) x = 20, 20, 30), and (c) x = (-12, -12, 30).

Figure 6.--Continued

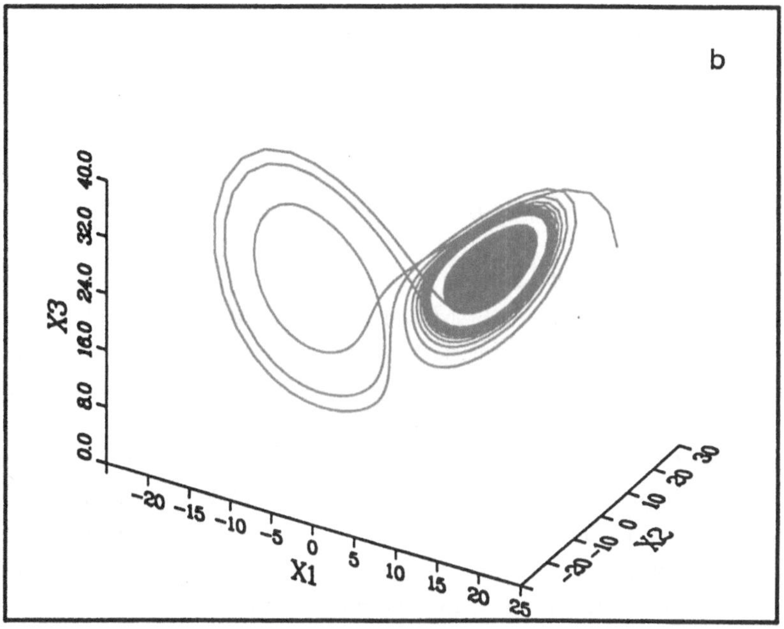

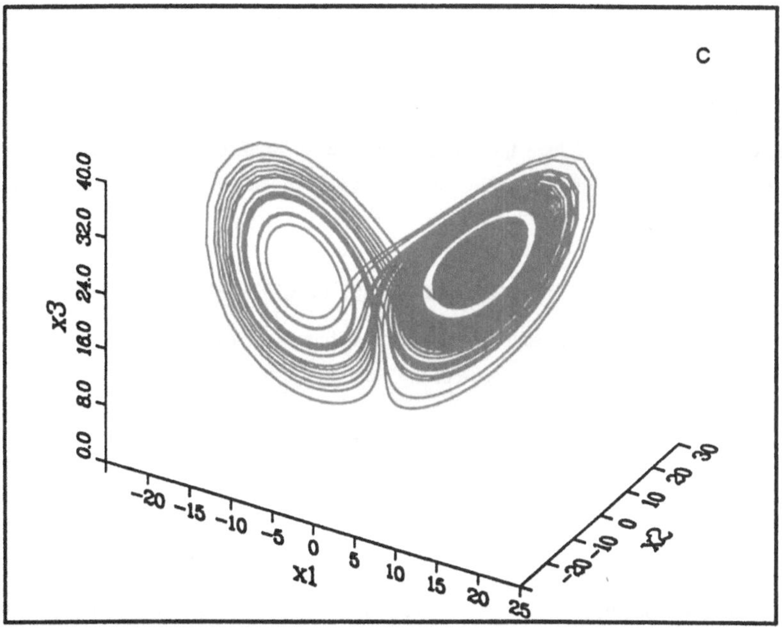

obtain chaotic transients of varying degree, depending on the initial conditions. In Figure 6a, the trajectory appears to wander for a while around the wrong equilibrium point before settling down to a slow spiral around the correct one. In Figure 6b, the trajectory first enters an apparent unstable orbit about the correct equilibrium point, then moves to an orbit about the wrong equilibrium point before finally settling in to the long-term stable orbit about the correct equilibrium point. Figure 6c illustrates the phenomenon known as "preturbulence" (Kaplan and Yorke, 1979), or as "meta-stable chaos" (Yorke and Yorke, 1979). In this case, the initial orbit appears to be chaotic; however, after a sufficiently long time period, the trajectory ultimately settles down to its slow decay to the proper equilibrium point. Even under this very weak feedback law, the system appears to be stable from every point in state space.

Thus, under linear feedback control, the behavior of the flow determined by the Lorenz equations may change dramatically, depending on the gain used in the feedback control law. The system may appear quite deterministic, or with chaotic transients. This is, of course, true of the uncontrolled Lorenz system as well. For example, by changing r, the Lorenz system can be made stable. Sparrow (1982) changed the r value to 14 and obtained the "stable manifold" of C1. Figure 7 illustrates a typical trajectory for this case. This is somewhat similar to Figure 5; however, by changing r, one also changes the equilibrium point from C1 = (8.4852, 8.4852, 27) to C1 = (5.8878, 5.8878, 13). Thus, while the Lorenz system can be made deterministic and stable by an appropriate change in parameters, one cannot necessarily have the system go to the desired point in state space that is achievable under feedback control. When Sparrow changed r to 22.4, he also got a typical "preturbulent" trajectory [see p. 31 of Sparrow (1982)], as illustrated in Figures 8a and 8b. In each of these cases, the system ultimately ends up at an equilibrium point. However, as these two figures illustrate, the final equilibrium state of the system is still uncertain. Figures 8a and 8b differ only in starting conditions. The trajectory of Figure 8a ends up at C1 = (7.5542, 7.5542, 21.4) and the trajectory of Figure 8b ends up at C2 = (-7.5542, -7.5542, 21.4), neither of which are at the desired controlled equilibrium point C1 = (8.4852, 8.4852, 27).

With control, we can produce the interesting effect of long-term determinism to a specified final state with apparent short-term chaotic motion. Such motion might be useful in certain stealth applications.

BOUNDED BANG-BANG CONTROL FOR THE LORENZ SYSTEM

We will now explore the possibility of using nonlinear control techniques to stabilize the Lorenz system (4)-(6). We will do this by means of the Controllability Minimum Principle:

If $u(x) \in U$ is an admissible control law that generates a trajectory $x(t)$ satisfying

$$\frac{dx}{dt} = f[x, u(x)] \tag{27}$$

such that for some time interval $0 \le t \le t_f$, $x(t)$ lies in the boundary of the controllable set to a specified target, then there must exist a non-zero N-dimensional vector $\lambda(t)$ such that for all $t \in [0, t_f]$

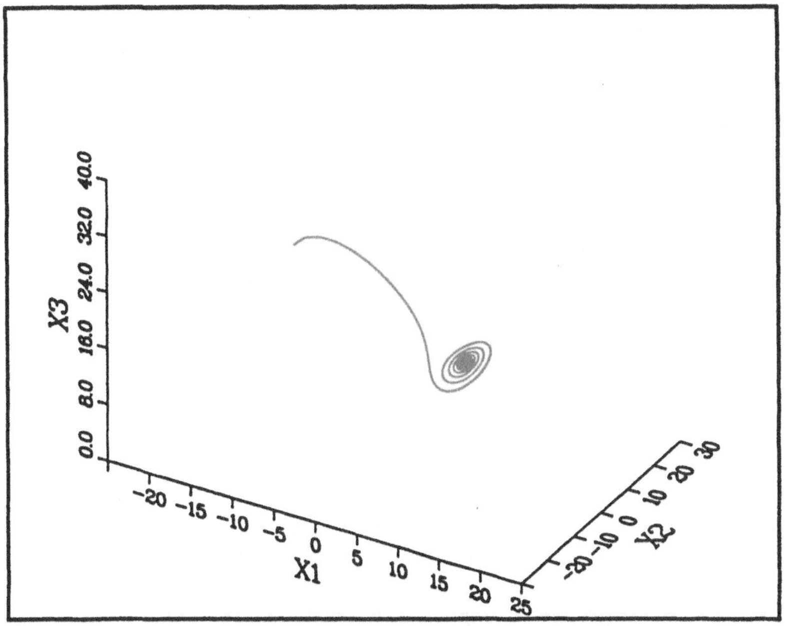

Figure 7. The uncontrolled Lorenz system with $\sigma = 10$, $s = 8/3$, and $r = 14$ starting from $x = (-10, -10, 30)$.

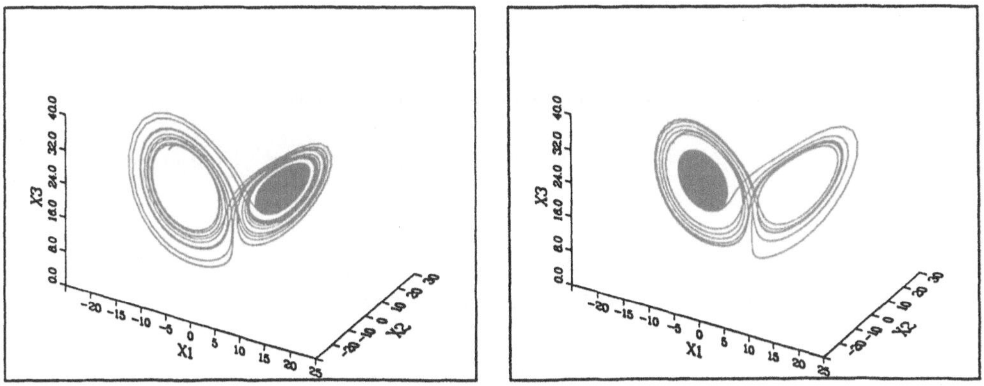

a

b

Figure 8. The uncontrolled Lorenz system with $\sigma = 10$, $s = 8/3$, and $r = 22.4$ starting from (a) $x = (-10, -10, 30)$ and (b) $x = (-12, -12, 30)$.

$$H[x(t), u[x(t)], \lambda(t)] \leq H[x(t)\ u, \lambda(t)] = 0 \ . \tag{28}$$

For all $u \in U$, with $\lambda(t)$ satisfying the adjoint equations,

$$\left[\frac{d\lambda}{dt}\right]^T = -\frac{\partial H}{\partial x}, \tag{29}$$

where the function H is defined by

$$H(x, u, \lambda) \equiv \lambda^T f(x, u) \ . \tag{30}$$

A more complete discussion and proof of this principle is given by Grantham and Vincent (1975). It has been found that this principle may sometimes be used to generate relatively simple control laws for stabilizing nonlinear systems (Gayek and Vincent, 1985).

We will use the Controllability Minimum Principle to determine a switching surface for the control to drive the system to the target C1. For the Lorenz system subject to control (4)-(6), the H function is given by

$$H = \lambda_1(-\sigma x_1 + \sigma x_2) + \lambda_2(r x_1 - x_2 - x_1 x_3 + u) + \lambda_3(x_1 x_2 - b x_3) \tag{31}$$

and the adjoint equations (29) take on a particularly simple form, given by

$$\frac{d\lambda_1}{dt} = -\frac{\partial H}{\partial x_1} = \sigma\lambda_1 - (r - x_3)\lambda_2 - x_2\lambda_3 \tag{32}$$

$$\frac{d\lambda_2}{dt} = -\frac{\partial H}{\partial x_2} = -\sigma\lambda_1 + \lambda_2 - x_1\lambda_3 \tag{33}$$

$$\frac{d\lambda_3}{dt} = -\frac{\partial H}{\partial x_3} = x_1\lambda_2 + b\eta_3 \ . \tag{34}$$

We obtain the control by noting that u appears linearly in H, which we can rewrite as

$$H = \lambda_1(-\sigma x_1 + \sigma x_2) + \lambda_2(r x_1 - x_2 - x_1 x_3) + \lambda_3(x_1 x_2 - b x_3) + \lambda_2 u \ . \tag{35}$$

Since

$$\frac{\partial H}{\partial u} = \lambda_2 , \tag{36}$$

we can seek a switching surface by setting λ_2 and its first time derivative equal to zero. That is,

$$\frac{d\lambda_2}{dt} = -\sigma\lambda_1 + \lambda_2 - x_1\lambda_3 = 0 \ . \tag{37}$$

Setting $\lambda_2 = 0$ in (35) and (37), it then follows from the requirement that $H = 0$ and (37) that

$$\lambda_1(-\sigma x_1 + \sigma x_2) + \lambda_3(x_1 x_2 - b x_3) = 0 \tag{38}$$

$$-\sigma\lambda_1 + \lambda_2 - x_1\lambda_3 = 0 , \tag{39}$$

which has a non-zero solution for λ_1 and λ_3 if and only if

$$- x_1(- \sigma x_1 + \sigma x_2) + \sigma(x_1 x_2 - b x_3) = 0 . \tag{40}$$

Substituting $\sigma = 10$, $b = 8/3$, and $r = 28$ into (40), we obtain

$$x_1^2 - \frac{8}{3} x_3 = 0 . \tag{41}$$

This equation defines a switching surface in the state space. A closed-loop control that will drive this system to the equilibrium point is given by

$$u = - 10 \, sgn\left[x_1^2 - \frac{8}{3} x_3\right] . \tag{42}$$

The behavior of the system (4)-(9) under the control (42) is illustrated in Figure 9.

Because of the nature of the motion of the trajectories in the neighborhood of the equilibrium point C1 on the strange attractor, the above control law may be made to be active only in the neighborhood of this point. In particular, if we apply (42) only in the region defined by

$$(x_1 - 8.4852)^2 + (x_2 - 8.4852)^2 + (x_3 - 27)^2 \le R^2 , \tag{43}$$

with $u = 0$ elsewhere, it is still possible to control the system to the equilibrium point, as illustrated in Figure 10. For this case, $R = 5$. The nature of the chaotic motion on the attractor is such that, ultimately, the system is driven to within the ball about C1 as defined by (43). Since the controller can drive the system from any point in this ball to the equilibrium point, the system is ultimately stabilized from any point in state space. Until the system enters the ball, the motion is truly chaotic with the same Lyapunov exponent that the uncontrolled Lorenz system would have.

CONCLUSIONS

Using both linear theory and the Controllability Minimum Principle, we are able to design controllers for the Lorenz system subject to a control input. It was demonstrated that behavior of this particular chaotic system can indeed be tamed to the point that it would be indistinguishable from any other well-behaved linear or nonlinear system. However, what is of perhaps more interest is that, when using a weaker feedback control or less effective bang-bang control, we can introduce chaos or preturbulence with a well-defined long-term outcome. In other words, we are able to create a system with a short- (or intermediate-) term sensitive dependence on initial conditions, but with no such long-term dependence. There are other systems, such as the two-link pendulum operating under gravity with a small amount of friction, that will display this same characteristic. However, the resultant equilibrium position, the lowest energy state, is rather transparent and obvious. By adding a controller to this system, such as a torquer at the link between the two bobs, one should be able to achieve the same short-term chaos with long-term determinism to a higher energy equilibrium state.

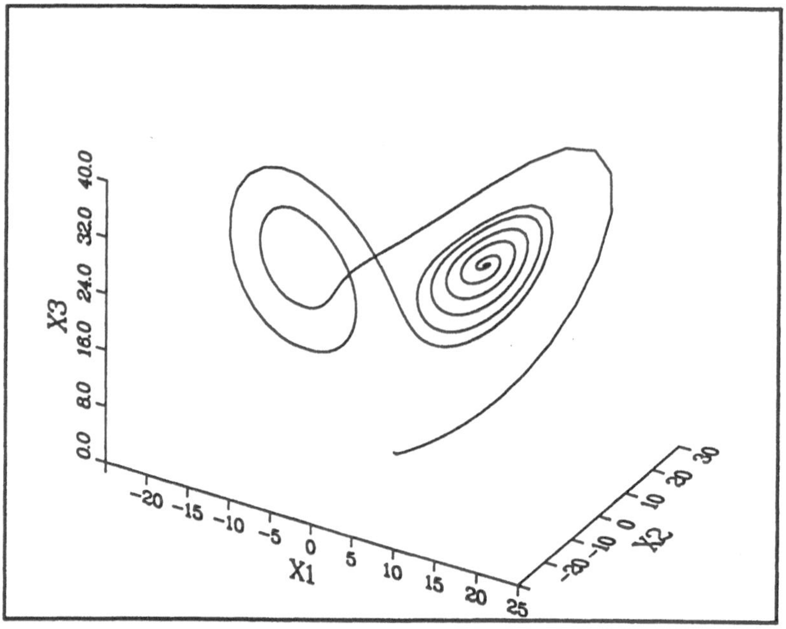

Figure 9. The Lorenz system under bang-bang control starting from $x = (0, 1, 0)$.

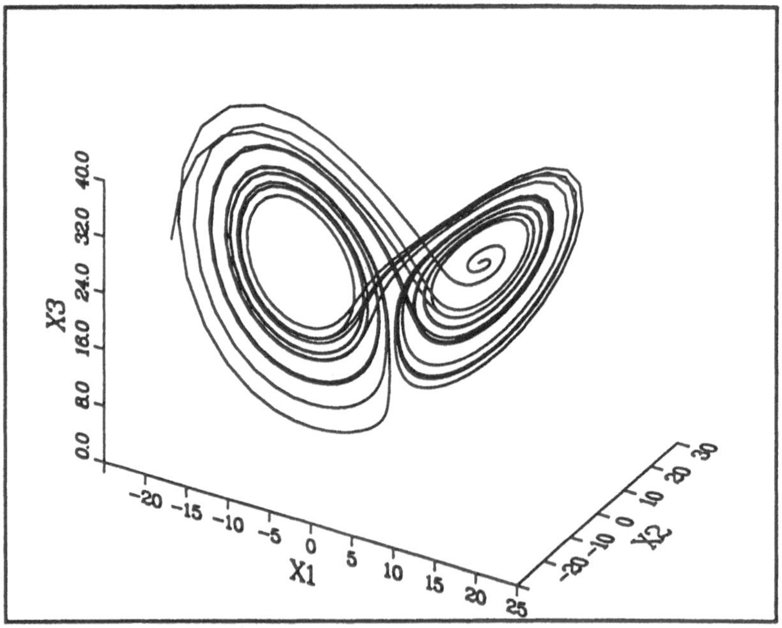

Figure 10. The Lorenz system under restricted bang-bang control starting from $x = (-20, -20, 30)$.

REFERENCES

J. Brindley and I. M. Motoz, "Lorenz attractor behavior in a continuously stratified baroclinic fluid," *Phys. Let.* vol. 77A pp. 441-444, 1980.

P. Ehrhard and U. Müller, "Dynamical behavior of natural convection in a single-phase loop," *J. Fluid Mech.* (in press).

J. E. Gayek and T. L. Vincent, "On the asymptotic stability of boundary trajectories, *Int. J. Control* vol. 41 pp. 1077-1086, 1985.

J. D. Gibbon, "Dispersive instabilities in nonlinear systems: The real and complex Lorenz Equations," in *Proc. Int. Symp. Synergetics*, Schlob Elmau, Bavaria, 1981.

J. D. Gibbon and M. J. McGuinness, "A derivation of the Lorenz equation for some unstable dispersive physical systems," *Phys. Let.* vol. 77A pp. 295-299, 1980.

W. J. Grantham and T. L. Vincent, "A controllability minimum principle," *J. Optimization Theory and Applications* vol. 17 pp. 93-114, 1975.

H. Haken, "Analogy between higher instabilities in fluids and lasers," *Phys. Let.* vol. 53A pp. 77-78, 1975.

J. L. Kaplan and J. A. Yorke, "Preturbulence: A regime observed in a fluid flow model of Lorenz," *Commun. Math. Phys.* vol. 67 pp. 93-108, 1979.

E. Knobloch, "Chaos in a segmented disc dynamo," *Phys. Let.* vol. 82A pp. 439-440, 1981.

E. N. Lorenz, "Deterministic non-periodic flow," *J. Atmos. Sci.* vol. 20 pp. 130-141, 1963.

M. V. R. Malkus, "Non-periodic convection at high and low Prandtl number," *Mémoires Société Royale des Sciences de Liége* Series 6 Vol. 4 pp. 125-128, 1972.

J. Pedlosky, "Limit cycles and unstable baroclinic waves," *J. Atmos. Sci.* vol. 29 p. 53, 1972.

J. Pedlosky and C. Frenzen, "Chaotic and periodic behavior of finite amplitude baroclinic waves," *J. Atmos. Sci.* vol. 37 pp. 1177-1196, 1980.

C. Sparrow, "The Lorenz equations: Bifurcations, chaos, and strange attractor," *Applied Mathematical Sciences* vol. 41, 1982.

E. D. Yorke and J. A. Yorke, "Metastable chaos: Transition to sustained chaotic behavior in the Lorenz model," *J. Stat. Phys.* vol. 21 pp. 263-277, 1979.

INTRODUCTION TO SUGGESTION THEORY

Po Lung Yu Dazhi Zhang

School of Business
University of Kansas
Lawrence, Kansas 66045-2003

ABSTRACT. For each decision problem there is a competence set consisting of ideas, knowledge, information and skills for its satisfactory solution. When the decision has *actually* acquired and mastered the competence set, he/she will be able to solve the problem; on the other hand, when the decision maker *thinks* he/she has already acquired and mastered the competence set as *perceived*, he/she will feel confident and comfortable making the decision and/or undertaking the challenge. Therefore, as consultant, how do we suggest to help the decision maker successfully expand his/her competence set with minimal time/cost is very important.

The proposed suggestion theory is based on competence set analysis which is new and based on a "set covering" concept instead of more traditional mathematical ordering and its maximization. The analysis may complement the existing methods for decision analysis.

1. INTRODUCTION

Why are some professional real estate agents *confident* and *quick* in making decisions in buying and selling houses? Probably because they think they have acquired and mastered the needed skills to make the deal.

Why might some teenagers also be confident and quick in making such a decision? Probably because some of the needed skills are unknown to them and consequently, they *think* they have the needed competence to make the decision.

For each problem, denoted by E (e.g., buying a house), there is a competence set consisting of ideas, knowledge, information and skills for its satisfactory solution. When the decision maker thinks that he/she has acquired and mastered the competence set as *perceived*, he/she will be confident and quick in making the decision. For the professional dealers, the competence set is acquired by hard working, practicing and learning. For some teenagers, the perceived competence set is very small and easily covered.

Through experience and learning, we consciously or implicitly have a perception of a collection of what it takes to successfully solve the problem or handle the event. The collection will be denoted by $Tr_t^*(E)$. We also have a perception of a collection of what skills, information or knowledge we have actually acquired for the problem E. This collection will be denoted by $Sk_t^*(E)$. In contrast, the truly needed competence set will be denoted by $Tr_t(E)$, it consists of skills, information or knowledge which is theoretically truly needed to successfully solve E; while the actually acquired skill set is denoted by $Sk_t(E)$. We use time subscript t when we want to emphasize that the competence sets are *evolving* and *expanding* with time. It can be dropped for notational simplicity when there is no confusion. Let us illustrate this concept by the following example.

John, a fresh college graduate, found a job at a midsize city, which was new to him. He wanted to buy a house (E). To successfully buy a house, he needed a set of information, knowledge and skills (competent set). Professional real estate agents usually have such a competent set, denoted by $Tr_t(E)$, which includes housing market information (price, location, house configurations, neighborhood impact, resale value, etc.) financing, packaging, quality assurance, and transaction procedure. As John explored more about house purchasing, his $Tr_t^*(E)$, $Sk_t^*(E)$ and $Sk_t(E)$ gradually or quickly expanded. Until his $Tr_t^*(E)$ and $Sk_t(E)$ are sufficiently expanded, he is not ready to make the purchase because his uncertainty and unknown are not yet clarified. Finally, at the time point, t_1, John found a reliable local agent to help him solve the problem.

Through discussion, interviewing, showing houses, and information, the agent could sense John's $Tr_{t1}^*(E)$, $Sk^*_{t1}(E)$ and $Sk_{t1}(E)$, and John could gradually understand and know $Tr_t(E)$, even vaguely or fuzzily. Over a sequence of interactions at the time points, $\{t_1, t_2, t_3 \dots\}$, John's $Tr_t^*(E)$, $Sk_t^*(E)$ and $Sk_t(E)$ were sufficiently expanded. His unknown and uncertainty about $Tr_t(E)$ were gradually clarified or reduced to a low level which he could easily handle. At that moment, John would be ready to make a decision by offering a financial package to buy his chosen house.

The following are worth mentioning.

(i) John's expansion of $Tr_t*(E)$, $Sk_t*(E)$ and $Sk_t(E)$ is highly dependent on his own knowledge base, memory and thinking process (i.e. his habitual domain) and on how credible he regards the information suggested by the agent.

(ii) To save his time and John's time in solving the problem, the agent's suggestions or idea must be comprehensible, credible and relevant. A suggestion or idea A (say, neighborhood impact) might be more easily digested by John than B (say, resale value) if A is "closer" to $Tr_{t_1}*(E)$ at time t_1. Thus the ordering of suggestion and "closeness" or *connectivity* of an idea/suggestion with the competence set are important. It is also to be noted that idea B can be more easily digested from idea A if B and A are more closely *connected.*

(iii) When a new set of ideas was accepted by John, he would integrate this new set of ideas with his existing ones. How he integrates will greatly affect his effectiveness in expanding his $Tr_t*(E)$, $Sk_t*(E)$ and $Sk_t(E)$. We shall discuss this shortly.

This paper serves as an introduction of the suggestion theory which is based on the *competence set analysis* (Yu 1990 and Yu and Zhang 1989) and how the introduced framework can be used to describe interesting aspects of decision making situations and characteristics. In the remainder of this paper, we will discuss decision cycles and learning processes (Section 2), effective suggestions (Section 3), and effective expansions (Section 4). Concluding remarks is given in Section 5.

2. DECISION CYCLES AND LEARNING PROCESSES

Recall from Section 1, that for each decision problem E, there is a competence set $Tr(E)$ consists of ideas/skills which are *truly needed* for solving problem E. The decision maker *perceived competence set* is denoted by $Tr*(E)$. When the decision maker believes that he/she has already acquired and mastered $Tr*(E)$, he/she feels comfortable and confident about making a decision. Otherwise, he/she hesitates to make a decision, especially when it involves high stakes. The perception, acquisition and mastering of $Tr*(E)$ thus plays an important role in determining how quickly and how effectively a decision is made and executed. These topics and their applications will be briefly discussed in this section (For detals, see Yu (1990) and Yu and Zhang (1989).) Specifically, we shall discuss decision cycles, core of habitual domains, learning processes and effective suggestions.

The period of time from the beginning (recognition) to the end (reaching a final decision) of a decision problem $\boldsymbol{E}$ is called the *decision cycle* of $\boldsymbol{E}$. For a simple grocery purchasing problem, the decision process is relatively simple and the decision cycle is short. For purchasing a house or preparing for a job interview, the corresponding decision process can be relatively more complex and the decision cycle can be relatively longer.

Notice that the decision cycle of problem $\boldsymbol{E}$ can be a function of how we perceive the problem's needed competence set $Tr^*(\boldsymbol{E})$ and how much we have acquired and mastered $Tr^*(\boldsymbol{E})$, in addition to the time constraints imposed on $\boldsymbol{E}$.

Given an event or a decision problem $\boldsymbol{E}$ which catches our attention at time t, the *propensity for an idea* **I** *to be activated* is denoted by $P_t(\mathbf{I},\boldsymbol{E})$. Like a conditional probability, we know that $0 \leq P_t(\mathbf{I},\boldsymbol{E}) \leq 1$, that $P_t(\mathbf{I},\boldsymbol{E}) = 0$ if **I** is unrelated to $\boldsymbol{E}$, and that $P_t(\mathbf{I},\boldsymbol{E}) = 1$ if **I** is automatically activated in the thinking process whenever $\boldsymbol{E}$ is presented.

Empirically, like with probability functions, $P_t(\mathbf{I},\boldsymbol{E})$ may be estimated by determining its *relative frequency*. For instance, if **I** is activated 7 out of 10 times whenever $\boldsymbol{E}$ is presented, then $P_t(\mathbf{I},\boldsymbol{E})$ may be estimated at 0.7. Probability theory and statistics can then be used to estimate $P_t(\mathbf{I},\boldsymbol{E})$.

Let us define the *α-core* of competence set for $\boldsymbol{E}$ at time t, denoted by $C_t(\alpha,\boldsymbol{E})$, to be the collection of the ideas or concepts that can be activated with a propensity larger than or equal to α. That is,

$$C_t(\alpha,\boldsymbol{E}) = \{ \mathbf{I} \mid P_t(\mathbf{I},\boldsymbol{E}) \geq \alpha \}.$$

By the *core of competence set for* $\boldsymbol{E}$, denoted by $C_t(\boldsymbol{E})$, we mean the collection of ideas or concepts that would almost surely be activated when $\boldsymbol{E}$ is presented. In other words, it is the α-core with $\alpha \to 1$. Sometimes, for convenience and to avoid confusion, the core of competence set may simply mean the α-core with a high value of α. Thus if **I** is an element of the core of competence set for $\boldsymbol{E}$, then $P_t(\mathbf{I},\boldsymbol{E})$ is large (close to the limit of 1) for most of time t when $\boldsymbol{E}$ is present.

Now recall that $Tr_t^*(E)$ is the perceived competence set for solving E, and the subscript "t" is used to emphasize its dynamics. Suppose that

$$Tr_t^*(E) \subseteq C_t(\alpha, E)$$

with a large value of α (that is, α is close to its upper limit 1). In this case the decision maker would feel comfortable with the problem and could solve it with a high degree of efficiency, because he/she has acquired and almost mastered $Tr_t^*(E)$.

Given a decision problem E and an idea or skill I of $Tr^*(E)$, suppose that $P_t(I,E) = 0$. That is, the decision maker does not associate I with E. Then implanting is needed. The purpose of implanting is to make a positive association between I and E. That is, $P_{t'}(I,E) > 0$ for some $t' > t$. This can be achieved through teaching, suggestion and/or training.

Once the idea I is implanted, $P_t(I,E)$ can be positive, yet still low. In order for "I" to have an impact on the decision maker, it needs to be high enough. To achieve this goal, we need to nurture the idea using training, practice and rehearsing. Like seedlings of a tree, without nurturing, the newly implanted ideas will wither and disappear. It is to be noted that "experiencing" and "self-suggesting", in addition to information inputs, are two important ways to strengthen our circuit patterns of new ideas. Our mind may not distinguish the sources. Both physical experience and mental exercise (or suggestion) are important in the nurturing process.

Through repeated practice and nurturing, a new idea "I" could gradually become an element of the core on the decision problem E. Thus, the propensity of activation of I is very high or, $P_t(I,E) \to 1$. That is, whenever our attention is paid to E, "I" would be almost surely activated. When we reach this stage for "I", we say that "I" is a *habituating element* of the competence sets.

Finally, we notice that the learning process of implanting, nurturing and habituating is not only applicable to self-learning, but also for suggestions to other people and/or training other people to acquire and expand the competence set.

3. EFFECTIVE SUGGESTIONS

A suggestion, S, is a set of ideas and/or operators. To help the decision maker reach decisions quickly (i.e., for the decision maker to have a shorter decision cycle), effective suggestions are extremely important.

From the dynamic behavior mechanism and habitual domain analysis (refer to Yu (1990) for details), we have the following observations:

(1) A suggestion can catch the attention only if it can create a relatively high level of charge on the decision maker. Although he/she has innate needs for external information, if the suggestion is unrelated to his/her charge structure (which is the collection of charges created by various events, see Yu (1990) for details), it will be most likely ignored or neglected.

(2) A suggestion cannot easily have a relatively high level of charge on the decision maker when the decision maker is preoccupied by other significant events (such as a grave illness or pressure to meet an important deadline unrelated to the suggestion).

(3) A suggestion can be more easily accepted, if it already exists in the memory of the decision maker (thus only "retrieving" is needed for the acceptance, "encoding" is not needed), or it is *closely connected* to the significant memory of the decision maker (because of association and analogy in information processing). That is, a suggestion can be easily accepted if it can broadly and deeply activate the circuit patterns of the receiver's memory. Thus we define

$$\mathrm{Conn}_1(S, \mathrm{Sk}(E)) = \sup \{P(I,S) \mid I \in \mathrm{Sk}(E)\}$$

and

$$\mathrm{Conn}_2(S, \mathrm{Sk}(E)) = \int_{\mathrm{Sk}(E)} P(I,S)\, \mu(dI).$$

They are called respectively the *type I and type II connectivity of a suggestion S with the receiver's competence set* Sk(E), where P(I,S) is the activation propensity function as defined in Section 2.

(4) A suggestion can receive a long duration of attention time if it can create a relatively high level of charge for a long duration, which occurs when the suggestion implants enthusiasm and confidence in the decision maker to achieve his/her burning desired goals. (Thus the suggestion not only creates a high level of charge, but also creates confidence for the release of charge).

(5) Usually a suggestion perceived as highly related to important life goals of the decision maker, can easily obtain the attention of the decision maker because of (2)-(4).

(6) If an accepted suggestion contains new ideas or knowledge, especially when S has a large intersection with the competence sets, the new ideas or

knowledge will be integrated with the existing memory, and the memory will be expanded. Furthermore, $Tr^*(E)$ and $Sk(E)$ will also be expanded. Yet they can be gradually stabilized, unless a new set of suggestions is encountered.

Now we assume that $Tr_t(E)$ is fairly stable. Let $Tr_t^*(E)$, $Sk_t^*(E)$ and $Sk_t(E)$ simply be represented by $Comp_t(E)$. Let $\{S_1, S_2, \ldots\}$ be a sequence of suggestions by the expert (S_i may be overlapping, i.e. $S_i \cap S_j \neq \emptyset$ and $S_i \neq S_j$). Let $Comp_0 = Comp_{t0}(E)$ be the initial competence set and for k=1,2,... , $Comp_k$ be the $Comp_t(E)$ after integrating S_k with $Comp_{k-1}$. Then a *successful suggestion program* is a sequence of suggestion $S = \{S_1, S_2, \ldots\}$ so that as $Comp_k$, k=1,2,..., consecutively expanding, there is a finite number m so that $Comp_m \supseteq Tr_t(E)$.

By assigning time and cost into S, one can study effective suggestion programs to create a successful program which minimizes time and cost. This is what we are going to discuss in the next section.

4. EFFECTIVE EXPANSIONS

Given a decision problem E. Recall that $Sk(E)$ and $Tr(E)$ denote respectively the actually acquired skill set and the truly needed competence set. Note that $Tr(E)\backslash Sk(E)$ is the set of ideas which are needed but not yet acquired. We want to cover this set with minimal time/cost. Let X be a large set containing both $Sk(E)$ and $Tr(E)$. For example, X can be the potential domain (Yu 1985, 1990) if it contains $Tr(E)$ or, simply set $X = Sk(E) \cup Tr(E)$.

Let f(a,b) be the time or costs needed for reaching idea b from idea a or a from b. Assume that X is discrete finite. If t is symetric and the time/costs needed to reach an idea is independent on the skill set, then we can apply the *minimal spanning tree method*, which is a variation of the so-called *shortest route problem* and is well known in network analysis. Here the nodes are the elements (ideas) in $Tr(E)\backslash Sk(E)$ and the distance between each pair of nodes is the time/costs needed to reach one from the other which is given by f(a,b).

In the above discussion, we assumed that t is symetric and the time/costs needed to reach an idea is independent on the skill set. If this is not the case, we can modify the approach as follows. First we define an *expansion process* as a path Γ in X which spans $Tr(E)$. Specifically, an ordered sequence $(x_1, x_2, \ldots, x_n)$ where $x_i \in X$, $x_i \neq x_j$, $i \neq j$, $i,j=1,2,\ldots,n$ such that

$$\{x_1, x_2, \ldots, x_n\} = X.$$

Assume that the *time/costs needed to reach idea x given current actual skill set A* is known, which is denoted by F(A,x). Then for any expansion process $\Gamma = (x_1, x_2, ..., x_n)$, the total time needed in the expansion process is given by

$$\Sigma\{F(Sk(E)\cup\{x_1,x_2,...,x_{i-1}\}, x_i) \mid i=1,2,...,n\}.$$

The *optimal* expansion process will be the one such that the above expression is minimized.

5. CONCLUDING REMARKS

The basic concepts of suggestion theory are briefly schetched in this paper. The paradigm may complement the existing literature on decision analysis.

A major common interest in the literature of traditional decision analysis is to represent and formulate the decision problems in terms of mathematical ordering (utility function or preference structures) over the feasible choices which are involved with random outcomes. Thus, it is important to know axiomatically under what conditions such a representation or some special form of representation is feasible and meaningful. (For instance, see Fishburn (1970) and Keeney and Raiffa (1976) and quotes therein.)

Our framework, however, is based on a set covering concept. When the DM's skill set contains or nearly contains the needed competence set, he/she is confident and comfortable in making an effective decision. We emphasize effective decision making, not the optimal choice. Furthermore, in our formulation, the skill set is allowed to expand with suggestions and experience. Thus it is a dynamic model allowing the relevant competence sets to evolve and change with time and situation. Enthusiasm, confidence, attention allocation, connectivity of ideas, implanting, nurturing, habituating, integration, decision instinct (concepts emerge from the core of competence sets), etc., are some important concepts which play an important role in effective decision making for nontrivial problems and are hardly discussed in the literature.

Future effort should be directed to the implementation of the proposed paradigm to real life decision problems. Given decision problem E, we wish to create for E a decision support system. The input data of the system are divided into two parts, the inside information, i.e., the decision maker's competence set, and the outside information, which includes the time constraint imposed on problem E, maximal effort we (the consultants) can put.

The output of the system should include such information as "Is the problem feasible?", "What should we suggest to the decision maker?", "How do we suggest?", "What are the possible outcomes (more precisely, what will be the expected competence set) by executing a special suggestion program?", and "When will the decision maker be ready (confident and capable) to solve problem E?".

We imagine that there should be such a system, which is able to provide with the decision maker an effective suggestion program when his/her competence set has been input. Therefore, as long as the decision maker gives the system his/her "changes" regularly, the system will continuously suggest him/her what to do.

REFERENCES

Chan, S. J. and Yu, P. L. (1985) Stable Habitual Domains: Existence and Implications, Journal of Mathematical Analysis and Applications, Vol.110, No.2, pp.469-482.

Chankong, V. and Haimes, Y.Y. (1983) Multiobjective Decision Making : Theory and Methodology, North-Holland, New York, N. Y.

Fishburn, P. C. (1970) Utility Theory for Decision Making, Wiley, New York, NY.

Hogarth, R. (1987) Judgment and Choice, John Wiley and Sons, New York, NY, 1987.

Isaacs, R. (1965) Differential Games, John Wiley and Sons, New York.

Janis, I. L., and Mann, L.(1977) Decision Making, A Psychological Analysis of Conflict, Choice and Commitment, The Free Press, New York.

Keeney, R. L., and Raiffa, H.(1976) Decision with Multiple Objectives: Preferences and Value Tradeoffs, Wiley, New York, NY..

Newell, A. and Simon, H. A. (1972) Human Problem Solving, Prentice Hall Inc., Englewood Cliffs, NJ.

Yu, P. L. (1988) Effective Decision Making Using Habitual Domain Analysis, Tutorial Lecture delivered at ORSA/TIMS Joint National Meeting, Denver, Oct 23-26.

Yu, P. L. (1985) Multiple Crireria Decision Making: Concepts, Techniques and Extensions, Plenum, New York.

Yu, P. L. (1990) Forming Winning Strategies - an integrated theory of habitual domains Springer-Verlag, Heidelberg.

Yu, P. L. and Huang, S. D. (1987) Knowing People and Making Strategical Decisions, China Coal Industry Publishing House, Beijing, China. (In Chinese).

Yu, P. L. and Zhang, D. (1989) Competence Set Analysis for Effective Decision Making, Control: Theory and Advanced Technology, Vol.5, No.4, pp. 523-547.

STABILITY OF CONTROLLERS WITH ON-LINE COMPUTATIONS

Pedro J. Zufiria[1] *and Ramesh S. Guttalu*[2]

Department of Mechanical Engineering
University of Southern California
Los Angeles, CA 90089-1453

Abstract: The dynamical systems theory developed in [4, 7, 8, 9] is applied to the stability analysis of control systems in which the feedback control law requires in real time the solution of a set of nonlinear algebraic equations. Since small sampling period is assumed, the stability and performance of the controlled process can be studied with a continuous-time formulation. A singularly perturbed system is used to model the combined dynamics of the system being controlled and a numerical iterative algorithm required to compute the control law. An updating control procedure has been proposed based on the iterative nature of the control algorithm. The results obtained in [9] regarding the behavior of a dynamical system that models the numerical algorithms lead to a considerable simplification in the analysis. For the case of control problem involving inverse kinematics, the numerical algorithm which solves for inverse kinematics can be considered as an observer (or an estimator) of the state space variables. The study provides an estimate of the required speed of computations to preserve the stability of the controller.

1. INTRODUCTION

In references [4, 7, 8, 9], global behavior of a class of dynamical systems of the form

$$\dot{x} = -J^{-1}(x)f(x), \quad x \in \mathbf{R}^{\mathrm{N}}, \; f \in C^1, \; J(x) = \nabla_x f(x) \tag{1}$$

is studied. In addition, some discrete formulations obtained by numerically integrating the dynamical system (1) are considered in [9]. These discrete formulations can be represented, in general, as discrete dynamical systems (or maps) of the form

$$x_{n+1} = G(x_n), \quad x \in \mathbf{R}^{\mathrm{N}}, \quad G : \mathbf{R}^{\mathrm{N}} \to \mathbf{R}^{\mathrm{N}}, \quad n \in \mathbf{Z} \tag{2}$$

Both types of systems (1) and (2) are typical formulations for determining the roots of the vector function f. A study of these dynamical systems sheds light on the behavior of some of the standard numerical algorithms employed to find the roots of f. Note that one-step Euler numerical integration of the continuous-time system (1) leads to the Newton method, which is an iterative technique or discrete system of the form (2).

---

[1] Research Associate
[2] Assistant Professor

The study of the dynamical systems (1) and (2) also finds potential application in control theory. One such application is considered here. We analyze problems where the computation of a control law needs primarily a solution of a system of nonlinear algebraic equations in real time. The algorithm that solves the algebraic system is referred to as *solving algorithm*. A classical analysis of the performance of these control designs requires that a solution of the algebraic system be determined in negligible real time, without affecting the performance of the controller. On the other hand, if the number of algebraic equations to be solved is considerably large or the dynamics of the system to be controlled are fast enough, then the dynamics of the solving algorithm should be taken into account. The time taken by the computer to calculate the control law has been considered in the literature on classical control of linear time-invariant systems where pure delay units are used to characterize the computation time [2]. The nonlinear case presents two main difficulties. One of them is how to characterize the dynamics of the solving algorithm. The other relates to the combined analysis of the dynamics of the system being controlled and the solving algorithm.

A characterization of the dynamics of algorithms for solving an algebraic system has been developed in [9]. Continuous-time as well as discrete-time formulations are considered in [9]. In this paper, the continuous-time formulation is employed to characterize both the solving algorithm and the system being controlled, based on the following assumptions. First, the sampling period is smaller than the characteristic time of evolution of the dynamics of the system being controlled. In addition the sampling period should also be smaller or of the same order of magnitude of the iteration time of the numerical algorithm. Finally, the control law is updated for every sampling instant with the iterate value that is being employed in the iterative algorithm at that sampling instant.

If the dynamics of the numerical algorithm are faster than the dynamics of the system being controlled, the combined analysis of both systems can be performed via a singular perturbation formulation, see [3, 5, 6]. The singularly perturbed system can be divided into two subsystems. The first one, called the *reduced order system*, represents the dynamics of the system being controlled. The second system, named *boundary layer system*, models the "fast" dynamics of the solving algorithm. Since they are coupled, both the systems can be studied simultaneously with the singular perturbation formulation.

The organization of the paper is as follows. In Section 2, a background of some known results concerning singular perturbed systems is provided. An implementation of a controller requiring on-line computations is presented in Section 3. A singularly perturbed system is used in Section 4 to model the system being controlled and the on-line controller. The theory developed in [9] simplifies considerably the analysis. The results are applied to obtain stability results for tracking control problems. Based on the same singularly perturbed formulation, some stability results are also obtained in Section 5 for controllers requiring computation of inverse kinematics. Examples provided here illustrate the utility of theoretical results. Concluding remarks appear in Section 6.

2. BACKGROUND

We begin by providing some of the main results regarding the singularly perturbed systems. For linear time-invariant systems the following result holds, see [5].

Theorem 1 *Consider the linear time-invariant singularly perturbed system given by*

$$\dot{x} = A_{11}x + A_{12}z, \quad x \in \mathbf{R}^N, \; z \in \mathbf{R}^M \tag{3}$$

$$\epsilon \dot{z} = A_{21}x + A_{22}z, \quad \epsilon > 0 \tag{4}$$

where $A_{11}, A_{12}, A_{21}, A_{22}$ are constant matrices of appropriate dimension. Let

$$A_0 = A_{11} - A_{12}A_{22}^{-1}A_{21}$$

If A_{22}^{-1} exists, and if A_0 and A_{22} are Hurwitz matrices, then there exists an $\epsilon^* > 0$ such that for all $\epsilon \in (0, \epsilon^*]$ the system (3-4) is asymptotically stable.

Note that this theorem only provides information about the existence of values of ϵ that guarantee asymptotic stability of the system. If the system (3-4) is a linear approximation of a nonlinear system at a given asymptotically stable equilibrium point, then no conclusions about the associated domains of attraction can be drawn. The following result provides some information about stability and domains of attraction for nonlinear systems, see [5].

Theorem 2 *Consider the singularly perturbed system*

$$\dot{x} = f(x,z), \quad f : \mathbf{R}^N \times \mathbf{R}^M \to \mathbf{R}^N \tag{5}$$

$$\epsilon \dot{z} = g(x,z), \quad g : \mathbf{R}^N \times \mathbf{R}^M \to \mathbf{R}^M \tag{6}$$

Let $B_x \in R^N$ and $B_z \in R^M$ denote closed sets. Let $\Psi(\cdot)$ and $\Phi(\cdot)$ be two scalar functions of vector argument which vanish only when their argument is zero (these are referred as comparison functions*). Suppose that the following assumptions are satisfied:*

1. *The origin is an isolated equilibrium of (5-6) in $B_x \times B_z$ and $z = h(x)$ is the unique root of the equation $0 = g(x,z)$ for (x,z) in $B_x \times B_z$.*

2. *There exists a Lyapunov function candidate $W(x,z)$ such that for all $(x,z) \in B_x \times B_z$ the following conditions are satisfied*

   (i) $W(x,z) > 0, \quad \forall z \neq h(x)$ *and* $W(x,h(x)) = 0$

   (ii) $(\nabla_z W) \cdot g(x,z) \leq -\alpha_2 \Phi^2(z - h(x)), \; \alpha_2 > 0$

   (iii) $(\nabla_x W) \cdot f(x,z) \leq \gamma \Phi^2(z - h(x)) + \beta_2 \Psi(x) \Phi(z - h(x))$

3. *There exists a Lyapunov function $V(x)$ that for all $(x,z) \in B_x \times B_z$ satisfies*

   (i) $(\nabla_x V) \cdot f(x,h(x)) \leq -\alpha_1 \Psi^2(x), \; \alpha_1 > 0$

(ii) $(\nabla_x V)\cdot[f(x,z)-f(x,h(x))] \le \beta_1\Psi(x)\Phi(z-h(x))$

Then the origin is an asymptotically stable equilibrium of the singularly perturbed system (5-6) for all $\epsilon \in (0,\epsilon^*)$, *where*

$$\epsilon^* = \frac{\alpha_1\alpha_2}{\alpha_1\gamma+\beta_1\beta_2} \tag{7}$$

Moreover, for every number $\rho \in (0,1)$ *the function*

$$\nu(x,z) = (1-\rho)V(x) + \rho W(x,z), \quad 0<\rho<1 \tag{8}$$

is a Lyapunov function for all $\epsilon \in (0,\epsilon_\rho)$, *where* $\epsilon_\rho \le \epsilon^*$ *is given by*

$$\epsilon_\rho \triangleq \frac{\alpha_1\alpha_2}{\alpha_1\gamma + \frac{1}{4(1-\rho)\rho}[(1-\rho)\beta_1+\rho\beta_2]^2} \tag{9}$$

The parameter ρ defines a family of Lyapunov functions and its selection may depend on different criteria such as the desired shape of the level curves or the upper bound ϵ_ρ. The upper bound has a maximum value ϵ^* which corresponds with the value $\rho^* = \frac{\beta_1}{\beta_1+\beta_2}$. Based on the Lyapunov function, we can define the set

$$S \triangleq \{(x,z) \mid \nu(x,z) \le c\} \subset B_x \times B_z, \quad c>0 \tag{10}$$

which is included in the domain of attraction of the equilibrium.

One notes that in the singularly perturbed system (5-6), the reduced system corresponds to

$$\dot{x} \;=\; f(x,h(x)), \quad f:\mathbf{R}^{\mathbf{N}}\times\mathbf{R}^{\mathbf{M}} \to \mathbf{R}^{\mathbf{N}} \tag{11}$$

and the fast system corresponds to

$$\epsilon\dot{z} \;=\; g(x,z), \quad g:\mathbf{R}^{\mathbf{M}}\times\mathbf{R}^{\mathbf{N}} \to \mathbf{R}^{\mathbf{M}} \tag{12}$$

where x is considered as a constant because of its slow time variation when compared with that of z for a small enough $\epsilon > 0$.

One can show that the assumptions made in the above theorem are mild and that there is a wide class of systems for which they hold as shown in the following lemma which is taken from [6].

Lemma 1 *Let* f, g *and* h *be continuously differentiable in* $\tilde{B}_x \times \tilde{B}_z$. *Suppose that the reduced system is exponentially stable such that* $\|x(t)\| \le c_3\|x(0)\|e^{-c_4 t}$ *whenever* $x(0)\in\tilde{B}_x$. *Let* $B_x \subseteq \tilde{B}_x$ *be a set that every trajectory starting in* B_x *remains inside* $\tilde{B}_x$. *For every* $x\in\tilde{B}_x$, *suppose that the equilibrium point* $\overline{z}=h(x)$ *of the "fast" system is inside* B_z *and is exponentially stable uniformly in* x *such that* $\|z(\tau)\| \le c_5\|z(0)-h(x)\|e^{-c_6\tau}$ *whenever* $z(0)\in\tilde{B}_z$. *Let* $B_z\subseteq\tilde{B}_z$ *be a set for which every trajectory starting in* B_z *remains inside* $\tilde{B}_z$. *Then the conditions of the above theorem are satisfied in* $B_x \times B_z$.

The proof of the lemma is based on the construction of conceptual Lyapunov functions $V(x)$ and $W(x,z)$ given by

$$V(x) = \int_0^T \|\mathcal{X}(t,x)\|^2 dt$$
$$W(x,z) = \int_0^T \|\mathcal{S}(\tau,z;x) - h(x)\|^2 d\tau$$

where $\mathcal{X}(t,x)$ is the trajectory of the reduced system starting at an initial point $x(0)$ and $\mathcal{S}(\tau,z;x)$ is the trajectory of the fast system starting at an initial point $z(0)$. Consequently, the above lemma is mainly of a theoretical value and does not provide a convenient way to determine some values for the constants $\alpha_1, \alpha_2, \beta_1, \beta_2$ and γ.

In the following sections, we will apply the above theorems to a special class of singularly perturbed systems which model dynamical systems with control laws requiring on-line solutions of coupled nonlinear algebraic equations. For some cases one is able to provide practical forms of Lyapunov functions.

3. AN "UPDATING CONTROL" PROCEDURE

Due to the digital nature of practical controllers, one can define a measurement sampling period which defines how frequently measurements are performed (at given sampling instants). These sampling instants usually define also the times at which the control law must be updated. The classical implementation of control laws with on-line computations updates the control value after the algorithm has obtained a new solution for the control value. Hence, this procedure assumes that the control algorithm is fast enough to provide a new control value before a new sampling instant arrives. It is known from digital control theory that the sampling period has to be chosen small enough when compared to the characteristic times of evolution of the system and controller dynamics. A problem arises whenever the computation of the control law requires more time than the sampling period.

We propose an alternative approach here for updating the control for the case in which the control value is obtained by a solving algorithm. Suppose that the numerical algorithm that computes the control law is based on an iterative procedure. Then whenever a sampling instant is reached, even if the algorithm did not find a solution, one can apply a control law based on the current iterate value available at that instant of time. This iterate value can be considered as an estimate of the exact solution. Under this control procedure, and assuming that the sampling period is bounded above by the order of magnitude of one iteration time, the algorithm that computes the control law can be seen as a dynamical system which is coupled with the system being controlled. One can expect the solving algorithm to be fast when compared with the system under control. Therefore, we can use a singularly perturbed formulation of the total system which is composed of a reduced order system and a fast system. The reduced order system represents the dynamical system being controlled (with the corresponding control law). The fast subsystem represents the numerical algorithm that is used to solve for the control law. As shown in [9], the evolution of the numerical algorithm can be associated with a *conceptual* time of evolution of a dynamical system. Hence, the parameter ϵ associated with the singularly perturbed system can be interpreted as

$$\epsilon = \frac{T_R}{T_C} \tag{13}$$

where T_R is the real computer time needed to compute a conceptual evolution time T_C in the solving algorithm. In other words, ϵ represents the ratio of the real time units of the system being controlled to the conceptual time units of the algorithm.

4. COMPUTED IMPLICIT MODEL REFERENCE CONTROL

In this section we consider control algorithms where the control is defined in an implicit manner so that the system follows a model reference system. Note that the term *implicit* here refers to the fact that the control law is given in an implicit function formulation. This concept is different to the one in direct adaptive control where the estimates of the parameters of the system are used implicitly. In direct adaptive control, the expression of the control is given explicitly but not as a function of the estimates of the parameters of the system (i.e. the control law is computed directly without need of a previous explicit estimation of such parameters).

Making use of the singularly perturbed formulation, we now analyze the stability of some of these control algorithms. Consider the system

$$\dot{x} = f(x,u), \quad f : \mathbf{R^N} \times \mathbf{R^N} \to \mathbf{R^N} \tag{14}$$

where $f(0,0) = 0$. We want $x(t)$ to follow the trajectory of the system given by

$$\dot{x} = g(x) \tag{15}$$

for which the origin is asymptotically stable and there is a positive definite function $V(x)$ for which $(\nabla_x V)\cdot g \leq -\alpha_1 \Psi^2(x)$, $\alpha_1 > 0$, where $\Psi(\cdot)$ is a comparison function. To characterize the algorithm that computes the control, suppose that

$$h(x,u) = f(x,u) - g(x) \tag{16}$$

and that there exists a control $u^*(x)$ which is the unique root of $h(x,u) = 0$ in a given region of study. Then one can model the algorithm that solves for $u^*(t)$ via the evolution equation for u given by

$$\epsilon\dot{u} = -(\nabla_u h(x,u))^{-1} h(x,u) = -(\nabla_u f(x,u))^{-1}(f(x,u) - g(x)) \tag{17}$$

which is of the form (1). The value of ϵ is given by the ratio defined by (13) which is the ratio of the real time units of the dynamics of (14) to the conceptual time units of the algorithm described by (17).

4.1 A LINEARIZATION APPROACH

Equations (14) and (17) when linearized around the solution $x = 0$ and $u = 0$ lead to the linear singularly perturbed system

$$\Delta\dot{x} = \nabla_x f(0,0)\Delta x + \nabla_u f(0,0)\Delta u \tag{18}$$

$$\Delta\dot{u} = -[(\nabla_u f(0,0))^{-1}(\nabla_x f(0,0) - \nabla_x g(0))]\Delta x - \Delta u \tag{19}$$

Note that

$$A_{22} = -I, \quad A_0 = \nabla_x g(x) \tag{20}$$

If the origin of the reference system (15) is exponentially stable, then by the definition of $g(x)$, A_0 and A_{22} are both Hurwitz matrices. Applying Theorem 2, we deduce that the system (18-19) is asymptotically stable for small enough ϵ.

Note that when the perturbed system is very close to the solution ($x = 0$, $u = 0$) the model (17) of the numerical algorithm (fast subsystem) reduces to a linear exponentially stable system which has an eigenvalue -1 with multiplicity N. Theoretically, this system would take infinite time to reach the equilibrium point.

In practice, a numerical implementation is suitably modeled by a discrete system. Some of the discrete forms applied to this linear system can provide convergence in a finite number of steps (e.g. in one step for the case of Euler one-step integration representing the standard Newton method). Then the rate of convergence of the algorithm is not very well modeled by a continuous-time formulation (in fact this would correspond with the case $\epsilon = 0$ because only one iteration is needed). This is another aspect which limits the applicability of the linearization approach.

4.2 A NONLINEAR SYSTEM APPROACH

Equations (14) and (17) define a nonlinear singularly perturbed system. Based on Theorem 2, suppose that $\Phi(\cdot)$ is another comparison function and that the following assumptions are met:

$$-2h^T h \le -\alpha_2 \Phi^2(u - u^*(x)), \ \alpha_2 > 0 \tag{21}$$

$$2h^T(\nabla_x h) \cdot f \le \gamma \Phi^2(u - u^*(x)) + \beta_2 \Psi(x)\Phi(u - u^*(x)) \tag{22}$$

$$(\nabla_x V) \cdot h(x,u) \le \beta_1 \Psi(x)\Phi(u - u^*(x)) \tag{23}$$

Then the controller guarantees asymptotic stability of the origin for (14) and (17) provided that the solving algorithm represented by equation (17) is fast enough so that $\epsilon \in [0, \epsilon^*)$ where ϵ^* is given by (7). The Lyapunov function for the entire singularly perturbed system has the form

$$\nu(x,u) = \rho V(x) + (1-\rho)h^T(x,u)h(x,u), \quad 0 < \rho < 1$$

Note that the whole set of conditions of the theorem for nonlinear systems reduce to three simple ones, as stated in (21-23), for this case.

In [9], it has been shown that the trajectories of the dynamical system defined by (1) may have pathological dynamics near singular manifolds. The singular manifolds for the system (17) are defined by $det(\nabla_u h^T(x,u^*(x))) = 0$. One may erroneously think that these manifolds may appear in the region of study and hence, may jeopardize the results obtained above. The following lemma proves that this is not the case and strengthens the validity of previous results.

Lemma 2 *The condition (21) is satisfied in a region of the state space where no singular manifold is present.*

Proof: Consider the Taylor expansion around $u = u^*(x)$:

$$\begin{aligned} -2h^T h &= -2((u-u^*(x))^T(\nabla_u h(x,u^*(x)))^T \nabla_u h(x,u^*(x))(u-u^*(x)) \\ &+ \mathcal{O}(\|u-u^*(x)\|^3) \end{aligned}$$

For the assumption (21) to be satisfied, it is required that

$$(\nabla_u h(x,u^*(x)))^T \nabla_u h(x,u^*(x)) > 0$$

which implies that $det(\nabla_u h(x,u^*(x))) \neq 0$. □.

This result indicates that α_2 is a critical constant in the sense that for $\alpha_2 > 0$ to be satisfied, the analysis must be restricted to a region where no singular manifold exists. In addition, the value of $\epsilon(\alpha_2)$ represents a tradeoff between the size of the region where Theorem 2 is applicable and the requirements to be met by the fast subsystem parameter ϵ. The larger the region the smaller the parameter ϵ (or faster the subsystem) must be.

4.3 EXAMPLE 1

To illustrate results obtained so far, consider the following control problem

$$\dot{x} = u + (1+u^4)x, \quad x \in \mathbf{R}, \ u \in \mathbf{R} \tag{24}$$

for which it is difficult to obtain an analytical expression for the control law. Suppose we want the system (24) to follow the reference system

$$\dot{x} = g(x) = -x, \quad x \in \mathbf{R} \tag{25}$$

then, the equation determining the fast subsystem is given by

$$\epsilon\dot{u} = -\frac{u + (1+u^4)x + x}{1+4u^3 x} \tag{26}$$

(a) Linear approach

A linearization of (24-26) in the neighborhood of the equilibrium $x = 0$ and $u = 0$ gives the linear time-invariant (second-order) singularly perturbed system

$$\begin{aligned} \Delta\dot{x} &= \Delta x + \Delta u \\ \epsilon\Delta\dot{u} &= -2\Delta x - \Delta u \end{aligned}$$

whose eigenvalues are

$$\lambda_{1,2} = \frac{1}{2\epsilon}\left\{-(1-\epsilon) \pm \sqrt{(1-\epsilon)^2 - 4\epsilon}\right\} \tag{27}$$

The linearized singularly perturbed system is asymptotically stable for $\epsilon < 1$. This provides an upper bound on ϵ but no information about the domain of attraction associated with the origin $x = 0$ and $u = 0$ of the original nonlinear system (24-26).

A delay model is employed in the classical analysis of the influence of computer time on the control of linear time-invariant systems. Suppose that the computer takes a time T to compute the control law. The transfer function of the whole system and controller would be

$$T_f = \frac{1}{s - 1 + 2e^{-T}} \tag{28}$$

which, via a Nyquist plot, implies that the system becomes unstable for $T > \ln(2) \approx 0.7$ s. In order to compare this absolute time result with the bound on the parameter ϵ obtained with the approach presented here, we need to explicitly take into account the computational details (method, computer features, etc). Suppose that the computer determines the control law via an iterative method that is modeled by the form (1) and that the sampling period is small enough. If the initial iterates are close to the solution, then the linear model can be considered as valid. The classical result obtained from a Nyquist plot would require the computer to find the roots (exact control law) in less than 0.7 s. On the other hand, suppose that we update the control with the current value of the iterate at every sampling instant whether or not a convergence to the root is achieved. Then, the stability of the origin of the total system (24-26) is guaranteed if the computer calculates the iterations equivalent to one logical second in less than one second of real time.

Note that, in the classical Newton method (with unit step size), one logical second corresponds to one iterate of the algorithm (that is, one iteration of the map (2)). Nevertheless, when the linear approximation is considered, the continuous-time model does not quite represent the time evolution of the algorithm. This is so because Newton method applied to a linear system converges in one iteration as opposed to the exponential decay of the continuous-time system.

(b) Nonlinear approach

More refined results relating the value of ϵ with the domain of attraction of the solution of the original nonlinear system can be obtained following the steps defined in the Theorem 2. After some algebra, we obtain the following quantities defined in that theorem:

$$B_x \times B_u = [-0.338, 0.338] \times [-0.88, 0.88]$$

$$W(x,u) = (xu^4 + u + 2x)^2$$

$$\Phi(y) = \Psi(y) = y$$

$$\|4u^3x\| \leq 1 - \sqrt{\tfrac{\alpha_2}{2}}, \quad 0 < \alpha_2 < 2$$

$$\gamma = 2(2 + 0.88^2)(2 - \sqrt{\tfrac{\alpha_2}{2}})^2 = 5.55(2 - \sqrt{\tfrac{\alpha_2}{2}})^2$$

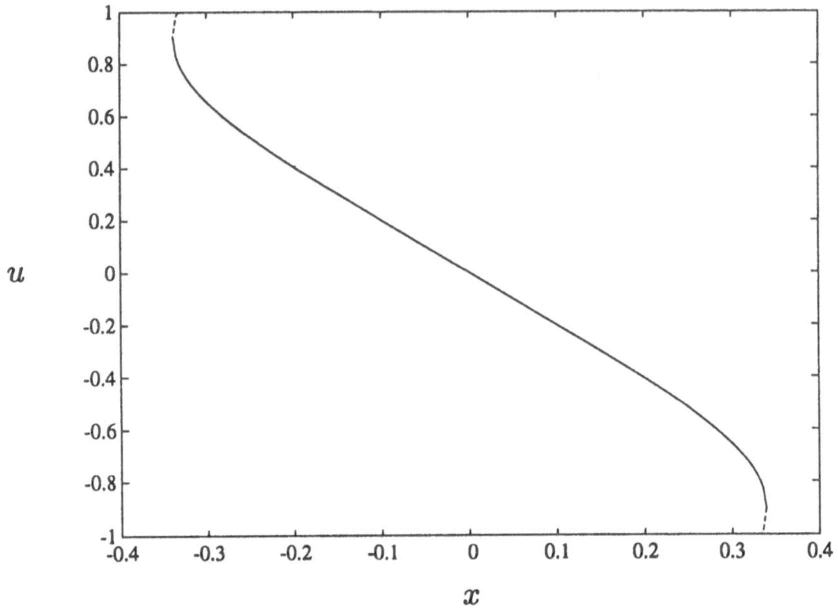

Figure 1: Solution curves for Example 1.

$$\beta_2 = 2(2 + 0.88^2)(2 - \sqrt{\tfrac{\alpha_2}{2}}) = 5.55(2 - \sqrt{\tfrac{\alpha_2}{2}})$$

$$\beta_1 = 2(2 - \sqrt{\tfrac{\alpha_2}{2}})$$

$$\epsilon^* = \frac{\alpha_2}{4(2+0.88^2)(2-\sqrt{\frac{\alpha_2}{2}})} = \frac{\alpha_2}{11.09(2-\sqrt{\frac{\alpha_2}{2}})}$$

$$\rho^* = 0.2649$$

Here $0 < \alpha_2 < 2$ is to be chosen. The singular manifold of the fast subsystem is defined by the equation $1 + 4u^3x = 0$ for $\alpha_2 = 0$. For a fixed value of α_2, we obtain an associated domain of attraction where asymptotic stability of the solution is guaranteed if $\epsilon < \epsilon^*(\alpha_2)$.

Figure 1 shows the solutions of $h(x, u) = 0$ which provide an ideal control law u as a function of the state x. For each value of x, up to four possible solutions for u may exist. Two of these appear in continuous and dashed lines, respectively, within the region of the figure. Figure 2 shows the regions of stability as defined in (10) which are delineated by the Lyapunov level curves. The estimates of the domain of attraction are delimited by $B_x \times B_u$ and parameterized by α_2. The dotted line corresponds to $\alpha_2 = 0.2$, the dashed line to $\alpha_2 = 1.0$, and the dashed-dotted line to $\alpha_2 = 1.8$. The line defined by the symbol "*" represents singular manifolds.

Simulation result for different values of ϵ is shown in Figure 3 for initial conditions $x(0) = 0.3$ and $u(0) = 0.8$. It can be seen that if $\epsilon > \epsilon^*(\alpha_2)$ asymptotic stability is not guaranteed and that

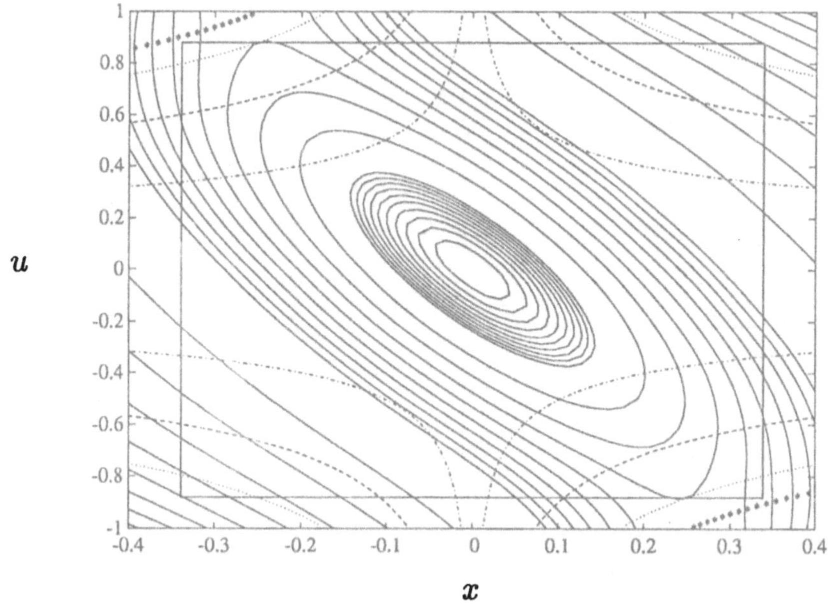

Figure 2: Level curves of the Lyapunov function $V(x)$ as a function of α_2 for Example 1.

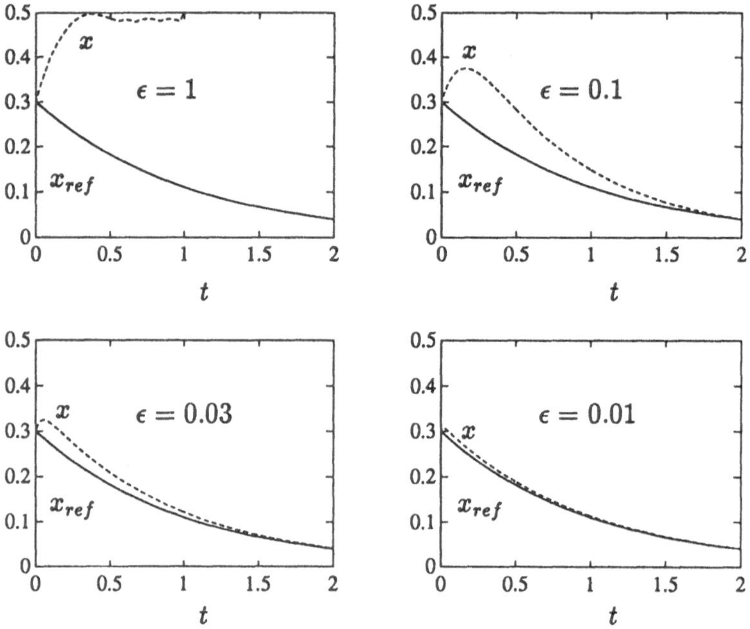

Figure 3: Simulation of trajectories of Example 1 for different values of ϵ.

the system can become unstable. If $\epsilon < \epsilon^*(\alpha_2)$, asymptotic stability is guaranteed. Smaller the ϵ better is the convergence meaning that the size of the guaranteed region of convergence becomes larger.

5. CONTROL BASED ON INVERSE KINEMATICS

The updating control law developed earlier and the results on singularly perturbed systems find potential application in control problems dealing with inverse kinematics. Consider the problem of controlling a mechanical system described in the form

$$\dot{\theta} = f(\theta, u), \quad u = u(\theta), \quad f : \mathbf{R^N} \times \mathbf{R^M} \to \mathbf{R^N} \tag{29}$$

The main difficulty here is in obtaining the feedback law $u(\theta)$ which arises due to the fact that θ is not directly measured, instead a related variable x is measured. In general, the kinematic relationship between coordinate variables is defined implicitly by

$$g(\theta, x) = 0, \quad g : \mathbf{R^N} \times \mathbf{R^N} \to \mathbf{R^N} \tag{30}$$

Suppose that we define a solution of (30) in the form $\theta = \theta(x)$. Then, one can define the following system in terms of the variables x and u:

$$\begin{aligned} \dot{x} &= h(x,u) = (\nabla_x g(\theta(x),x))^{-1}\nabla_\theta g(\theta(x),x) f(\theta,u), \quad h: \mathbf{R^N} \times \mathbf{R^M} \to \mathbf{R^N} \\ u &= u(\theta(x)) \end{aligned} \tag{31}$$

Without loss of generality, we suppose that $\theta = 0$ is an equilibrium point of (29) and that this corresponds to $x = 0$ in (30) so that $g(0,0) = 0$. Note that while the measured vector is x, the control is defined in terms of θ which is a vector of joint coordinates. Suppose that there exists a control law $u = u(\theta)$ that makes the origin of the system (29) an asymptotically stable equilibrium point. Then, the corresponding point $x_e = 0$ and $g(0, x_e) = 0$ will also be an asymptotically stable equilibrium point of (31) under the control law $u = u(\theta(x)) = u(x)$. To characterize the algorithm that computes the control law, we define $\theta(x)$ as the unique root of (30) for a given measurement x in a specified region of study (i.e. a neighborhood of $\theta = 0$ and $x = 0$). Then, an iterative algorithm to solve for $\theta(x)$ can be characterized via an evolution equation for $\hat{\theta}$, the estimate of $\theta(x)$, given by

$$\epsilon\dot{\hat{\theta}} = -(\nabla_{\hat{\theta}} g(\hat{\theta},x))^{-1} g(\hat{\theta},x) \tag{32}$$

Here, again ϵ characterizes the ratio between the logical time of evolution of the solving algorithm and the real time of evolution of the system dynamics. If we consider either equation (29) or (31) where $u = u(\hat{\theta})$ together with (32), we can define a singularly perturbed system either as

$$\begin{aligned} \dot{\theta} &= f(\theta, u(\hat{\theta})) \\ \epsilon\dot{\hat{\theta}} &= -(\nabla_{\hat{\theta}} g(\hat{\theta}, x(\theta)))^{-1} g(\hat{\theta}, x(\theta)) \end{aligned} \tag{33}$$

or as

$$\begin{aligned} \dot{x} &= h(x,u) = (\nabla_x g(\theta(x),x))^{-1}\nabla_\theta g(\theta(x),x) f(\theta,u(\hat{\theta})), \\ \epsilon\dot{\hat{\theta}} &= -(\nabla_{\hat{\theta}} g(\hat{\theta},x))^{-1} g(\hat{\theta},x) \end{aligned} \tag{34}$$

We consider a region of analysis where a diffeomorphism between θ and x is defined implicitly by (30).

5.1 LINEARIZATION APPROACH

The linearized form around $\theta = 0$ and $\hat{\theta} = 0$ of the formulation given by (33) is

$$\Delta\dot{\theta} = \nabla_\theta f(\theta, u(\hat{\theta}))\Delta\theta + \nabla_{\hat{\theta}} f(\theta, u(\hat{\theta}))\Delta\hat{\theta} \tag{35}$$

$$\epsilon\Delta\dot{\hat{\theta}} = -(\nabla_{\hat{\theta}} g(\hat{\theta}, x(\theta))^{-1}\nabla_\theta g(\hat{\theta}, x(\theta))\Delta\theta - \Delta\hat{\theta} \tag{36}$$

for which we obtain $A_0 = \nabla_\theta f(\theta, u(\theta))$ and $A_{22} = -1$, both of which are Hurwitz matrices whenever the control stabilizes the origin of the mechanical system exponentially. Hence, for small enough ϵ, the singularly perturbed system will also have an asymptotically stable equilibrium point.

5.2 NONLINEAR APPROACH

In order to obtain some information regarding the domain of attraction associated with the equilibrium point, we need to use the nonlinear approach. For definiteness, consider the formulation given by equation (33). Suppose that there is a positive definite function $V(\theta)$ satisfying $(\nabla_\theta V)\cdot f(\theta,u(\theta)) \le -\alpha_1\Psi^2(\theta)$, $\alpha_1 > 0$. The same can be concluded about formulation (34) by considering the implicit function theorem and regular partial derivatives of g at $x = 0$ and $u = 0$. Then, there exists a positive definite function $V(x)$ satisfying $(\nabla_x V)\cdot h(x,u(x)) \le -\alpha_1\Psi_1^2(x)$, $\alpha_1 > 0$, where $\Psi_1(\cdot)$ is a comparison function. Assume that $\Phi(\cdot)$ is another comparison function and that, referring to the theorem for nonlinear systems, the following assumptions are met in a neighborhood of the origin in the state space region $B_\theta \times B_{\hat\theta}$:

$$-2g^T(\hat\theta,x(\theta))\cdot g(\hat\theta,x(\theta)) \le -\alpha_2\Phi^2(\hat\theta-\theta),\ \alpha_2>0 \tag{37}$$

$$2g^T(\nabla_\theta g(\hat\theta,x(\theta)))\cdot f(\theta,u(\hat\theta)) \le \gamma\Phi^2(\hat\theta-\theta)+\beta_2\Psi(\theta)\Phi(\hat\theta-\theta) \tag{38}$$

$$(\nabla_\theta V)\cdot[f(\theta,u(\hat\theta))-h(\theta,u(\theta))] \le \beta_1\Psi(\theta)\Phi(\hat\theta-\theta) \tag{39}$$

Then the controller guarantees asymptotic stability of the origin for (29) and (32) provided that the algorithm is fast enough so that $\epsilon\in[0,\epsilon^*)$ where ϵ^* is given by (7). The Lyapunov function has the form

$$\nu(\theta,\hat\theta)=\rho V(\theta)+(1-\rho)g^T(\hat\theta,x(\theta))g(\hat\theta,x(\theta)),\ 0<\rho<1$$

The singular manifolds are defined by $det\{\nabla_{\hat\theta} g(\hat\theta,x(\theta))\}=0$. One can again prove that Lemma 2 is true and that the condition given by (37) is satisfied in a region where no singular manifold is present.

5.3 EXPLICIT FORM OF TRANSFORMATION

A general implicit form relating the coordinates x and θ has been previously considered. We now consider the inverse kinematics problem where the transformation of coordinates is defined explicitly as

$$x = A(\theta) \tag{40}$$

Here, an analytical expression for the inverse function $\theta = A^{-1}(x)$ is not available. As before, we can work with either θ or x as the state variable. If we define a region of study in θ as B_θ, then the corresponding region for x is given by

$$B_x=\{x \mid x=A(\theta),\ \theta\in B_\theta\} \tag{41}$$

For purpose of simplification, let the initial guesses be confined to the same region $B_{\hat\theta}=B_\theta$. Suppose that (40) is a diffeomorphism in the region of study $B_\theta\times B_{\hat\theta}$. Then, there is a unique relation between the two sets of variables θ and $\hat\theta$ defined by $g(\hat\theta,x(\theta))=0$ and the first condition of Theorem 2 is satisfied. Let

$$g(\hat\theta,x(\theta))=A(\hat\theta)-x=A(\hat\theta)-A(\theta) \tag{42}$$

Singularly perturbed system (33) now takes the form

$$\begin{aligned} \dot{\theta} &= f(\theta, u(\hat{\theta})) \\ \epsilon\dot{\hat{\theta}} &= -(\nabla_{\hat{\theta}} A(\hat{\theta}))^{-1}(A(\hat{\theta}) - A(\theta)) \end{aligned} \tag{43}$$

Note that without loss of generality $\theta = 0$ and $\hat{\theta} = 0$ is the equilibrium point of the singularly perturbed system. For a tracking control problem, one may apply a control law such that x^r (or equivalently θ^r) is the new equilibrium point. In that case one can define

$$\overline{x} = x - x^r, \quad \overline{\theta} = \theta - \theta^r \tag{44}$$

$$\overline{x} = A(\overline{\theta}) \tag{45}$$

where $B_{\overline{x}}$ or $B_{\overline{\theta}}$ define a neighborhood around $\overline{x} = 0$ or $\overline{\theta} = 0$. Note that the Jacobian of the transformation must be analyzed in a neighborhood of x^r (or equivalently near θ^r).

A linearization approach leads to

$$\begin{aligned} \Delta\dot{\theta} &= \nabla_{\theta} f(\theta, u(\hat{\theta}))\Delta\theta + \nabla_{\hat{\theta}} f(\theta, u\hat{\theta})\Delta\hat{\theta} \\ \epsilon\Delta\dot{\hat{\theta}} &= \Delta\theta - \Delta\hat{\theta} \end{aligned} \tag{46}$$

For a nonlinear analysis, let

$$\begin{aligned} J &= \nabla_{\theta} A(\theta) \\ K &= \nabla_{\hat{\theta}} f(\theta, u(\hat{\theta})) \\ L &= \nabla_{\theta} V(\theta) \\ \sigma_{\max}(\cdot) &= \textit{Largest singular value} \\ \sigma_{\min}(\cdot) &= \textit{Smallest singular value} \end{aligned}$$

Then the conditions mentioned in Theorem 2 can be put in the following form:

(a)

$$\begin{aligned} -2g(\hat{\theta}, x(\theta))^T g(\hat{\theta}, x(\theta)) &= -2(A(\hat{\theta}) - A(\theta))^T(A(\hat{\theta}) - A(\theta)) \\ &\leq -2\sigma_{\min}^2(J)\|\hat{\theta} - \theta\|^2 \leq -\alpha_2\Phi^2(\hat{\theta} - \theta), \; \alpha_2 > 0 \end{aligned}$$

(b)

$$\begin{aligned} 2g^T(\nabla_{\theta} g(\hat{\theta}, x(\theta)))f(\theta, u(\hat{\theta})) &= 2(A(\theta) - A(\hat{\theta}))^T J(\theta) f(\theta, u(\hat{\theta})) \\ &= 2(A(\theta) - A(\hat{\theta}))^T J(\theta)[f(\theta, u(\hat{\theta})) - f(\theta, u(\theta))] \\ &+ 2(A(\theta) - A(\hat{\theta}))^T J(\theta) f(\theta, u(\theta)) \\ &\leq 2\sigma_{\max}^2(J)\sigma_{\max}(K)\|\hat{\theta} - \theta\|^2 \\ &+ 2\sigma_{\max}(J)\|J(\theta)f(\theta, u(\theta))\| \cdot \|\hat{\theta} - \theta\| \\ &\leq 2\sigma_{\max}^2(J)\sigma_{\max}(K)\|\hat{\theta} - \theta\|^2 \\ &+ 2\sigma_{\max}^2(J)\|f(\theta, u(\theta))\| \; \|\hat{\theta} - \theta\| \\ &\leq \gamma\Phi^2(\hat{\theta} - \theta) + \beta_2\Psi(\theta)\Phi(\hat{\theta} - \theta) \end{aligned}$$

(c)

$$\begin{aligned}(\nabla_\theta V)[f(\theta,u(\hat{\theta}))-f(\theta,u(\theta))] &= L(\theta)[-f(\theta,u(\hat{\theta}))+f(\theta,u(\theta))] \\ &\leq \sigma_{\max}(K)\|L(\theta)\|\ \|\hat{\theta}-\theta\| \\ &\leq \beta_1\Psi(\theta)\Phi(\hat{\theta}-\theta)\end{aligned}$$

Here, $\sigma_{\min}(\cdot)$ and $\sigma_{\max}(\cdot)$ are evaluated at a point $(\theta,\hat{\theta}) \in B_\theta \times B_{\hat{\theta}}$.

To obtain some bounds (and to satisfy above inequalities), one must find maxima of singular values or related expressions in any pair $(\theta,\hat{\theta}) \in B_\theta \times B_{\hat{\theta}}$:

$$\max_{(\hat{\theta},\theta)\in B_\theta\times B_{\hat{\theta}}} \Omega(\hat{\theta}) = \max\,[\max_{\hat{\theta}\in B_{\hat{\theta}}}\Omega(\hat{\theta}), \max_{\theta\in B_\theta}\Omega(\theta)] \tag{47}$$

Here, Ω is any expression to be maximized (e.g. a singular value of a matrix). If the region $B_{\hat{\theta}}$ is chosen equal to B_θ, then the maximization problem reduces to one variable. Hence, we can now define

(a) $\alpha_2 = 2\min_{(\hat{\theta},\theta)\in B_\theta\times B_{\hat{\theta}}} \sigma_{\min}(J) = 2\min_{\hat{\theta}\in B_{\hat{\theta}}} \sigma_{\min}(J(\hat{\theta}))$

(b) $\gamma = 2\max_{\hat{\theta}\in B_{\hat{\theta}}}(\sigma^2_{\max}(J)\sigma_{\max}(K))$

(c) $\beta_2 = 2\max_{\hat{\theta}\in B_{\hat{\theta}}} \sigma^2_{\max}(J)$, $\beta_1 = \max_{\hat{\theta}\in B_{\hat{\theta}}} \sigma_{\max}(K)$

$$\alpha_1 = \min_{\theta\in B_\theta} \frac{\min[\|L(\theta)\|, \|f(\theta,u(\theta))\|]}{\max[\|L(\theta)\|, \|f(\theta,u(\theta))\|]} = \min_{\hat{\theta}\in B_{\hat{\theta}}} \frac{\min[\|L(\hat{\theta})\|, \|f(\hat{\theta},u(\hat{\theta}))\|]}{\max[\|L(\hat{\theta})\|, \|f(\hat{\theta},u(\hat{\theta}))\|]}$$

We also define the functions $\Phi(\theta)$ and $\Psi(\theta)$ to be

$$\begin{aligned}\Phi(\theta) &= \|A(\theta)\| \\ \Psi(\theta) &= \max[\|L(\theta)\|, \|f(\theta,u(\theta))\|]\end{aligned}$$

such that $\Psi(\theta)$ for $\theta \in B_\theta$ satisfies

$$\begin{aligned}&\sigma^2_{\max}(J)\|f(\theta,u(\theta))\| \leq \beta_2\Psi(\theta) \\ &\sigma_{\max}(K)\|L(\theta)\| \leq \beta_1\Psi(\theta) \\ &L(\theta)f(\theta,u(\theta)) \leq -\alpha_1\Psi^2(\theta)\end{aligned}$$

and the region of validity $(\hat{\theta},\theta) \in B_\theta \times B_{\hat{\theta}}$ is determined by the value of α_2. Notice that for $\alpha_2 > 0$, no region with singular manifold can be included.

The Lyapunov function is given by

$$\nu(\theta,\hat{\theta}) = \rho V(\theta) + (1-\rho)(A(\hat{\theta})-A(\theta))^T(A(\hat{\theta})-A(\theta)),\ 0<\rho<1 \tag{48}$$

and the level curves are in general $(2N-1)$ dimensional manifolds defined by $\nu(\hat{\theta},\theta) = c > 0$ which can be located by employing the singular manifold algorithm provided in [9]. One can successively check the level curves $\nu(\theta,\hat{\theta}) = c$ associated with different values of c and find the curve with the

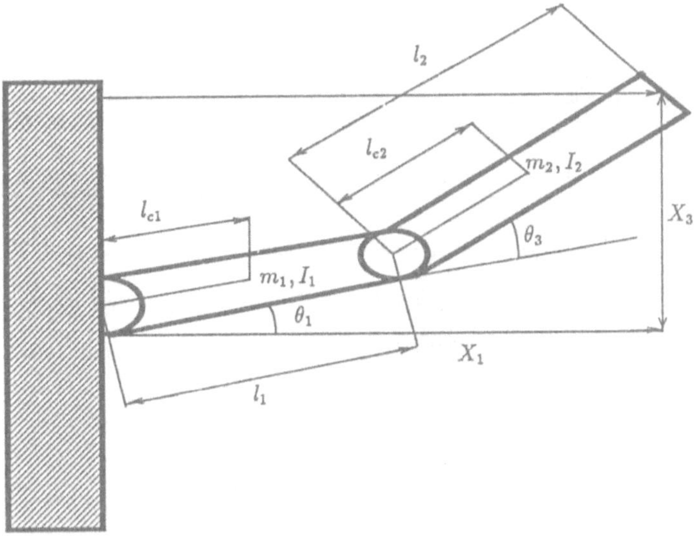

Figure 4: Two-link robot manipulator of Example 2.

highest value of c that can be included in $B_\theta \times B_{\dot{\theta}}$. Alternatively, one can calculate the minimum of c on the boundary of the region of study via a nonlinear programming algorithm. In order to get estimates of domains of attraction we choose sets as defined in (10).

Note that the results obtained by this procedure provide sufficient conditions for stability and they can be quite conservative depending on the margins required when satisfying the inequalities. In general, the updating control procedure is quite robust in providing stability for a larger range of initial conditions and parameter values.

5.4 EXAMPLE 2

Consider the two-link robot arm shown in Figure 4. Suppose that an arbitrary torque is applied at each of the joints. We define the following two related sets of state space variables:

$$x = \begin{bmatrix} x_1 \\ \dot{x}_1 \\ x_3 \\ \dot{x}_3 \end{bmatrix} = \begin{bmatrix} x_1 \\ x_2 \\ x_3 \\ x_4 \end{bmatrix}, \quad \theta = \begin{bmatrix} \theta_1 \\ \dot{\theta}_1 \\ \theta_3 \\ \dot{\theta}_3 \end{bmatrix} = \begin{bmatrix} \theta_1 \\ \theta_2 \\ \theta_3 \\ \theta_4 \end{bmatrix} \tag{49}$$

Referring to [1], the dynamics including the control are given by

$$\begin{aligned}
\dot{\theta}_1 &= \theta_2 \\
\dot{\theta}_2 &= \frac{hH_{33}\theta_4^2 + 2hH_{33}\theta_2\theta_4 - H_{33}G_1 + hH_{13}\theta_2^2 + H_{13}G_3 + H_{33}u_1 - H_{13}u_2}{H_a} \\
\dot{\theta}_3 &= \theta_4 \\
\dot{\theta}_4 &= \frac{-hH_{11}\theta_2^2 - H_{11}G_3 - hH_{13}\theta_4^2 - 2hH_{13}\theta_2\theta_4 + H_{13}G_1 - H_{13}u_1 + H_{11}u_2}{H_a}
\end{aligned}$$

where the control vector u represents the torque applied to the joints and

$$\begin{aligned}
H_{11} &= m_1 l_{c1}^2 + I_1 + m_2[l_1^2 + l_{c2}^2 + 2l_1 l_{c2}\cos\theta_3] + I_2 \\
H_{33} &= m_2 l_{c2}^2 + I_2 \\
H_{13} &= m_2 l_1 l_{c2}\cos\theta_3 + m_2 l_{c2}^2 + I_2 \\
H_a &= H_{11} + H_{33} + H_{12}^2 \\
H_b &= H_{11} + H_{33} - H_{12}^2 \\
h &= m_2 l_1 l_2 \sin\theta_3 \\
G_1 &= m_1 l_{c1}\cos\theta_1 + m_2 g[l_{c2}\cos(\theta_1+\theta_3) + l_1\cos\theta_1] \\
G_3 &= m_2 l_{c2} g\cos(\theta_1+\theta_3)
\end{aligned}$$

A simple control can be defined as a function of the joint angles and velocities. This control law is chosen to cancel nonlinearities and to follow a prescribed reference path θ_{1R} and θ_{3R}. It is given by

$$\begin{aligned}
u(1) &= -h\theta_4^2 - 2h\theta_2\theta_4 + G_1 + \frac{H_a H_{11}}{H_b}[k_1(\theta_1 - \theta_{1R}) + k_2\theta_2] + \frac{H_a H_{13}}{H_b}[k_3(\theta_3 - \theta_{3R}) \\
&+ k_4\theta_4] \\
u(2) &= h\theta_2^2 + G_3 + \frac{H_a H_{13}}{H_b}[k_1(\theta_1 - \theta_{1R}) + k_2\theta_2] + \frac{H_a H_{33}}{H_b}[k_3(\theta_3 - \theta_{3R}) + k_4\theta_4]
\end{aligned}$$

Therefore, the dynamics of the controlled system are given by

$$\begin{aligned}
\dot{\theta}_1 &= \theta_2 \\
\dot{\theta}_2 &= k_1(\theta_1 - \theta_{1R}) + k_2\theta_2 \\
\dot{\theta}_3 &= \theta_4 \\
\dot{\theta}_4 &= k_3(\theta_3 - \theta_{3R}) + k_4\theta_4
\end{aligned}$$

Suppose that only displacement and velocity of the end of the arm is measured. The kinematic relation between the variables θ and x is given by

$$x = A(\theta) \tag{50}$$

where the transformation function $A(\theta)$ is given by

$$A(\theta) = \begin{bmatrix} l_1\cos\theta_1 + l_2\cos(\theta_1+\theta_3) \\ -l_1\theta_2\sin\theta_1 - l_2(\theta_2+\theta_4)\sin(\theta_1+\theta_3) \\ l_1\sin\theta_1 + l_2\sin(\theta_1+\theta_3) \\ l_1\theta_2\cos\theta_1 + l_2(\theta_2+\theta_4)\cos(\theta_1+\theta_3) \end{bmatrix} \tag{51}$$

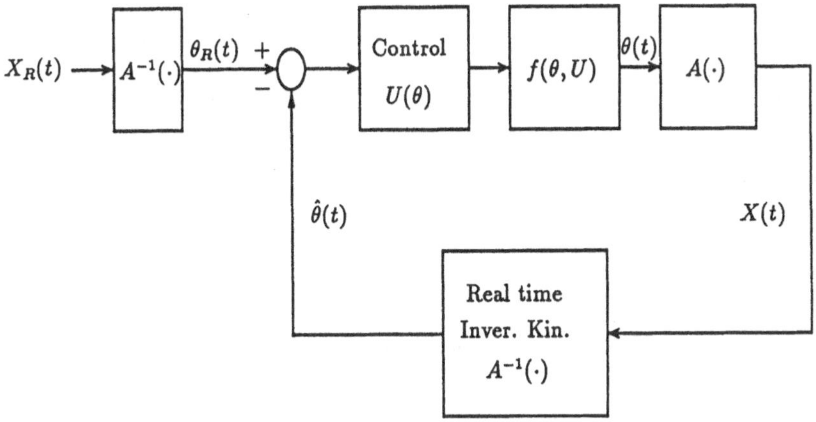

Figure 5: Block diagram of controller and robot manipulator for Example 2.

Here, the controller must solve for θ as a function of the measured vector x. The general structure of the whole system is shown in Figure 5. Hence, for the analysis, one can employ the singularly perturbed system formulation given in (33). The region of evolution $B_{\hat{\theta}} \times B_{\theta}$ must be chosen so that the conditions for applying the theorem for nonlinear systems are satisfied. From the condition of uniqueness of the solution, it is easy to check that the range of variation of θ_3 must satisfy $0 < \theta_3 < \pi$ or $-\pi < \theta_3 < 0$. Also $\alpha_2 > 0$ must be satisfied. Lemma 2 requires that no point with singular Jacobian $\nabla_{\theta} A(\theta)$ be included. After some algebra, the determinant of the Jacobian is given by

$$det(\nabla_{\theta} A(\theta)) = l_1^2 l_2^2 \sin^2 \theta_3 \tag{52}$$

Hence, we again conclude that the region of study must be a subset of either the region

$$B_{\hat{\theta}} = B_{\theta} \subset \{(x, \theta) | 0 < \theta_3 < \pi\}$$

or the region

$$B_{\hat{\theta}} = B_{\theta} \subset \{(x, \theta) | -\pi < \theta_3 < 0\}$$

In addition, one may expect that stability may also depend on the velocity of the arm and some limits must be imposed on the range of values attainable by θ_2 and θ_4. For example, let us chose two regions of study in the form

$$B_{\dot{\theta}} = B_{\theta} = \{\theta \mid -0.5 < \theta_2 < 0.5,\ 0.3\pi < \theta_3 < 0.7\pi, -0.5 < \theta_4 < 0.5\}$$

and

$$B_{\dot{\theta}} = B_{\theta} = \{\theta \mid 1.0 < \theta_2 < 1.0,\ 0.2\pi < \theta_3 < 0.8\pi, 1.0 < \theta_4 < 1.0\}$$

where $-\pi < \theta_1 < \pi$ for all the cases. Let the physical parameters be

$$\begin{aligned} m_1 &= m_2 = 1\ Kg,\ \ l_1 = l_2 = 1\ m \\ l_{c1} &= l_{c2} = \frac{1}{2}\ Kg - m^2,\ \ I_1 = I_2 = \frac{1}{12}\ Kg - m^2 \end{aligned}$$

and the control law parameters be

$$\begin{aligned} k_1 = k_3 = -1 \\ k_2 = k_4 = -2 \end{aligned}$$

so that

$$\begin{aligned} V(\theta) &= (\theta - \theta_R)^T P(\theta - \theta_R) \\ &= (\theta - \theta_R)^T \begin{bmatrix} \frac{3}{2} & \frac{1}{2} & 0 & 0 \\ \frac{1}{2} & \frac{1}{2} & 0 & 0 \\ 0 & 0 & \frac{3}{2} & \frac{1}{2} \\ 0 & 0 & \frac{1}{2} & \frac{1}{2} \end{bmatrix} (\theta - \theta_R) \end{aligned} \tag{53}$$

After some involved algebra[3] one can obtain analytical expressions for J, K and L. A nonlinear programming approach has been employed to obtain the values of the constants $\alpha_1, \alpha_2, \beta_1, \beta_2$ and γ. The value of c that determines the highest level Lyapunov curve that can be fitted in the regions of study is obtained by calculating the minimum of the Lyapunov function on the boundary of the region of study, also making use of a nonlinear programming algorithm. The parameter values obtained for the two different regions considered are shown in the following table.

Velocity Range Displacement Range	$-0.5 < \theta_{2,4} < 0.5$ $0.3\pi < \theta_3 < 0.7\pi$	$-1.0 < \theta_{2,4} < 1.0$ $0.2\pi < \theta_3 < 0.8\pi$
α_1	0.4030772	0.1740433
α_2	0.1532493	0.0237532
β_1	37.0494662	87.7230993
β_2	13.8329242	33.4230647
γ	512.5024606	2931.9748298
ρ^*	0.7281392	0.72410959
ϵ^*	0.0002399	0.0000012

[3]The computer algebra was performed by Macsyma.

For the case when θ_R is a function of time (i.e. the problem of tracking a given trajectory) we impose that $\theta_R(t) \in B_\theta$ so that the state variables of the system will remain in the region of study. In addition, the maximum value of the constant c that one can choose in equation (10) depends on θ_R. In order to guarantee stability of the tracking problem, we choose the minimum of all the possible c values when $\theta_R(t)$ is restricted to evolve within a given region. The following table shows different values of c obtained for two different regions where $-\frac{\pi}{4} < \theta_{1R}(t) < \frac{\pi}{4}$ and $\theta_{3R}(t)$ is allowed to evolve.

Velocity Range Displacement Range	$-0.5 < \theta_{2,4} < 0.5$ $0.3\pi < \theta_3 < 0.7\pi$	$-1.0 < \theta_{2,4} < 1.0$ $0.2\pi < \theta_3 < 0.8\pi$
$0.35\pi < \theta_{3R} < 0.65\pi$	0.0014541	0.0097079
$0.45\pi < \theta_{3R} < 0.55\pi$	0.0160932	0.0347817

To preserve stability of the entire system, the initial states and estimates must satisfy certain conditions. For instance, suppose that the initial conditions θ_i and initial estimate $\hat{\theta}_i$ satisfy $\nu(\theta_i, \hat{\theta}_i) < 0.0014541$ and they are within the first region under study. The results provided in the tables imply that stability is preserved whenever the computer performs a logical time unit integration (or one iteration in the case of standard Newton technique) in less than 0.23 ms as determined by ϵ^*, and the tracking reference signal remains within $-\frac{\pi}{4} < \theta_{1R}(t) < \frac{\pi}{4}$ and $0.35\pi < \theta_{3R}(t) < 0.65\pi$. In the same way, for the second region of study for which initial conditions θ_i and initial estimate $\hat{\theta}_i$ satisfy $\nu(\theta_i, \hat{\theta}_i) < 0.0347817$ and the tracking signal is within $-\frac{\pi}{4} < \theta_{1R}(t) < \frac{\pi}{4}$ and $0.45\pi < \theta_{3R}(t) < 0.55\pi$, stability is preserved when the computer performs a logical time unit integration in less than 1.2 μs.

Several simulations of the singularly perturbed system have been performed. It can be demonstrated that the above results are very conservative for this problem in which stability is guaranteed for a larger range of initial conditions and for larger values of ϵ. Figure 6 presents the time evolution of the joint displacements $\theta_1(t)$ and $\theta_3(t)$ of the mechanical system under study. The continuous lines represent an ideal evolution of the system in the case when the joint variables can be directly measured for control feedback. The reference point to be tracked is $\theta_{1R} = 0.5$ and $\theta_{3R} = \frac{\pi}{2}$ and the the initial conditions chosen for the system variables are $\theta_1 = \theta_2 = \theta_4 = 0$ and $\theta_3 = \frac{\pi}{2}$. Figure 6(a) illustrates one case in which initial guesses are chosen purposely to be $\hat{\theta}_2 = \hat{\theta}_4 = 0$, $\hat{\theta}_1 = \frac{\pi}{2}$ and $\hat{\theta}_3 = -\frac{\pi}{2}$ to generate a pathological behavior. These values of $\hat{\theta}$ correspond to an alternate solution associated with the same value of the measurement vector x as the real mechanical system joint variables. In other words, $x = A(\theta) = A(\hat{\theta})$ but $\theta \neq \hat{\theta}$. In this case, the inverse kinematics solver locates an undesired solution leading to pathological behavior. Note that, of course, the initial conditions and guesses do not satisfy sufficiency conditions obtained above to guarantee stability of the entire system. The rest of the figures correspond to initial guesses $\hat{\theta}_2 = \hat{\theta}_4 = 0$ and $\hat{\theta}_1 = \hat{\theta}_3 = 0.5$ and for different values of ϵ. Note that the stability behavior is good and it improves as ϵ becomes smaller.

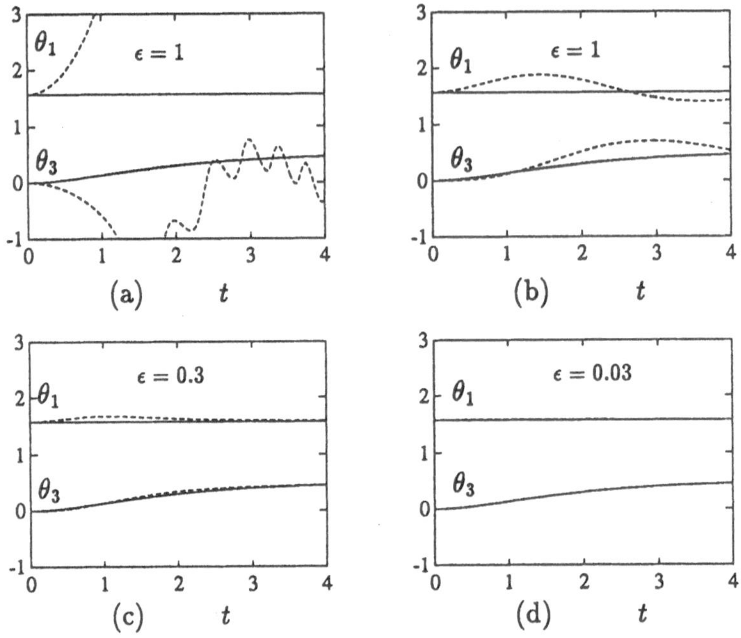

Figure 6: Simulation of trajectories of Example 2 for different values of ϵ and initial guesses.

The results obtained above are very conservative. This comes from the fact that the theorem based on the use of Lyapunov functions provides sufficient conditions for stability via some inequalities. In addition, this paper has been focused on methods of analysis of systems and the synthesis of the controllers was not explicitly addressed. Note that one could improve the above results by choosing different control laws. Theoretically speaking, an optimal control problem can be formulated where the cost function depends on the perturbation parameter (ϵ is to be maximized) and the estimates of the domains of attraction (also to be increased in size).

6. CONCLUDING REMARKS

Some applications of the analysis of the dynamical system (1) to control theory have been developed. An updating control procedure has been proposed based on the iterative nature of the control algorithm. A singularly perturbed system formulation has been used to model the combined dynamics of the system being controlled and a numerical iterative algorithm which is required to compute the control law. This approach can also be viewed as a combined controller and an observer (or an estimator) formulation. The results of [9] regarding the behavior of dynamical system that models these algorithms lead to a considerable simplification in the analysis. For the case of inverse kinematics control, it leads to a formulation in which the numerical algorithm that solves inverse kinematics can be considered as an observer (or an estimator) of the state variables. The study provides an estimate of the required speed of computations to preserve the stability of the controller. In general such results are very conservative and improvement is needed in that direction by obtaining less conservative bounds for the theorem based on the Lyapunov functions and by choosing proper control laws. Additional work is also needed for relaxing the assumption made in the modeling process. A combination of continuous-time and discrete systems (differential-difference equations) in a singularly perturbed structure, if possible, may lead to more accurate models.

Acknowledgments

The work reported here is partially supported by a grant from the National Science Foundation Grant. P. J. Zufiria also thanks the support of a Formación de Postgrado fellowship of the Programa Nacional de F.P.I. provided by the D.G.I.C.T. of the Ministerio de Educación y Ciencia of Spain.

REFERENCES

[1] H. Asada and J. E. Slotine, *Robot Analysis and Control,* John Wiley & Sons, 1986.

[2] R. C. Dorf, *Modern Control Systems,* Addison-Wesley, 1986.

[3] L. T. Grujić, *Uniform asymptotic stability of non-linear singularly perturbed general and large-scale systems,* Int. J. Control, Vol. 33, No. 3 pp. 481-504, 1981.

[4] R. S. Guttalu and P. J. Zufiria, *On a class of nonstandard dynamical systems: Singularity issues,* Advances in Controls and Dynamic Systems (Editor: C.T. Leondes), Vol. 34, 1990.

[5] P. Kokotović, H. K. Khalil and J. O'Reilly, *Singular Perturbation Methods in Control: Analysis and Design,* Academic Press, London, 1986.

[6] A. Saberi and H. Khalil, *Quadratic-type Lyapunov functions for singularly perturbed systems,* IEEE Trans. Aut. Control, Vol. AC-29, No. 6, pp. 542-550, 1984.

[7] P. J. Zufiria and R. S. Guttalu, *On an application of dynamical system theory to determine all the zeros of a vector function.* J. Math. Anal. and Appl. (accepted for publication).

[8] P. J. Zufiria and R. S. Guttalu, *A computational method for locating all the roots of a vector function.* J. Appl. Math. Comp. (accepted for publication).

[9] P. J. Zufiria, *Global behavior of a class of nonlinear dynamical systems: Analytical, computational and control aspects.* Ph.D. Dissertation, University of Southern California, 1989.

Lecture Notes in Control and Information Sciences

Edited by M. Thoma and A. Wyner

Vol. 97: I. Lasiecka/R. Triggiani (Eds.)
Control Problems for Systems
Described by Partial Differential Equations
and Applications
Proceedings of the IFIP-WG 7.2
Working Conference
Gainesville, Florida, February 3-6, 1986
VIII, 400 pages, 1987.

Vol. 98: A. Aloneftis
Stochastic Adaptive Control
Results and Simulation
XII, 120 pages, 1987.

Vol. 99: S. P. Bhattacharyya
Robust Stabilization Against
Structured Perturbations
IX, 172 pages, 1987.

Vol. 100: J. P. Zolésio (Editor)
Boundary Control and Boundary Variations
Proceedings of the IFIP WG 7.2 Conference
Nice, France, June 10-13, 1987
IV, 398 pages, 1988.

Vol. 101: P. E. Crouch,
A. J. van der Schaft
Variational and Hamiltonian
Control Systems
IV, 121 pages, 1987.

Vol. 102: F. Kappel, K. Kunisch,
W. Schappacher (Eds.)
Distributed Parameter Systems
Proceedings of the 3rd International Conference
Vorau, Styria, July 6–12, 1986
VII, 343 pages, 1987.

Vol. 103: P. Varaiya, A. B. Kurzhanski (Eds.)
Discrete Event Systems:
Models and Applications
IIASA Conference
Sopron, Hungary, August 3-7, 1987
IX, 282 pages, 1988.

Vol. 104: J. S. Freudenberg/D. P. Looze
Frequency Domain Properties of Scalar
and Multivariable Feedback Systems
VIII, 281 pages, 1988.

Vol. 105: Ch. I. Byrnes/A. Kurzhanski (Eds.)
Modelling and Adaptive Control
Proceedings of the IIASA Conference
Sopron, Hungary, July 1986
V, 379 pages, 1988.

Vol. 106: R. R. Mohler (Editor)
Nonlinear Time Series and
Signal Processing
V, 143 pages, 1988.

Vol. 107: Y. T. Tsay, L.-S. Shieh, St. Barnett
Structural Analysis and Design
of Multivariable Systems
An Algebraic Approach
VIII, 208 pages, 1988.

Vol. 108: K. J. Reinschke
Multivariable Control
A Graph-theoretic Approach
274 pages, 1988.

Vol. 109: M. Vukobratović/R. Stojić
Modern Aircraft Flight Control
VI, 288 pages, 1988.

Vol. 110: B. J. Daiuto, T. T. Hartley,
S. P. Chicatelli
The Hyperbolic Map and Applications
to the Linear Quadratic Regulator
VI, 114 pages, 1989

Vol. 111: A. Bensoussan, J. L. Lions (Eds.)
Analysis and Optimization
of Systems
XIV, 1175 pages, 1988.

Vol. 112: Vojislav Kecman
State-Space Models of Lumped
and Distributed Systems
IX, 280 pages, 1988

Vol. 113: M. Iri, K. Yajima (Eds.)
System Modelling and Optimization
Proceedings of the 13th IFIP Conference
Tokyo, Japan, Aug. 31 – Sept. 4, 1987
IX, 787 pages, 1988.

Vol. 114: A. Bermúdez (Editor)
Control of Partial Differential Equations
Proceedings of the IFIP WG 7.2
Working Conference
Santiago de Compostela, Spain, July 6–9, 1987
IX, 318 pages, 1989

Vol. 115: H.J. Zwart
Geometric Theory for Infinite
Dimensional Systems
VIII, 156 pages, 1989.

Vol. 116: M.D. Mesarovic, Y. Takahara
Abstract Systems Theory
VIII, 439 pages, 1989

Lecture Notes in Control and Information Sciences

Edited by M. Thoma and A. Wyner

Vol. 117: K.J. Hunt
Stochastic Optimal Control Theory
with Application in Self-Tuning Control
X, 308 pages, 1989.

Vol. 118: L. Dai
Singular Control Systems
IX, 332 pages, 1989

Vol. 119: T. Başar, P. Bernhard
Differential Games and Applications
VII, 201 pages, 1989

Vol. 120: L. Trave, A. Titli, A. M. Tarras
Large Scale Systems:
Decentralization, Structure Constraints
and Fixed Modes
XIV, 384 pages, 1989

Vol. 121: A. Blaquière (Editor)
Modeling and Control of Systems
in Engineering, Quantum Mechanics,
Economics and Biosciences
Proceedings of the Bellman Continuum
Workshop 1988, June 13–14, Sophia Antipolis, France
XXVI, 519 pages, 1989

Vol. 122: J. Descusse, M. Fliess, A. Isidori,
D. Leborgne (Eds.)
New Trends in Nonlinear Control Theory
Proceedings of an International
Conference on Nonlinear Systems,
Nantes, France, June 13–17, 1988
VIII, 528 pages, 1989

Vol. 123: C. W. de Silva, A. G. J. MacFarlane
Knowledge-Based Control with
Application to Robots
X, 196 pages, 1989

Vol. 124: A. A. Bahnasawi, M. S. Mahmoud
Control of Partially-Known
Dynamical Systems
XI, 228 pages, 1989

Vol. 125: J. Simon (Ed.)
Control of Boundaries and Stabilization
Proceedings of the IFIP WG 7.2 Conference
Clermont Ferrand, France, June 20–23, 1988
IX, 266 pages, 1989

Vol. 126: N. Christopeit, K. Helmes
M. Kohlmann (Eds.)
Stochastic Differential Systems
Proceedings of the 4th Bad Honnef Conference
June 20–24, 1988
IX, 342 pages, 1989

Vol. 127: C. Heij
Deterministic Identification
of Dynamical Systems
VI, 292 pages, 1989

Vol. 128: G. Einarsson, T. Ericson,
I. Ingemarsson, R. Johannesson,
K. Zigangirov, C.-E. Sundberg
Topics in Coding Theory
VII, 176 pages, 1989

Vol. 129: W. A. Porter, S. C. Kak (Eds.)
Advances in Communications and
Signal Processing
VI, 376 pages, 1989.

Vol. 130: W. A. Porter, S. C. Kak,
J. L. Aravena (Eds.)
Advances in Computing and Control
VI, 367 pages, 1989

Vol. 131: S. M. Joshi
Control of Large Flexible Space Structures
IX, 196 pages, 1989.

Vol. 132: W.-Y. Ng
Interactive Multi-Objective Programming
as a Framework for Computer-Aided Control
System Design
XV, 182 pages, 1989.

Vol. 133: R. P. Leland
Stochastic Models for Laser Propagation
in Atmospheric Turbulence
VII, 145 pages, 1989.

Vol. 134: X. J. Zhang
Auxiliary Signal Design in Fault
Detection and Diagnosis
XII, 213 pages, 1989

Vol. 135: H. Nijmeijer, J. M. Schumacher (Eds.)
Three Decades of Mathematical System Theory
A Collection of Surveys at the Occasion of the
50th Birthday of Jan C. Willems
VI, 562 pages, 1989

Vol. 136: J. Zabczyk (Ed.)
Stochastic Systems and Optimization
Proceedings of the 6th IFIP WG 7.1
Working Conference,
Warsaw, Poland, September 12–16, 1988
VI, 374 pages. 1989

Lecture Notes in Control and Information Sciences

Edited by M. Thoma and A. Wyner

Vol. 137: S. L. Shah, G. Dumont (Eds.)
Adaptive Control Strategies for Industrial Use
Proceedings of a Workshop
Kananaskis, Canada, 1988
VI, 360 pages. 1989

Vol. 138: D. C. McFarlane, K. Glover
Robust Controller Design Using Normalized Coprime Factor Plant Descriptions
X, 206 pages. 1990

Vol. 139: V. Hayward, O. Khatib (Eds.)
Experimental Robotics I
The First International Symposium
Montreal, June 19-21, 1989
XVII, 613 pages. 1990

Vol. 140: Z. Gajic, D. Petkovski, X. Shen
Singularly Perturbed and Weakly Coupled Linear Control Systems
A Recursive Approach
VII, 202 pages. 1990

Vol. 141: S. Gutman
Root Clustering in Parameter Space
VIII, 153 pages. 1990

Vol. 142: A. N. Gündeş, C. A. Desoer
Algebraic Theory of Linear Feedback Systems with Full and Decentralized Compensators
V, 176 pages. 1990

Vol. 143: H.-J. Sebastian, K. Tammer (Eds.)
System Modelling and Optimization
Proceedings of the 14th IFIP Conference
Leipzig, GDR, July 3-7, 1989
X, 960 pages. 1990

Vol. 144: A. Bensoussan, J. L. Lions (Eds.)
Analysis and Optimization of Systems
Proceedings of the 9th International Conference, Antibes, June 12-15, 1990
XII, 992 pages. 1990

Vol. 145: M. B. Subrahmanyam
Optimal Control with a Worst-Case Performance Criterion and Applications
X, 133 pages. 1990

Vol. 146: D. Mustafa, K. Glover
Minimum Entropy H_∞ Control
IX, 144 pages. 1990

Vol. 147: J. P. Zolésio (Ed.)
Stabilization of Flexible Structures
Third Working Conference, Montpellier, France, January 1989
V, 327 pages, 1990

Vol. 148: In preparation

Vol. 149: K. H. Hoffmann, W. Krabs (Eds.)
Optimal Control of Partial Differential Equations
Proceedings of the IFIP WG 7.2 International Conference, Irsee, April 9-12, 1990
VI, 245 pages. 1991

Vol. 150: L. C. G. J. M. Habets
Robust Stabilization in the Gap-topology
IX, 126 pages. 1991

Vol. 151: J. M. Skowronski, H. Flashner, R. S. Guttalu (Eds.)
Mechanics and Control
Proceedings of the 3rd Workshop on Control Mechanics
in Honor of the 65th Birthday of George Leitmann
January 22-24, 1990, University of Southern California
IV, 497 pages. 1991